Advances in Intelligent Systems and Computing

Volume 1046

The series "Advances in Intelligent Systems and Computing" contains publications on theory, applications, and design methods of Intelligent Systems and Intelligent Computing. Virtually all disciplines such as engineering, natural sciences, computer and information science, ICT, economics, business, e-commerce, environment, healthcare, life science are covered. The list of topics spans all the areas of modern intelligent systems and computing such as: computational intelligence, soft computing including neural networks, fuzzy systems, evolutionary computing and the fusion of these paradigms, social intelligence, ambient intelligence, computational neuroscience, artificial life, virtual worlds and society, cognitive science and systems, Perception and Vision, DNA and immune based systems, self-organizing and adaptive systems, e-Learning and teaching, human-centered and human-centric computing, recommender systems, intelligent control, robotics and mechatronics including human-machine teaming, knowledge-based paradigms, learning paradigms, machine ethics, intelligent data analysis, knowledge management, intelligent agents, intelligent decision making and support, intelligent network security, trust management, interactive entertainment, Web intelligence and multimedia.

The publications within "Advances in Intelligent Systems and Computing" are primarily proceedings of important conferences, symposia and congresses. They cover significant recent developments in the field, both of a foundational and applicable character. An important characteristic feature of the series is the short publication time and world-wide distribution. This permits a rapid and broad dissemination of research results.

**** Indexing: The books of this series are submitted to ISI Proceedings, EI-Compendex, DBLP, SCOPUS, Google Scholar and Springerlink ****

More information about this series at http://www.springer.com/series/11156

Radek Silhavy · Petr Silhavy ·
Zdenka Prokopova
Editors

Intelligent Systems Applications in Software Engineering

Proceedings of 3rd Computational Methods in Systems and Software 2019, Vol. 1

Editors
Radek Silhavy
Faculty of Applied Informatics
Tomas Bata University in Zlin
Zlin, Czech Republic

Petr Silhavy
Faculty of Applied Informatics
Tomas Bata University in Zlin
Zlin, Czech Republic

Zdenka Prokopova
Faculty of Applied Informatics
Tomas Bata University in Zlin
Zlin, Czech Republic

ISSN 2194-5357 ISSN 2194-5365 (electronic)
Advances in Intelligent Systems and Computing
ISBN 978-3-030-30328-0 ISBN 978-3-030-30329-7 (eBook)
https://doi.org/10.1007/978-3-030-30329-7

This Springer imprint is published by the registered company Springer Nature Switzerland AG
The registered company address is: Gewerbestrasse 11, 6330 Cham, Switzerland

Preface

This book constitutes the refereed proceedings of the Computational Methods in Systems and Software 2019 (CoMeSySo 2019), held on October 2019.

CoMeSySo 2019 conference intends to provide an international forum for the discussion of the latest high-quality research results in all areas related to cybernetics and intelligent systems. The addressed topics are the theoretical aspects and applications of Intelligent Systems Applications in Software Engineering. The papers address topics as software engineering cybernetics and automation control theory, econometrics, mathematical statistics or artificial CoMeSySo 2019 have received (all sections) 140 submissions, 84 of them were accepted for publication.

The volume Intelligent Systems Applications in Software Engineering brings the discussion of new approaches and methods to real-world problems. Furthermore, the exploratory research that describes novel approaches in software engineering and informatics in the scope of the intelligent systems is presented.

The editors believe that readers will find the following proceedings interesting and useful for their research work.

July 2019

Radek Silhavy
Petr Silhavy
Zdenka Prokopova

Organization

Program Committee

Program Committee Chairs

Petr Silhavy	Department of Computers and Communication Systems, Faculty of Applied Informatics, Tomas Bata University in Zlin, Czech Republic
Radek Silhavy	Department of Computers and Communication Systems, Faculty of Applied Informatics, Tomas Bata University in Zlin, Czech Republic
Zdenka Prokopova	Department of Computers and Communication Systems, Tomas Bata University in Zlin, Czech Republic
Krzysztof Okarma	Faculty of Electrical Engineering, West Pomeranian University of Technology, Szczecin, Poland
Roman Prokop	Department of Mathematics, Tomas Bata University in Zlin, Czech Republic
Viacheslav Zelentsov	Doctor of Engineering Sciences, Chief Researcher of St. Petersburg Institute for Informatics and Automation of Russian Academy of Sciences (SPIIRAS), Russian Federation
Lipo Wang	School of Electrical and Electronic Engineering, Nanyang Technological University, Singapore
Silvie Belaskova	Head of Biostatistics, St. Anne's University Hospital Brno, International Clinical Research Center, Czech Republic

International Program Committee Members

Pasi Luukka	North European Society for Adaptive and Intelligent Systems & School of Business and School of Engineering Sciences Lappeenranta University of Technology, Finland
Ondrej Blaha	Louisiana State University Health Sciences Center New Orleans, New Orleans, USA
Izabela Jonek-Kowalska	Faculty of Organization and Management, The Silesian University of Technology, Poland
Maciej Majewski	Department of Engineering of Technical and Information Systems, Koszalin University of Technology, Koszalin, Poland
Alena Vagaska	Department of Mathematics, Informatics and Cybernetics, Faculty of Manufacturing Technologies, Technical University of Kosice, Slovak Republic
Boguslaw Cyganek	DSc, Department of Computer Science, University of Science and Technology, Krakow, Poland
Piotr Lech	Faculty of Electrical Engineering, West Pomeranian University of Technology, Szczecin, Poland
Monika Bakosova	Institute of Information Engineering, Automation and Mathematics, Slovak University of Technology, Bratislava, Slovak Republic
Pavel Vaclavek	Faculty of Electrical Engineering and Communication, Brno University of Technology, Brno, Czech Republic
Miroslaw Ochodek	Faculty of Computing, Poznan University of Technology, Poznan, Poland
Olga Brovkina	Global Change Research Centre, Academy of Science of the Czech Republic, Brno, Czech Republic
Elarbi Badidi	College of Information Technology, United Arab Emirates University, Al Ain, United Arab Emirates
Gopal Sakarkar	Shri. Ramdeobaba College of Engineering and Management, Republic of India
V. V. Krishna Maddinala	GD Rungta College of Engineering and Technology, Republic of India

Anand N. Khobragade (Scientist) — Maharashtra Remote Sensing Applications Centre, Republic of India
Abdallah Handoura — Computer and Communication Laboratory, Telecom Bretagne, France

Organizing Committee Chair

Radek Silhavy — Tomas Bata University in Zlin, Faculty of Applied Informatics
email: comesyso@openpublish.eu

Conference Organizer (Production)

OpenPublish.eu s.r.o.
Web: http://comesyso.openpublish.eu
Email: comesyso@openpublish.eu

Conference Web site, Call for Papers:

http://comesyso.openpublish.eu

Contents

Predictive Controller Based on Feedforward Neural Network with Rectified Linear Units

Petr Dolezel(✉), Daniel Honc, and Dominik Stursa

Faculty of Electrical Engineering, University of Pardubice,
Studentska 95, 532 10 Pardubice, Czech Republic
{petr.dolezel,daniel.honc,dominik.stursa}@upce.cz
http://www.fei.upce.cz

Abstract. This paper deals with a nonlinear Model Predictive Control with a special form of the process model. Controller uses for the prediction purposes a locally valid linear sub-models. The sub-models are obtained from a neural model with the rectifier activation function in hidden neurons. Simulation example is given to demonstrate proposed solution - neural model design and predictive controller application.

Keywords: Process control · Piecewise-linear neural model · Model predictive control

1 Introduction

Automatic process control is a wide and complex branch of industrial control systems theory. It includes classical approaches (logic control, PID control), as well as modern and more advanced algorithms (internal model control, adaptive control, model reference control, predictive control, fuzzy and neuro-fuzzy control, etc.). Therefore, many possibilities for process control system design are available. However, most of them consider the controlled process to be linear or close to linear. On the contrary, if the controlled process is characterized by the behavior typified by nonlinearities, nonlinear system control theory needs to be applied and there exist far fewer approaches to implement, in comparison to linear process control.

This paper deals with one particular subset of nonlinear process control. In order to be more specific, a nonlinear process, which is able to be approximated by a piecewise-linear model, is considered here. Piecewise-linear model is a model compounded of a finite set of linear sub-models, where each sub-model is valid in a distinguished region of state space. Piecewise-linear models have received a lot of attention recently for their equivalence to regular classes of the systems [1], but mainly for their practicality. This structure of a piecewise-linear model can be dealt with using techniques which were originally proposed for linear systems. Besides, some specialized tools for a piecewise-linear model analysis were published, too [2].

R. Silhavy et al. (Eds.): CoMeSySo 2019, AISC 1046, pp. 1–12, 2019.
https://doi.org/10.1007/978-3-030-30329-7_1

Identification of the nonlinear system using a piecewise-linear model is, in most cases, not an elementary problem. In a particular case, when the number of regions and, consequently, the number of linear submodels is known, the issue of identification is transformed into a classical linear model identification. Hence, the parameters of the sub-models are estimated from the corresponding input-output data. Nevertheless, the significant challenge occurs, when the number of sub-models is not known. Many contributions dealing with the identification of piecewise-linear models (or more general piecewise-affine systems) have been proposed over time; Vidal et al. propose an algebraic approach [3], while others use clustering approach [4] or the bounded error method [5].

Once a piecewise-linear model of the nonlinear system is designed, it can be advantageously applied in stability investigation, prediction, fault diagnosis and, apparently, in process control. In 2012, a novel way of piecewise-linear model design was proposed [6]. Authors used a special topology of feed-forward neural network, which could ingenuously provide a piecewise-linear model of nonlinear system to any degree of accuracy. This approach was then used for nonlinear system control using PID controller [7,8], and tested in industrial environment [9]. As the last contribution to this topic, the paper [10] introduced a predictive controller design for the mentioned approach.

It should be emphasized that in [6], the authors strictly use the symmetric linear saturated activation function for neurons in a hidden layer of a neural network. Clearly, the used activation function should be piecewise-linear in order to provide the expected behavior, but many other activation functions, which fulfill this requirement, are available in addition to the the symmetric linear saturated activation function. As a natural choice, a rectifier activation function is suggested. This function was considerably discussed in [11] with biological motivation. Later, in 2011, Glorot et al. [12] demonstrated the power of rectifiers in training of deep networks. The aim of this paper is to present the first tests of the rectifier activation function used in a piecewise-linear neural model as introduced in [6]. The tests are based on the predictive controller design for the nonlinear process, as described in [10].

The paper is organized as follows. In Sect. 2, the aim of this paper is properly formulated. Then, the particular predictive controller, which fits to the problem defined in Sect. 2, is suggested (Sect. 3) and its features are demonstrated on a simulated process (Sect. 4).

2 Problem Formulation

A novel and computationally simple technique providing a piecewise-linear model of a nonlinear process is presented in [6]. It works as illustrated in Fig. 1.

The deterministic form of the piecewise-linear model used in Fig. 1 is a set of following difference equations.

$$\begin{aligned} A^1(q^{-1})y(k) &= B^1(q^{-1})u(k) + c^1, \text{ if } \mathbf{x} \in \mathbf{X_1} \\ A^2(q^{-1})y(k) &= B^2(q^{-1})u(k) + c^2, \text{ if } \mathbf{x} \in \mathbf{X_2} \\ &\vdots \\ A^R(q^{-1})y(k) &= B^R(q^{-1})u(k) + c^R, \text{ if } \mathbf{x} \in \mathbf{X_R} \end{aligned} \tag{1}$$

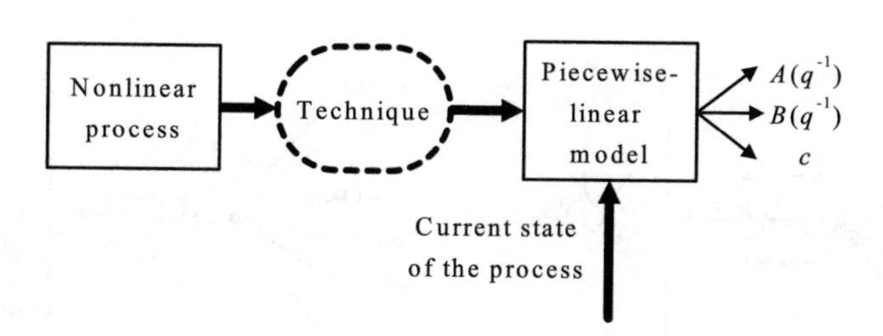

Fig. 1. Technique.

In Eq. (1), k is a discrete time slot, $y(k)$ is the output of the model, $u(k)$ is the input to the model and q denotes the forward shift operator, i.e. $q^{-1}y(k) = y(k-1)$. The vector $\mathbf{x}$ defines the current state vector of the model, $\mathbf{x} = [u(k-1), \cdots, u(k-m), y(k-1), \cdots, y(k-n)]^{\mathrm{T}}$. Then, $\mathbf{X} = \bigcup_{i \in \{1,2,\cdots\},R} \mathbf{X_i}$ denotes the state space partition into closed regions. Moreover, $A^i(q)$, $B^i(q)$, $i = 1, 2, \cdots, R$, are the auto-regression and correlation polynomials (filters), which, together with the constant c^i, determine the linear sub-model valid for the particular region i. The polynomials are defined as

$$B^i(q^{-1}) = [0 + b_1^i q^{-1} + b_2^i q^{-2} + ... + b_m^i q^{-m}], \tag{2}$$

$$A^i(q^{-1}) = [1 + a_1^i q^{-1} + a_2^i q^{-2} + ... + a_n^i q^{-n}], m \leq n. \tag{3}$$

where m is the order of the polynomial $B^i(q^{-1})$ and n is the order of the polynomial $A^i(q^{-1})$.

The main idea of the proposed approach is smart and simple determination of $A^i(q)$, $B^i(q)$ using a special topology of the artificial neural network - see esp. [13] for the detailed description of the technique. In simple words, only linear or piecewise-linear elements are used in the neural model of the process and, therefore, the neural model can be transformed into the structure (1). In [6,13], the symmetric linear saturated activation function is used as the piecewise-linear element - see (4).

$$y_{\text{out}} = \begin{cases} 1 \text{ for } & y_a > 1, \\ y_a \text{ for } & -1 \leq y_a \leq 1, \\ -1 \text{ for } & y_a < 1. \end{cases} \tag{4}$$

In Eq. (4), y_{out} is the output from the activation function and y_a is the input.

Another and computationally even simpler activation function is proposed in the contribution. Based on [12], the rectifier activation function is considered instead of the symmetric linear saturated activation function - see (5).

$$y_{\text{out}} = \max\left(0, y_a\right) \tag{5}$$

While one symmetric linear saturated activation function divides the state space to three linear subregions, the rectifier activation function provides only two subregions, as shown in Fig. 2.

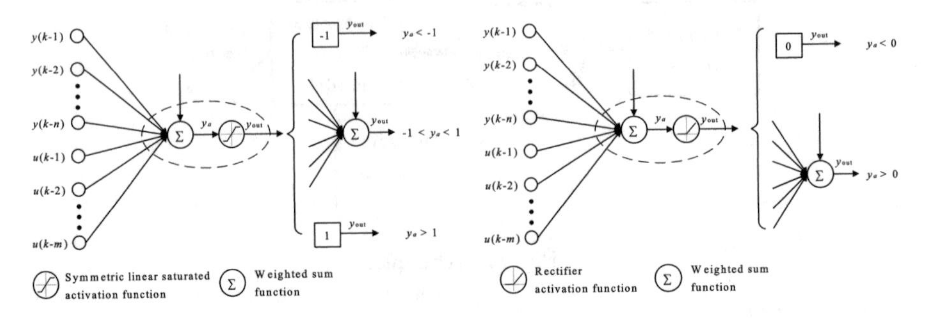

Fig. 2. Neuron with symmetric linear saturated activation function (left) and rectifier activation function (right).

The neural model with R neurons and with the rectifier activation function in the hidden layer provides 2^R linear sub-models, where each sub-model is valid in a distinguished region of the state space.

This approach is tested on application of nonlinear model predictive control. The derivation of the predictive controller, considering the piecewise-linear model (1), is shown in the next section.

3 Predictive Controller

Model predictive control is an advanced control technique frequently used in industry. Model predictive controllers minimize weighted future control errors and control afford taking into account constraints and using concept of the receding horizon. The comprehensive review of the predictive control approaches can be found in [14,15] or [16].

In our approach, the cost function is defined as follows.

$$J = \sum_{j=1}^{N} \left(\hat{y}(k+j) - w(k+j)\right)^2 + r \sum_{j=1}^{N-1} \left(u(k+j) - u(k+j-1)\right)^2, \tag{6}$$

where k is the current discrete time slot, $y(k)$ is the output of the process to be controlled, $w(k)$ is its reference value, $\hat{y}(k+j)$ is its predicted value j time slots ahead, $u(k)$ is the current control action, N is the length of the prediction and control horizon and r is the control action change penalization ($r > 0$). Tuning of this parameter affects the dynamics of the controller.

For further manipulation, it is convenient to transform the cost function to its matrix form. Thus,

$$J = (\mathbf{y} - \mathbf{w})^{\mathbf{T}} (\mathbf{y} - \mathbf{w}) + \mathbf{u}^{\mathrm{T}}\mathbf{R}\mathbf{u}, \tag{7}$$

where $\mathbf{y} = \begin{bmatrix} \hat{y}(k+1) \\ \hat{y}(k+2) \\ \vdots \\ \hat{y}(k+N) \end{bmatrix}$, $\mathbf{w} = \begin{bmatrix} w(k+1) \\ w(k+2) \\ \vdots \\ w(k+N) \end{bmatrix}$, $\mathbf{u} = \begin{bmatrix} u(k) \\ u(k+1) \\ \vdots \\ u(k+N-1) \end{bmatrix}$, and $\mathbf{R} = \begin{bmatrix} r & -r & 0 & \cdots & 0 \\ -r & 2r & -r & \cdots & 0 \\ 0 & -r & 2r & \cdots & 0 \\ \vdots & \vdots & \vdots & \ddots & \vdots \\ 0 & 0 & 0 & \cdots & r \end{bmatrix}$.

The aim is to calculate a control action course $u(k), u(k+1), \cdots, u(k+N-1)$ which ensures the cost function (6) to be minimal. In the following paragraphs, solution is shown for the piecewise-linear model (1), where $m = 2, n = 2$. The generalization, however, is obvious.

Let us suppose an ith linear sub-model of the piecewise-linear model (1) as follows. The upper indexes are removed for the simplicity of the notation.

$$y(k) + a_1 y(k-1) + a_2 y(k-2) = b_1 u(k-1) + b_2 u(k-2) + c. \tag{8}$$

According to the previous equation, the following set of prediction equations can be used.

$$\hat{y}(k+1) + a_1 y(k) + a_2 y(k-1) = b_1 u(k) + b_2 u(k-1) + c, \tag{9}$$

$$\hat{y}(k+2) + a_1 \hat{y}(k+1) + a_2 y(k) = b_1 u(k+1) + b_2 u(k) + c, \tag{10}$$

$$\hat{y}(k+3) + a_1 \hat{y}(k+2) + a_2 \hat{y}(k+1) = b_1 u(k+2) + b_2 u(k+1) + c, \tag{11}$$

$$\vdots$$

$$\hat{y}(k+N) + a_1 \hat{y}(k+N-1) + a_2 \hat{y}(k+N-2) = b_1 u(k+N-1) + b_2 u(k+N-2) + c. \tag{12}$$

The symbol $\hat{y}$ means that the value is predicted by using the model, while y means measured value of the system output.

Again, the set of the equations above is transformed into a matrix form. Note that the predicted and measured terms of the equation are separated.

$$\mathbf{A}_p \mathbf{y} = \mathbf{B}_p \mathbf{u} + \mathbf{A}_m \begin{bmatrix} y(k) \\ y(k-1) \end{bmatrix} + \mathbf{B}_m u(k-1) + \mathbf{C}_m c, \tag{13}$$

where $\mathbf{A}_p = \begin{bmatrix} 1 & 0 & 0 & \cdots & 0 \\ a_1 & 1 & 0 & \cdots & 0 \\ a_2 & a_1 & 1 & \cdots & 0 \\ \vdots & \vdots & \vdots & \ddots & \vdots \\ 0 & 0 & 0 & \cdots & 1 \end{bmatrix}$, $\mathbf{B}_p = \begin{bmatrix} b_1 & 0 & 0 & \cdots & 0 \\ b_2 & b_1 & 0 & \cdots & 0 \\ 0 & b_2 & b_1 & \cdots & 0 \\ \vdots & \vdots & \vdots & \ddots & \vdots \\ 0 & 0 & 0 & \cdots & b_1 \end{bmatrix}$, $\mathbf{A}_m = \begin{bmatrix} -a_1 & -a_2 \\ -a_2 & 0 \\ 0 & 0 \\ \vdots & \vdots \\ 0 & 0 \end{bmatrix}$, $\mathbf{B}_m = \begin{bmatrix} b_2 \\ 0 \\ 0 \\ \vdots \\ 0 \end{bmatrix}$, $\mathbf{C}_m = \begin{bmatrix} 1 \\ 1 \\ 1 \\ \vdots \\ 1 \end{bmatrix}$.

Predictor equation can be written as

$$\mathbf{y} = \mathbf{A}_p^{-1}\mathbf{B}_p\mathbf{u} + \mathbf{A}_p^{-1}\mathbf{A}_m \begin{bmatrix} y(k) \\ y(k-1) \end{bmatrix} + \mathbf{A}_p^{-1}\mathbf{B}_m u(k-1) + \mathbf{A}_p^{-1}\mathbf{C}_m c, \tag{14}$$

and after introducing matrices $\mathbf{G}$ and $\mathbf{F}_p$ we get

$$\mathbf{y} = \mathbf{G}\mathbf{u} + \mathbf{F}_p\mathbf{x}_p, \tag{15}$$

where $\mathbf{G} = \mathbf{A}_p^{-1}\mathbf{B}_p$, $\mathbf{F}_p$ consists of three matrices in a row, i.e. $\mathbf{F}_p = [\mathbf{A}_p^{-1}\mathbf{A}_m \;\; \mathbf{A}_p^{-1}\mathbf{B}_m \;\; \mathbf{A}_p^{-1}\mathbf{C}_m]$ and data vector $\mathbf{x}_p = \left[y(k)\; y(k-1)\; u(k-1)\; c\right]^{\mathrm{T}}$.

The predictor is composed of two parts - the term $\mathbf{Gu}$ is called the forced response and the term $\mathbf{F}_p\mathbf{x}_p$ is named as the free response. The free response is the system response assuming that the current and future control actions are zero - influence of the initial conditions. The forced response is the system response to the nonzero current and future control actions. The controller is able to affect only the forced response part of the predictor. For the simplicity of the notation, the free response is labeled as $\mathbf{f}$ in the following equations; i.e. $\mathbf{f} = \mathbf{F}_p\mathbf{x}_p$.

Using (15), the future course of the system output $\mathbf{y}$ can be predicted with respect to the future control action course $\mathbf{u}$. Thus, cost function in Eq. (7) can be written as follows.

$$J = (\mathbf{Gu} + \mathbf{f} - \mathbf{w})^{\mathrm{T}} (\mathbf{Gu} + \mathbf{f} - \mathbf{w}) + \mathbf{u}^{\mathrm{T}}\mathbf{Ru}. \tag{16}$$

Now, using some basic matrix algebra operations, the cost function can be rewritten as

$$\begin{gathered} J = \left(\mathbf{u}^{\mathrm{T}}\mathbf{G}^{\mathrm{T}} + \mathbf{f}^{\mathrm{T}} - \mathbf{w}^{\mathrm{T}}\right)(\mathbf{Gu} + \mathbf{f} - \mathbf{w}) + \mathbf{u}^{\mathrm{T}}\mathbf{Ru} = \\ \mathbf{u}^{\mathrm{T}}\mathbf{G}^{\mathrm{T}}\mathbf{Gu} + \mathbf{u}^{\mathrm{T}}\mathbf{G}^{\mathrm{T}}\mathbf{f} - \mathbf{u}^{\mathrm{T}}\mathbf{G}^{\mathrm{T}}\mathbf{w} + \mathbf{f}^{\mathrm{T}}\mathbf{Gu} + \mathbf{f}^{\mathrm{T}}\mathbf{f} - \mathbf{f}^{\mathrm{T}}\mathbf{w} - \mathbf{w}^{\mathrm{T}}\mathbf{Gu} - \mathbf{w}^{\mathrm{T}}\mathbf{f} + \mathbf{w}^{\mathrm{T}}\mathbf{w} + \mathbf{u}^{\mathrm{T}}\mathbf{Ru}. \end{gathered} \tag{17}$$

The previous equation can be formally simplified as

$$J = \mathbf{u}^{\mathrm{T}}\mathbf{Hu} + \mathbf{u}^{\mathrm{T}}\mathbf{g} + \mathbf{g}^{\mathrm{T}}\mathbf{u} + k = \mathbf{u}^{\mathrm{T}}\mathbf{Hu} + 2\mathbf{u}^{\mathrm{T}}\mathbf{g} + k, \tag{18}$$

where $\mathbf{H} = \mathbf{G}^{\mathrm{T}}\mathbf{G} + \mathbf{R}$, $\mathbf{g} = \mathbf{G}^{\mathrm{T}}(\mathbf{f} - \mathbf{w})$ and $k = (\mathbf{f} - \mathbf{w})^{\mathrm{T}}(\mathbf{f} - \mathbf{w})$.

The solution of the cost function minimization (18) can be found e.g. by following derivation.

$$J = \left(\mathbf{u} + \mathbf{H}^{-1}\mathbf{g}\right)^{\mathrm{T}} \mathbf{H} \left(\mathbf{u} + \mathbf{H}^{-1}\mathbf{g}\right) - \mathbf{g}^{\mathrm{T}}\mathbf{H}^{-1}\mathbf{g} + k. \tag{19}$$

Now, assuming that $\left(\mathbf{H}^{-1}\right)^{\mathrm{T}} = \mathbf{H}^{-1}$ and in unconstrained case, the cost function optimum (19) can be found analytically. The future course of the optimal control actions is

$$\mathbf{u} = -\mathbf{H}^{-1}\mathbf{g} = \left(\mathbf{G}^{\mathrm{T}}\mathbf{G} + \mathbf{R}\right)^{-1} \mathbf{G}^{\mathrm{T}} \left(\mathbf{w} - \mathbf{F}_p \mathbf{x}_p\right), \tag{20}$$

or

$$\mathbf{u} = \mathbf{L} \left(\mathbf{w} - \mathbf{F}_p \mathbf{x}_p\right), \tag{21}$$

where $\mathbf{L} = \left(\mathbf{G}^{\mathrm{T}}\mathbf{G} + \mathbf{R}\right)^{-1} \mathbf{G}^{\mathrm{T}}$.

Since only the current control action is required, the term (21) can be simplified to

$$u(k) = \mathbf{l} \left(\mathbf{w} - \mathbf{F}_p \mathbf{x}_p\right), \tag{22}$$

where $\mathbf{l}$ is the first row of the matrix $\mathbf{L}$.

The predictive controller (22) with the piecewise-linear neural model of the process can be illustrated as seen in Fig. 3.

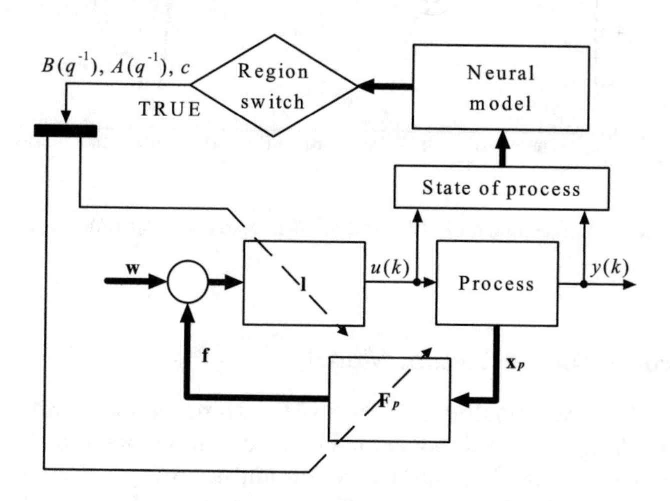

Fig. 3. Control loop.

4 Sample Application

As an application for the proposed approach, the following issue is dealt with - model predictive control simulation of the second order nonlinear process.

4.1 Nonlinear Process Model

As the controlled process, we consider a Hammerstein system consisting of a static nonlinearity in series with a second order linear system (23). The response of the process to the step function is shown in Fig. 4 in order to demonstrate the features of the controlled process.

$$\begin{aligned} u^* &= \tan(u^3), u \in [-1; 1], \\ y(k) - 1.7y(k-1) + 0.8y(k-2) &= 0.2u^*(k-1) - 0.1^*u(k-2). \end{aligned} \tag{23}$$

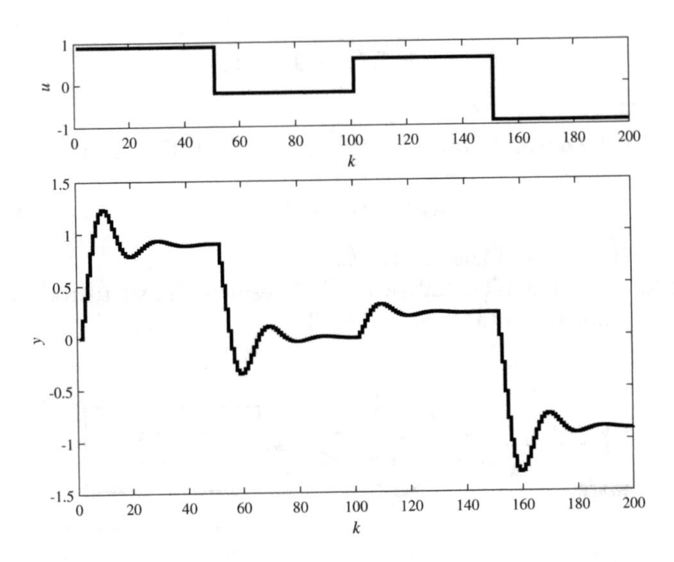

Fig. 4. Response of the controlled process to step function.

4.2 Piecewise-Linear Neural Model

In order to divide this nonlinear process (23) into set of linear subsystems, it is necessary to design a neural model of the process, where neurons in the hidden layer contain a rectifier activation function and the output neuron contains a linear (identical) activation function. This procedure involves training set acquisition, neural network training and pruning, and neural model validating. As this sequence of processes is illustrated closely in many other publications [17,18], it is not referred here in detail.

Eventually, the neural model with the structure shown in Fig. 5 is designed. The neural network used inside this model consists of four inputs ($m = 2, n = 2$), four neurons with rectifier activation functions in hidden layer, and one output neuron with linear (identical) activation function. It means that the whole state space of the controlled process can be divided up to 16 subregions.

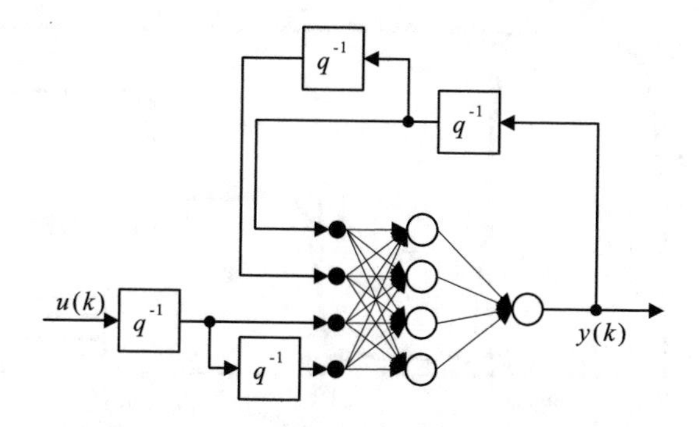

Fig. 5. Designed neural model.

4.3 Control Experiment

The piecewise-linear neural model provides the parameters a_1, a_2, b_1, b_2 and c of the linear sub-model, which is currently valid. Using these parameters, the vector $\mathbf{l}$ and the matrix $\mathbf{F}_p$ of the predictive controller can be determined and the control action $u(k)$ can be computed. Apparently, the current state of the system as well as the future course of the reference variable must be known.

Control response with the simulated process (23) is computed for the control horizon $N = 10$ and the control action penalization $r = 0.2$. The responses are shown in Fig. 6. The process output y and its reference value w are situated at the top part of the figure, while the control action is at the bottom. In addition, the regions, in which particular linear models are activated, are numbered. The numbering corresponds with the Table 1.

Table 1. Linear submodels used for the predictive controller, see Fig. 6.

Number	a_1	a_2	b_1	b_2	c
1	−1.8090	0.8614	0.2898	−0.1692	−0.0634
2	−1.6441	0.7746	0.2745	−0.0443	0.1121
3	−1.7011	0.8040	0.1194	−0.0943	0.0002
4	−1.7289	0.8313	0.2555	−0.0813	−0.0846

According to Fig. 6, the control response is quick and smooth. Switching between the linear sub-models does not bring any significant deterioration. The small steady-state control error is caused by the combination of the imperfect model of the process and a missing integral action of the predictive controller.

The control response can be tuned by the control action penalization r. In Fig. 7, the control responses for several values of r are plotted, keeping other parameters unmodified.

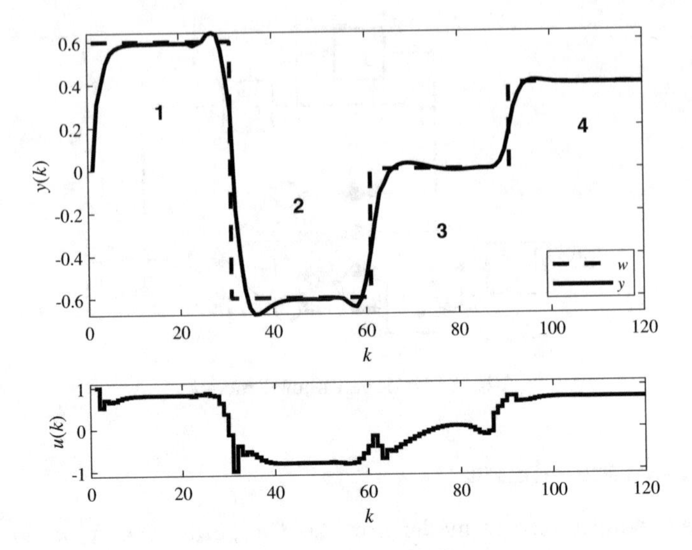

Fig. 6. Control response.

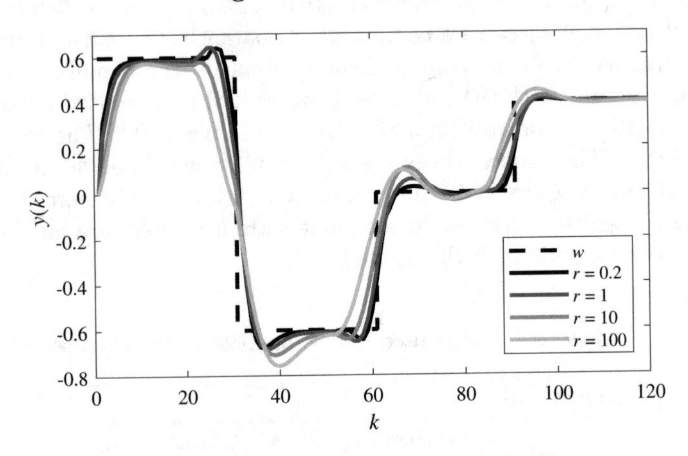

Fig. 7. Control responses for various values of penalization r.

5 Conclusion

As shown in this paper, a locally valid linear sub-model provided by a piecewise-linear neural model can be obtained not only by using the symmetric linear saturated activation function in hidden neurons, but also by the rectifier activation function. Such a model can be used for the process control design using the predictive controller in a very suitable way. The predictive controller used in this contribution deals with a particular linear sub-model provided by a piecewise-linear neural model in a very satisfactory way, as it is shown in the illustrative example. The controller design method was kept as simple as possible to keep the contribution clear and easy to follow. However, the controller design can be modified with many interesting features such as an integral action, prediction error evaluation, constraint respecting, multi-variable control, etc.

Acknowledgments. The work has been supported by SGS grant at Faculty of Electrical Engineering and Informatics, University of Pardubice, Czech Republic. This support is very gratefully acknowledged.

References

1. Heemels, W.P.M.H., Schutter, B.D., Bemporad, A.: On the equivalence of classes of hybrid dynamical models. In: Proceedings of the 40th IEEE Conference on Decision and Control (Cat. No.01CH37228), vol. 1, pp. 364–369 (2001)
2. Hedlund, S., Johansson, M.: PWL Tool, a Matlab Toolbox for Piecewise Linear System. Technical report, Department of Automatic Control, Lund Institute of Technology (LTH) (1999)
3. Vidal, R., Soatto, S., Ma, Y., Sastry, S.: An algebraic geometric approach to the identification of a class of linear hybrid systems. In: Proceedings of 42nd IEEE Conference on Decision and Control, 2003, vol. 1, pp. 167–172, December 2003
4. Ferrari-Trecate, G., Muselli, M., Liberati, D., Morari, M.: A clustering technique for the identification of piecewise affine systems. Automatica **39**(2), 205–217 (2003)
5. Bemporad, A., Garulli, A., Paoletti, S., Vicino, A.: Data classification and parameter estimation for the identification of piecewise affine models. In: 43rd IEEE Conference on Decision and Control, 2004. CDC, vol. 1, PP. 20–25, December 2004
6. Dolezel, P., Taufer, I.: Piecewise-linear artificial neural networks for PID controller tuning. Acta Montanist. Slovaca **17**(3), 224–233 (2012)
7. Dolezel, P., Skrabanek, P.: Piecewise-linear neural network - possible tool for non-linear process control. In: International Conference on Soft Computing MENDEL, pp. 245–250 (2012)
8. Dolezel, P., Rozsival, P., Mariska, M., Havlicek, L.: PID controller design for nonlinear oscillative plants using piecewise linear neural network. In: 2013 International Conference on Process Control (PC), pp. 19–24, June 2013
9. Dolezel, P., Gago, L.: Piecewise linear neural network for process control in industrial environment. In: 2016 17th International Carpathian Control Conference (ICCC), pp. 161–165 (2016)
10. Honc, D., Dolezel, P., Gago, L.: Predictive control of nonlinear plant using piecewise-linear neural model. In: 2017 21st International Conference on Process Control (PC), pp. 161–166, June 2017
11. Hahnioser, R., Sarpeshkar, R., Mahowald, M., Douglas, R., Seung, H.: Digital selection and analogue amplification coexist in a cortex- inspired silicon circuit. Nature **405**(6789), 947–951 (2000)
12. Glorot, X., Bordes, A., Bengio, Y.: Deep sparse rectifier neural networks. J. Mach. Learn. Res. **15**, 315–323 (2011)
13. Dolezel, P., Heckenbergerova, J.: Computationally simple neural network approach to determine piecewise-linear dynamical model. Neural Netw. World **27**(4), 351–371 (2017)
14. Camacho, E.F., Bordons, C.A.: Model Predictive Control in the Process Industry. Springer, New York (1997)

15. Rossiter, J.: Model-Based Predictive Control: A Practical Approach. CRC Press, London (2003)
16. Wade, C.: Model Predictive Control: Theory Practices and Future Challenges. Nova Science Publishers, New York (2015)
17. Haykin, S.: Neural Networks: A Comprehensive Foundation. Prentice Hall, New Jersey (1999)
18. Nguyen, H., Prasad, N., Walker, C.: A First Course in Fuzzy and Neural Control. Chapman and Hall/CRC, London (2002). ISBN: 1584882441

Using Growing Neural Gas Networks for Clustering of Web Data as a Foundation for Marketing Automation in Brick-and-Mortar Retailing

Felix Weber(✉)

University of Duisburg-Essen, Essen, Germany
felix.weber@icb.uni-due.de

Abstract. Even though more than 90 percent of retail executives agree to personalization in marketing being a top priority for them, only a handful deliver on this aspect. In brick-and-mortar retailing marketing actions that are conducted for large-scale grid based on an arbitrarily set distribution area - not tailored to single stores. This is mainly due to the lack of data and missing processes for automatization of marketing management. Even though the collection of data is easy to establish, the resulting data quality and condition is a problem for many clustering algorithms. Even methods of Machine Learning that are considered to be more robust to noise have issues with the unpredictable nature of such data collection, as they rely on a predefined structure. This paper presents a method for collecting a large amount of relevant location information for a major German brick-and-mortar retailer from online data sources and mapping them to a grid model of Germany at the level of 1 km^2. The clustering is conducted using methods of Machine Learning, in particular Growing Neural Gas (GNG), a neural network variation. The resulting store clusters are then tested for the purpose of location specific marketing. The GNG adapts good to the noisy data and the practical quality is feasible.

Keywords: Artificial intelligence · Neural network · Growing neural gas · Retail · Marketing automation

1 Introduction

Technology has finally advanced to the point where the largest datasets can be processed for marketing in a way that is both meaningful to customers and profitable for the company. However, even though more than 90 percent of retail executives agree to personalization in marketing being a top priority for them, only a handful deliver on this aspect. In particular, for brick-and-mortar retailers, having a personalized approach towards their customers is laborious. Due to the historically low need and the particular setting of retailing in physical stores, the customers are largely unknown and unidentifiable to the retail companies. Just the use of customer/loyalty cards or coupons poses a possibility for customer identification, but even this simple method is largely unused among the German grocery retailers. The marketing actions are not tailored to

R. Silhavy et al. (Eds.): CoMeSySo 2019, AISC 1046, pp. 13–27, 2019.
https://doi.org/10.1007/978-3-030-30329-7_2

single stores and the corresponding catchment area, but rather for a large-scale grid based on an arbitrarily set distribution area. This is the result of several realities, but mainly, the needed data is largely not available, and no processes for automatization of marketing management have been put in place so far. The low utilization of methods from the fields of advanced analytics and artificial intelligence [1] is also part of a larger lack of knowledge in these fields within the domain of retailing. The collection of external data, especially from the web, could be a way of developing a more personalized understanding and thus marketing to stationary customers. Even though the collection of possible valuable data is rather straightforward, the resulting data quality and condition is a problem for many clustering algorithms. Even methods of Machine Learning that are considered to be more robust to noise have issues with the unpredictable nature of such data collection. For example, the mapping capability of Artificial Neural Networks (ANN) is dependent on their predefined structure (i.e., the number of layers or the number of hidden units); for an unpredictable and fast changing data collection, this imposes a great effort for adaptation.

In this paper, we present a method for collecting a large amount of relevant location information for a major German brick-and-mortar retailer from online data sources, including internal location data (like assortment, sales history, pricing history), competition, social structure (from income to housing situation), and demographics. This data set is mapped to a grid model of Germany at the level of 1 km-squares. Based on this granular data, we build a clustering model of similar stores operated by the retail chain. The clustering is conducted using methods of Machine Learning, in particular Growing Neural Gas (GNG), a neural network variation. The learning algorithm of GNG is also called incremental learning which characteristically modifies the network topology in the learning process. This results in a certain independence from any assumptions made on the data structure and relationship in advance.

The resulting store clusters are then tested for the purpose of location specific marketing. This separation of the marketing actions away from nationwide marketing campaigns towards clusters of stores is seen as an improvement towards the optimization (for example A-B-testing can be conducted between stores within the same class), and finally, complete automatization may occur. Finally, the results are discussed under consideration of the current practice and experience of the responsible local marketing managers.

1.1 Location in Brick-and-Mortar Retailing

The selection of a suitable shop location is one of the most important decisions in stationary retailing. The term "location" is used in the trading literature in two respects. On the one hand, it is the geographical location of a retail company where it combines its production factors and where contact with consumers is to be established. On the other hand, it is about all decisions and questions concerning the placement of the central warehouses. This article deals only with the sales locations in stationary retail trade (also shops, business premises, shopping centers, etc.). The location of retail establishments whose activity is sales-oriented and whose sales areas are all characterized by relatively narrow limits must be clearly determined by the primacy of sales. Location decisions have ongoing and lasting effects on the sales result. The property of the location influencing the sales result is thus undisputed [2], so location in the retail trade—subject to the possibility of free choice being available—must be regarded as a sales policy instrument. Its importance in comparison with the other sales policy instruments is all the greater when its duration and intensity are much greater than those of all other instruments. Free and advertising policy measures, assortment policy, and service policy, for example, have more or less narrow lateral limits. They can be continuously adapted to market requirements, which also means that a measure that does not correspond to the market can be revised and changed, in most cases, without consequences that endanger the existence of the company. The constructive character of a location policy decision and the level of the associated investments, on the other hand, prohibit a short-term revision, since location policy misalignment can, at the same time, lead to the closure of the business. Within the framework of the marketing mix, location policy plays a dominant role, as it modifies the other sub-politics. Above all, assortment and price policies are closely correlated with the location quality. For example, a high-quality city location requires a different assortment level and assortment dimensions than a peripheral location or a location in a suburban center; the proximity of a department store will also determine the price level and assortment of a specialist shop.

This paper assumes that the location choice of the retail company is not influenceable and has already been implemented. So, the setting describes a situation in which the store locations are present and already operating. Moreover, the current marketing process uses the set locations as the basis for the local marketing efforts. With the store locations, the catchment area is defined, and this catchment area is the segmentation basis for all marketing efforts. A further segmentation, the narrowing down of the catchment area into different subsegments, is currently not part of the most marketing processes in brick-and-mortar retailing.

1.2 Marketing Segmentation and Catchment Area

Catchment Area of Brick-and-Mortar Retailers. Marketing decisions are delimited by the catchment area of the retailer's own location and also those of the companies that may be competitors. This is necessary because the competition-important competitors also cover the overlapping catchment area from a geographical point of view, and the market potential must therefore be calculated differently. The catchment area of a location comprises the entirety of the locations (apartments or offices) of all potential customers. Several methods are available for determining the catchment area. They are systematized differently in the literature. Generally, there are two distinguishable models: macro- and micro-market areas [3, 4]. A macro-market area is the sales area of an entire town, a historically grown retail settlement, or a shopping center. The demarcation of the market area at this aggregation level does not take company-specific factors into account and considers the overall attractiveness of the retail agglomeration. A micro-market area is the sales area of a single retail business. Here, company-specific characteristics, such as the type of enterprise, are included in the analysis. In principle, the requirements for the delimitation of a macro-market area are likely to be lower than for a micro-market area. This is because micro-market areas require the demand to be known, while the definition of the macro-market area only includes the demands attributable to the whole area, independently of how demand is distributed between competing stores and businesses.

Marketing and Market Segmentation. The market segmentation concept was first developed in the 1950s for consumer goods marketing and has now been an integral part of marketing science for many years [5]. In the concrete design of individual business relationships, it is now an economic imperative to evaluate relationships with customers and customer groups in order to allocate scarce resources, such as marketing and sales budgets as well as visiting hours in sales (which, in turn, can be expressed in monetary units) to their most productive use. The information on the optimal design of marketing and sales policy is important for entrepreneurial practice, not least because the optimal allocation of scarce resources to customers offers considerably higher potential for profit increases than the determination of the optimal total amount [6]. Market segmentation can be regarded as the process of dividing a large market into smaller groups or clusters [7, 8].

The level of detail of segmentation can range from classic segment marketing, to niche marketing, to individual marketing. The recommended procedure for segmenting markets is often described in the literature using the model "Segmenting, Targeting, Positioning" (STP) [9]. This approach divides the process of market segmentation into three main steps, which must be carried out in chronological order. The first step is the actual segmentation, i.e., the division of the overall market into individual segments through the use of suitable segmentation variables. These segments optimally represent buyer groups that are as clearly distinguishable as possible and are each to be addressed with a range of services or a specific marketing mix specially tailored to them. In order to assess the opportunities in each of the sub-markets, a supplier must now evaluate the attractiveness of the segments and, on the basis of this evaluation, determine the ones it wishes to serve. In the third step, a positioning concept is developed for each target market in order to establish a sustainable, competitive position and this is signaled to demanders. Although segmentations in service markets and retail trade are also treated separately in some cases, there are hardly any specific segmentation criteria or approaches for the latter two areas. Rather, researchers suggest that the concepts developed for B2C physical goods markets can also be used for end-user services and for retail businesses [9]. Several advantages can be derived from market segmentation. The most important benefit is that decision makers can target a smaller market with greater precision. This allows resources to be deployed more wisely and efficiently. In addition, market segmentation forges closer relationships between the customers and the company. Furthermore, the results of market segmentation can be used for decision makers to determine the particular competitive strategies (i.e., differentiation, low cost, or focus strategy).

1.3 Classical Clustering Approaches and Growing Neural Gas

Cluster Analysis is a technique used to divide a set of objects into k groups so that each group is uniform in regard to certain attributes based on the specific criteria. The cluster analysis aims to make it a popular marketing segmentation tool. Clustering algorithms can, in general, be classified as partitioning methods (e.g., k-means), hierarchical methods (e.g., agglomerative approach), density-based methods (e.g., Gaussian mixture models), and grid-based methods (e.g., self-organizing maps) [10, 11].

Heuristic Approaches. The most commonly used way to define groups of store clusters is the manual heuristic approach. Based on a few descriptive categories, the clusters are delimited from each other. Mostly, the resulting clusters follow a practical pattern related to the organizational setting of the company. For example, the number of groups follow geographical or regional aspects. Other groupings could follow the logic of different store types or differences within the assortments or store layout.

K-Means. The basic idea of the agglomerative procedures is the successive combination of the most similar observation units [12]. Once units have been combined into aggregates, they are no longer re-sorted into different aggregates during the agglomeration process but are combined as a whole in subsequent steps to form larger aggregates. This creates the hierarchical system of aggregates described above, which forms a coarse sequence of partitions of the sample. K-means, on the other hand, optimizes a given partition by a series of repositioning steps of individual observations from one aggregate to another. The number of aggregates remains unchanged. The optimality criterion is a measure of the heterogeneity of aggregates and for partitions, namely, the sum of quadrilateral. This measure is a so-called index. The smaller this index is, the more homogeneous the aggregates are and the better they can be interpreted as clusters. We are therefore looking for the partition with the smallest index, given by the number of aggregates.

Kohonen Networks or Self-organizing Maps (SOM). Kohonen networks are a type of unsupervised learning algorithm that uncovers structures in data. They are also called self-organizing maps (SOM) [13]. The input vectors are compared with the weight vectors. As a rule, the Euclidean distance is the measure of similarity. The neuron with the smallest distance or the highest similarity to the input pattern wins and receives the entire activation. The weighting of the winning neuron in the input layer is modified so that the similarity increases further. Geometrically, the algorithm shifts the weight vector in the direction of the input vector. So far, the Kohonen network works like other clustering methods. In order to generate the topological structure, the weight vectors of the neighboring neurons of the winner also change. This requires a definition of neighborhoods that implement different functions, such as the Gauss function or the cosine. These functions provide a measure of the distance of each neuron in the Kohonen layer from the winning neuron, which influences the intensity of the weight change. The closer a neuron is to the activated neuron, the more strongly its weight vector is adapted. The vectors of very close neurons are thus always shifted in similar directions. This creates clusters in which similar patterns are mapped [13].

Problems with the Classical Clustering Approaches. Within data analysis, the problems are reduced to a set of the most informative features which should describe the objects best. This kind of abstraction is needed to use these smaller sets of characteristics to compose a vector in some multidimensional space of features. Then, a certain kind of similarity between these vectors is measured. Such similarity can be measured quantitatively by the closeness (proximity) of vectors to each other in the space. The closeness between these vectors within that space is then used to determine the similarity; this is usually understood as the Euclidean distance or the Mahalanobis distance if there is a correlation between the vectors. With Big Data, the amount of input data is now much larger, as more information and data can be recorded and used. The attributes to describe an object or problem are now becoming much more available. Mitsyn and Ososkov [14] pointed out that the major difficulties in solving this kind of problems are as follows:

- the number of measurements to be processed become extremely large (upto 106 and more);
- new data sources and measurements become dynamically available over time;
- the space used for grouping the input data into regions has many more dimensions;
- no preliminary information is available about the number and locations of the sought for regions.

At this point, the neural gas algorithms come into play. Unlike the Kohonen networks or SOM, neural gas is intended for data clustering and not for data visualization so that it can use an optimal data lattice and is not limited by any topological restrictions. In terms of quantization error or classification, therefore, better results can often be obtained [15]. The ability of methods associated with machines to learn and recognize peculiar classes of objects in a complex and noisy space of parameters and to learn the hidden correlations between parameters of objects has been shown to be particularly suitable for the problem of neural networks and neural gas classification [16].

Neural Gas and Growing Neural Gas. Neural Gas (NG) is based on works by Martinetz and Schulten [17]. It is a single-layered neural net that is trained with an unsupervised learning process. The weights of the neurons correspond to vectors in the input space. Each neuron represents a subspace of the input space in which all data points to its weight vector have a smaller Euclidean distance than to the weight vectors of the other neurons. The Growing Neural Gas (GNG) is an incremental neural network that can be used with both a supervised and an unsupervised learning procedure depending on the problem. It was developed in the early nineties by Fritzke [18]. The number of layers of adaptable weights depends on the training algorithm used. Monitored learning requires two layers since its basic structure is strongly based on RBF networks. In contrast, unsupervised training requires only one layer, analogous to the carbon maps and the neural gas. The goal of an unsupervised GNG is to represent an input data distribution with minimal quantization error; in contrast, a supervised GNG should classify input data with as little error as possible. The neurons of the GNG layer each represent a subspace of the input space in which all data points have a smaller Euclidean distance to their weight vector than to the weight vectors of other neurons. I.e., a Voronoi decomposition of the input space takes place. Furthermore, the neurons of the GNG layer correspond to nodes in an undirected graph, whose edges are constructed dependent on the neighborhoods of the weight vectors in the input space. The incremental architecture of the Growing Neural Gas allows an optimal adaptation of the network topology and the structure of the graph to the respective problem. For this purpose, starting from a minimal number, successive neurons can be inserted into the GNG layer, but also implicitly removed. The criteria that control network growth depend on the basic paradigm (supervised or unsupervised). Accumulator variables and neighborhood relations in the input space are decisive for the growth process. Neurons are inserted where the accumulated error is greatest. This reduces the total error of the network. The structure of the graph is continuously adapted. All edges have an aging parameter. Edges that have reached a maximum age are deleted. Nodes (or neurons) that have no more edges are also removed. An explicit mechanism for deleting nodes is not required.

Growing Neural Gas for Clustering. The extension and application of GNG following the initial idea [17] are manifold. GNG is also used for a wide range of application scenarios like soft competitive learning [19]. All recent developments in GNN, including supervised learning and unattended learning, are examined in Xinjian, Guojian [20].The suitability of GNG for clustering is show in a wide range of different domains, even for astronomical images [21] or clustering of data stream [22].

1.4 Research Design

Regarding the Above Outlined Work, the Following Research Questions are Proposed:

(RQ1) Is it Possible to Gather Enough Data from the Web to Build Store Clusters? It is not clear whether it is feasible to rely on freely available data collected from the Internet in order to build store clusters for the case-study company. Major marketing research companies have business models built around the fact that retail companies currently solely rely on their data for such decisions. We aim to investigate whether this dependence is justified or whether more independence and thus a potential for cost savings and a competitive advantage is possible.

(RQ2) Can the Neural Gas Algorithm Produce Suitable Store Clusters that Have High Intra-class and Low Inter-class Similarity? As outlined above, many algorithms and methods already exist in the literature for clustering. Our resulting clusters need to be tested against standard quality measures for clusters in order to judge the usefulness of the proposed method. The ability to adapt to changing data inputs and being robust against noise is not enough to justify the algorithm's use in practice.

(RQ3) Are the Resulting Store Clusters Suitable for Segmenting Marketing Activities and Thus Use as the Foundation for Marketing Automation in Brick-and-Mortar Retailing? The resulting clusters need to be tested not only against a mathematically theory-based valuation but also against the practical usability and usefulness within the current marketing practice.

2 The Data and Sources

Several authors [22, 19] described the availability of relevant information in price management, an integral part of the decisions within the marketing mix, as the basic prerequisite for effective design. A fundamental distinction can be made between internal information (on products, customers, costs, branches) and external information (on the market structure, the competitive environment or the target group and customer behavior), as well as primary statistical and secondary statistical information [23], whereby the integration of external and internal information is one of the most serious challenges in marketing operations [24]. Primary statistics is the form of (statistical) data collection that takes place specifically and exclusively for statistical purposes, e.g., a population census. Secondary statistics is the term used to describe a form of statistical collection that consists essentially of the transfer of data that were not originally collected for statistical purposes, e.g., the transfer of accounting data for statistical purposes (Table 1).

Table 1. Overview of used data sources

Type	Source	Collected information
Direct available (primary statistics)	Internal ERP system	Master data
	Bureau for census 2011	1 km grid, inhabitants, foreigners, gender, age, household size, vacancy rate, living space, average living space
	Federal statistical office	Average household income
	Local registration offices	Average household income
Web scraping (secondary statistics)	Websites	Master data, event data, competition data
	Twitter API	Ratings, sentiments
	Facebook API	Ratings, sentiments
	Google places API	Ratings, competitions
	Real estate platforms (partly API)	Average rental rate, real estate quality

The use of primary statistical information sources usually has the advantage that this data is already available because it was originally collected for a purpose other than price management. On the other hand, for the collection of secondary statistical information, separate processes must be set up or purchased as an external service.

The direct available data comes from several sources that provide prepared and cleansed data that is easily available and mostly identifiable by geo-coordinates or name tags (cities or other geographical markers). The standard source in current retail marketing practice is the procurement of data from major marketing and market-research companies, such as The Nielsen Company. By default, the data is only available at the level of the 5-digit postal code areas and municipalities.

The internal ERP system already offers the locations and rudimentary information about the operating stores. This includes the name, location, type, sales data, and assortment.

The Bureau of the Census in Germany offers a collection of information collected within the last census in Germany, which took place in 2011.

Each inhabitant is assigned to an address and thus to a grid cell with a side length of 1 km. Members of the Bundeswehr, the police authorities, and the Foreign Service working abroad as well as their resident families are not taken into account in this evaluation.

Persons with a non-German nationality are counted as foreigners. When classifying citizenship, a distinction is made between persons with German and non-German citizenship. Persons with German nationality are regarded as Germans, irrespective of the existence of other nationalities.

The gender of each person was recorded as “male” or “female” in the 2011 census. No further specifications are planned, as this corresponds to the data provided by the residents’ registration offices on 9 May 2011.

The average age of the population (in years) is the ratio of the sum of the age in years of the total population to the total population per grid cell.

The average household size is the ratio of the number of all persons living in private households to the total number of private households per km^2 and is based on the data from the 2011 census. In this context, secondary residents are also taken into account. A private household consists of at least one person. This is based on the “concept of shared living”. All persons living together in the same dwelling, regardless of their residential status (main/secondary residence), are considered members of the same private household, so there is one private household per occupied dwelling. Persons in shared and institutional accommodation are not included here, only persons with their own managed household.

The vacancy rate (apartments) is the ratio of vacant apartments to all inhabited and vacant apartments per grid cell in percent. The following dwelling are not taken into account: holiday and leisure apartments, diplomatic apartments/apartments of foreign armed forces, and commercially used apartments. The calculation is made for dwellings in residential buildings (excluding dormitories) and is based on data from the previous census in Germany (2011).

Living space is the floor area of the entire apartment in m^2. The apartment also includes rooms outside the actual enclosure (e.g., attics) as well as basement and floor rooms developed for residential purposes. For the determination of the floor space, the rooms are counted as the following:

- Full: the floor areas of rooms/room parts with a clear height of at least 2 metres;
- Half: the floor areas of rooms/room parts with a clear height of at least 1 m but less than 2 metres; unheatable winter gardens, swimming pools and similar rooms closed on all sides; normally a quarter but not more than half: the areas of balconies, loggias, roof gardens, and terraces.

The average living space per inhabitant is the ratio of the total living space of inhabited dwellings in m^2 to the total number of persons in inhabited dwellings per grid cell and is based on data from the 2011 census, excluding diplomatic dwellings/ apartments of foreign armed forces, holiday and leisure dwellings, and commercially used dwellings. The calculation is made for dwellings in residential buildings (excluding dormitories).

The average living space per dwelling is the ratio of the total living space of the dwellings in m^2 to the total number of dwellings per grid cell and is based on the data of the 2011 census. Not included are diplomatic dwellings/dwellings of foreign armed forces and commercially used dwellings. The calculation is made for dwellings in residential buildings (excluding dormitories).

The average income or median income in a society or group is the level of income from which the number of households (or persons) with lower incomes is equal to that of households with higher incomes. The median thus defines the average income.

The ratings from mostly anonymous web users are collected through the Twitter, Google Places and Facebook APIs. Here, a numerical rating and a textual review of each place is available. It is important to note that these reviews are preselected and censored, as all major platforms use automatic spam detection methods to delete reviews that are likely to be spam. Based on the collected review data, the sentiments are calculated and stored on an individual and aggregated level for each place.

From several different event and news platforms, such as Eventbrite, Eventime, or local newspaper announcements, the information about local events with their dates, locations, and categories are scraped and matched to the grid-map.

The same process is incorporated for competition data. This includes local shops, restaurants and other types of business. Here, all information available from the several platforms is consolidated to a single record. The records are then matched to the list of shops within the geographical distance.

From a set of different German real estate platforms, like immowelt or Immobilienscout24, the average rental rate and the real estate quality rating for each grid-map are gathered. Here, the street names within the grind are used to calculate an overall average value for both of these metrics.

The Grid. All this data was collected and then added to the core concept of a grid-based map of Germany on the granularity level of 1 km^2.

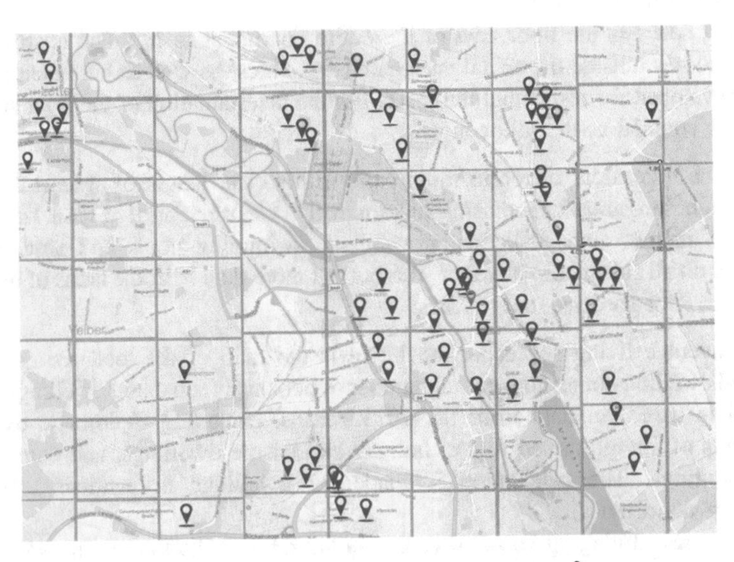

Fig. 1. Visualization of the base map (1 km^2)

The above image shows the visualization of a random example area. Each of the squares in this grid includes the following data: grid-id, geographical location (latitude and longitude), number of inhabitants, number of foreigners, gender share, average age, household size, vacancy rate, living space, average living space, average household income, average household income, average rental rate, and real estate quality. The shops, competitors, and events are then matched by their location to a single grid-id (Fig. 1).

3 Implementation

To collect the data from the diverse range of websites, several scraping bots were implemented. Web services are the de facto standard for the integration of data and services. There are, however, data integration scenarios that web services cannot fully cover. A number of Internet databases and tools do not support web services, and existing web services do not meet any user data requirements. As a result, web data scraping, one of the oldest web content extraction techniques, is still able to provide a valuable service for a wide range of applications, ranging from simple extraction robots to online meta servers. As much of the needed data in this project were not directly available, simply due to the fact that no API exists, most of the data were collected by scraping bots (independent computer program for collecting data from a certain website or set of similar websites). Here, a single bot is set up, using the programming language Python, for each data source that is automatically searching and extracting the needed data. As a subsequent step, the data are cleansed and validated. Finally, the data are exported to the central SAP HANA in-memory database. For the data sources with a public accessible API available, the same procedure is set up. All bots are built to regularly check on possible updates and data changes. Using the native Streaming Server this process could also be set up in real-time as the events (data) enters the system [25], due to the nature of the relative stable data set used here, we waived that idea for the initial prototype.

Based on the data available from the central database, the bot for the Growing Neural Gas is established. At this stage, the newly available data are considered to be added to the existing network and the cluster is re-evaluated.

The central database is not solely used as data storage but also for sentiment analysis of the collected review data sets (Fig. 2).

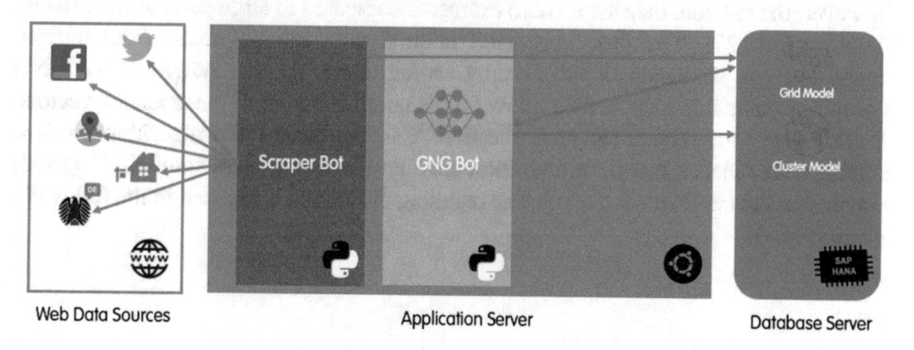

Fig. 2. System architecture of the implementation

Growing Neural Gas Algorithm and Approach. Following [18] the GNG algorithm is constructed presented by the following pseudocode:

1. Start with two units a and b at random positions w_a and w_b in R^n. Matrices must also store connection information and connection age information. In addition, each neuron has an error counter, representing the cumulative error.
2. Generate an input vector ξ according to $P(\xi)$.
3. Find the nearest unit s1 and the second-nearest unit s2.
4. Increment the age of all connection emanating s1. Also sum up the error counter with $\Delta error(s1) = ||ws1 - \xi||^2$.
5. Compute the displacement of s1 and its direct topological neighbors in the direction of ξ with $\Delta w_{s1} = \varepsilon_b(\xi - w_{s1})$; $\Delta w_n = \varepsilon_n(\xi - w_n)$. Whereby $\varepsilon_b\, \varepsilon_n \in (0, 1)$ represents parameters for adjusting the movement.
6. Reset the connections age, if s1 and s2 were connected, otherwise create one. Delete all connections with an age higher than the predefined maximal age amax. Also delete neurons which have no emanating connections. Increment the age of all emanating connection of s1 by one.
7. Every λ-iteration locate the neuron q with the greatest value of its error counter and its topological neighbor with the highest error counter f. Create a new neuron r between q and f with $w_r = \left((w_q + w_f)/2\right)$. Also create connections among r, q and f and delete the original connection between q and f. Reduce the error counter of q and f by a multiplication with α. Initialize the error counter of the new neuron with the value of f.
8. Reduce the error counter of all neurons by multiplication with δ and continue with step one.

4 Findings and Discussion

Regarding the first research question "Is it possible to gather enough data from the web to build store clusters?" (**RQ1**) a comparison to related literature is expedient. The majority of literature regarding the decision about the location selection of new retail stores, a clear distinction between internal and external (trade area) variables can be made. Regarding the present data set, it becomes clear that internal data is contempt. However, the external data set is more extensive compared to large parts of the existing literature [26, 27]. The less considered internal data can be added from internal resources in an extension of the existing setting. Here, the advantages of the GNG algorithms come into play as they allow for a smooth adaption to the extended vectors.

To answer the second research question "Can the Neural Gas algorithm produce suitable store clusters that have high intra-class and low inter-class similarity?" (**RQ2**) a mathematical check is conducted. The resulting cluster are visualized in the following image (Fig. 3).

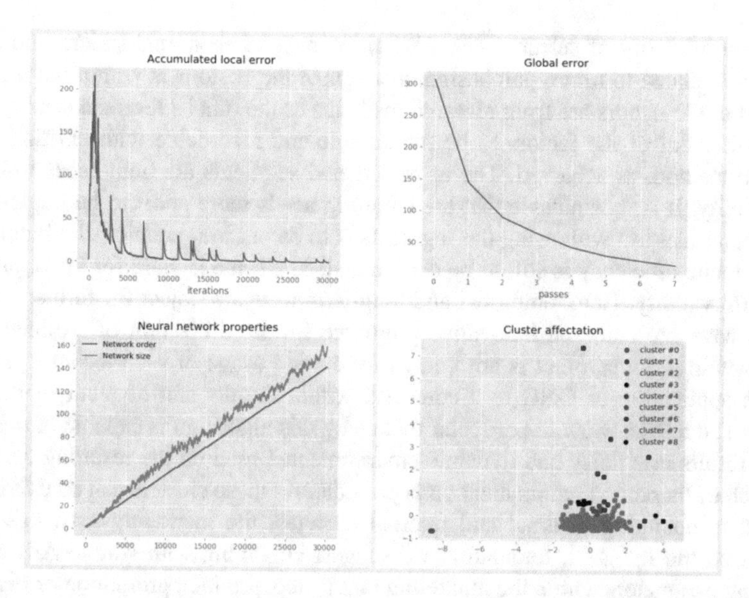

Fig. 3. Two-dimensional representation of the dataset and resulting clusters: the accumulated and global error, the network order and size and a high-level visualization of the resulting clusters

A total of 9 different cluster out of 3761 shops seems low at the first sight but for marketing praxis in retailing that is a quite feasible. As most of the changes within the marketing mix require also physical modifications in the store (change of assortment or placement of products).

So, besides the mathematical quality, that is prescribed by the GNG, the usability for the domain of brick-and-mortar-retailing is feasible.

To answer the latest question "Are the resulting store clusters suitable for segmenting marketing activities and thus use as the foundation for marketing automation in brick-and-mortar retailing?" (RQ3) we follow a guideline developed by Kesting and Rennhak [9].

Criteria for segmentation must meet certain conditions. The literature [28, 29] usually places six requirements on them, one of the purposes of which is to ensure that market partitioning is expedient (see table below) (Table 2).

Table 2. Requirements for segmentations in literature

Requirements for segmentations	
Relevance on purchase behavior	Suitable indicators for future purchasing behavior
Measurability (operationality)	Measurable and recordable with existing market research methods
Accessibility and accessibility	Ensuring a targeted approach to the formed segments
Capacity to act	Ensuring the targeted use of marketing instruments
Economic efficiency	The benefits of the survey should be greater than the costs involved
Temporal stability	Long-term validity of the information collected by means of the criteria

As the majority of selected variables are related to economic (micro and macro) factors are related to future purchasing behavior of the customers within the segments. As most of the factors are from research institutes or the official German census bureau the requirement of the factors to be measurable and recordable with existing market research methods is achieved. The to be targeted segments are both accessibility and accessibility. It is indisputable that the targeting needs more effort to be implemented. As an empirical test with using the segments is to be put into practice, the judgment of the economic efficiency is still to be determined. The same is valid for the request of a long-term validity. Here, empirical and long-term tests are object for further research.

We have presented an innovative prototype for the utilization of artificial intelligence within a domain that is not known for a wide usage of such technologies [30].

The application of GNG to a core task within the domain of retail marketing is unique. The results show in particular that using this algorithm is capable of delivering usable results in a noisy and dynamic environment. Based on the resulting clusters an approach of marketing automatization is possible. As these clusters can be dynamically adapted to changing external and internal changes, the marketing activities can be tailored on the fly. As a foundation we suggest beginning with simple A/B-tests for changing parameters within the marketing mix (price, product, promotion or place) and the derivation of tailored marketing activities based on that.

References

1. Weber, F., Schütte, R.: A domain-oriented analysis of the impact of machine learning—the case of retailing. Big Data Cogn. Comput. **3**(1), 11 (2019)
2. Kari, M., Weber, F., Schütte, R.: Datengetriebene Entscheidungsfindung aus strategischer und operativer Perspektive im Handel. HMD Praxis der Wirtschaftsinformatik (2019)
3. Schöler, K.: Das Marktgebiet im Einzelhandel: Determinanten, Erklärungsmodelle u. Gestaltungsmöglichkeiten d. räumlichen Absatzes. Duncker & Humblot (1981)
4. Schröder, H.: Handelsmarketing Methoden und Instrumente im Einzelhandel, 1st edn. Redline Wirtschaft, München (2002)
5. Wedel, M., Kamakura, W.A.: Market segmentation: Conceptual and Methodological Foundations, vol. 8. Springer, New York (2012)
6. Doyle, P., Saunders, J.: Multiproduct advertising budgeting. Mark. Sci. **9**(2), 97–113 (1990)
7. Smith, W.R.: Product differentiation and market segmentation as alternative marketing strategies. J. Mark. **21**(1), 3–8 (1956)
8. Weinstein, A.: Market segmentation: Using Niche Marketing to Exploit New Markets. Probus Publishing, hicago (1987)
9. Kesting, T., Rennhak, C.: Marktsegmentierung in der deutschen Unternehmenspraxis. Springer, Hidelberg (2008)
10. Huang, J.-J., Tzeng, G.-H., Ong, C.-S.: Marketing segmentation using support vector clustering. Expert Syst. Appl. **32**(2), 313–317 (2007)
11. Jiang, H., Kamber, M.: Data Mining: Concept and Techniques, pp. 26–78. Morgan Kaufmann Publishers Inc., San Francissco (2001)
12. MacQueen, J.: Some methods for classification and analysis of multivariate observations. In: Proceedings of the Fifth Berkeley Symposium on Mathematical Statistics and Probability, Oakland, CA, USA (1967)

13. Kohonen, T.: Self-organization and Associative Memory, vol. 8. Springer, Heidelberg (2012)
14. Mitsyn, S., Ososkov, G.: The growing neural gas and clustering of large amounts of data. Opt. Mem. Neural Netw. **20**(4), 260–270 (2011)
15. Cottrell, M., et al.: Batch and median neural gas. Neural Netw. **19**(6–7), 762–771 (2006)
16. Brescia, M., et al.: The detection of globular clusters in galaxies as a data mining problem. Mon. Not. R. Astron. Soc. **421**(2), 1155–1165 (2012)
17. Martinetz, T., Schulten, K.: A "neural-gas" network learns topologies (1991)
18. Fritzke, B.: A growing neural gas network learns topologies. In: Advances in Neural Information Processing Systems (1995)
19. Chaudhary, V., Ahlawat, A.K., Bhatia, R.: Growing neural networks using soft competitive learning. Int. J. Comput. Appl. **21**(3), 1–6 (2011). (0975-8887)
20. Xinjian, Q., Guojian, C., Zheng, W.: An overview of some classical growing neural networks and new developments. In: 2010 2nd International Conference on Education Technology and Computer (2010)
21. Angora, G., et al.: Neural gas based classification of globular clusters. In: International Conference on Data Analytics and Management in Data Intensive Domains. Springer (2017)
22. Ghesmoune, M., Lebbah, M., Azzag, H.: A new growing neural gas for clustering data streams. Neural Netw. **78**, 36–50 (2016)
23. Hartmann, M.: Preismanagement im Einzelhandel. 1. (ed.) Gabler Edition Wissenschaft. Dt. Univ.-Verl, Wiesbaden (2006)
24. Diller, H., Preispolitik. 3. (ed.). Kohlhammer, Stuttgart (2000)
25. Weber, F.: Streaming analytics—real-time customer satisfaction in Brick-and-Mortar retailing. In: Cybernetics and Automation Control Theory Methods in Intelligent Algorithms. Springer, Cham (2019)
26. Mendes, A.B., Themido, I.H.: Multi-outlet retail site location assessment. Int. Trans. Oper. Res. **11**(1), 1–18 (2004)
27. Themido, I.H., Quintino, A., Leitão, J.: Modelling the retail sales of gasoline in a Portuguese metropolitan area. Int. Trans. Oper. Res. **5**(2), 89–102 (1998)
28. Meffert, H., Burmann, C., Kirchgeorg, M.: Marketing Grundlagen marktorientierter Unternehmensführung, Konzepte, Instrumente, Praxisbeispiele. 9. Auflage ed., Wiesbaden: Gabler (2000)
29. Freter, H.: Marktsegmentierung (= Informationen für Marketing-Entscheidungen). Stuttgart (1983)
30. Weber, F., Schütte, R.: State-of-the-art and adoption of artificial intelligence in retailing. Digital Policy, Regulation and Governance. https://doi.org/10.1108/dprg-09-2018-0050

Development Method of Building a Modular Simulator of Quantum Computations and Algorithms

Vyacheslav Guzik, Sergey Gushanskiy, and Viktor Potapov(✉)

Department of Computer Engineering, Southern Federal University, Taganrog, Russia
{vfguzik, smgushanskiy, vpotapov}@sfedu.ru

Abstract. The article assumes a description of the fundamentals of the theory of quantum computing in the field of quantum algorithms. A universal concept of a quantum algorithm is given, and the time of operation of the algorithm with the determination of the probability of a particular result at the output is theoretically described. A method for constructing a modular simulator of quantum computations and algorithms, its architecture and the interactions of its various components is considered. The paper developed a method for constructing a quantum algorithm for graph interpretation, which is a study of the relationship between classical and quantum elements and concepts. An algorithm for graph interpretation and elimination (reduction) of graph vertices is built, and a method of paralleling an undirected graph model by fixing the values of graph vertices is implemented. The advantage of this strategy is that all these assessments can be carried out in parallel. In this paper, an assessment was made of the complexity of a particular algorithm based on the complexity function and a universal formula for calculating it was derived. The basics of developing quantum algorithms are described in accordance with specific software for implementing quantum algorithms and the stages of their development. Quantum algorithms involve the use of vector and matrix algebra. In accordance with this, "quantum" software is defined, including: a quantum intermediate representation of information, a quantum language of physical operations, and a quantum assembler.

Keywords: Quantum register · Quantum computer simulator · Complex plane · Qubit · Quantum algorithm · Phase amplitude

1 Introduction

Recently, there has been a rapid increase interest in quantum computing devices [1]. The use of quantum computers allows to significantly increase the speed of solving computational problems and, most importantly, exponentially increase the speed of solving NP-complete problems [2] that can be solved on classical machines in an unacceptable time. The relevance of the research topic is due to the development of this direction in the quantum world and its enormous importance in the development and implementation of quantum calculators, because without modeling the operation of

R. Silhavy et al. (Eds.): CoMeSySo 2019, AISC 1046, pp. 28–36, 2019.
https://doi.org/10.1007/978-3-030-30329-7_3

quantum algorithms [3], their results become difficult, and sometimes impossible at all, quantum computing technology – quantum computers and algorithms, quantum cryptanalysis [4]. What will be another step forward in the research of the elementary theoretical base of a quantum computing device and, as a result, the practical, physical implementation of this device, not to mention the increase in time and, accordingly, performance due to the reduction [5] of the graph [6] vertices.

Work is based on the use of quantum mechanical phenomena such as superposition [7] and entanglement [8] to convert input data into output, which can actually provide effective performance by 3–4 orders of magnitude higher than any modern computing device, which will solve the above and others tasks in natural and accelerated time scale.

2 The Algorithm for the Execution of the Quantum Computer Simulator (QCS) and the Interactions of Its Various Components

Consider the algorithm for the execution of QCS and interactions of its various components: framework, core [11], quantum gates [12], quantum circuits [13], libraries of basic quantum algorithms, interfaces for programming languages, an educational development environment, and a user interface.

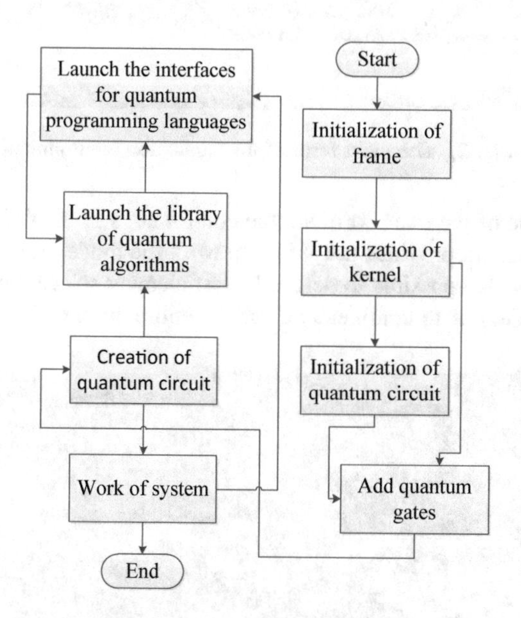

Fig. 1. Initialization of the QCS

The interdependence of the QCS parts is extremely high. The algorithm for the execution of the modeling environment and the interaction of its components is shown in Fig. 1. The framework is the structure of the QCS, uniting its various parts and components. Initialization of the framework (shell) leads to the initialization of the modeling core, which sets the number of qubits in the modeling environment. This data is needed to initialize the quantum scheme and display data on quantum bits from the core. Next, you will need to initialize the user interface [14], where most of the work on the computer will occur. In a parallel or sequential way, you can choose one of the alternatives for further work (Fig. 2).

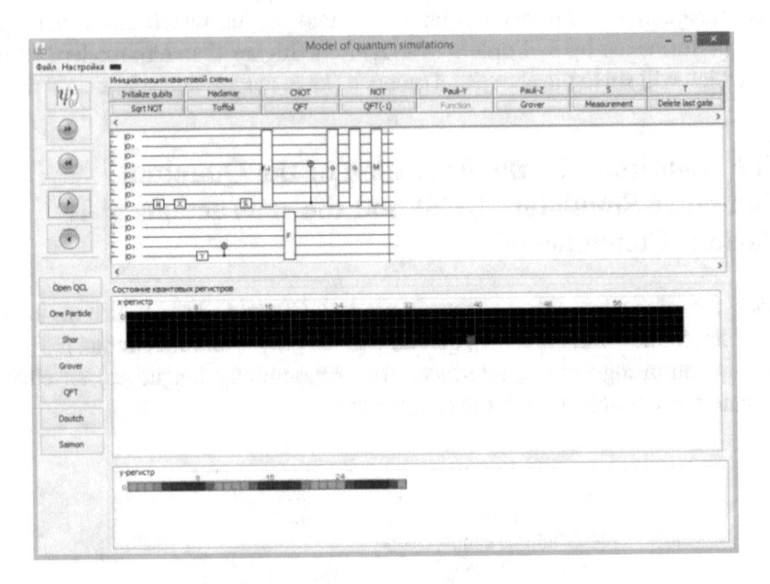

Fig. 2. The main form of the simulation environment

On the left side of the main form are the control keys of the QCS. This area of the simulation environment provides the ability to work the model in an automatic or step-by-step mode, it is also possible to delete the last element selected and recorded in the quantum circuit (QC) or to completely clear the entire circuit.

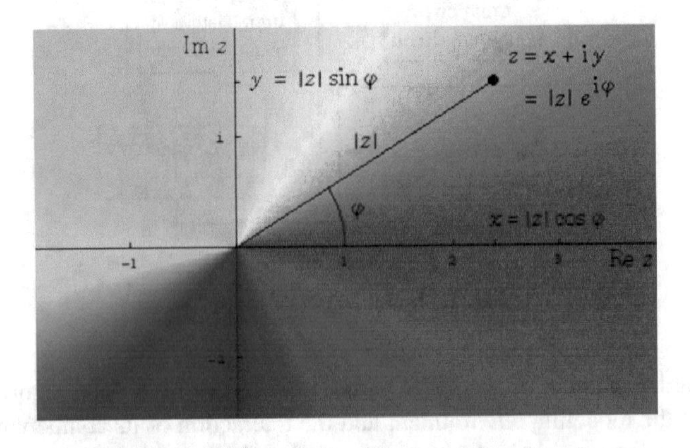

Fig. 3. Colour card register; complex plane

In the upper part is the navigation menu, with which the control and configuration of the QCS are carried out, in the center, there is a set of quantum logic gates, and at the bottom is a color complex diagram of quantum states of the X and Y registers. The result of the simulation of the QCS is reflected in the form of a color set of pixels [15]. In Fig. 3 describes a colored map in the complex plane [16]. The hue color depends on the phase [17] value ψ. A positive value is represented by a red pixel color, a negative one – a light blue. A register [18] of quantum bits of dimension n is given as 2 to the power n of basic quantum states [19]. Each of these states is represented by a single square (pixel), the phase amplitude of which reflects its hue.

If the quantum bit |j> takes the amplitude an in the interval (0, 1], then the j-th pixel is a red hue. The whole set of qubits has amplitudes Z = 0 and is coloured dark.

3 The Concept of a Quantum Algorithm

At the moment there are a large number of quantum algorithms that solve a wide variety of tasks, but there is no universal concept (definition) of a quantum algorithm that is not based on the properties of a particular quantum algorithm. In Fig. 4 represents one cycle of accessing a quantum resource. Here, using the pulses of a classical computer, a multitude of qubits S_k of a universal basis is created, then a pure state $|0^n\rangle$ is prepared. This is followed by various unitary transformations and measurement in the computational basis. The final stage is the processing of measurement results by classical means – x.

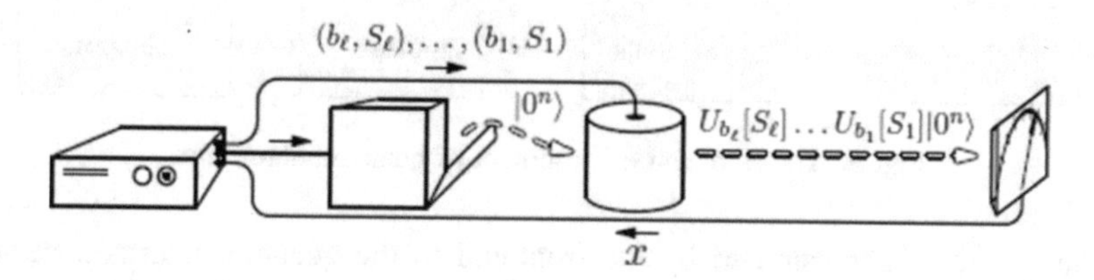

Fig. 4. The cycle of reference to the quantum resource

Thus, we give the definition of a quantum algorithm. The quantum algorithm Q is a classical algorithm A, which can be implemented on classical computing devices and implements the description of the quantum scheme C_x in a universal finite basis that implements the operator U_x on n_x qubits and the description of the result register S_x. The running time of the algorithm at input x is nothing more than the sum of the running time of algorithm A and the size of the circuit C_x. If it is necessary to implement a quantum algorithm Q, which will calculate f(x) with an error probability $\varepsilon < ½$, then the algorithm $Q_s^{'}$ will work as follows:

1. Repeat the Q algorithm S times.
2. The result is the y value that occurred most often.

The software implementation of classical algorithms differs from similar quantum development, due to the existence of a quantum informatics paradigm that requires a shift towards paradoxes. The development environment of quantum algorithms must be capable of transforming high-level quantum programs into error-resistant implementations in various quantum environments. At the same time, it should contain programming languages, compilers, optimizers, simulators, debuggers and other tools with resistance to correcting quantum errors (Fig. 5). Currently, quantum programming languages are divided into languages aimed at practical application (simulation of quantum mechanical systems, programming of quantum circuits, etc.); languages of analysis of quantum algorithms. Languages of the second kind are used when it is impossible to formally prove the algorithm's correctness/efficiency, since it can be based on unproven mathematical assumptions or heuristic methods. The algorithm can be implemented in a high-level programming language independent of quantum technology (for example, Quipper).

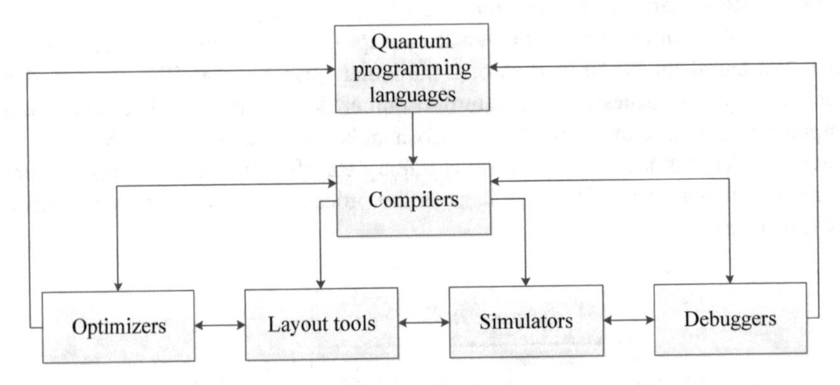

Fig. 5. Development environment of quantum algorithms

The program is transmitted by the front-end to the quantum intermediate representation in accordance with the quantum scheme. At the same time, the presence of grammar and an optimizer that performs code conversion and optimization is necessary. During the conversion, a description of quantum circuits in the quantum assembler language (QASM) occurs. Next, the optimizer transforms the QASM program into instructions of the quantum language of physical operations (QPOL). The developed algorithm is based on the tensor reduction of the network, in which the width of the graph will be the dominant factor in determining the time and computational complexity.

The algorithm involves the following set of steps:

1. Build a simplified undirected graph.
2. Divide the whole task into many subtasks, keeping the time of each subtask as short as possible. The basic idea is to remove the most expensive nodes by parallelizing their values.
3. Use this estimate to determine the effective order of eliminating the remaining nodes in each subtask.

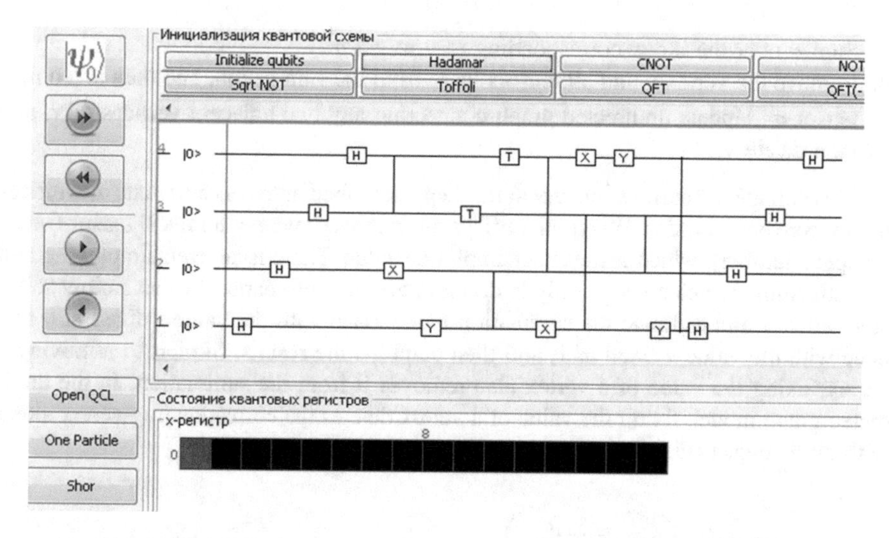

Fig. 6. Example of scheme C, which is estimated using the graph model

The simulator (Fig. 6) is based on a complex undirected graph model. This is essentially a variant of the tensor network proposed by Markov and Shi, but which also uses diagonal gates. The graph can be simplified if the tensor operators turn out to be diagonal. For example, if the nodes u and v are connected by a one-qubit diagonal gate, then one member in the integral of the Feynman trajectory can remain in the graph only when the corresponding label we choose satisfies u = v. Therefore, the two nodes u and v can be combined together. Similarly, the tensors on the left side of Fig. 7 (highlighted in the background) correspond to their graph interpretations on the right.

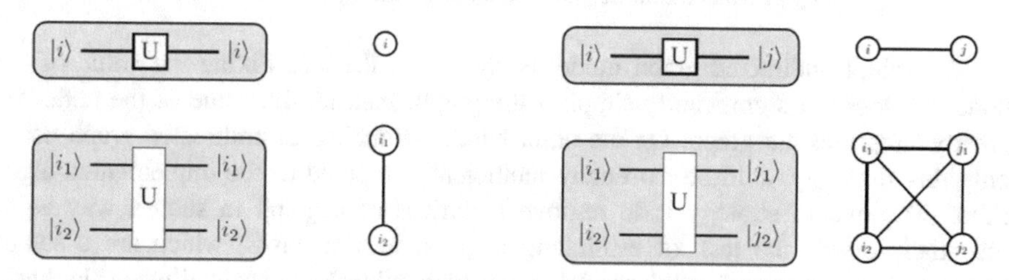

Fig. 7. One- and two-qubit diagonal and non-diagonal gates and their graph interpretation

The sum estimation algorithm is similar to the algorithm in [27] and continues by eliminating one vertex at a time. Typically, this process results in a tensor of rank greater than 2. To eliminate the binary vertex $v = i_k^j$:

1. Find the set of tensors $T_v = \{\tau | v \to \tau\}$ and the set of vertices $V_v = \bigcup_{\tau \in T_v} \{v' | v' \to \tau\}$.
2. Multiply all tensors $\tau \to T_v$ to obtain a new rank tensor σ and index $|V_v|$ vertices in $|V_v|$.

3. Sum σ over the index corresponding to v to get σ'.
4. Remove the vertex v and all tensors in T_v from the summation, and then add a new tensor σ'. Update undirected graph connecting any two adjacent vertices of v, and then delete v.

To estimate the entire sum, repeat the steps described above to eliminate all vertices in any convenient order. When all vertices are removed, we get a rank 0 tensor (i.e., a complex number), which is the exact amplitude value. There is an even simpler method for estimating, which consists only in dividing the sum into parts. We can simply select any vertex v and estimate the summation twice, once with the value v fixed at 0 and once with the value v fixed at 1, and then combine the results. Similar to removing a vertex, fixing the value of a vertex also removes it from the summation. In the undirected graph model, fixing the value of a vertex deletes the corresponding vertex along with all its edges (Fig. 8).

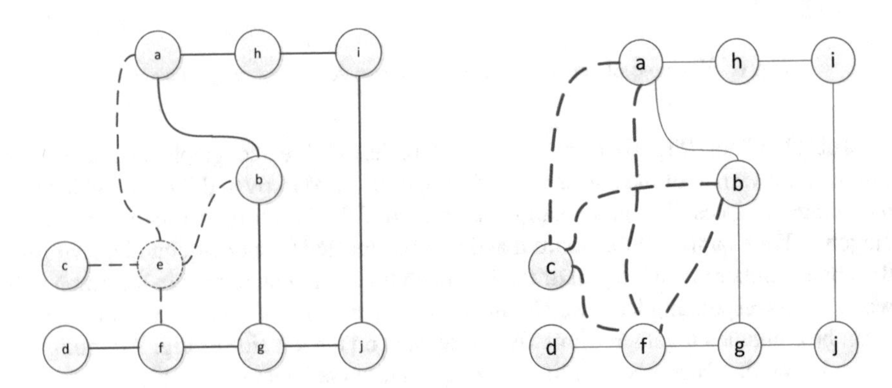

Fig. 8. Illustration of parallelization by fixing the node value

The initial undirected graph model is shown on the left. Fixing the value of the node "c" does not significantly simplify the graph. Instead, the value of the node "e" greatly simplifies the graph. On the other hand, estimating an undirected graph using only this strategy would be extremely inefficient compared to the implemented algorithm. The overall strategy is to remove t vertices of a graph in such a way as to essentially divide the task of estimating a graph into subtasks, which are a set of vertices of an undirected graph model. Then each subtask is executed using the basic algorithm. Of course, this is valid only if the cost of estimating the resulting graph is significantly lower than the cost of estimating the original graph. Further, it is possible to optimize computational processes in a quantum region by translating the reduced graph into a quantum scheme.

By the complexity of the algorithm, we mean the number of "elementary" operations performed by the algorithm to solve a specific problem in a given formal system. When using algorithms for solving practical problems, we are faced with the problem of a rational choice of a problem solving algorithm. The solution to the problem of choice is connected with the construction of a system of comparative assessments, which relies heavily on the formal model of the algorithm. A specific problem is given by N words of memory, so the input to the algorithm $N_\beta = N * \beta$ information qubits. The tool that implements the algorithm consists of M machine instructions for β_M qubits of $M_\beta = M * \beta_M$ information qubits. Obviously, for different applications, the weights of resources will be different, which leads to the following comprehensive assessment of the algorithm:

$$\varphi_\gamma = c_1 * F_a(N) + c_2 * M_\beta + c_3 * S_d + c_4 * S_\gamma \quad (1)$$

Where c_i – resource weights; S_d – memory for storing intermediate results; S_γ – memory for the organization of the computational process.

All weights were set in accordance with the importance and significance of a resource for the successful implementation of the algorithm and the operation of the device as a whole. $F_a(N)$ is the number of "elementary" operations performed by the algorithm to solve a specific problem in a given formal system.

4 Conclusion

The method for constructing a quantum algorithm for graph interpretation that is described in this paper is the study of the relationship between classical and quantum elements and concepts. The advantage of this strategy is that all these assessments can be carried out in parallel. Intuitively, deleting vertices with a high degree of accuracy should simplify the graph further; however, even a vertex with a small degree can be a "key vertex" that contains large parts of the graph together. To choose which vertices to remove, we use a greedy strategy based on the estimated time value of the basic algorithm. Thus, we obtain an increase in time and, consequently, productivity due to the reduction of graph vertices. The paper developed a method for constructing a quantum algorithm for graph interpretation, which is a study of the relationship between classical and quantum elements and concepts. An algorithm for graph interpretation and elimination (reduction) of graph vertices is built, and a method of paralleling an undirected graph model by fixing the values of graph vertices is implemented. Quantum algorithms are already beginning to be embodied in actually functioning experimental devices, and quantum computing is a fairly developed area of knowledge.

Acknowledgments. This work was carried out within the State Task of the Ministry of Science and Higher Education of the Russian Federation (Project part No. 2.3928.2017/4.6) in Southern Federal University.

References

1. Sukachev, D.D., Sipahigil, A., Lukin, M.D.: Silicon-vacancy spin qubit in diamond: a quantum memory exceeding 10 ms with single-shot state readout. Phys. Rev. Lett. **119**(22), 223602 (2017)
2. Lukin, M.D.: Probing many-body dynamics on a 51-atom quantum simulator. Nature **551** (7682), 579 (2013)

3. Potapov, V., Gushansky, S., Guzik, V., Polenov, M.: Architecture and software implementation of a quantum computer model. In: Advances in Intelligent Systems and Computing, vol. 465, pp. 59–68. Springer (2016)
4. Raedt, K.D., Michielsen, K., De Raedt, H., Trieu, B., Arnold, G., Richter, M., Lippert, T., Watanabe, H., Ito, N.: Massively parallel quantum computer simulator. Comput. Phys. Commun. **176**, 121–136 (2007)
5. Boixo, S., Isakov, S.V., Smelyanskiy, V.N., Babbush, R., Ding, N., Jiang, Z., Martinis, J.M., Neven, H.: Characterizing quantum supremacy in near-term devices. arXiv preprint arXiv: 1608.00263 (2016)
6. Stierhoff, G.C., Davis, A.G.: A history of the IBM systems Journal. IEEE Ann. Hist. Comput. **20**(1), 29–35 (1998)
7. Lipschutz, S., Lipson, M.: Linear Algebra (Schaum's Outlines), 4th edn. McGraw Hill, New York (2009)
8. Collier, D.: The Comparative Method. In: Finifter, A.W. (ed.) Political Sciences: The State of the Discipline II, pp. 105–119. American Science Association, Washington, DC (1993)
9. Vectorization.https://en.wikipedia.org/w/index.php?title=Vectorization&ldid=829988201
10. Williams, C.P.: Quantum gates (chap. 2). In: Explorations in Quantum Computing. Texts in Computer Science, pp. 51–122. Springer (2011)
11. Olukotun, K.: Chip Multiprocessor Architecture – Techniques to Improve Throughput and Latency. Morgan and Claypool Publishers, San Rafael (2007)
12. Potapov, V., Guzik, V., Gushanskiy, S., Polenov, M.: Complexity estimation of quantum algorithms using entanglement properties. In: Informatics, Geoinformatics and Remote Sensing (Proceedings of 16-th International Multidisciplinary Scientific Geoconference, SGEM 2016, Bulgaria), vol. 1, pp. 133–140. STEF92 Technology Ltd. (2016)
13. Inverter (logic gate). https://en.wikipedia.org/w/index.php?title=Inverter_(logic_gate) &oldid = 844691629
14. Lachowicz, P.: Walsh – Hadamard Transform and Tests for Randomness of Financial Return-Series. http://www.quantatrisk.com/2015/04/07/walsh-hadamard-transform-python-tests-for-randomness-of-financial-return-series/ (2015)
15. Potapov, V., Gushanskiy, S., Guzik, V., Polenov, M.: The computational structure of the quantum computer simulator and its performance evaluation. In: Software Engineering Perspectives and Application in Intelligent Systems. Advances in Intelligent Systems and Computing, vol. 763, pp. 198–207. Springer (2019)
16. Zwiebach, B.: A First Course in String Theory. Cambridge University Press, Cambridge (2009)
17. Potapov, V., Gushanskiy, S., Samoylov, A., Polenov, M.: The quantum computer model structure and estimation of the quantum algorithms complexity. In: Advances in Intelligent Systems and Computing, vol. 859, pp. 307–315. Springer (2019)
18. Universe of Light: What is the Amplitude of a Wave? Regents of the University of California. http://cse.ssl.berkeley.edu/light/measure_amp.html
19. Sternberg, R.J., Sternberg, K.: Cognitive Psychology, 6th edn. Wadsworth, Cengage Learning (2012)

A Technique of Workload Distribution Based on Parallel Algorithm Structure Ontology

A. B. Klimenko[1(✉)] and I. B. Safronenkova[2]

[1] Scientific Research Institute of Multiprocessor Computer Systems, Southern Federal University, 2, Chehova Street, 347928 Taganrog, Russia
anna_klimenko@mail.ru

[2] Federal Research Centre the Southern Scientific Centre of the Russian Academy of Sciences, 41, Chehova Street, 344006 Rostov-on-Don, Russia

Abstract. A workload distribution problem is quite topical nowadays. A large number of applications, which function in distributed environments, uses various techniques of a workload relocation. Yet very few studies consider the workload relocation in fog-computing environment emphasizing the increase of a search space and the distances between the computational nodes. In this paper a new workload distribution technique, based on ontological analysis of algorithm structures and the available resources is presented. The aim is to limit and reduce the search space of the workload distribution problem. Such strategy decreases the time of workload location and so decreases the time needed to solve the general computational task of application.

Keywords: Ontology · Workload distribution problem · Workload relocation · Distributed systems · Parallel algorithm · Fog computing · Cloud computing

1 Introduction

A workload distribution problem is topical nowadays, because of ubiquitous usage, including numerous ICSs, CSs, DCAD, etc. One of the most important criteria of such systems is their performance, which depends on many factors including the speed of workload distribution problem solving. The time needed for this procedure can be long enough, so it affects the computational process.

There exist numerous methods of workload distribution, including the ones with the reliability function emphasized [1–4]. However, only few of them take into account the "fog-computing" infrastructure peculiarities. They are:

- the number of nodes in the "fog" layer;
- the instability of nodes;
- the heterogeneity of nodes and communication channels in terms of performance and capacity;
- the distances between computational nodes, which affect the time of data transmission through the network.

Moreover, there must be the constrained set of nodes for the tasks distribution, and it is highly desirable to reduce this set. A "cloud" presupposes a limited set of nodes

R. Silhavy et al. (Eds.): CoMeSySo 2019, AISC 1046, pp. 37–48, 2019.
https://doi.org/10.1007/978-3-030-30329-7_4

and is based on the "closed" system principle. Outside the "cloud" we face quite large number of devices, which can be unstable [5].

Consequently, the issue of search space limitation forming for workload distribution problem solving is quite important for "fog-computing" concept. Such search space limitation is reached with appropriate nodes pre-selection.

In this paper a technique of workload distribution based on parallel algorithm structure ontology is proposed. Domain ontology and a new technique, which allows to choose the nodes for workload distribution in a "fog" layer, has been developed. The quantitative analysis of workload distribution techniques was conducted.

2 The Technique of Workload Relocation Based on Local Device Group

Consider a boundary forming technique for workload relocation problem solving, which is based on local device groups [6].

Given the graph of a computational problem *G*, which is split into two subgraphs, *G'* and *G"*. *G'* is considered to be bound to the computational nodes (CNs) of the network segment *P'*, while *G"* is still performed on the network segment *P"*, as is shown in Fig. 1.

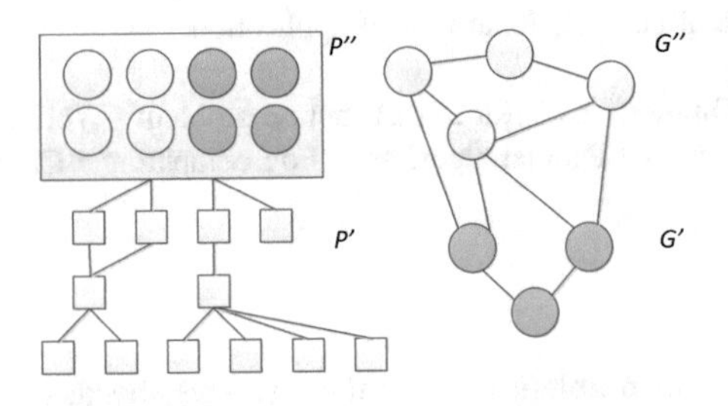

Fig. 1. Subtasks distribution for the offloading computational model.

Consider the term "local device group" (LDG) as a set of devices, which are:

- interconnected by the high-velocity communication channels without any transitional nodes;
- participated in computational task *G* solving;

It must be mentioned, that the local groups can intersect and have the same nodes.

We consider the example, when the initial location of tasks is a "cloud", so the problem is to relocate some tasks to the "fog".

As far as *G'* is considered to be brought to a "fog" layer, there must be a node in a "cloud" layer (*P"*), which is responsible for workload relocation. Consider the LDG, which solves the *G"*. In LDG a leader must be elected by any of existing methods [7]. When the leader has been elected, it sends the request for resources to its LDG (see Fig. 2).

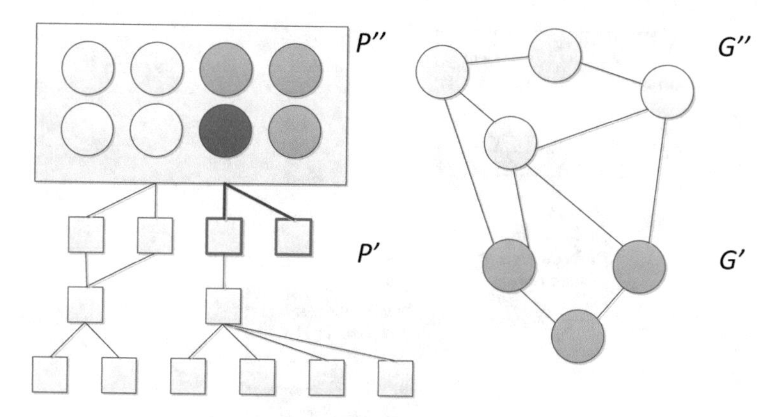

Fig. 2. The elected leader sends the requests to its LDG.

If devices answer the request with agreement to allocate the resources needed for the subtasks to be relocated, the workload distribution problem is solved within the set of those devices. If the solution is acceptable, the tasks are linked to the devices. In case if the resources are insufficient, or the solution is unacceptable, the devices send the request to their LDGs (see Fig. 3).

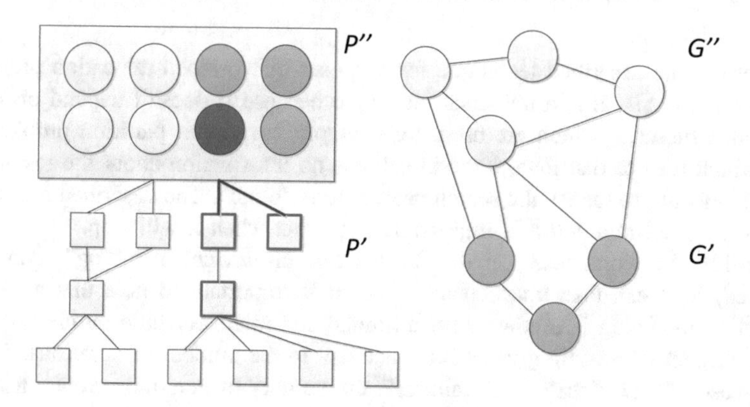

Fig. 3. The LDG enhancement procedure.

This procedure is iterative and eventually allows to form the set of nodes, which are capable of solving the computational subtasks. After the boundary conditions are formed, the problem of workload distribution is solved.

The scheme of the examined technique of workload relocated problem is presented below (see Fig. 4).

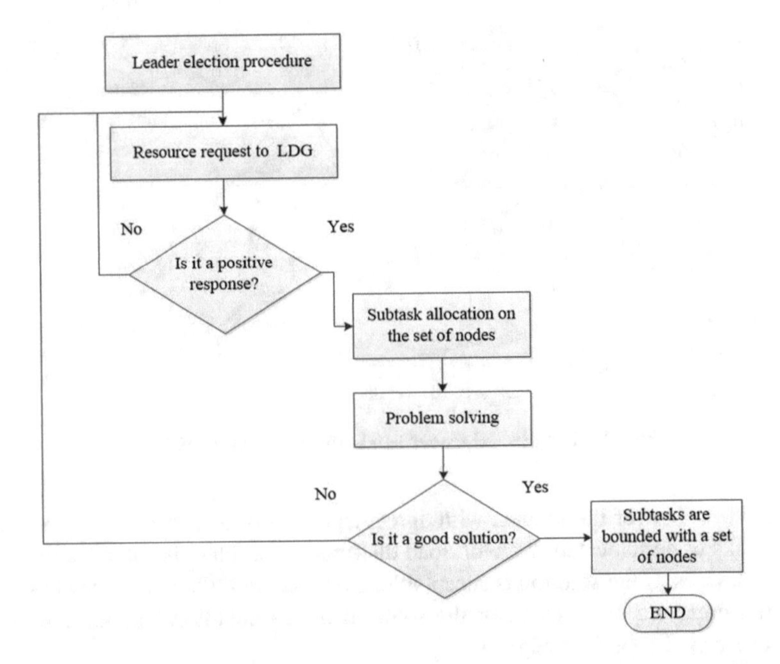

Fig. 4. The scheme of the LDG-based technique of workload relocation problem.

Consider the described technique for purposes of workload relocation problem in the field of DCAD. It is a universal technique because it doesn't depend on parallel algorithm structures, which are used, for example, for circuit partition problem. The node, which tries to distribute a workload, has no information about the resources of nodes in a local group. So, the search procedure is "blind". The described technique is an iterative procedure and it is impossible to predict when it will stop.

Consider the worst-case scenario. In this case all devices of a "fog" layer can be examined, so it can take unacceptable time. It is important to note that a workload distribution modeling is produced for a limited and relatively little number of nodes, but the number of modeling procedures is equal to the number of iterations. So, with the increase of task distribution "failures", the number of iterations grows. It is obviously not a proper solution.

As far as a limited number of known algorithms and their parallel implementations are used in modern DCAD, a following solution can be proposed.

If there is information about the structure of workload to be relocated and an opportunity to survey the fairly large number of "fog" nodes within the "fog" domain, it will be possible to choose best-fit nodes for workload distribution preliminary. In this situation we have restriction to a number of nodes, which are considered as candidates for workload distribution. By means of qualitative evaluation of relocating workload structure, the search space shrinks. It happens because we eliminate the nodes which aren't suitable for graph structure relocation.

3 A Technique of Workload Distribution Based on Parallel Algorithm Structure Ontology

In this paper a technique of workload distribution based on parallel algorithm structure ontology is proposed, which includes a procedure of ontological analysis of population algorithm structures and their parallel implementations, which are used for circuit partition problem solving in DCAD, and given information about available resources of "fog" nodes. The ontological analysis allows to decrease the number of candidate-nodes for computational task distribution according to the subtask algorithmic structure. The information about resources of "fog" layer allows to estimate resource availability and node remoteness from a "leader", which is responsible for collection, storage and updating of information.

Assume that:

1. Each node can provide information about its resource availability and remoteness from a "leader"
2. A set of "fog" nodes is sufficiently large but limited (we consider the problem within the "fog" domain, for example)
3. There is a finite set of parallel algorithm partition ways and it can be described with the help of ontology.

According to the item 3 from the assumption list, a set of algorithm partition ways is finite. For compliance with such an assumption it is necessary to impose restrictions on the ways of parallel population algorithm structures. In this case each parallel model is characterized by the individual set of partition ways.

According to the existing classification of parallel models of population algorithms, there are [8]:

- Global model;
- Island model;
- Diffusion model.
- Population algorithms, which use a global model, are focused on parallel computing by "master-slave" type [9]. In this case the following restrictions on the ways of graph partition are possible:
- "master" process relocation (see Fig. 5);

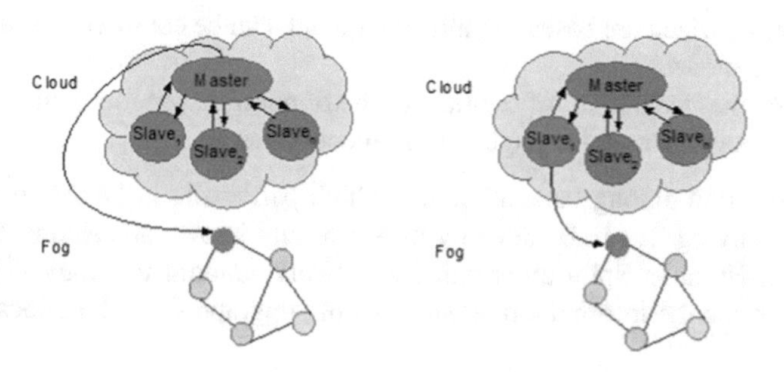

Fig. 5. The scheme of "master" and "slave" process relocation.

- “slave” process relocation (see Fig. 5);
- “master-slave” process relocation (see Fig. 6);

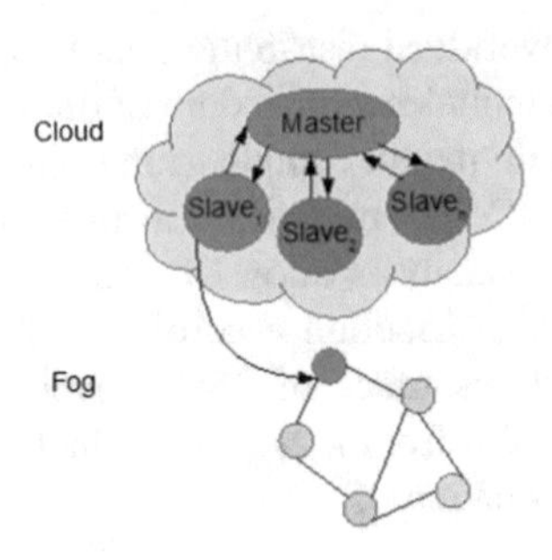

Fig. 6. The scheme of “master-slave” process relocation.

The other cases of relocation combinations are possible, but within this work they are not observed.

Migration algorithms, which are based on island model, are also called multiple-deme algorithms [10]. In this case it is sensible to formulate only one rule of graph partition – “island” (“population”) relocation (see Fig. 7).

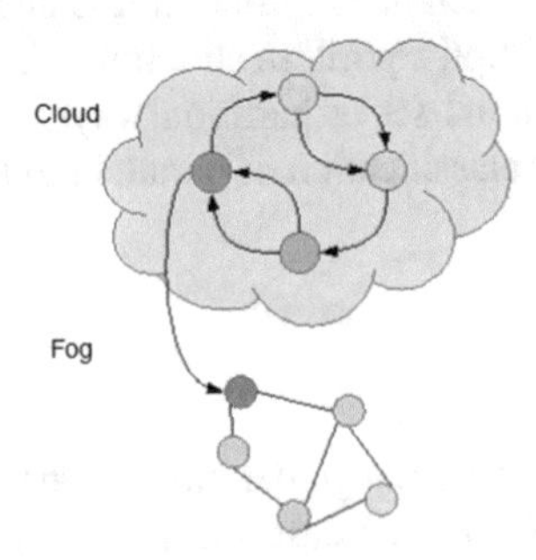

Fig. 7. The scheme of “island” process relocation.

Algorithms, which are based on diffusion model, can be considered as the particular case of island model.

Consider the technique of workload distribution based on parallel algorithm structure ontology. The main stages of the process are:

1. Leader election among “cloud” nodes, which participate in DCAD computational problem solving. It can be done by means of any known algorithms, for instance, Gallagher-Humblet-Spira algorithm for network random topology [7]. Herewith, the leader has the information about tasks of supgraph G’ to be relocated.

2. The leader node surveys the resources of the fog nodes in some predetermined neighborhood. The area to be explored must be large enough. The procedure begins from a leader-node and extends through the "fog" layer nodes. This procedure can be conducted with the help of echo wave algorithm [11]. When the procedure has been conducted, the leader node has the following information:

- resource status of the "fog" layer nodes;
- remoteness of the candidate nodes from a leader node.

Now we have information about the "fog" layer nodes. Then an ontological analysis should be conducted. On its basis the set of candidate nodes is reduced.

The ontology of parallel algorithm structure formalizes the following knowledge:

- the class of a parallel algorithm [12]
- the model of parallelization,
- the description of graph, which is considered to be relocated: initial placement ("cloud" layer, user device, "fog" layer), the way of graph partition (restrictions were examined above).

The ontology of parallel algorithm structure and the ways of algorithm partition was developed in Protege 4.2 [13], and is shown below (see Fig. 8).

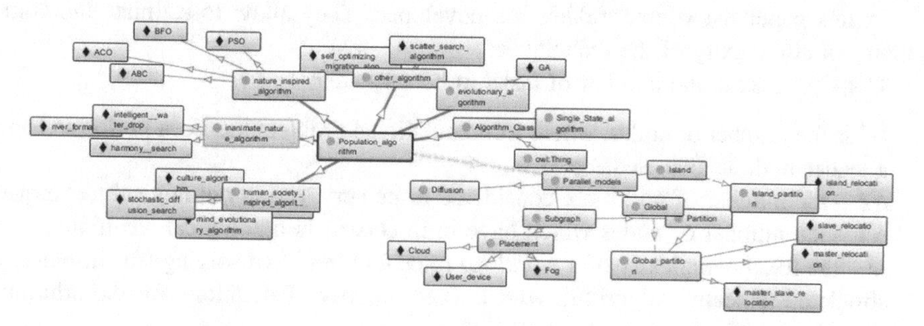

Fig. 8. The ontology of a parallel algorithm structure and its partition.

The choice of appropriate nodes for the workload distribution is made by the ontological analysis according to heuristics rules [14]. Then the simulation of computational task distribution takes place on the restricted set of "fog" nodes. The scheme of developed method is shown below (see Fig. 9).

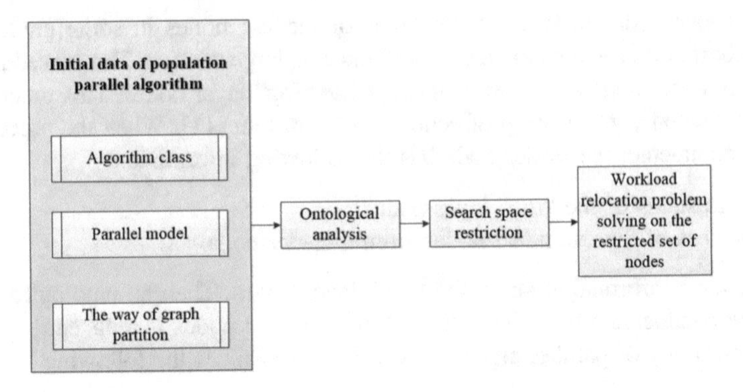

Fig. 9. The scheme of ontology-based method.

4 The Quantitative Analysis of Workload Distribution Techniques

Consider the efficiency of methods in the context of their computational workload. If the computational workload is high, the workload relocation problem needs more time to be solved.

In this paper the simple models are developed. They allow to estimate the computational complexity of the techniques.

Consider *task* is the number of tasks to be relocated;

N1 is the number of nodes, where a computational problem has been performed and a leader node is also elected;
N2 – the number of nodes are considered to be surveyed in ontological technique;
N3 – the number of nodes which have been chosen in ontological technique;
D – the fixed complexity of a workload relocation problem solving (for instance, a simulated annealing algorithm with a fixed number of iterations for the arbitrary size of a search space can be used;

In this case the complexity of a workload relocation problem solving procedure does not depend on the number of tasks and the number of nodes. In this case a decision quality can be low while the problem solving time is limited.

*N_* - a number of nodes, which are ready to give their resources and have been found on the basis of local group technique.

iteration – the number of iterations, which have been conducted on the basis of local group technique to reach a solution.

Consider the complexity of each procedure, which is performed within the ontology-based technique. At the beginning a problem is performed in a "cloud" on *N1* nodes. Then the complexity of election procedure is *kN1,* where *k* – is a ratio between the input data and a computational complexity.

The computational complexity models of workload distribution problem solving are presented below.

The first case: the complexity of workload distribution is fixed:

$$Eo = kN1 + 2 \cdot kN2 + D \tag{1}$$

$$Eg = kN1 + (kN\_ + D) \cdot iteration \tag{2}$$

where Eo – is a workload distribution complexity based on ontological technique;

Eg – is a workload distribution complexity based on LDG-based technique

The second case: the complexity of a workload relocation problem is proportional to a number of subtasks and a number of nodes:

$$Eo = kN1 + 2kN2 + kN3 \cdot task \tag{3}$$

$$Eg = kN1 + (kN\_ + kN\_ \cdot task) \cdot iteration \tag{4}$$

Consider the simulation results on the basis of developed models.

Parameter "iteration" is varied while other parameters are constant for the case when D is fixed. The diagram of workload relocation problem complexity is shown below (see Fig. 10).

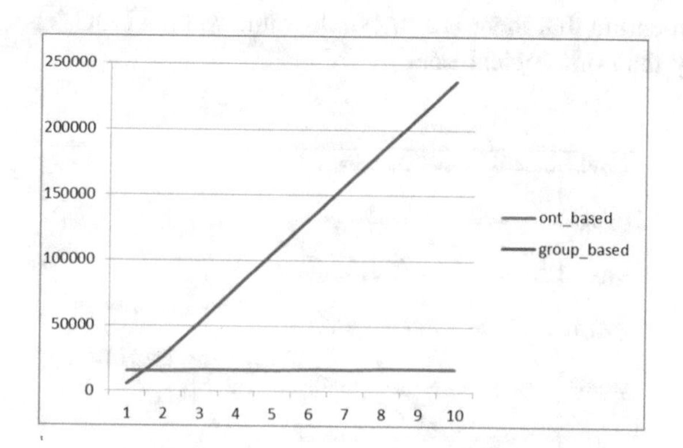

Fig. 10. The diagram of workload relocation problem complexity for "iteration" parameter is variable when D is fixed.

One can see that the complexity of LDG-based method grows proportionally to a number of iterations, while the complexity of ontology based method is a constant. Yet when a number of iterations is small, a LDG-based technique is more promising.

In the case when the complexity of a workload distribution problem is proportional to a number of subtasks and a number of nodes, the diagram behavior is changed, yet, we also can see an intersection between plots when a number of iterations is small (see Fig. 11).

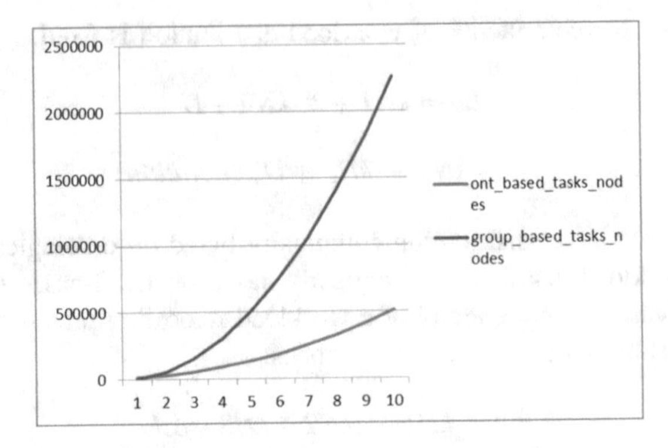

Fig. 11. The diagram of workload relocation problem complexity for the variable "iteration" and D parameters.

In the following figure parameter "task" is varied while other parameters are constants for the case of proportional complexity.

One can see that the computational complexity of ontological technique increases faster than LDG-technique in a Fig. 12. The last one has the linear nature. It is worthwhile to mention that there is a threshold value when a LDG-based technique has less complexity than ontological one.

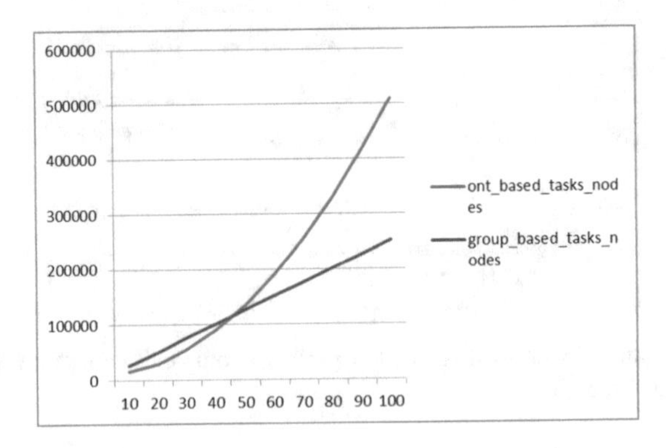

Fig. 12. The diagram of workload relocation problem complexity for the variable "task" and *D* parameters.

Consequently, in each of examined methods when "iteration" and "task" parameters are variable there are some threshold values when one of the method is more efficient for workload relocation problem solving.

5 Conclusion

A dynamic workload relocation problem in distributed CAD systems based on "fog-computing" paradigm requires new methods of solution. It is caused by some peculiar features of a "fog" layer infrastructure. On the one hand, there is a huge amount of "fog" nodes. On the other hand, a workload relocation problem belongs to combinatorial optimization problem and requires the search space to be constrained.

An ontology-based technique of workload relocation problem in the field of DCAD in "fog" environment was proposed in this paper. Boundary conditions are formed and search space is reduced with the help of ontological analysis and the heuristic rules application. Simple mathematical models for examined techniques (LDG-based and ontology-based) were developed to estimate the efficiency of the techniques.

The results of simulation are as follows: ontology-based technique does not depend on a number of iterations and can be more efficient in case of a large number of iterations in LDG-based technique. Yet, if the number of iterations are fixed and considerably small, there is a threshold value of a number of tasks, when the complexity of ontology-based technique is higher than the complexity of LDG-based technique.

So, the hybrid approach can be promising, combining the LDG-based and ontology-based techniques.

Acknowledgments. The paper has been prepared within the RFBR project 18-29-03229 and the GZ SSC RAS N GR of project AAAA-A19-119011190173-6.

References

1. Melnik, E.V., Klimenko, A.B., Klimenko, V.V., Taranov, A.Yu.: Distributed informational and control system configuration quality improving by the load balancing. Herald of computer and information technologies. Sci. Tech. Prod. Mon. J. **11**(49), 33–38 (2016)
2. Longbottom, R.: Computer System Reliability. Energoatomizdat, Moscow (1985)
3. Ivanichkina, L.V., Neporada, A.P.: The reliability model of a distributed data storage in case of explicit and latent disk faults. Tr. ISP RAN **27**(6), 253–274 (2015)
4. Haldar, V., Chakraborty, N.: Power loss minimization by optimal capacitor placement in radial distribution system using modified cultural algorithm. Int. Trans. Electr. Energy Syst. **25**, 54–71 (2015)
5. Cisco: Fog computing and the internet of things: extend the cloud to where the things are. https://www.cisco.com/c/dam/en_us/solutions/trends/iot/docs/computing-overview.pdf. Accessed 21 Apr 2019
6. Melnik, E.V., Klimenko, A.B., Ivanov, D.Ya., Gandurin, V.A.: Methods for providing the uninterrupted operation of network-centric data-computing systems with clusterization. Izv. SFedU. Eng. Sci. **12** (185) 71–84 (2016)
7. Distributed systems. http://pv.bstu.ru/networks/books/NetBook&Algoritms.pdf. Accessed 21 Apr 2019
8. El-Ghazali, T., et al.: Parallel approaches for multiobjective optimization. In: Branke, J., Deb, K., Miettinen, K., Słowiński, R. (eds.) Multiobjective Optimization. Lecture Notes in Computer Science, vol. 5252, pp. 349–372. Springer, Berlin (2008)

9. Glushan', V.M., Lavrik, P.V.: Distributed CAD. Architecture and possibilities. TNT, Stary Oskol (2015)
10. Karpenko, A.P.: Population algorithms for global continuous optimization. review of new and little-known algorithms. Inf. Technol. **7**, 1–32 (2017)
11. Wave algorithms and traversal algorithms. http://ccfit.nsu.ru/~tarkov/Dictributed%20algorithms/Books/Tel_Distributed_Algorithms.pdf. Accessed 21 Apr 2019
12. Karpenko, A.P.: The main essentials of population-based algorithms of global optimization. systematization experience. Internet-zhurnal « NAUKOVEDENIE ». vol. 9, no. 6. https://naukovedenie.ru/PDF/46TVN617.pdf. Accessed 21 Apr 2019
13. Protégé 4.2. https://protege.stanford.edu. Accessed 21 Apr 2019
14. Holod I.I.: Models and methods of distributed data mining parallel algorithm development: Author's abstract of Doctor of Engineering Science: 05.13.11. – St. Petersburg (2018)

Preprocessing CT Images for Virtual Reality Based on MATLAB

Jan Kohout(✉), Jan Egermaier, Nad'a Tylová, and Jan Mareš

Department of Computing and Control Engineering, University of Chemistry and Technology in Prague, Technická 1905/5, 166 28 Praha 6, Prague, Czech Republic
jan.kohout@vscht.cz

Abstract. 3D modeling and virtual reality is a modern branch with many interdisciplinary applications. These approaches are commonly used in technical praxis (computer vision, autonomous robotic systems or navigation in space). The aim of this paper is an extension of 3D modeling and virtualization methods to biomedical and radiological applications. We created a MATLAB GUI application for preprocessing computed tomography (CT) data to be able to select a part of the human body that a user can be interested in. Then, a 3D model is created and exported into a STL file. After final conversion, it can be shown in virtual reality.

Keywords: Computational tomography · Virtual reality · Preprocessing · CT · MATLAB

1 Introduction

Computed tomography (CT) and magnetic resonance imaging (MRI) are two of the most powerful imaging techniques. These modalities produce digital data, which can be modified to more precise diagnoses without further burdening the patient. A basic software is at disposal from business partners but it is often user-unfriendly and allows only several basic operations.

Modified and integrated data to 3D models and virtual reality are able to increase the number of patients with early diagnoses of different pathological findings [4]. Moreover, this data can allow the surgeon to navigate more precisely when resecting findings from parenchymal organs by providing information on the relationship between the finding and the rest of the organ [5,9]. Alongside similar elective procedures, this method in orthopedic oncology has aided in the planning of subsequent reconstructive procedures that aim to maintain or restore the functional integrity and load-bearing of the skeleton [3,8]. In analogy, the precise orientation is crucial for safe management of the deeply seated pathologies that are managed by otolaryngologist, head and neck surgeons [1].

What is more, all these employments are applicable and useful to the fields of emergency medicine and traumatology as well. In complex fractures with multiple fragments and their associated bone defects, the precision with which this

R. Silhavy et al. (Eds.): CoMeSySo 2019, AISC 1046, pp. 49–57, 2019.
https://doi.org/10.1007/978-3-030-30329-7_5

method defines the fracture field can not only optimize the pre-operative planning but it can also improve surgical orientation during the operation. Lastly, among the wide-ranging uses for the information gained by way of 3D modeling the musculoskeletal system and the human body, it has potential as an educational tool.

1.1 Virtual Reality

Standard CT scans display just two-dimensional sections of a patient body. Image processing that converts individual slices into a 3D model displays scan as a complex structure. A virtual reality (VR) headset is used to get a more realistic view of this 3D model. In contrast with a flat screen, a view angle of the VR headset is wider. For example, the view angle of the HTC Vive headset is about 110° [2,6].

Basic VR applications allow to user to rotate with a model, to zoom in and out and in some cases to grab the whole model. The user can observe the model from different angles. The HTC Vive[1] also allows a user to move around the virtual scene.

Advanced VR applications are able to display a cross-section of the 3D model in real time. This process requires the latest graphical card. With the latest hardware, it is possible to create a very realistic scene where shadows of the object are displayed and brightness and contrast could be set up by the user. These applications could be used for surgery planning [7,14].

2 Materials and Methods

Our first step how to show CT images (stored in DICOM format) in virtual reality is in MATLAB[2]. A scheme of the workflow is in a Fig. 1. Firstly, we need to load a DICOM[3] format and then to select a volume (part of a patient's body) that is interesting for us. For this reason, we created a simple GUI. Finally, a user can export an edited model in STL format. Steps how to get the model to virtual reality are described in Sect. 2.3.

2.1 MATLAB Preprocessing

Main steps in MATLAB shows Fig. 2. Function `dicomreadVolume` reads CT image data from DICOM files:

$$[\mathrm{V}, \mathrm{spatial}, \mathrm{dim}] = \mathrm{dicomreadVolume}(\mathrm{source}) \quad (1)$$

[1] https://www.vive.com/eu/.
[2] https://www.mathworks.com/products/matlab.html.
[3] https://www.dicomstandard.org/.

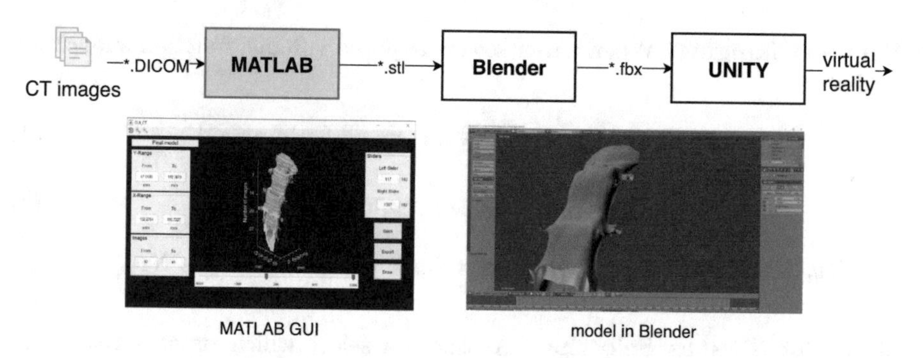

Fig. 1. A scheme of CT images preprocessing for virtual reality.

Information from DICOM. Information from DICOM structure are obtained by:

$$\text{info} = \text{dicominfo}(\text{filename}) \tag{2}$$

the most important are variables `SliceThickness` (thickness of one image slice in mm [12]) and `PixelSpacing` (a distance between centers of pixels – vertically, respectively horizontally – in mm [11]).

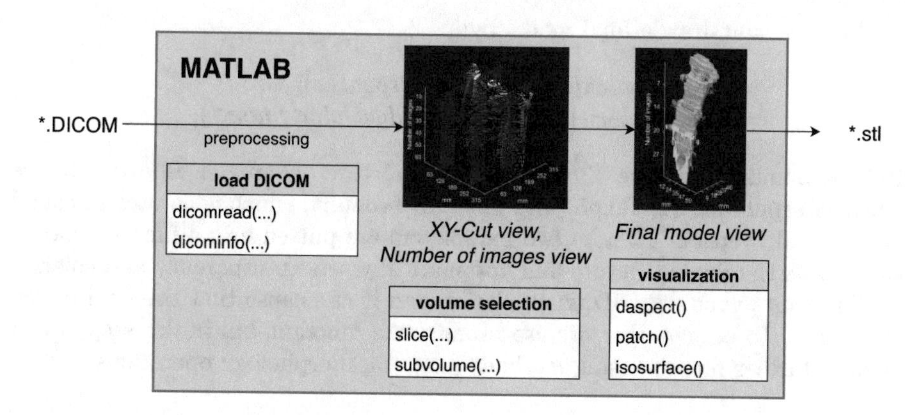

Fig. 2. A scheme of our preprocessing CT images in MATLAB.

Volume Selection. For volume selection, we use function `slice` (based on functions from [10]):

$$\text{slice}(\text{X}, \text{Y}, \text{Z}, \text{V}, \text{xslice}, \text{yslice}, \text{zslice}) \tag{3}$$

where X, Y, Z are coordinates from function `meshgrid`, V is 3D matrix with image data, *xslice*, *yslice* and *zslice* are planes that a user use for volume selection (drag and drop).

Volume Changing. When a user selects required volume, function `subvolume` is used:

$$\text{limits} = [\text{xmin}, \text{xmax}, \text{ymin}, \text{ymax}, \text{zmin}, \text{zmax}] \tag{4}$$

$$[\text{Nx}, \text{Ny}, \text{Nz}, \text{Nv}] = \text{subvolume}(\text{V}, \text{limits}) \tag{5}$$

Nx, Ny, Nz are coordinates of a new volume (saved in variable Nv).

Slider for Tissues Selection. A user can select which tissue want to see by setting required Hounsfield units values[4]. It can be done with a GUI slider (in Fig. 3 at the bottom). Selected range means which voxel values will not be deleted.

Drawing. To avoid axes deformations, it is necessary to use before drawing function:

$$\begin{array}{l} \text{daspect(ratio)} \\ \text{ratio} = [\text{x}, \text{y}, \text{z}] \end{array} \tag{6}$$

where x and y are set to 1 and z is equal to `PixelSpacing/SliceThickness` (see above).

Then we can draw a final model by:

$$\begin{array}{l} hiso = patch(isosurface(V, isovalue); , ... \\ 'FaceColor', [.90, .90, .90], 'EdgeColor', 'none'); \end{array} \tag{7}$$

It joins points in volume V with value equal to a parameter *isovalue*. It use linear interpolation (`interp3`) and function `isosurf`, which generates so called "faces" and "vertices". Those two parameters are putted into a function `patch` that draws the final structure and automatically sets transparency and colors.

Function `patch` uses 3D triangulation and it can cause that the final model is rastered. To be smoother, we use `isonormals` function, but in this step, it can be added other preprocessing methods (smooth, morphology operations, ...).

Export to STL. To be able to use our model in virtual reality, we use export to STL format by function `stlwrite` (from [13]).

2.2 MATLAB GUI Description

User interface is based on MATLAB's GUIDE[5]. It is splitted into three parts: XY-Cut view, Number of images view a Final model view (see Fig. 2).

[4] http://radclass.mudr.org/content/hounsfield-units-scale-hu-ct-numbers.

[5] GUI development environment.

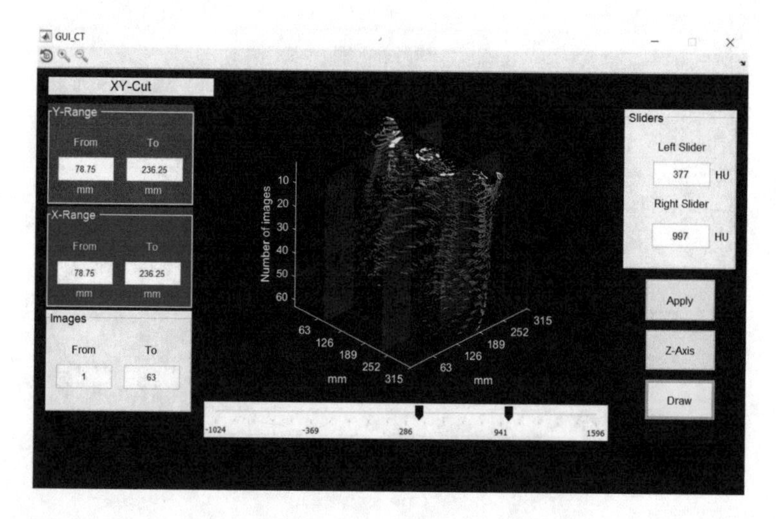

Fig. 3. XY-Cut view. Moving with red and blue planes, a user can specify an area of interest.

XY-Cut View. An undersampled (for fast displaying) 3D model is shown in a Fig. 3. Moving with color planes, a user can specify areas of interest – blue for the x-axis, red for the y-axis. For better navigation, it is possible to change slider (in Fig. 3 at the bottom) that selects a range of shown voxels based on their Hounsfield units (HU).

Button "Apply" switches GUI to Final model view, "Z-Axis" to Number of images view.

Number of Images View. It is used for a range selection in z-axis (number of images) – see Fig. 4.

Final Model View. A user can export the final model to STL.

2.3 Converting Models into Virtual Reality

A virtual scene with a 3D model is made in Unity editor[6]. Firstly, Blender3 editor[7] is used to convert STL to FBX (Unity format).

A basic Unity scene includes the environment, a floor, and light. Without this stuff, a user can be disoriented. It is expected that the user will be moving in the scene by walking. A model is imported to Unity as a new asset and placed to the center of the scene. A head position of the user and his moving are set as the scene properties. A VR headset allows to user to zoom and control a view angle in real time by moving in the scene.

[6] https://unity3d.com.

[7] https://www.blender.org/.

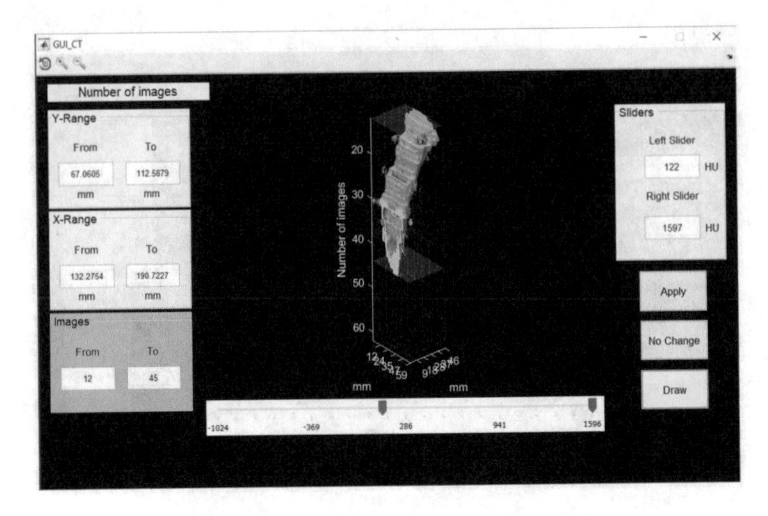

Fig. 4. Number of images view. Moving with green planes, a user can specify an area of interest.

2.4 Removing Artifacts in Blender

In Blender, it is also possible to edit a model and remove an artifact. [15] In medicine, common artifacts are rest of a treatment couch or discontinuity of tissue. To avoid artifacts from the treatment couch, it is necessary to select the closest part as possible. In some cases, the discontinuity of the model depends on the tissues properties and threshold settings. For example, it is difficult to distinguish cartilages connecting ribs to sternum from soft tissues.

The artifacts removing process starts with selecting an object of interest. In this step, it is possible to determine significant parts of the examined tissue, but it is strongly dependent on the user's experience. There is a function to select all together connected vertices and after inverting this selection, small objects can be removed from the model (it is because they are usually made by several vertices). It is necessary to balance between a real part of the tissue, the noise removing and model editing. After this preprocessing, the model is exported into the FBX format.

3 Results

The final visualization of a model prepared in our MATLAB GUI is displayed in Fig. 5. Blender smoothing function was used for adjustment. This function use rendering algorithms to make the model more realistic. The smoothing quality is influenced by scanning properties especially in direction of scanning.

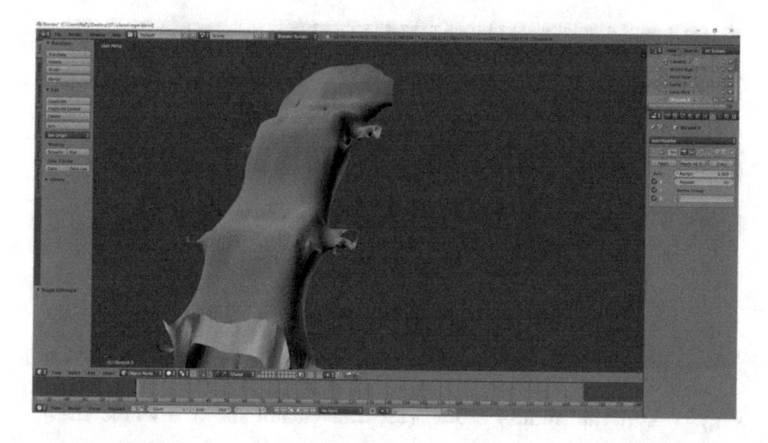

Fig. 5. A model exported by MATLAB GUI and adjusted in Blender.

3.1 Comparison with 3Dslicer

We used 3Dslicer[8] for comparison. It is a free open-source 3D medical reconstruction software that is able to export the STL model from DICOM data.

Fig. 6. Detail of final STL models comparison in Blender – preprocessed by our MATLAB GUI (left) and by 3Dslicer (right).

Exported STL model from the 3Dslicer and our MATLAB GUI are compared in a Fig. 6. Although both exported models have similar quality, our approach provides an option for volume selection and the process is automatized. Problematical parts of the models (edges between slices, the noise of unrecognized soft tissue) are noticeable in both figures. These issues could be solved in Blender as a part of preprocessing for converting to virtual reality (see Sect. 2.4 and Fig. 7).

[8] https://www.slicer.org/.

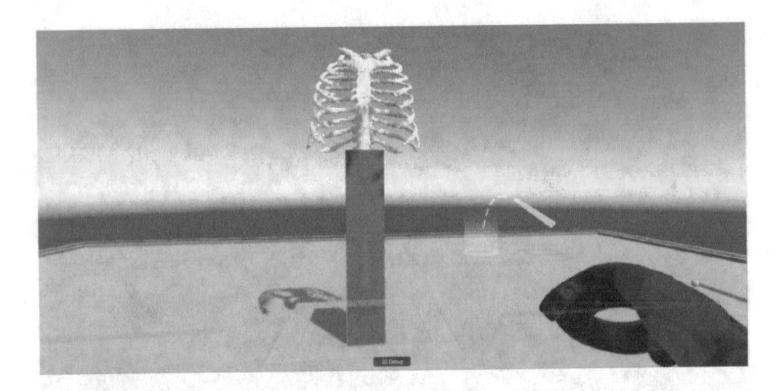

Fig. 7. A virtual reality model (for illustration shows a whole chest).

4 Conclusion

In this study, we developed a system for converting CT images into virtual reality model and compared 3Dslicer and our own approach.

Displaying a 3D model in virtual reality is more comfortable for a user. Therefore, we developed a system for preprocessing 3D medical models from CT images for virtual reality. We created our MATLAB GUI to be able to focus on some small parts of a body. The main advantages of this approach are:

1. modularity – other smoothing or reconstruction algorithms can be added
2. we are able to simply process only a selection of CT images (a crop in volume) which means fewer computer requirements
3. automatization of artifact removing
4. it can lead to further application in medicine, exported STL format is usable for 3D print

We compare our approach with 3Dslicer. A basic principle of both software is the same. The first step is to localize an area of interest and to set an optimal threshold of an examined tissue. 3Dslicer allows to clear artifacts, but it is necessary to clear them in each slide separately – it means that this operation is significantly time-consuming.

Future work will be to improve smoothing and noise filtering.

Acknowledgements. Financial support from specific university research (MSMT No 21-SVV/2019).

References

1. Bosc, R., Hersant, B., Carloni, R., et al.: Mandibular reconstruction after cancer: an in-house approach to manufacturing cutting guides. Int. J. Oral Maxillofacial Surg. **46**(1), 24–31 (2017)

2. Egger, J., Gall, M., Wallner, J., et al.: HTC Vive MeVisLab integration via OpenVR for medical applications. Plos One **12**(3), e0173972 (2017)
3. Glicksman, J.T., Reger, C., Parsher, A.K., et al.: Accuracy of computer-assisted navigation: significant augmentation by facial recognition software. Int. Forum Allergy Rhinol. **7**(9), 884–888 (2017)
4. Honingmann, P., et al.: A simple 3-dimensional printed aid for a corrective palmar opening wedge osteotomy of the distal radius. J. Hand Surg. Am. **41**(3), 464–9 (2016)
5. Chana-Rodriguez, F., et al.: 3D surgical printing and pre contoured plates for acetabular fractures. Injury **47**(11), 2507–2511 (2016)
6. King, F.: An immersive virtual reality environment for diagnostic imaging. A thesis (2015)
7. Torner, J., Gómez, S., Alpiste, F., Brigos, M.: Virtual reality application applied to biomedical models reconstructed from CT scanning. In: 24th Conference on Computer Graphics, Visualization and Computer Vision (2016)
8. Wang, S., Wang, L., Liu, Y., et al.: 3D printing technology used in severe hip deformity. Exp. Ther. Med. **14**(3), 2595–2599 (2017)
9. Xu, W., Zhang, X., Ke, T., Cai, H., Gao, X.: 3D printing-assisted preoperative plan of pedicle screw placement for middle-upper thoracic trauma: a cohort study. BMC Musculoskelet. Disord. **18**, 348 (2017)
10. Wang, G.: 3D CT/MRI images interactive sliding viewer. https://www.mathworks.com/matlabcentral/fileexchange/29134-3d-ct-mri-images-interactive-sliding-viewer?s_tid=prof_contriblnk
11. Pixel Spacing Attribute. DICOM Standard Browse. Innolitics. https://dicom.innolitics.com/ciods/ct-image/image-plane/00280030
12. Slice Thickness Attribute. DICOM Standard Browser. Innolitics. https://dicom.innolitics.com/ciods/ct-image/image-plane/00180050
13. Lohsen, W.G.: stlwrite - Write binary or ascii STL file. https://www.mathworks.com/matlabcentral/fileexchange/36770-stlwrite-write-binary-or-ascii-stl-file
14. Virtual medicine. https://www.medicinevirtual.com/
15. Visualizing MRI & CT Scans in Mixed Reality / VR / AR, Part 3: 3D Model Maker. https://www.andreasjakl.com/visualizing-mri-ct-scans-mixed-reality-vr-ar-part-3-3d-model-maker/

A PRP-HS Type Hybrid Nonlinear Conjugate Gradient Method for Solving Unconstrained Optimization Problems

Olawale J. Adeleke[1(✉)], Micheal O. Olusanya[2], and Idowu A. Osinuga[3]

[1] Department of Mathematics, College of Science and Technology, Covenant University, Ota, Nigeria
wale.adeleke@covenantuniversity.edu.ng
[2] Department of Information Technology, Faculty of Accounting and Informatics, Durban University of Technology, Durban, Republic of South Africa
[3] Department of Mathematics, College of Physical Sciences, Federal University of Agriculture, Abeokuta, Nigeria

Abstract. Many engineering problems that occur in real-life are usually constrained by one or more factors which constitute the basis for the complexity of obtaining optimal solutions. While some of these problems may be transformed to the unconstrained forms, there is a large pool of purely unconstrained optimization problems in engineering which have practical applications in the industry. One effective approach for solving this latter category of problems is the nonlinear conjugate gradient method (NCGM). Particularly, the NCGM uses an efficient recursive scheme to solve unconstrained optimization problems with very large dimensions. In this paper, a new hybrid NCGM is proposed based on the recent modifications of the Polak-Ribiére-Polyak (PRP) and Hestenes-Stiefel (HS) methods. Theoretical analyses and numerical computations using standard benchmark functions, as well as comparison with existing NCGM schemes show that the proposed PRP-HS type hybrid scheme is globally convergent and computationally efficient.

Keywords: Hybrid methods · Nonlinear conjugate gradient method · Unconstrained optimization problems · Descent property · Global convergence · Line search

1 Introduction

This study considers the nonlinear conjugate gradient method (NCGM), an iterative method for solving the unconstrained optimization problem of the form

$$\min_{x \in \mathbb{R}^n} f(x), \tag{1}$$

where the gradient of the continuous function $f : \mathbb{R}^n \rightarrow \mathbb{R}$ can be evaluated at every point The method and all its variants constitute a class of iterative scheme for solving

R. Silhavy et al. (Eds.): CoMeSySo 2019, AISC 1046, pp. 58–68, 2019.
https://doi.org/10.1007/978-3-030-30329-7_6

(1) effectively especially for large values of n. The NCGM scheme for (1) can be written as follows:

$$x_{k+1} = x_k + \alpha_k d_k, \tag{2}$$

where the stepsize, α_k, evaluated by a suitable line search process, and the search direction, d_k, is computed by the formula

$$\begin{cases} d_0 = g_0, & \text{for } k = 0 \\ d_k = -g_k + \beta_k d_{k-1}, & \text{for } k \geq 1, \end{cases} \tag{3}$$

where g_k is the gradient of f at x_k, β_k is the NCGM update parameter and is usually chosen so that the (2) and (3) together reduces to the linear form of the method when f is a strictly convex quadratic function and α_k is determined by a one-dimensional exact line search technique.

Some of the variants of NCGM include, Hestenes and Stiefel (HS) [1], Fletcher and Reeves (FR) [2], Polak, Ribieré and Polyak (PRP) [3, 4], Liu and Storey (LS) [5], Dai and Yuan (DY) [6] and Hager and Zhang (CG_DESCENT (N)) [7]. These methods are computed as follows: $\beta_k^{HS} = g_k^T y_k / d_{k-1}^T y_{k-1}$, $\beta_k^{FR} = \|g_k\|^2 \big/ \|g_{k-1}\|^2$, $\beta_k^{PRP} = g_k^T y_k \big/ \|g_{k-1}\|^2$, $\beta_k^{LS} = -g_k^T y_k / d_{k-1}^T g_{k-1}$, $\beta_k^{DY} = \|g_k\|^2 \big/ d_{k-1}^T y_{k-1}$ and $\beta_k^N = (y_k - 2d_{k-1} \left(\|y_k\|^2 \big/ d_{k-1}^T y_k\right))^T g_k / d_{k-1}^T y_k$, where $\|\cdot\|$ denotes the Euclidean norm and $y_k = g_k - g_{k-1}$.

These variants are equivalent for strictly convex functions and exact line search. However, for non-quadratic functions, the methods perform differently [8, 9]. Establishing the global convergence results of these methods usually requires that the stepsize α_k satisfies certain approximate line search criteria as most exact line search procedures are cost-ineffective. The most widely used of these conditions are the strong Wolfe-Powell inequalities given as follows:

$$\begin{cases} f(x_{k-1}) - f(x_k) \geq -\delta \alpha g_{k-1}^T d_{k-1} \\ \left|g_k^T d_{k-1}\right| \leq -\sigma g_{k-1}^T d_{k-1}, \end{cases} \tag{4}$$

with $0 < \delta < \sigma < 1$. The weaker version of this condition is obtained by combining the first part of (4) with

$$g_k^T d_{k-1} \geq \sigma g_{k-1}^T d_{k-1}. \tag{5}$$

For the NCGM and its variant to be considered efficient, the search direction d_k, must satisfy a property known as the descent property. That is, d_k making an angle of strictly less than $\pi/2$ radians with $-g_k$, will always guaranteed a decrease in f provided that α_k is sufficiently small. In other words, an NCGM algorithm is said to satisfy the descent property if

$$g_k^T d_{k-1} < 0, \forall k \geq 1. \tag{6}$$

A more natural way of ensuring descent for NCGM algorithms is through the use of sufficient descent property. This is given by

$$g_k^T d_{k-1} \leq -c\|g_{k-1}\|^2, \forall k \geq 1, \tag{7}$$

where c is a positive constant.

Each CG method has very striking features that makes it adaptable to some sets of unconstrained problems. For instance, the FR and DY methods have been identified as having the best convergence results (See [10] and [11]). However, for general objective functions, the two methods have poor computation power. Conversely, the HS and PRP methods have good computational strength even though they exhibit poor convergence results. These contrasting features have led to the development of hybrid methods which are constructed with the aim of overcoming existing deficiencies in two or more methods. For instance, a well-constructed hybrid method of FR and PRP should perform well computationally as well as yield good convergence properties. This is the main motivation behind this paper. In this paper, a new hybrid NCGM variant is proposed through the combination of recently proposed variants of the traditional PRP and HS methods. Even though the two methods are specifically known to perform well computation-wise, it is interesting to know that the hybrid methods proposed here exhibit strong global convergence properties.

Recently, Du et al. [12] proposed four modified NCGMs. Only two which are of interest are mentioned in this study. The first of these, a variant of the PRP method which shall denoted as DPRP, has its β_k value given as follows:

$$\beta_k^{DPRP} = g_k^T \left(g_k - \left(|g_k^T g_{k-1}| \middle/ \|g_{k-1}\|^2 \right) g_{k-1} \right) \middle/ \|g_{k-1}\|^2 \tag{8}$$

and the second, denoted as DHS is a variant of the HS method and has its β_k given as follows:

$$\beta_k^{DHS} = g_k^T \left(g_k - \left(|g_k^T g_{k-1}| \middle/ \|g_{k-1}\|^2 \right) g_{k-1} \right) \middle/ d_{k-1}^T y_k. \tag{9}$$

It was further established in their study that the DPRP and DHS methods possess the sufficient descent property (Eq. 7) and are globally convergent under strong Wolfe line search with $0 < \sigma < 1/4$ and $0 < \sigma < 1/3$, respectively. It was also shown that the two methods are always nonnegative under the same assumptions of strong Wolfe line search and the respective intervals of convergence above.

Based on the DPRP and DHS methods, in this paper, a mixed hybrid NCGM is proposed. The development of this hybrid method was motivated by the observation that the numerical results reported in Du et al. [12] for the DPRP and DHS methods revealed that the methods are not so efficient compared to some classical methods, especially the CG_DESCENT method. The objective of this study is therefore, to construct a computationally efficient NCGM by hybridizing the PRP, HS, DPRP and

DHS methods. The remaining parts of this paper are organized as follow. The proposed hybrid method and the corresponding algorithm is presented in Sect. 2, while the descent and global convergence properties are described in Sect. 3. In Sect. 4, the numerical computations as well as the discussion of results are reported. The concluding remarks are given in Sect. 5.

2 Proposed Hybrid CG Methods and Algorithm

In this section, the new hybrid method of PRP, HS, DPRP and DHS is presented. A suitable algorithm for implementing the method for unconstrained optimization test problems is also provided in this section. Motivated by the ideas of hybrid method construction in [8, 13, 16] and the mixed performance profile of these methods, the β_k of the hybrid of DPRP, DHS, PRP and HS, denoted as HNCG, is given as

$$\beta_k^{HNCG} = \left(\|g_k\|^2 - \max\left\{\left(|g_k^T g_{k-1}| / \|g_{k-1}\|^2\right) g_k^T g_{k-1}, g_k^T g_{k-1}\right\}\right) / \max\left\{\|g_{k-1}\|^2, d_{k-1}^T (g_k - g_{k-1})\right\}. \tag{10}$$

By observation, the HNCG method represented in (10) can be reduced to any of PRP, DPRP, HS and DHS. A suitable algorithm for this method is given as follows:

Algorithm 1: Hybrid HNCG Scheme

1: input $x_0 \in \mathbb{R}^n$, $\varepsilon \geq 0$, set $d_0 = -g_0$, $k = 0$;
2: obtain α_k so that (4) is satisfied;
3: while $\|g_k\| < \varepsilon$ do
4: $\quad y_k = g_k + g_{k-1}$;
5: $\quad \beta_k = \beta_k^{HNCG}$ (as in (10));
6: $\quad d_k = -g_k + \beta_k^{HNCG} d_{k-1}$;
7: $\quad x_{k+1} = x_k + \alpha_k d_k$;
8: $\quad k = k + 1$;
9: end

3 Descent and Global Convergence Properties of the HNCG Method

3.1 Descent Property of Algorithm 1

The descent property of d_k and the global convergence results for Algorithm 1 are presented in this section. The following result establishes the descent properties of the sequence of search directions generated by the Hybrid Algorithm. Interestingly, this result was obtained independent of any line search procedure.

Theorem 1: Suppose that $f(x)$ in (1) is a smooth function and d_k is generated by the Hybrid Algorithm. Then $g_k^T d_k < 0$ for each $k \geq 0$.

Proof: Clearly, $g_0^T d_0 = -\|g_0\|^2 < 0$ for $k = 0$. By assumption, let $g_k^T d_k < 0$ holds for $k \geq 1$. To show that $g_k^T d_k < 0$ for all k, the proof is divided into four different cases for each of the hybrid method. Note that if $\beta_k = 0$, then, from (3), $d_k = -g_k + \beta_k d_{k-1} = -g_k \Rightarrow g_k^T d_k = -\|g_k\|^2 < 0$. Thus, it is assumed that $\beta_k \neq 0$ in all the cases.

Case I: If $g_k^T g_{k-1} < \left(|g_k^T g_{k-1}| \big/ \|g_{k-1}\|^2\right) g_k^T g_{k-1}$ and $d_{k-1}^T(g_k - g_{k-1}) \geq \|g_{k-1}\|^2$, then from (10), $\beta_k^{HNCG} = \beta_k^{DHS}$. Notably, since $\|g_{k-1}\|^2 > 0$, the denominator of β_k^{DHS}, that is, $d_{k-1}^T(g_k - g_{k-1}) > 0$.. Hence, starting with (3), we have

$$\begin{aligned} g_k^T d_k = g_k^T(-g_k + \beta_k^{DHS} d_{k-1}) &= \left(\|g_k\|^2 g_{k-1}^T d_{k-1} \big/ d_{k-1}^T(g_k - g_{k-1})\right) \\ &\quad - \left(\left(|g_k^T g_{k-1}| \big/ \|g_{k-1}\|^2\right) \cdot g_k^T g_{k-1} \big/ d_{k-1}^T(g_k - g_{k-1})\right) \cdot g_k^T d_{k-1} \\ &< \left(\|g_k\|^2 g_{k-1}^T d_{k-1} \big/ d_{k-1}^T(g_k - g_{k-1})\right) < 0. \end{aligned} \quad (11)$$

In (11), the first equality was obtained by finding the inner product of (3) and substituting β_k^{DHS} for β_k. The first inequality is obvious because $\left(|g_k^T g_{k-1}| \big/ \|g_{k-1}\|^2\right) > 1$ from the first assumption above, while the second was obtained from the fact that $g_{k-1}^T d_{k-1} < 0$.

Case II: If $g_k^T g_{k-1} < \left(|g_k^T g_{k-1}| \big/ \|g_{k-1}\|^2\right) g_k^T g_{k-1}$ and $d_{k-1}^T(g_k - g_{k-1}) < \|g_{k-1}\|^2$, then (10) yields $\beta_k^{HNCG} = \beta_k^{DPRP}$. Notice that the second assumption implies that $g_k^T d_{k-1} - \|g_{k-1}\|^2 < g_{k-1}^T d_{k-1}$. Thus, starting from (3), we obtain the following result:

$$\begin{aligned} g_k^T d_k &= g_k^T(-g_k + \beta_k^{DPRP} d_{k-1}) = \left(\left(g_k^T d_{k-1} - \|g_{k-1}\|^2\right) \|g_k\|^2 \big/ \|g_{k-1}\|^2\right) \\ &\quad - \left(\left(|g_k^T g_{k-1}| \big/ \|g_{k-1}\|^2\right) \cdot g_k^T g_{k-1} \big/ \|g_{k-1}\|^2\right) \cdot g_k^T d_{k-1} \\ &< \left(\left(g_k^T d_{k-1} - \|g_{k-1}\|^2\right) \|g_k\|^2 \big/ \|g_{k-1}\|^2\right) < \left(\|g_k\|^2 \big/ \|g_{k-1}\|^2\right) \cdot g_{k-1}^T d_{k-1} < 0. \end{aligned} \quad (12)$$

Case III: If $g_k^T g_{k-1} \geq \left(|g_k^T g_{k-1}| \big/ \|g_{k-1}\|^2\right) g_k^T g_{k-1}$ and $d_{k-1}^T(g_k - g_{k-1}) \geq \|g_{k-1}\|^2$, then from (10), $\beta_k^{HNCG} = \beta_k^{HS}$. Proceeding from the inner product of (3) with g_k gives

$$\begin{aligned} g_k^T d_k = g_k^T\left(-g_k + \beta_k^{HS} d_{k-1}\right) &= \left(\|g_k\|^2 g_{k-1}^T d_{k-1} \Big/ d_{k-1}^T (g_k - g_{k-1})\right) \\ &\quad - \left(g_k^T g_{k-1} \cdot g_k^T d_{k-1} / d_{k-1}^T (g_k - g_{k-1})\right) \\ &< \left(\|g_k\|^2 g_{k-1}^T d_{k-1} \Big/ d_{k-1}^T (g_k - g_{k-1})\right) < 0 \end{aligned} \tag{13}$$

Case IV: If $g_k^T g_{k-1} \geq \left(|g_k^T g_{k-1}| \Big/ \|g_{k-1}\|^2\right) g_k^T g_{k-1}$ and $d_{k-1}^T (g_k - g_{k-1}) < \|g_{k-1}\|^2$, then from (10), $\beta_k^{HNCG} = \beta_k^{PRP}$. The second assumption in this case allows us to set $g_k^T d_{k-1} - \|g_{k-1}\|^2 < g_{k-1}^T d_{k-1}$ and beginning from (3) as in other cases, we obtain

$$\begin{aligned} g_k^T d_k = g_k^T\left(-g_k + \beta_k^{PRP} d_{k-1}\right) &< \left(\|g_k\|^2 \left(g_k^T d_{k-1} - \|g_{k-1}\|^2\right) \Big/ \|g_{k-1}\|^2\right) \\ &< \left(\|g_k\|^2 \Big/ \|g_{k-1}\|^2\right) \cdot g_{k-1}^T d_{k-1} < 0 \end{aligned} \tag{14}$$

The descent property is satisfied in (11)–(14). Hence, the sequence of search directions generated by the HNCG method satisfies the descent property.

3.2 Global Convergence Property of Algorithm 1

The following important result is a consequence of Theorem 1 and is very important in establishing the global convergence of Algorithm 1. The result is stated as follows without proof.

Lemma 2: The inequality $0 \leq \beta_k \leq \frac{g_k^T d_k}{g_{k-1}^T d_{k-1}}$ always hold for every $k \geq 1$, where $\beta_k = \beta_k^{HNCG}$.

Next, the global convergence result for Algorithm 1 is presented. The result is established under the following assumptions.

Assumption:

1. Bound on the Objective Function: $f(x)$ is bounded from below on the level set $\Omega = \{x \in \mathbb{R}^n : f(x) \leq f(x_0)\}$, where x_0 is an initial guessed point.
2. Lipschitz Condition: within some neighbourhood N of Ω, $f(x)$ is continuously differentiable, and its gradient is Lipschitz continuous, that is, for all $x, y \in N$, there exists a constant $L \geq 0$ such that

$$\|g(x) - g(y)\| \leq L\|x - y\|. \tag{15}$$

3. Let an iterative scheme of the form (2) where d_k is a descent direction and α_k satisfies the Wolfe line search conditions (first part of 4) and (5). If Assumptions 1 and 2 hold, then,

$$\sum_{k=0}^{\infty}\left(\left(g_k^T d_k\right)^2 \Big/ \|d_k\|^2\right) < +\infty \tag{16}$$

Assumption 3 is the well-known Zoutendijk condition, the proof of which may be found in [11]. Using Lemma 2 and these assumptions, the following global convergence result of Algorithm 1 is presented.

Theorem 3: Let d_k be generated by the iterative rules (2)–(3) with β_k computed as in (10). If Assumptions (1)–(3) hold, then, $\lim_{k\to\infty} \inf\|g_k\| = 0$.

Proof: Suppose by contradiction that the conclusion does not hold, that is, $\lim_{k\to\infty} \inf\|g_k\| \neq 0$ However, since $\|g_k\| > 0$, there exists a constant $\varsigma > 0$ such that $\|g_k\| \geq \varsigma, \forall k$.

Starting with (3), the following holds,

$$\|d_k\|^2 = \beta_k^2\|d_{k-1}\|^2 - 2d_k^T g_k - \|g_k\|^2. \tag{17}$$

Dividing both sides of (17) by $\left(d_k^T g_k\right)^2$ and using Lemma 2, the following was obtained

$$\begin{aligned}\left(\|d_k\|^2 \Big/ \left(d_k^T g_k\right)^2\right) &= \left(\|d_{k-1}\|^2 \Big/ \left(d_{k-1}^T g_{k-1}\right)^2\right) - \left(2/d_k^T g_k\right) - \left(\|g_k\|^2 \Big/ \left(d_k^T g_k\right)^2\right) \\ &= \left(\|d_{k-1}\|^2 \Big/ \left(d_{k-1}^T g_{k-1}\right)^2\right) - \left((1/\|g_k\|) + (\|g_k\|/d_k^T g_k)\right)^2 + \left(1 \Big/ \|g_k\|^2\right) \\ &\leq \left(\|d_{k-1}\|^2 \Big/ \left(d_{k-1}^T g_{k-1}\right)^2\right) + \left(1 \Big/ \|g_k\|^2\right).\end{aligned} \tag{18}$$

Since the expression on the left side of (18) is the same as $1 \Big/ \left(d_1^T g_1\right)^2$, Eq. (18) gives

$$\left(\|d_{k-1}\|^2 \Big/ \left(d_{k-1}^T g_{k-1}\right)^2\right) \leq \sum_{i=1}^{k}\left(1 \Big/ \|g_i\|^2\right) \leq k/\varsigma, \forall k \tag{19}$$

Equation (19) implies that $\sum_{k\geq 1}\left(g_k^T d_k\right)^2 \Big/ \|d_k\|^2 = \infty$, which contradicts (16). □

4 Numerical Implementation of Algorithm 1

In this section, the numerical implementation of Algorithm 1 is presented. The indicators for comparing the algorithm with other existing methods are: the number of iterations (Table 1), the CPU time of executing the algorithm (Table 2), the optimal values of the objective function (Table 3) and the infinite norm of the gradient (Table 4). All the test functions were drawn from the Andrei [14], many of which are also in the CUTEr library described in Bongartz et al. [15]. The algorithm was implemented on Matlab R2016a installed on a PC with Windows 10 OS and 2G RAM. For all the twenty selected problems, the dimension n is set to 5000 and 10000.

Table 1. Numerical test results indicating the number of iterations

Function name	Dim	DPRP	DHS	CGD	HNCG
EBD-1	5000	60	61	107	72
	10000	71	101	59	67
Diagonal-4	5000	91	8	52	18
	10000	91	29	53	18
EH	5000	140	95	33	29
	10000	105	77	33	30
GR	5000	39	56	57	64
	10000	39	56	57	64
EWH	5000	37	150	83	63
	10000	37	150	83	63
ET	5000	3	10862	180	628
	10000	3	13652	176	692
PQD	5000	3179	26	1590	37
	10000	965	25	483	36
LIARWHD	5000	18	21	30	21
	10000	18	22	31	23
QUARTC	5000	185	181	8	191
	10000	148	146	8	6
GWH	5000	37	150	84	63
	10000	37	150	88	63

The preliminary results show that the proposed hybrid method is efficient and competes very well with existing methods. For instance, by observing Table 1 where the numeral values of the numbers of iterations were reported, HNCG produces better results that the CG_DESCENT and DHS methods (HNCG vs. CG_DESCENT (14 to 6), HNCG vs. DPRP (8 to 12), HNCG vs. DHS (11 to 8, with one tie)). The same is true for the CPU time in Table 2 (HNCG vs. CG_DESCENT (15 to 5), HNCG vs. DPRP (10 to 10), HNCG vs. DHS (11 to 9)). The following acronyms were used in the tables: EBD = Extended Block Diagonal-1; EH = Extended Himmelblau; GR = Generalized Rosenbrock; EWH = Extended White & Holst; ET = Extended Tridiagonal; PQD = Perturbed Quadratic Diagonal; GWH = Generalized White & Holst; Dim = Problem dimension; CGD = CG_DESCENT.

Table 2. Numerical test results indicating the CPU time of Algorithm 1

Function name	Dim	DPRP	DHS	CGD	HNCG
EBD-1	5000	1.000	0.452	0.824	0.459
	10000	1.619	0.693	0.386	0.638
Diagonal-4	5000	1.530	0.112	0.492	0.161
	10000	2.212	0.348	0.606	0.207
EH	5000	2.215	0.890	0.264	0.242
	10000	2.468	1.137	0.405	0.375
GR	5000	0.429	0.540	0.651	0.611
	10000	0.686	0.865	1.086	0.972
EWH	5000	0.563	2.005	1.237	0.909
	10000	0.939	3.379	2.154	1.379
ET	5000	0.094	72.565	1.774	4.089
	10000	0.182	136.981	2.324	7.749
PQD	5000	24.268	0.113	6.145	0.129
	10000	16.431	0.216	4.162	0.280
LIARWHD	5000	0.125	0.131	0.212	0.340
	10000	0.475	0.318	0.463	0.343
QUARTC	5000	0.423	0.395	0.050	0.430
	10000	0.661	0.498	0.097	0.064
GWH	5000	0.359	1.145	0.660	0.484
	10000	0.649	2.153	1.305	0.875

Table 3. Numerical test results indicating the values of the objective function

Function name	Dim	DPRP	DHS	CGD	HNCG
EBD-1	5000	3.85e−15	6.79e−14	1.93e−16	4.03e−15
	10000	7.58e−15	2.57e−14	8.71e−14	2.46e−14
Diagonal-4	5000	5.58e−16	NaN	7.34e−16	5.89e−16
	10000	1.12e−15	5.10e−15	6.60e−16	1.18e−15
EH	5000	3.37e−15	3.29e−15	6.57e−15	4.24e−15
	10000	6.11e−15	1.19e−15	1.31e−14	7.34e−15
GR	5000	3.99e+00	3.99e+00	3.99e+00	3.99e+00
	10000	3.99e+00	3.99e+00	3.99e+00	3.99e+00
EWH	5000	1.24e−16	3.07e−16	5.99e−17	2.12e−16
	10000	1.24e−16	3.07e−16	5.99e−17	2.12e−16
ET	5000	0.00e+00	9.91e−09	6.58e−09	8.82e−09
	10000	0.00e+00	1.25e−08	4.34e−10	1.11e−08
PQD	5000	3.60e−16	6.43e−17	3.60e−16	1.67e−16
	10000	9.37e−17	4.98e−18	9.37e−17	6.00e−17
LIARWHD	5000	1.62e−23	1.20e−22	1.05e−22	4.83e−22
	10000	3.23e−23	0.00e+00	2.93e−23	2.32e−23
QUARTC	5000	9.13e−11	9.17e−11	6.61e−15	9.12e−11
	10000	7.28e−11	7.23e−11	1.32e−14	4.03e−12
GWH	5000	1.24e−16	3.07e−16	7.16e−17	2.12e−16
	10000	1.24e−16	3.07e−16	9.22e−17	2.12e−16

Table 4. Numerical test results indicating the values of norm of gradient

Function name	Dim	DPRP	DHS	CGD	HNCG
EBD-1	5000	1.97e−07	7.92e−07	8.15e−08	3.58e−07
	10000	4.93e−07	3.41e−07	8.26e−07	8.56e−07
Diagonal-4	5000	6.88e−07	NaN	7.93e−07	6.25e−07
	10000	9.73e−07	8.82e−07	7.52e−07	8.83e−07
EH	5000	8.43e−07	6.82e−07	6.84e−07	7.28e−07
	10000	9.50e−07	3.83e−07	9.67e−07	6.19e−07
GR	5000	7.29e−07	9.37e−07	8.22e−07	9.00e−07
	10000	7.29e−07	9.37e−07	8.22e−07	9.00e−07
EWH	5000	8.30e−07	9.85e−07	6.01e−07	8.97e−07
	10000	8.30e−07	9.85e−07	6.01e−07	8.97e−07
ET	5000	0.00e+00	9.96e−07	9.21e−07	9.92e−07
	10000	0.00e+00	9.98e−07	4.10e−07	9.94e−07
PQD	5000	9.84e−07	4.16e−07	9.84e−07	6.71e−07
	10000	9.85e−07	2.27e−07	9.85e−07	7.88e−07
LIARWHD	5000	1.62e−07	4.41e−07	4.13e−07	8.85e−07
	10000	4.58e−07	0.00e+00	4.36e−07	3.88e−07
QUARTC	5000	9.94e−07	9.97e−07	7.80e−10	9.92e−07
	10000	9.97e−07	9.92e−07	1.56e−09	1.14e−07
GWH	5000	8.30e−07	9.85e−07	6.63e−07	8.97e−07
	10000	8.30e−07	9.85e−07	7.16e−07	8.97e−07

5 Conclusion

This study proposed a new mixed hybrid conjugate gradient method for solving large-scale unconstrained optimization problems. The method was constructed using ideas from previously constructed hybrid method and taking as base methods two recently proposed methods of the Polak-Ribiére-Polyak (PRP) and Hestenes-Stiefel (HS) family of methods. Analyses revealed that the method satisfies the descent condition and is globally convergent. Numerical experiments showed that hybrid methods is computationally efficient compared to non-hybrid methods. As part of future work, the proposed method will be tested against existing hybrid methods using larger set of test functions.

References

1. Hestenes, M.R., Stiefel, E.L.: Method of conjugate gradients for solving linear systems. J. Res. Nat. Bur. Standards **49**, 409–436 (1952)
2. Fletcher, R., Reeves, C.: Function minimization by conjugate gradients. Comput. J. **7**, 149–154 (1964)
3. Polak, E., Ribieré, G.: Note sur la convergence de directions conjugeés. Rev. Francaise Informat Recherche Operationelle, 3e Année, vol. 16, pp. 35–43 (1969)

4. Polyak, B.T.: The conjugate gradient method in extreme problems. USSR Comput. Math. Math. Phys. **9**, 94–112 (1969)
5. Liu, Y., Storey, C.: Efficient generalized conjugate gradient algorithms, part 1: theory. J. Optim. Theory Appl. **69**, 129–137 (1991)
6. Dai, Y.H., Yuan, Y.: A nonlinear conjugate gradient method with a strong global convergence property. SIAM J. Optim. **10**, 177–182 (1999)
7. Hager, W.W., Zhang, H.: A new conjugate gradient method with guaranteed descent and an efficient line search. SIAM J. Optim. **16**(1), 170–192 (2005)
8. Adeleke, O.J., Osinuga, I.A.: A five-term hybrid conjugate gradient method with global convergence and descent properties for unconstrained optimization problems. Asian J. Sci. Res. **11**, 185–194 (2018)
9. Hager, W.W., Zhang, H.: A survey of nonlinear conjugate gradient methods. Pac. J. Optim. **2**, 35–58 (2006)
10. Al-Baali, M.: Descent property and global convergence of the Fletcher-Reeves method with inexact line search. IMA J. Numer. Anal. **5**(1), 121–124 (1985)
11. Dai, Y.H., Yuan, Y.: An efficient hybrid conjugate gradient method for unconstrained optimization. Ann. Oper. Res. **103**, 33–47 (2001)
12. Du, X., Zhang, P., Ma, W.: Some modified conjugate gradient methods for unconstrained optimization. J. Comput. Appl. Math. **305**, 92–114 (2016)
13. Jian, J., Han, L., Jiang, X.: A hybrid conjugate gradient method with descent property for unconstrained optimization. Appl. Math. Model. **39**, 1281–1290 (2015)
14. Andrei, N.: An unconstrained optimization test functions collection. Adv. Mod. and Optim. **10**(1), 147–161 (2008)
15. Bongartz, I., Conn, A.R., Gould, N.I.M., Toint, P.L.: CUTE: constrained and unconstrained testing environments. ACM Trans. Math. Software **21**, 123–160 (1995)
16. Oladepo, D.A., Adeleke, O.J., Ako, C.T.: A mixed hybrid conjugate gradient method for unconstrained engineering optimization problems. In: Silhavy, R. (eds.) Cybernetics and Algorithms in Intelligent Systems, CSOC2018 2018. AISC, vol. 765. Springer, Cham (2019)

Efficient Sensory Data Transformation: A Big Data Approach

Akram Pasha[1(✉)] and P. H. Latha[2]

[1] REVA University, Bengaluru, India
akramyelandur@gmail.com
[2] Sambhram Institute of Technology, Bengaluru, India
phdlatha2017@gmail.com

Abstract. Big Data Analytics has immensely solved complex problems involving massive and complex data. Data filtering plays a vital role in helping the analysis processes to perform analytics with ease and in precise form. The data variety and veracity are some of the major problems that add enough complexity to data, creating the overall hindrance to an effective big data analytics process. The proposed work targets the data staging phase in a big data classification to tackle variety and veracity problems found in massive sensory data. The study adopts a novel approach for data storage, then designs and implements a simple algorithm to perform data transformation. The concept of cloud storage-bucket is used for effective storage and transformation of sensory data. Such an analytical approach is proven to retain the capability of an effective and faster data transformation. The algorithm performs conversion of unstructured data to semi-structured data in the first stage, then converts the semi-structured data to structured data in the second stage, and finally stores the resulting structured data into virtually localized distributed storage. The outcome of this study offers faster response time and higher data purity for data transformation process of data staging phase in any big data analytics application.

Keywords: Big data · Cloud computing · Data aggregation · Data mining · Wireless sensor network · Unstructured data

1 Introduction

The term 'big data' has opened several challenges for the organizations, by making the storage of data and its analysis a foundation for rest of the challenges encountered in big data [1,2]. However, the advent of cloud environments have eased the challenges of storage for most of the big data problems having massive size, but has drastically expanded the demands of data analysis [3]. Performing efficient analytical operations on big data still remains an unsolved challenge in big data [4,5] due to complexity of data.

R. Silhavy et al. (Eds.): CoMeSySo 2019, AISC 1046, pp. 69–82, 2019.
https://doi.org/10.1007/978-3-030-30329-7_7

1.1 Challenges of Big Data

Big data gets its name due to various characteristics of data denoted as V's of big data. Among many views of **'V's**, $5Vs$ are most commonly observed characteristics in existing massive data. The quantity of the data being originated is represented as **V**olume while diversified data originated from multiple discrete sources in different variants of formats is represented by **V**ariety. The degree of ambiguity involved in data is represented by **V**eracity while the speed of data arrival is represented as **V**elocity of data, and the key benefit or **V**alue obtained out of data makes it a big data problem [6, 7]. Many studies [8, 9] have claimed that Internet-of-Things (IoT) applications generate voluminous unstructured data which forms the basis for any big data analytics application. Many industries have started investing high capitals to investigate big data analytics over IoT-generated sensory data [10]. Research initiative towards such forms of complex data leads to evolution of new arena of analytical services as well as application associated with sophisticated analytics. The core contributions of such analytical approach is that it enabled enhanced decision making in businesses by cutting down the cost incurred in business processes. The major challenges [11] of big data analytics as seen in the literature can be enumerated as under:

- the investigation of storage mechanisms for voluminous unstructured data (issue of *Volume*);
- the investigation of security, privacy preserving and purity mechanisms for data retrieval (issue of *Veracity*);
- integration or aggregation of the data originating from heterogeneous origins and resulting in various formats (issue of *Variety*);
- extracting the valuable information from the massive data (issue of *Value*);
- developing parallel architectures for processing such data in real time and near-real time (issue of *Velocity*);
- development of visualizing tools to rapidly gain insights from the data;
- most importantly, the investigation of efficient data analytics models for such data.

1.2 Contributions of the Proposed Work

The present research archives has witnessed various approach to sort out the problems associated with big data as well as it also presented discussion towards various issues to be solved, which is yet an open-end problem. The major objective of this proposed work is to develop an efficient scheme for pre-processing big sensory data resulting in a clean data suitable for easy mining.

The following are the significant contributions of the proposed work:

- it offers overall cost effective data transformation without using any external tool or any form of third party agents,
- it offers conversion of unstructured data to semi-structured data in a simplified manner,
- it offers capability to mine key attributes of sensory data,

- it makes the knowledge delivery process human readable by using grammar-based syntax matching with the strings,
- it allows users to create cloud storage-based buckets, an initiative to promote ownership of data and to assist in distributed positioning of mined data, and
- it assists in an effective data fusion process by minimizing errors, re-positing data in distributed locations with higher degree of data purity, and reduced response time.

The proposed manuscript is organized in the following structure: Sect. 2 discusses the recent studies conducted using the big data approaches to address the data variety and data veracity problems in big data. Discussion of proposed solution and approach is carried out in Sect. 3 with illustration of the algorithm design and execution. In Sect. 4, experimental results are discussed, followed by a brief description of a "Threat for validity" that lists the major assumptions made in this study. Section 5, concludes the work proposed by enumerating the major contributions of this paper and the future directions of the work.

2 Related Work

Many studies pertaining to address the big data based approaches on sensory data analytics are conducted in the past. The work proposed in [12], designed the recommendation system using big data approach for a IoT-based smart home energy management system. Business Intelligence (BI) and proprietary software products are reportedly utilized in this study without focusing on the issues related to variety and veracity in sensory data generated from IoT. Software packages are susceptible to skip or pay less attention on the lower level analytical operation/s performed on such data. Usage of fog intelligence was proposed in the work of [13] to improve analytical performance for sensory big data. The authors have used statistical approach for obtaining sensory data without focusing at data variety and veracity issues in big data. Concentric framework of computation was proposed in the work of [13] and [14] for big data analytics. The work carried out in [15] has adopted dynamic data of mobile sensor focusing on an effective data management as the core goal of the study. In the work of [16], clustering approach was used to perform data fusion focussing on the overall performance of analytical operation. The approach presented in [17] has addressed the complexity issues associated with energy as well as precision factor connected with dataset. The work of [18], presents the data analysis by incorporating the hybridized algorithms and distributed storage models. In [19], an effort was made to analyze big data generated from mobile sensors. Large scale tempo-spatial streams were analyzed using data ingestion approach in their work. Similar research can be seen in the work of [20–22] to perform data aggregation phase using big data approach. The work published in [23] the radiation signals were analyzed using big data approach.

To summarise, efforts have been made in past in the direction of sophisticated mining for enhancing security [24, 25], to perform clustering using principal component analysis [26] and to enhance the decision making [27] (a case study of avionics), to analyse the weather information [28], to perform ecological monitoring [30].

The research studies conducted in the direction of sensory big data analytics have witnessed one or more of the following limitations:

- hypothetical modeling approach is used while addressing the problems associated with an application [10,13,18];
- Emphasized over improving the performance ignoring the adoption of practical constraints associated with particular sensor network or even IoT [11,12,28]; and
- Utilized of various existing propriety tools for carrying out analytical operation over big data [14,15,18].

As observed in [31], the HACE theorem states that the Heterogeneous Autonomous Complex and Evolving (HACE) data is having large volume of heterogeneous data coming from autonomous sources. The sources of data are distributed across many clusters and the control is essentially decentralized. These characteristics of big data poses extreme challenges in identifying and extracting the useful insights from data. Addressing the data complexity in a cost effective manner still remains a challenge that needs high attention in big data analytics domain. Ensuring the security of data is yet another challenge on top of data complexity for sensory data generated and collected from any IoT-enadled WSN. The data stored on cloud storage requires secure access mechanisms for effective data analytics in such environments [29]. The literature studies conducted reveals that no single joint implementation in big data analytics exists that address all the issues of big data (5 Vs). In the current study, the effort has been made to address variety and veracity problems jointly using big data approach. The subsequent section introduces proposed work and elaborates the step-by-step procedure of algorithm design.

3 Proposed Work

Big data analytics in its most generalized view involves 4 important phases of development to address the major issues; data sources phase addressing data volume and integration issues, content format phase addressing data variety issue, data staging phase addressing the data veracity issue and data processing phase addressing all the issues. Such a stage-wise classification of big data is as shown in the Fig. 1. The staging process of data is one significant phase in big data classification filters the data helping the overall big data analytics operation. It includes the three sub-phases; data cleaning, data transformation and data normalization. Data transformation refers to normalizing the data followed by transforming the data for supporting analytical operation in subsequent phases of big data analytics application. The work proposed in this paper tackles the data transformation problem in an intermediate data staging phase of any big data analytics application. The proposed system shown in Fig. 2 uses basic analytical approach to focus at data transformation process in particular. The conversion

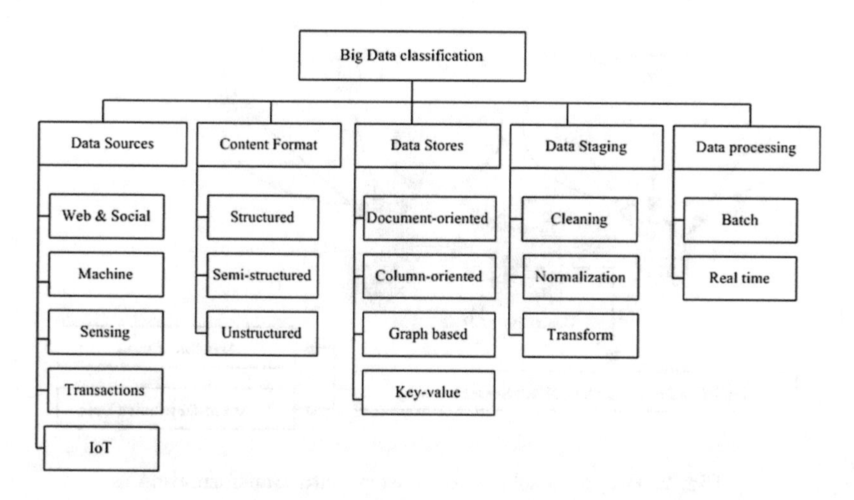

Fig. 1. Big data classification [32]

of the unstructured to structured data is carried out linearly in this process with an aid of a simple mining operation. In the first part of the mining operation the algorithm extracts the significant attributes from sensory data and then converts it to a structured form suitable for analytics in the subsequent phases of any big data analytics application. Recently, there is a report of beneficial feature of cloud-based services over conventional data processing platforms offering higher scope of big data [33]. The algorithm proposed in this work uses the concept of cloud storage-bucket. Buckets are the data structures that are used basically to retain the data that is highly user specific. Any sensory data that we intend to store in cloud storage must be contained in a bucket. The buckets organize the sensory data and control the overall access to the stored sensory data. The fact that the bucket creation and deletion is bounded with limitations, the design of storage applications favor intensive object operations, and as a result, only a few bucket operations are required to be performed relatively. Additionally, the cloud storage-bucket mechanism proposed in this work allows the user to localize the distributed cloud-storage buckets as per their requirements while re-posting the mined data.

Methodology: The proposed solution targets to resolve the issues associated with data variety and data veracity problems associated with the massive unstructured sensory data. Figure 3 represents the initial implementation phase of proposed system. The algorithm design is carried out by considering the text mining as the core approach to extract the inputs from big data and process them. In the first step, it creates the massive unstructured sensory data with a pre-specified number of attributes. In the second step, the most valuable attributes are extracted from the unstructured sensory data. In the third step, the transformation of unstructured data to semi-structured data is performed.

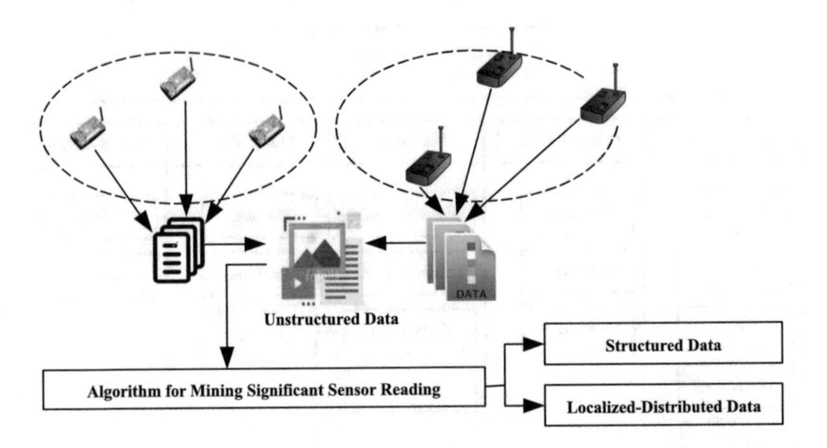

Fig. 2. Proposed scheme of sensory data transformation

The fourth and the last step performs the transformation of semi-structured to structured data, suitable to be re-posited on a distributed storage. The pseudo code form of an algorithm in its simplest form is depicted in Algorithm 1.

Creation of Attributes and Their Values for Unstructured Sensory Data: To store the information that is highly important, the list of attributes are created. Each row of a sensory data is stored in the list: $a_1, a_2, \ldots, a_n$. The number of attributes to be used depends on the kind of dataset analyzed. Thus, n, the number of attributes can be set to maximum value based on the kind of dataset analyzed. However, in the current study for simplicity the value of n is set to a fixed value, 9. The attributes are grouped as

$$g_i = a_1, a_2, , ..., a_9 \tag{1}$$

Correspondingly, each of the group is represented in a database $db1$ as $g_1, g_2, \ldots, g_m$. Hence,

$$G_{db1} = g_1, g_2, \ldots, g_m \tag{2}$$

In the above expression, cumulative group number is represented as m that is present with an inequality of $m > n$. The number of groups m is much higher than the number of attributes.

Extraction of Most Valuable Attributes in Sensory Data: Each of attributes are acquired, discretisized and stored in cloud storage to ensure that the accuracy of mined sensory data is enhanced. The whole process of extracting the most valuable attributes in sensory data is obtained algorithmically.

Transformation of Unstructured Data to Semi-structured Data: The algorithm is made to leverage the semi-structured data before it converts it to

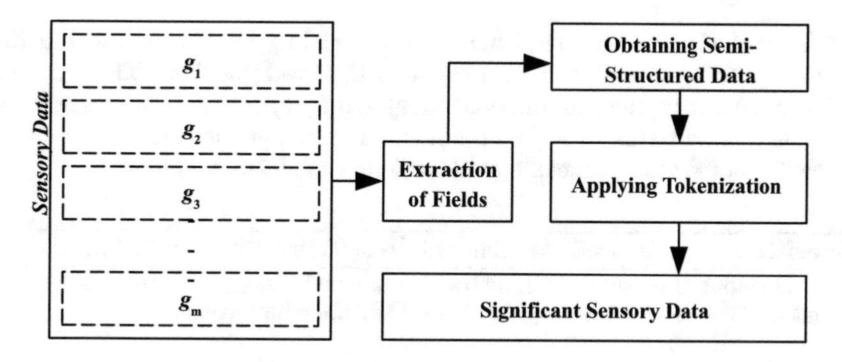

Fig. 3. Preliminary phase of ESDT-ABDA implementation

structured data, facilitating the performance of intermediate stage-wise analysis with ease. This method of conversion of unstructured data to semi-structured data reduces the inherent data complexity involved in sensory data eventually.

Transformation of Semi-structured to Structured Data: The concept of tokenization is used for extracting the attributes from the semi-structured data. Finally, the proposed system considers sensory data in the form of text assists in rapidly understanding the extracted structured sensory data at the end.

The above operations are carried out stage-wise for extracting significant sensor reading that takes the input of D_i (Input of Sensory Data) and m (Number of cloud-storage buckets). In the pipeline, the algorithm produces the final output of the function O_4, the transformed sensory data, and the output of function O_c, the data with certainty. The pseudo code representation of sequence of analytical operations performed on any sensory data set $x \epsilon D_i$ is abstractly depicted in Algorithm 1.

The step-by-step discussion of Algorithm 1 is as under.

- After taking the input $x \epsilon D_i$ of unstructured sensory data *(Line-1)*, the algorithm initially looks for the word, sentences, and paragraph by exploring the position of single space, period, and new line character.
- A simple and integrated group function $g(x)$ is developed for this purpose to collect all the distinctive fields *(Line-2)*.
- The algorithm considers all the counted attributes for the given database *(Line-3)*.
- A specific function $\theta_1(x)$ is developed which relates all the values associated with the attributes with grammatical syntax in order to offer better interpretation of the text-based sensory readings *(Line-4)*.
- The next function $\theta_2(x)$ is developed for obtaining the semi-structure form of unstructured input data *(Line-5)*.
- A specific template is then developed using the same fields and all the collected information associated in the attributes is also collected. This mechanism is called as data transformation capable of representing the unstructured data in its semi-structured form.

– The next step of the algorithm calls for reading all the groups and their corresponding attributes with respect to their cell positions. This operation is carried out by the two functions $\theta_3(x)$ and $\theta_4(x)$. These functions associate the extracted terms with the grammatical syntax and extract the significant readings of sensors respectively *(Line-6 and Line-7)*.

Algorithm 1. Pseudocode for Efficient Data Transformation algorithm

Input: D_i(Input of sensory data), m(Number of cloud-storage buckets).
Output: O_4(Transformed sensory data) and O_c(Data with certainty).

1: Initialize D_i (Input of sensory data), m (Number of cloud-storage buckets) and n with 9.
2: $F = g_1(D_i)$
3: **for** $j = 1$ to n **do**
4: $O_1 = Apply\ \theta_1(x)\ on\ f \mid f \subset F$
5: $O_2 = Apply\ \theta_2(x)\ on\ f$
6: $O_3 = Apply\ \theta_3(x)\ on\ f$
7: $O_4 = Apply\ \theta_4(x)\ on\ f$
8: **end for**
9: $O_c = Apply\ \theta_5(x)\ on\ O_4$

The methodology shown in Algorithm 1 ensures that when highly unstructured data is owing to the problem of data variety, it is capable of identifying this problem and offers a simple solution to convert the unstructured data to structured data. In the process of conversion, it also extracts the significant sensory readings which can be treated as mined data. The linear steps of algorithm discussed so far offer simple solution by eliminating the problem of data variety for the given unstructured data. However, it doesn't address the problem of data veracity yet in the same scan. Thus, to address the problem of data veracity, it is essential for any cloud environment to have distinctive hold of the information associated with the distributed storage model targeted to store semi-structured data to be re-posited eventually to a different location. This is achieved finally by developing a fifth function $\theta_5(x)$ in pipeline, which is fed with the output O_4 resulting from the application of $\theta_4(x)$.

The proposed system applies a unique and simplified strategy to address the problem of data veracity mainly associated with precision and trustworthiness of the data. A function is developed for addressing the problems of data unworthiness or uncertainty factor caused due to data lineage, biases, bugs, noise, data security, abnormalities, complexities or hardware problems with sources, human error, out-of-data, and falsification. As these factors result in a data that cannot be trusted for its reliability, it is essential to ensure that the data doesn't suffer after normalization. The proposed system constructs virtual directories that are essentially meant for re-positing the extracted mined information in highly distributed manner so that distributed mining algorithm could be further investigated on top of it. This is achieved by constructing a cloud storage-based

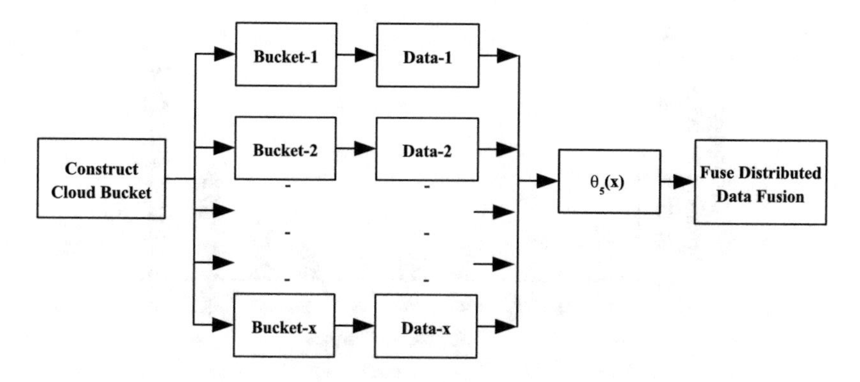

Fig. 4. Scheme for mitigating data veracity problem

bucketing system, which is a form of a dedicated virtual storage over cloud. This mechanism used allows the mined information to re-posit discretely and uniquely in cloud storage-based buckets.

The Fig. 4 shows that there are x-number of cloud storage-buckets being created, which allows exactly x-number of mined data to be re-posited uniquely in different locations. All the indexes of data repositories are carried out in dynamic order and can be effectively maintained over any cloud cluster retaining the location information in encoded form. The proposed system then applies a specific function $\theta_5(x)$ to fetch data from each dynamically constructed cloud storage-bucket and accesses respective data sources. All the fetched data sources will be then subjected to the similar process of elimination of data variety problems as discussed in the algorithm, so that cleansed data, suitable for analysis could be obtained in the end. The major contribution of the proposed algorithm is that it performs the effective and efficient data processing by combating the problems of data variety and veracity together that are commonly encountered, in any big data analytics application.

4 Experimentation and Discussion of Results

The proposed model is simulated and testified on the synthetic data, modified from standard publicly available Hadoop Illuminated test dataset. The formats of the data are $XLS/CSV/JSON$ with each data to possess approximately 5000 rows and more than 11 columns with headers. The data is programmatically incorporated with the heterogeneity and veracity problems. A synthetic database of sensory application is constructed for testifying the proof of the proposed algorithm. The scripting of the proposed logic is carried out using MATLAB. The performance parameters considered for evaluating the effectiveness of adopted technique are mainly the computational time and data quality. Following are the core findings of the proposed system.

The graphical outcome of Fig. 5 highlights that data transformation time increases with increase in size of the input data over 10 different evaluation

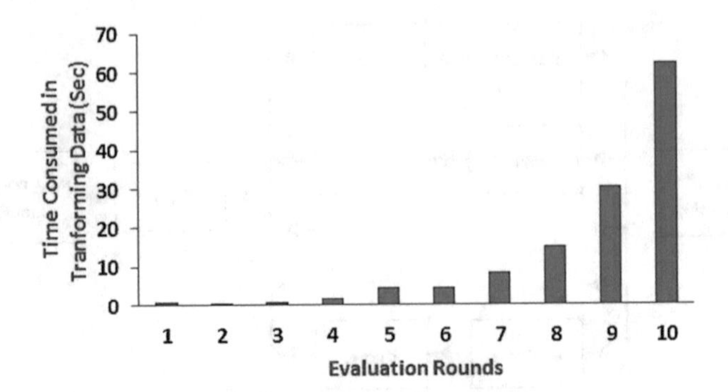

Fig. 5. Analysis of time consumption in transforming data: (a) Time consumption in seconds

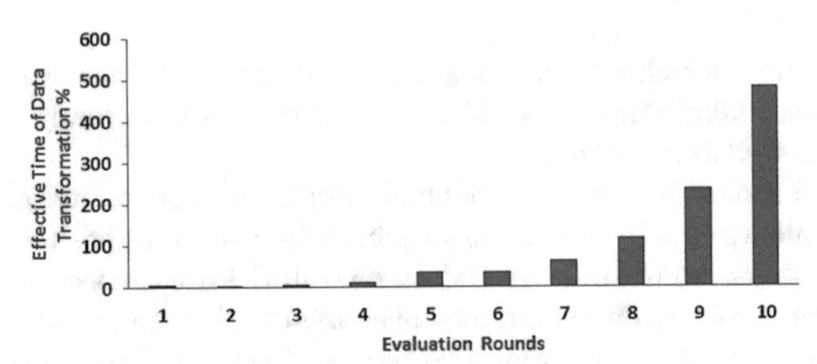

Fig. 6. Analysis of time consumption in transforming data: (b) Percentage of effective time

rounds. Figure 6 shows effective data transformation time where proposed system is visibly found to offer nearly consistent time consumption for more than half the rounds of evaluation and then it peaks up to higher level owing to increase of incoming data traffic. It clearly shows that there is no preliminary sluggishness in processing the sensory big data and can control the data processing time to a larger extent. However, transformation process is an important part of the overall proposed analytical operation. Without this process it is not feasible to transform unstructured data to semi-structured data. At the same time, the processing time in our proposed system denotes the time required to obtain the significant information from the sensory data, which we refer here as mined information. This fact can be assessed by comparing data transformation duration with the mining duration as the former is just one part of the operation in performing mining operation. The outcome obtained is basically a mined result against the given set of sensory data.

Figure 7 highlights that the effective data transformation time is always found to be under significant control as against mining. The height of bars for mining remains tall implying that there are many other processes running after performing transformation operation for the sensory big data. It also implies that there is no abnormal fluctuation for mining duration too, and the increment is too

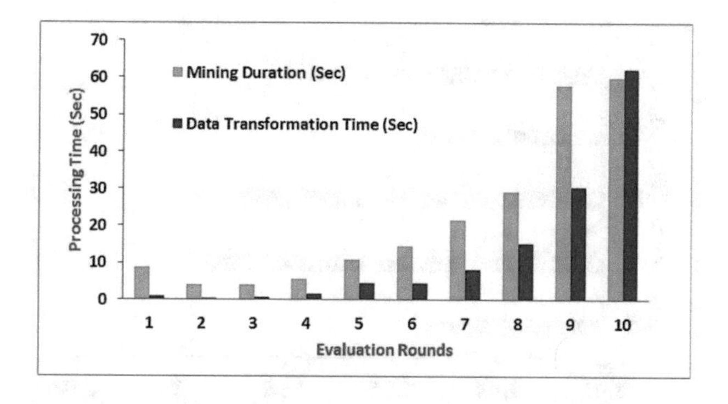

Fig. 7. Analysis of processing time

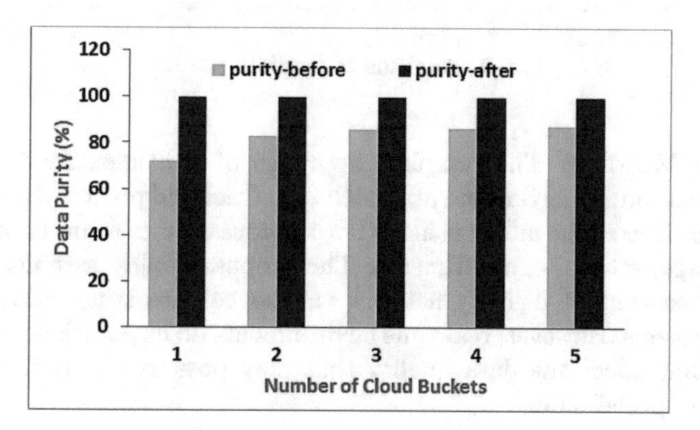

Fig. 8. Analysis of data purity

smooth and predictive in nature. Data purity is another performance parameter considered for analyzing the effectiveness of proposed work. The study shows that there are different degrees of data purity prior to performing implementation. The method used computes the degree of completeness of data which acts as the measure of data variety function shown in the Fig. 8. The average data purity obtained prior to data variety function implementation is approximately 83%, while after applying data variety function, the data purity is found to be reaching perfectly 100%.

Finally, the proposed system is assessed for prediction time as shown in Fig. 9, which is overall time of evolving with final structured data to resolve the problem associated with data veracity. The overall prediction time of proposed system is approximately 1.97 s. The earlier works have been either reported to have used complex functionality, or have used off-the shelf business analytical tools, or any such tools that are reported of yielding error-prone information in big data. It overcomes such problems and offers a simple and efficient analytical operation, and retains a better balance between response time of analytical operation and superior quality of data.

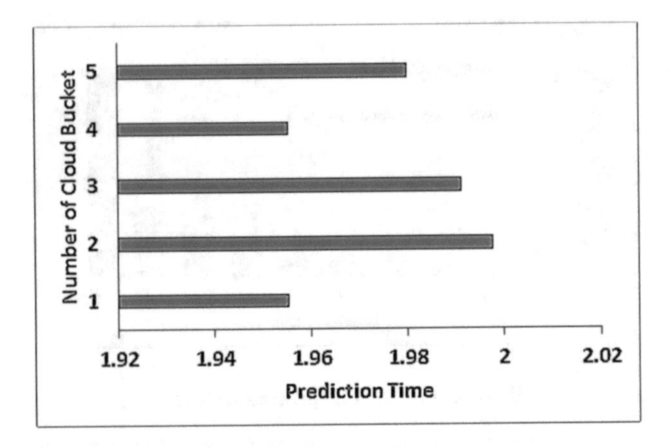

Fig. 9. Analysis of prediction time

Threat to Validity: The complete approach of implementing the proposed logic uses text mining as the core approach to extract and process the inputs from the big data. Hence, the model is likely not to process other forms of inputs in the form of image, speech, or multimedia. The proposed study assumes that there is a dedicated content delivery network and hence data is not affected by any channel problems. However, real-time environments do have such channel related problems that affect the data quality that may pose as threat to operational feature of proposed model.

5 Conclusion and Future Enhancement

Proposed approach presents a very simplified technique of data transformation and data localization in distributed cloud storage-based model. The methodology adopted uses an analytical approach to solve data variety and data ambiguity issues connected with sensory data. The experimental work shows that the significant control over the effective data transformation time in comparison with traditional mining operations can be performed during data transformation. The data purity after applying the data variety function is found to be reaching perfectly 100%. The overall prediction time is found to be almost instantaneous. The analytical operation designed is not only simple in its implementation and efficient but also retains a better balance between response time of analytical operation and superior quality of data. The future work will be to focus on: extending the similar framework to address data value problem, to work out heuristic or meta-heuristic approaches to collectively address the 3 Vs of big data in unstructured massive sensory data. Evaluating the security metrics with data security concerns on cloud storage-bucket remains yet another direction of research in this domain.

References

1. Trovati, M., Hill, R., Zhu, S., Liu, L.: Big-Data Analytics and Cloud Computing. Springer, Heidelberg (2015)
2. Mazumder, S., Bhadoria, R.S., Deka, G.C.: Distributed Computing in Big Data Analytics: Concepts, Technologies and Applications. Springer, Heidelberg (2017)
3. Zanoon, N., Al-Haj, A., Khwaldeh, S.M.: Cloud computing and big data is there a relation between the two: a study. Int. J. Appl. Eng. Res. **12**(17), 6970–6982 (2017)
4. Marjani, M., Nasaruddin, F., Gani, A., Karim, A., Hashem, I.A.T., Siddiqa, A., Yaqoob, I.: Big IoT data analytics: architecture, opportunities, and open research challenges. IEEE Access **5**, 5247–5261 (2017)
5. Lv, Z., Song, H., Basanta-Val, P., Steed, A., Jo, M.: Next-generation big data analytics: state of the art, challenges, and future research topics. IEEE Trans. Ind. Inform. **13**(4), 1891–1899 (2017)
6. Marr, B.: Big Data: Using SMART Big Data, Analytics and Metrics to Makebetter Decisions and Improve Performance. Wiley, Hoboken (2015)
7. Li, K.C., Jiang, H., Zomaya, A.Y.: Big Data Management and Processing. CRC Press, Cambridge (2017)
8. Puthal, D.: Lattice-modelled information flow control of big sensing data streams for smart health application. IEEE Internet of Things J. (2018)
9. Han, G., Guizani, M., Lloret, J., Chan, S., Wan, L., Guibene, W.: Emerging trends, issues, and challenges in big data and its implementation toward future smart cities: part 2. IEEE Commun. Mag. **56**(2), 76–77 (2018)
10. Habibzadeh, H., Boggio-Dandry, A., Qin, Z., Soyata, T., Kantarci, B., Mouftah, H.T.: Soft sensing in smart cities: handling 3Vs using recommender systems, machine intelligence, and data analytics. IEEE Commun. Mag. **56**(2), 78–86 (2018)
11. Raafat, H.M., Hossain, M.S., Essa, E., Elmougy, S., Tolba, A.S., Muhammad, G., Ghoneim, A.: Fog intelligence for real-time IoT sensor data analytics. IEEE Access **5**, 24062–24069 (2017)
12. Al-Ali, A., Zualkernan, I.A., Rashid, M., Gupta, R., Alikarar, M.: A smart home energy management system using IoT and big data analytics approach. IEEE Trans. Consum. Electron. **63**(4), 426–434 (2017)
13. ur Rehman, M.H., Ahmed, E., Yaqoob, I., Hashem, I.A.T., Imran, M., Ahmad, S.: Big data analytics in industrial IoT using a concentric computing model. IEEE Commun. Mag. **56**(2), 37–43 (2018)
14. Yang, C., Puthal, D., Mohanty, S.P., Kougianos, E.: Big-sensing-data curation for the cloud is coming: a promise of scalable cloud-data-center mitigation for next-generation IoT and wireless sensor networks. IEEE Consum. Electron. Mag. **6**(4), 48–56 (2017)
15. Zhang, D., He, T., Lin, S., Munir, S., Stankovic, J.A.: Taxi-passenger-demand modeling based on big data from a roving sensor network. IEEE Trans. Big Data **3**(3), 362–374 (2017)
16. Din, S., Ahmad, A., Paul, A., Rathore, M.M.U., Jeon, G.: A cluster-based data fusion technique to analyze big data in wireless multi-sensor system. IEEE Access **5**, 5069–5083 (2017)
17. Cheng, S., Cai, Z., Li, J., Gao, H.: Extracting kernel dataset from big sensory data in wireless sensor networks. IEEE Trans. Know. Data Eng. **29**(4), 813–827 (2017)
18. Ebner, K., Buhnen, T., Urbach, N.: Think big with big data: identifying suitable big data strategies in corporate environments. In: 47th Hawaii International Conference on System Sciences (HICSS), pp. 3748–3757. IEEE (2014)

19. Hu, G., Zhang, X., Duan, N., Gao, P.: Towards reliable online services analyzing mobile sensor big data. In: IEEE International Conference on Web Services (ICWS), pp. 849–852. IEEE (2017)
20. Ren, M., Li, J., Guo, L., Li, X., Fan, W.: Distributed data aggregation scheduling in multi-channel and multi-power wireless sensor networks. IEEE Access (2017)
21. Takaishi, D., Nishiyama, H., Kato, N., Miura, R.: Toward energy efficient big data gathering in densely distributed sensor networks. IEEE Trans. Emerg. Top. Comput. **2**(3), 388–397 (2014)
22. Karim, L., Al-kahtani, M.S.: Sensor data aggregation in a multi-layer big data framework. In: 7th IEEE Annual Conference on Information Technology, Electronics and Mobile Communication (IEMCON), pp. 1–7. IEEE (2016)
23. Jeong, M.H., Sullivan, C.J., Wang, S.: Complex radiation sensor network analysis with big data analytics. In: IEEE Nuclear Science Symposium and Medical Imaging Conference (NSS/MIC), pp. 1–4. IEEE (2015)
24. Kandah, F.I., Nichols, O., Yang, L.: Efficient key management for big data gathering in dynamic sensor networks. In: International Conference on Computing, Networking and Communications (ICNC), pp. 667–671. IEEE (2017)
25. Zhu, C., Shu, L., Leung, V.C., Guo, S., Zhang, Y., Yang, L.T.: Secure multimedia big data in trust-assisted sensor-cloud for smart city. IEEE Commun. Mag. **55**(12), 24–30 (2017)
26. Li, J., Guo, S., Yang, Y., He, J.: Data aggregation with principal component analysis in big data wireless sensor networks. In: 12th International Conference on Mobile Ad-Hoc and Sensor Networks (MSN), pp. 45–51. IEEE (2016)
27. Miao, W., Zheng, D., Hangyu, G., Tao, Y.: Research on big data management and analysis method of multi-platform avionics system. In: 16th International Conference on Computer and Information Science (ICIS), pp. 757–761. IEEE (2017)
28. Onal, A.C., Sezer, O.B., Ozbayoglu, M., Dogdu, E.: Weather data analysis and sensor fault detection using an extended IoT framework with semantics, big data, and machine learning. In: IEEE International Conference on Big Data, pp. 2037–2046. IEEE (2017)
29. Latha, P., Vasantha, R.: MDS-WLAN: maximal data security in wlan for resisting potential threats. Int. J. Electr. Comput. Eng. **5**(4), 859 (2015)
30. Wiska, R., Habibie, N., Wibisono, A., Nugroho, W.S., Mursanto, P.: Big sensor-generated data streaming using Kafka and impala for data storage in wireless sensor network for CO_2 monitoring. In: International Workshop on Big Data and Information Security (IWBIS), pp. 97–102. IEEE (2016)
31. Wu, X., Zhu, X., Wu, G.Q., Ding, W.: Data mining with big data. IEEE Trans. Knowl. Data Eng. **26**(1), 97–107 (2014)
32. Hashem, I.A.T., Yaqoob, I., Anuar, N.B., Mokhtar, S., Gani, A., Khan, S.U.: The rise of big data on cloud computing: review and open research issues. Inf. Syst. **47**, 98–115 (2015)
33. Cai, H., Xu, B., Jiang, L., Vasilakos, A.V.: IoT-based big data storage systems in cloud computing: perspectives and challenges. IEEE Internet of Things J. **4**(1), 75–87 (2017)

Software Architecture as a Thinging Machine: A Case Study of Monthly Salary System

Sabah Al-Fedaghi(✉) and Majd Makdessi

Computer Engineering Department, Kuwait University, P.O. Box 5969 Safat, 13060 Kuwait City, Kuwait
sabah.alfedaghi@ku.edu.kw,
majd.makdessi@grad.kw.edu.kw

Abstract. Software architecture is concerned with the study of a software system's high-level structures and involves the design and production of such structures. According to experts n the field, software teams struggle to communicate software architecture. To communicate the software architecture of a software system, developers use diagrams that tends to be a confused mess of boxes and features, unstable notation, ununiform naming, unlabeled relationships, generic terminology, missing technology decisions, mixed abstractions, etc. This paper proposes a new diagrammatic representation, called a thinging machine (TM), as a methodology in software architecture. We show that a TM can express situations that arise in practice and is therefore an alternative to tools currently in use. The general aim is to provide alternative representations that may be used to develop more refined tools in this field of study.

Keywords: Software systems architecture · System modeling · Conceptual model · Diagrammatic representation

1 Introduction

Software architecture concerns the study of a software system's high-level structures and involves the design and production such structures. The term "architecture," as used here, is a metaphor, analogous to the architecture of a building, which is considered a blueprint for a system [1–3].

Each structure comprises software elements and relations among them. Software architecture notions include components that deliver application functionality and connectors (e.g., procedure call, pipe, and shared memory) that specify interactions among components.

Architecture system-level design is concerned with topics that include

- The main functional components and how they fit together
- The interaction of these components with one another and with the outside world
- System-level managed, stored, and presented information
- Physical hardware and software elements that support the system
- System-level operational features and capabilities
- System-level development, testing, support, and training environments [4]

R. Silhavy et al. (Eds.): CoMeSySo 2019, AISC 1046, pp. 83–97, 2019.
https://doi.org/10.1007/978-3-030-30329-7_8

Software architecture is used for parallel development, integration, and testing of software components. It is also a valuable tool to facilitate understanding of the "big-picture" of a system.

1.1 Problem and Proposed Solutions

Software teams struggle to communicate software architecture (see Fig. 1). According to Brown [5],

> Ask a software developer to communicate the software architecture of a software system using diagrams and you'll likely get a confused mess of boxes and lines … inconsistent notation (colour coding, shapes, line styles, etc.), ambiguous naming, unlabeled relationships, generic terminology, missing technology choices, mixed abstractions, etc.

Brown [5] then went to Google to see what a software architecture diagram looked like. He found diagrams (see Fig. 2) modeling all of the problems we see in hand-written diagrams.

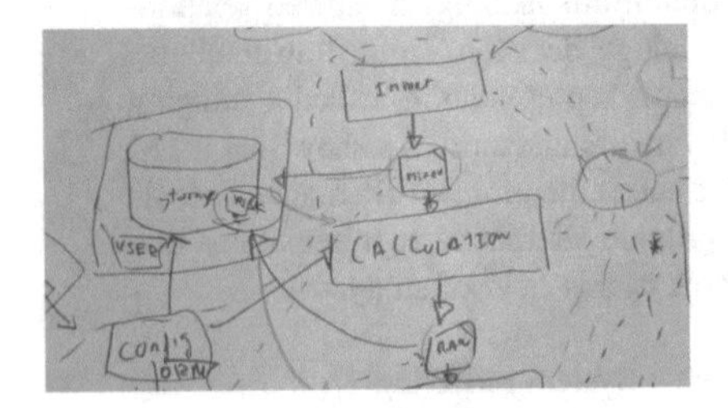

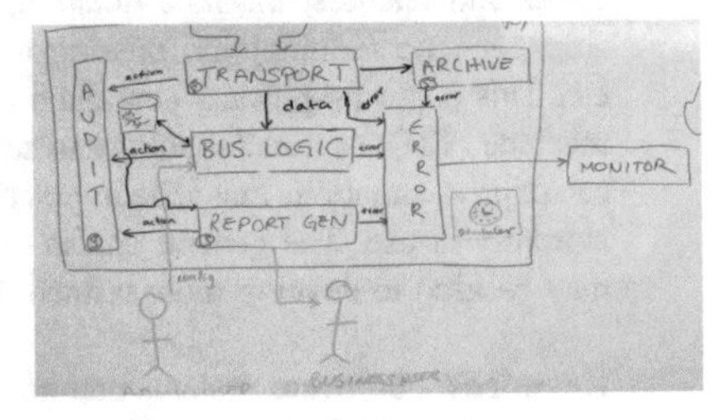

Fig. 1. Samples of communication of architecture (Adapted from [5]).

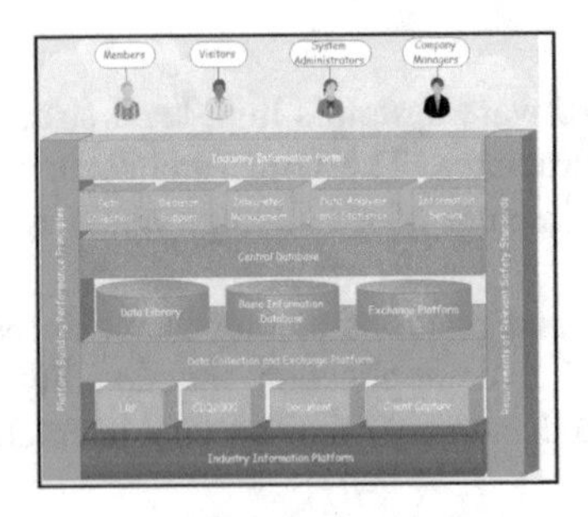

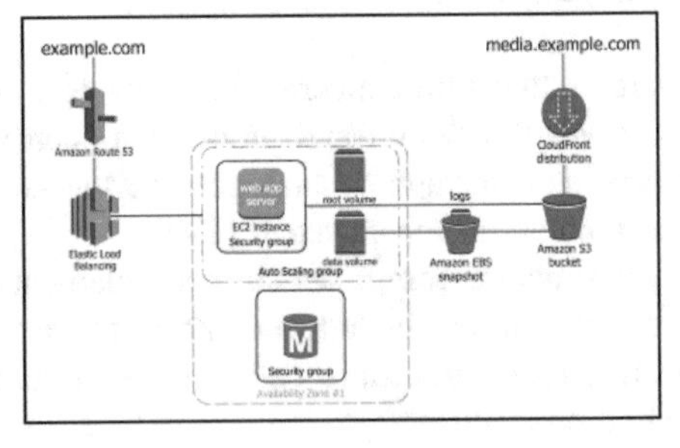

Fig. 2. Samples of architecture diagrams in Google (partial).

According to Brown [5], as an industry, we have Unified Modeling Language (UML), ArchiMate, and SysML, but asking whether these provide an effective way to communicate software architecture is often irrelevant because many teams have already abandoned in favor of much simpler "boxes and line" diagrams.

Rozanski [4] stated that "it is not possible to capture the functional features and quality properties of a complex system in a single comprehensible model that is understandable by and of value to all stakeholders." Rozanski [4] advocated representing complex systems in a manageable way that is comprehensible by a range of business and technical stakeholders, such as approaching the problem from different perspectives (e.g., contextual, functional, deployment, or concurrent points of view).

Brown [5] proposed a model based on abstractions "that reflect how software architects and developers think about and build software." The model consists of static structures that are understandable in terms of containers, components, and code (see Fig. 3).

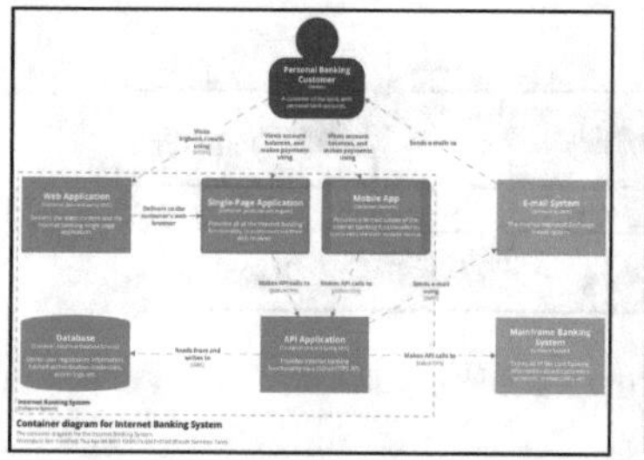

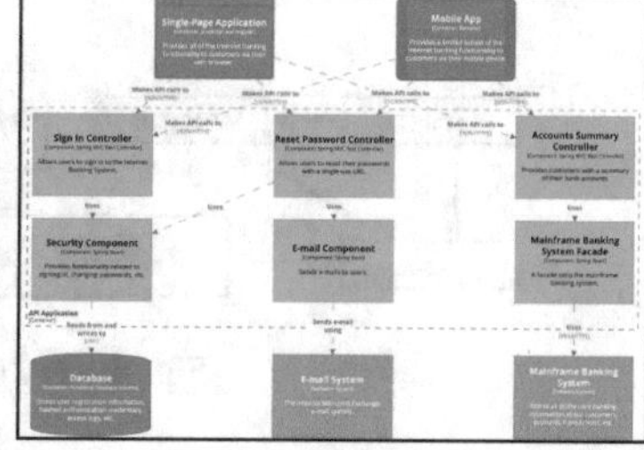

Fig. 3. Samples of abstractions: a container (left) and a component (right) (partial figures from [5]).

1.2 Thinging Machine as a Basis for Modeling Software Architecture

In this paper, we propose the adoption of a conceptual modeling method called a thinging machine (TM) that views all elements of software systems architecture in terms of a single notion: a flow machine. The TM model serves this purpose at various levels of detail. The method is applied to the real case study of a monthly salary system to demonstrate that the viability of TM as a methodology in this context.

To establish the bona fides of TM as a research framework, Sect. 2 reviews previous uses of the TM in several published papers [6–14].

2 Thinging Machine

"Philosophical work is not only intrinsically important, but it can also stand up in terms of some of the more established research metrics to other types of IS [information systems] research" [15]. Furthermore, Hassan et al. [15] stated, "Most research in IS will either begin with qualities and characteristics of objects of study already laid out by other disciplines, with very little contemplation of the metaphysics of 'information' or of 'system'."

We adopt a conceptual model that is centered on things and machines in a system. The philosophical foundation of this approach is based on Heidegger's notion of thinging [16]. Thus, instead of perceiving things (to be defined later) as everyday objects, we conceptualize them, pursuant to Heidegger's view, in their interrelation to the surrounding environment. In Heidegger's terms, the environment refers to the place where things reside. Conceptualization, as will become clearer later, denotes the production of a description of the observed system.

From this perspective, things are, each in its own way, "equipment" (Heideggerian term), such that they are not put to just any specific use, but are, rather, part of the complex mesh of things that make up the whole system of interwoven components. In a TM, we capture this equipment feature of things through the notion of machinery. A thing is an (abstract) machine. Things form a mesh of interacting machines in their residence in the environment (which becomes a grand machine). A TM is a basic type of thing/machine, as shown in Fig. 4.

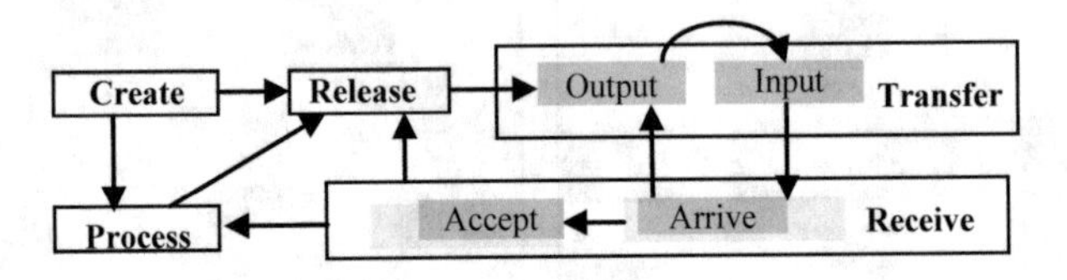

Fig. 4. A flow machine.

The flow of things in a TM refers to the exclusive conceptual movement among the five operations (stages) as shown in Fig. 4. A thing is what is created, processed, released, transferred, and/or received in a machine.

Accordingly, the stages of a TM can be described as operations that transform, modify, etc. things, either in the abstract sense or in the "concrete" sense. They are briefly described as follows.

Arrive: A thing flows into a new machine (e.g., packets arrive at a port in a router).

Accept: A thing enters a TM after arrival (we will assume that all arriving things are accepted); hence, we can combine arrive and accept to engender the **receiving** stage.

Release: A thing is marked ready for transfer outside the machine (e.g., in an airport, passengers wait to board after passport clearance).

Process: A thing is changed by modifying its description rather than the nature of the thing itself (e.g., an apple after it is peeled). However, if we (the crafter of the description) consider the peeled apple to be a new thing, then the processing of an apple triggers the creation of a peeled apple.

Create: A new thing is born in a machine (e.g., a forward packet is generated in a machine router).

Transfer: A thing is inputted or outputted in/out of a machine.

A TM includes one additional notation—triggering (denoted by a dashed arrow)—that initiates a new flow (e.g., a flow of electricity triggers a flow of air). Multiple machines can interact with each other through flows, or by triggering stages.

3 Software Architecture

The following example is adapted from [2]. Assume one is required to draw a bar chart for stock data representing two components: CollectStockData (collects stock data to a file) and DisplayStockData (draws a graph based on the file). A coordinator system acts as a connector to invoke these two components, as shown in Fig. 5.

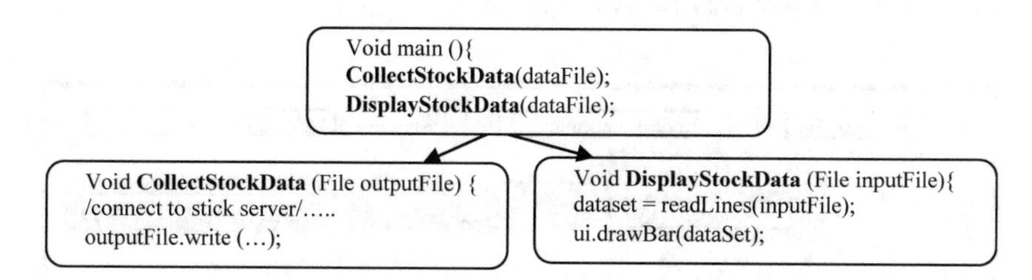

Fig. 5. An example of connector.

We claim that architectural notions, such as "a connector" deserve a conceptual description above programming language. Figure 6 shows the corresponding TM representation of this example. The figure reflects a conceptual picture, and therefore does not use such notations (e.g., semicolons) as are found in a programming language. The main machine connects three components: File1 (circle 1), CollectStockData (2), and DisplayStockData (3).

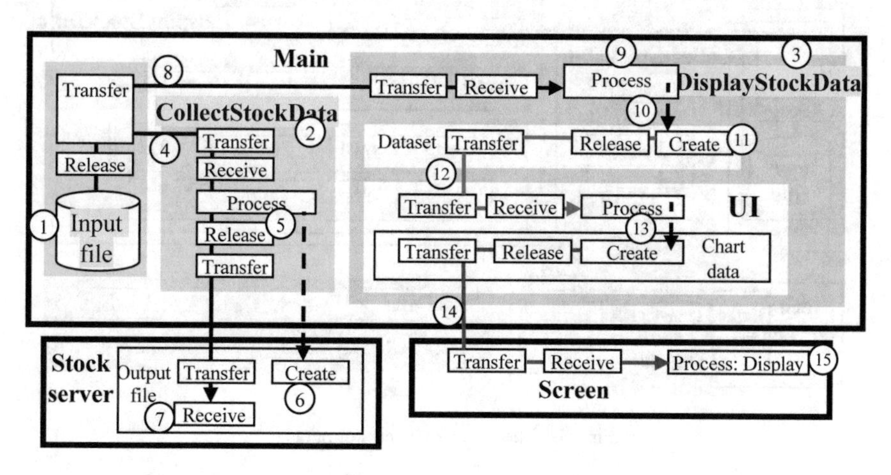

Fig. 6. The TM representation of the connector.

- The data flow from the input file to CollectStockData (4) where it is processed (5) to create the output file in the stock server (6) that is transferred to the input file (7).
- The data also flow (8) to DisplayStockData, where the information is processed (9) to trigger (10) the creation (11) of the data set.

- The data set flows to the user interface (UI) function (12), where it is processed to create the chart data (13) that flows (14) to the screen to be displayed (15).

We use the notion of an event to model the behavior of the connector system. An event is a TM machine that is formed from time, region, and the event itself as submachines. For example, the event called "Data" is processed in CollectStockData and an Output file is created, as modelled in Fig. 7. However, for simplicity's sake, we will represent the event only by its region.

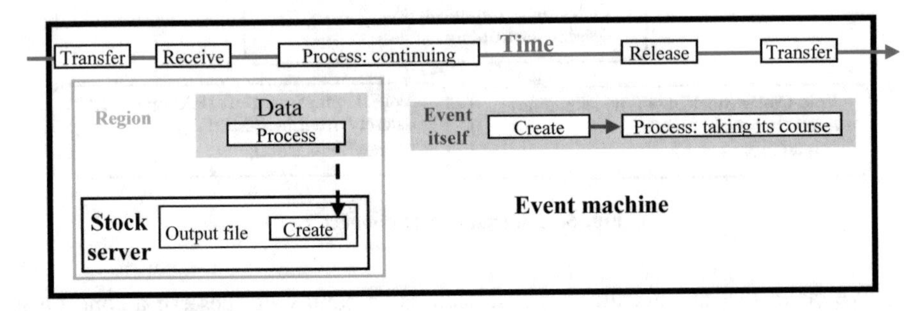

Fig. 7. The TM representation of the event called "Data" is processed in CollectStockData and the Output file is created.

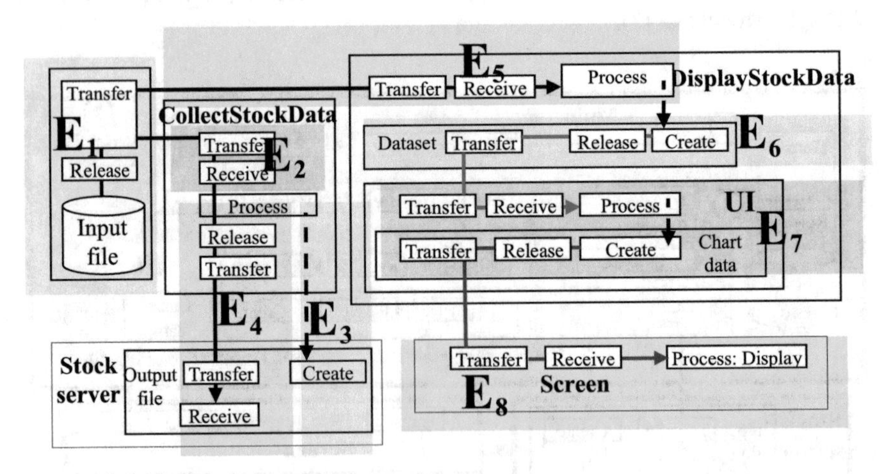

Fig. 8. The events of the connector.

The behavior of the connector can be described by identifying its events (see Fig. 8) as follows:

Event 1 (E_1): Data flow from the input file.
Event 2 (E_2): Data are received by CollectStockData.
Event 3 (E_3): Data are processed in CollectStockData and an output file is created.

Event 4 (E_4): Data are processed in CollectStockData and data flow to the output file flows in the stock server.
Event 5 (E_5): Data are received and processed in DisplayStockData.
Event 6 (E_6): Data set is created.
Event 7 (E_7): The data set flows to the UI, where it is processed and chart data is created.
Event 8 (E_8): Chart data flow to the screen, where it is displayed as a chart.

Note that the TM diagram can be presented at various levels of granularity. For example, the directions of the arrows can reflect the connection and the flow among machines, thus we can remove the release, transfer, and receive stages if there is no functionality involved in these stages. Figure 6 can be simplified, as shown in Fig. 9. The diagram can be simplified further, as shown in Fig. 10. These simplifications are each based on the original TM diagram, just as simplified engineering diagrams (block diagrams in electrical engineering) are not arbitrarily simplified, but are instead based on a complete blueprint of the system (e.g., electrical circuitry).

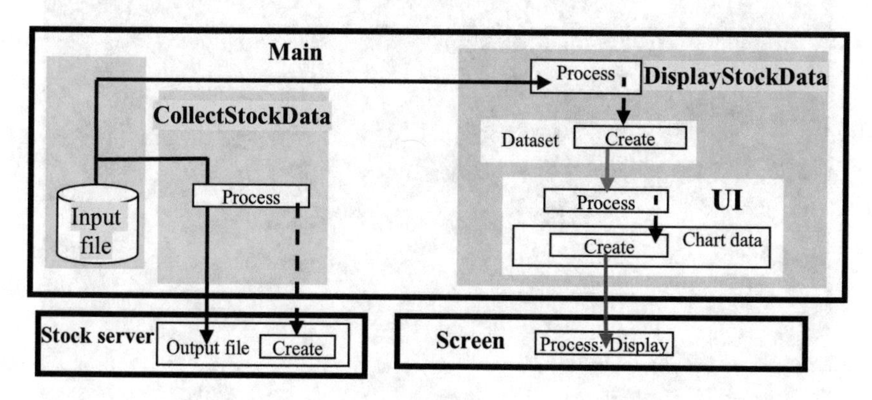

Fig. 9. Simplification of the TM representation of the connector.

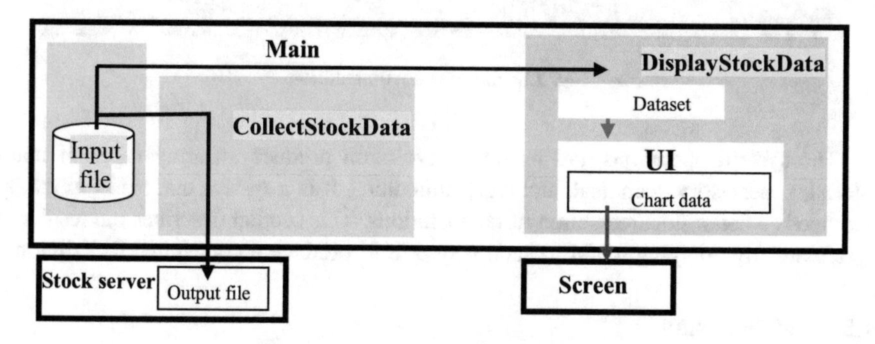

Fig. 10. Further simplification of the TM representation of the connector.

4 Case Study: Monthly Salary Software System

This case study models a current software system in a private software enterprise. One aspect of the software system involves issuing monthly salaries to the employees of the client organization. In this case study, the firm that created the software system has used an ad hoc technique that resulted in skilled software developers building the monthly salary system piece-by-piece over several years. No current documentation exists, even though the software manager has drawn flowcharts showing the full description of the processes Underpinning the operation of the monthly salary system as shown in Fig. 11.

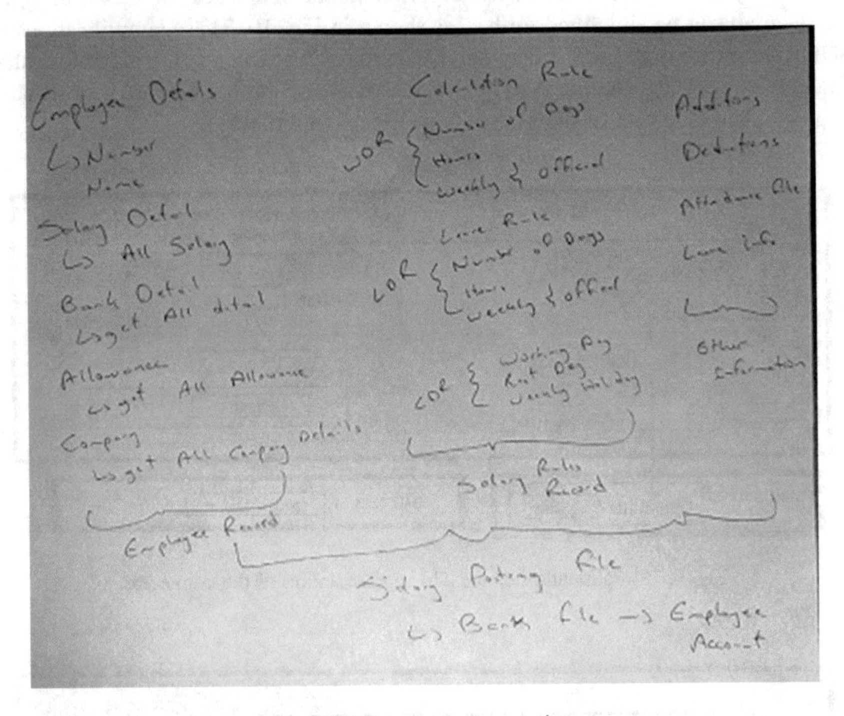

Fig. 11. Current documentation.

The system operations and its future evolution produce an array types of managerial, supervisory, technical, and legal difficulties. It is a system that exists in reality and needs a better documentation of its operations. This section describes the results of systematically applying a TM to such a system to produce a conceptual description.

4.1 TM Modeling

Discussion with developers of the software system resulted in the TM description shown in Fig. 12, which represents the software system architecture, as described below.

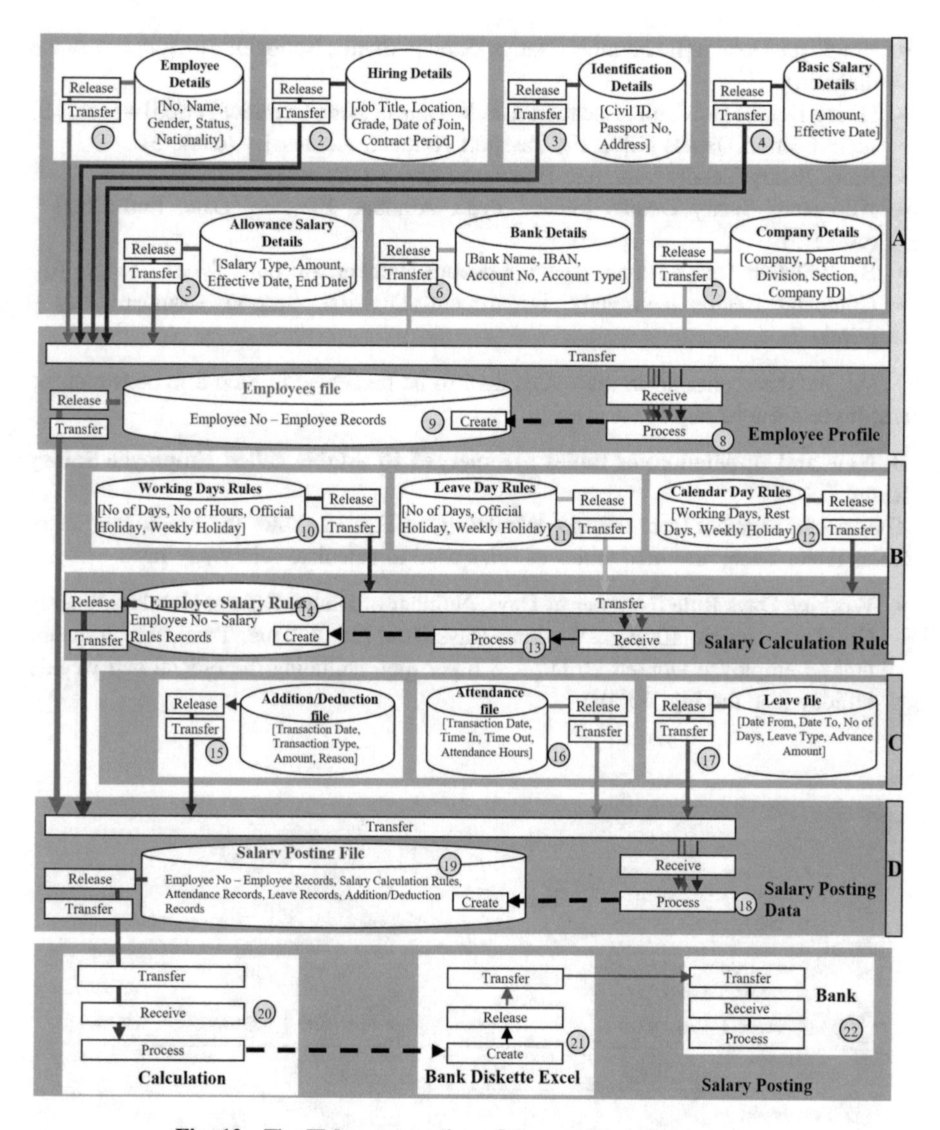

Fig. 12. The TM representation of the monthly salary system.

4.2 Static TM Description

The execution of the monthly salary system is described in several parts, as shown in Fig. 11. The employee data involve each employee's identification number as a key and each employee's other relevant details.

A. Monthly updating of data

The upper segment labelled A in Fig. 12 shows the monthly updating of data. New and updated data flow to seven files as follows:

- Employee Details [ID Number, Name, Gender, Status, Nationality] – (circle 1 in the diagram)
- Hiring Details [Job Title, Location, Grade, Date of Hire, Contract Period] – (circle 2)
- Identification Details [Civil ID, Passport Number, Address] – (circle 3)
- Basic Salary Details [Amount, Effective Date] – (circle 4)
- Allowance Salary Details [Salary Type, Amount, Effective Date, End Date] – (circle 5)
- Bank Details [Bank Name, IBAN, Account Number, Account Type] – (circle 6)
- Company Details [Company, Department, Division, Section, Company ID] – (circle 7)

All the data collected in these files flow to be processed in Step 8 to construct the **employee's file** in step 9.

B. New and updated rules tables are merged in a table called Employee Salary Rules

The second segment from the top, labelled B, in Fig. 12 shows the Employee Salary Rules table. There are three tables of rules used to calculate salary pay rates.

- Working Days Rule [Number of Days, Number of Hours, Official Holiday, Weekly Holiday] – (circle 10; e.g., Working Days Rule [26, 8, True, True], which means that an employee worked 26 Days, 8 h per day, including the official and weekly holiday [see the Fig. 13(a)]).

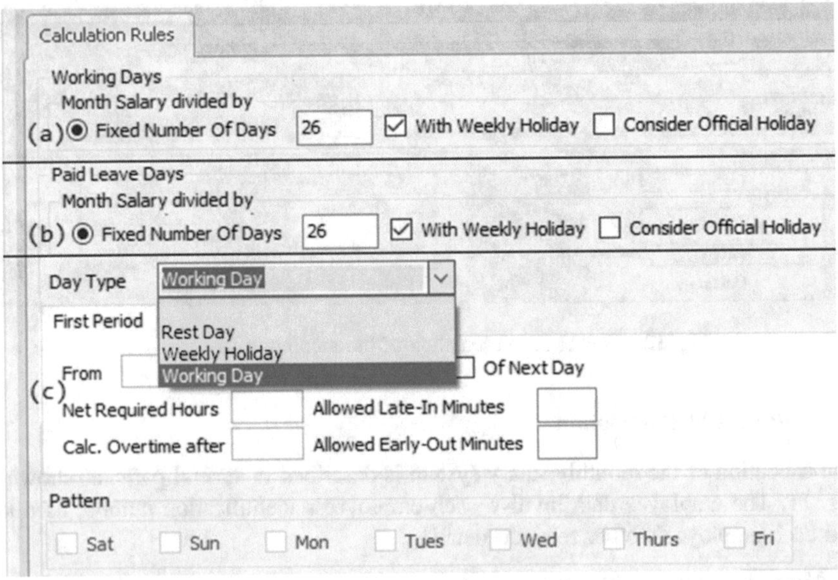

Fig. 13. Employee Salary Rules.

- Leave Day Rule [Number of Days, Official Holiday, Weekly Holiday] – (circle 11; e.g., leave rules are usually the same as working days rules, but companies have the ability to change them accordingly [see Fig. 13(b)])
- Calendar Day Rule [Working Days, Rest Days, Weekly Holiday] – (circle 12; e.g., companies can put their preferred working day rules, rest day rules, and weekly holiday rules [see Fig. 13(c)])

These rules (circles 10 to 12 in Fig. 12) are processed (circle 13) to construct the **Employee Salary Rules** (14).

C. Addition/Deduction, Attendance and Leave files

The third segment from the top labelled C in Fig. 12 shows the Addition/Deduction and Attendance and Leave files. To start the monthly salary posting, we must collect the following data files.

(1) The addition/deduction records (circle 15 in Fig. 12; e.g., a given employee may have some additions and deduction added to his or her salary, such as external payments and penalty deductions).
(2) The employee attendance records for the month (circle 16; e.g., each employee's attendance is recorded for deductions if he or she is late or absent).
(3) The employee leave records (17; e.g., the leave record for each employee is recorded because the attendance record shows absences, but when an employee is on leave, no penalties are applied).

D. Grand Salary Posting file

The segment labeled D in Fig. 12 shows the Grand Salary file. All the data in the three files in sections A, B, and C, as described above, are processed (18) to produce the Grand Salary Posting file (19). Thus, each employee has a grand record constructed from his or her records in A, B, and C.

The rest of the processes in the system are as follow:

- The grand salary posting file (19) flows to the calculation module (20) that calculates the monthly employee pay rate.
- The employee posting calculations are exported to bank diskettes such as an Excel sheet (21) that is used to transfer the salary amounts to the employees' bank accounts at the bank (22).

Accordingly, the employees are able to view all the information processed in Step 20.

4.3 Behavior Modeling

The events of interest can now be identified as follows (see Fig. 14):

Event 1 (E_1): Latest *Employee Details* have been prepared.
Event 2 (E_2): Latest *Hiring Details* have been prepared.
Event 3 (E_3): Latest *Identification Details* have been prepared.
Event 4 (E_4): Latest *Basic Salary Details* have been prepared.
Event 5 (E_5): Latest *Allowance Salary Details* have been prepared.

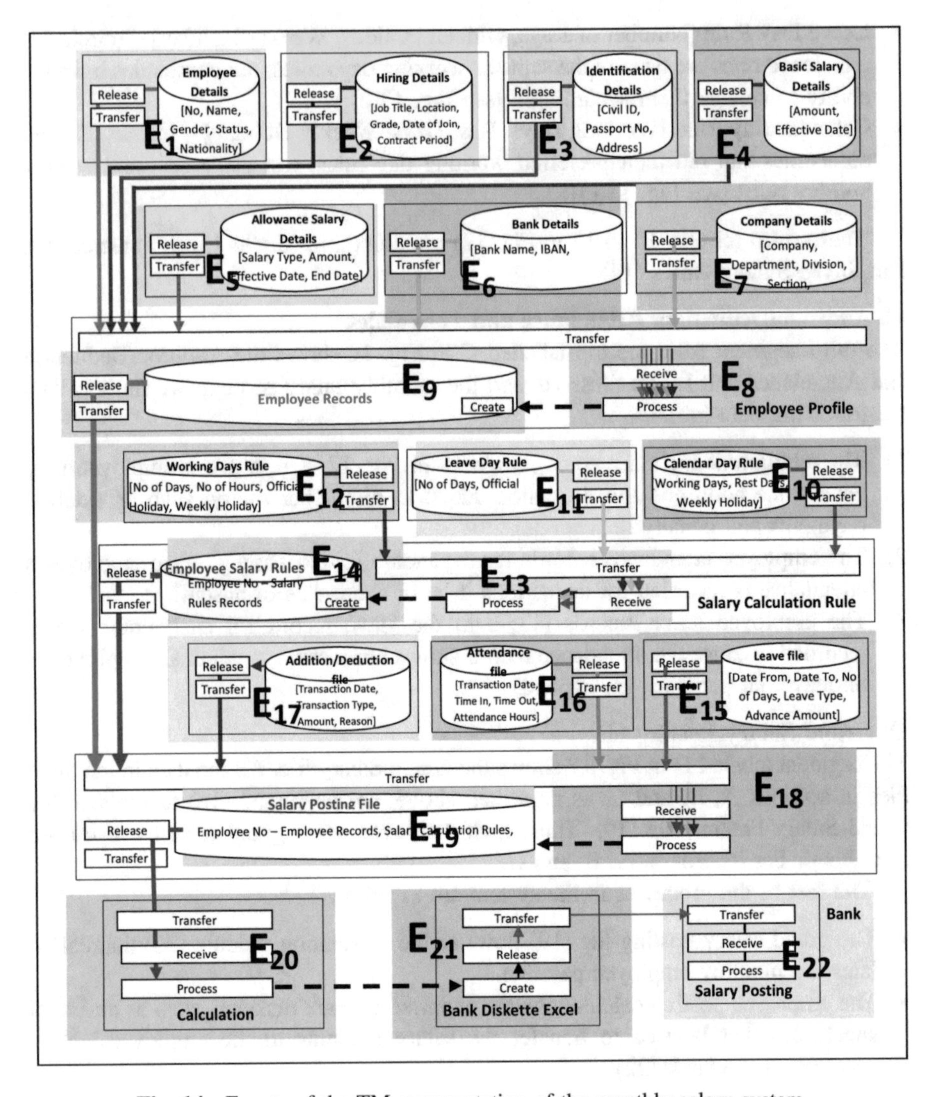

Fig. 14. Events of the TM representation of the monthly salary system.

Event 6 (E_6): Latest *Bank Details* have been prepared.
Event 7 (E_7): Latest *Company Details* have been prepared.
Event 8 (E_8): Previous files have been processed successfully.
Event 9 (E_9): *Employee files* have been created.
Event 10 (E_{10}): Latest *Working Days Rules* have been prepared.
Event 11 (E_{11}): Latest *Leave Day Rules* have been prepared.
Event 12 (E_{12}): Latest *Calendar Day Rules* have been prepared.
Event 16 (E_{16}): Latest *Attendance files* have been prepared.
Event 17 (E_{17}): Latest *Addition/Deduction files* have been prepared.

Event 18 (E_{18}): Data from *Employees files*, *Employee Salary Rules*, *Leave files*, *Attendance files*, and *Addition/Deduction files* have been processed successfully.
Event 19 (E_{19}): *Grand salary posting file* has been created.
Event 20 ($E1_{20}$): Salaries are calculated.
Event 21 ($E1_{21}$): Bank Diskette is created.
Event 22 (E_{22}): Salaries are posted.

Figure 15 shows the chronology of events in the monthly salary system.

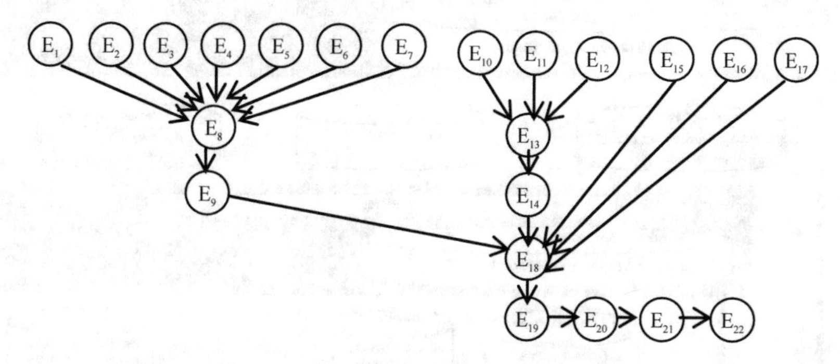

Fig. 15. Chronology of events in the monthly salary system.

4.4 Simulation

The TM diagram can be used as a conceptual model in simulation. Specifically, TM operational semantics uses events to define fine-grained activities, resulting in an integration of the static domain description and the dynamic chronology of events. We claim that the TM methodology provides a more complete specification, suitable for the static domain model and dynamic aspects a simulation requires.

To focus the incorporation of a TM into the simulation, without loss of generality, we will discuss flowcharting in the simulation language arena. An arena flowchart is generated by a cognitive agent, and its success depends on how well the designer divides the system's events and projects them in the flowchart. An arena flowchart captures purely arbitrary information because the magnitude of depicted events depends on the subjectivity of the flowchart designer.

By contrast, a TM diagram is an objective, seizure f all *elementary* events (i.e., the create, process, release, transfer, and receive steps of the modelled system). Therefore, if we desire to subjectively achieve a coarse description of events, we can start with a TM diagram and identify event boundaries from elementary events until we attain the required level of granularity. Consequently, the flowthing machine description of events can produce diagrams in which elementary events form more complex events.

4.5 Simplification

As mentioned above, the TM representation may be simplified to any required level of granularity, based on the original TM description. For example, Fig. 16 was produced from the original TM representation of the monthly salary system as follows:

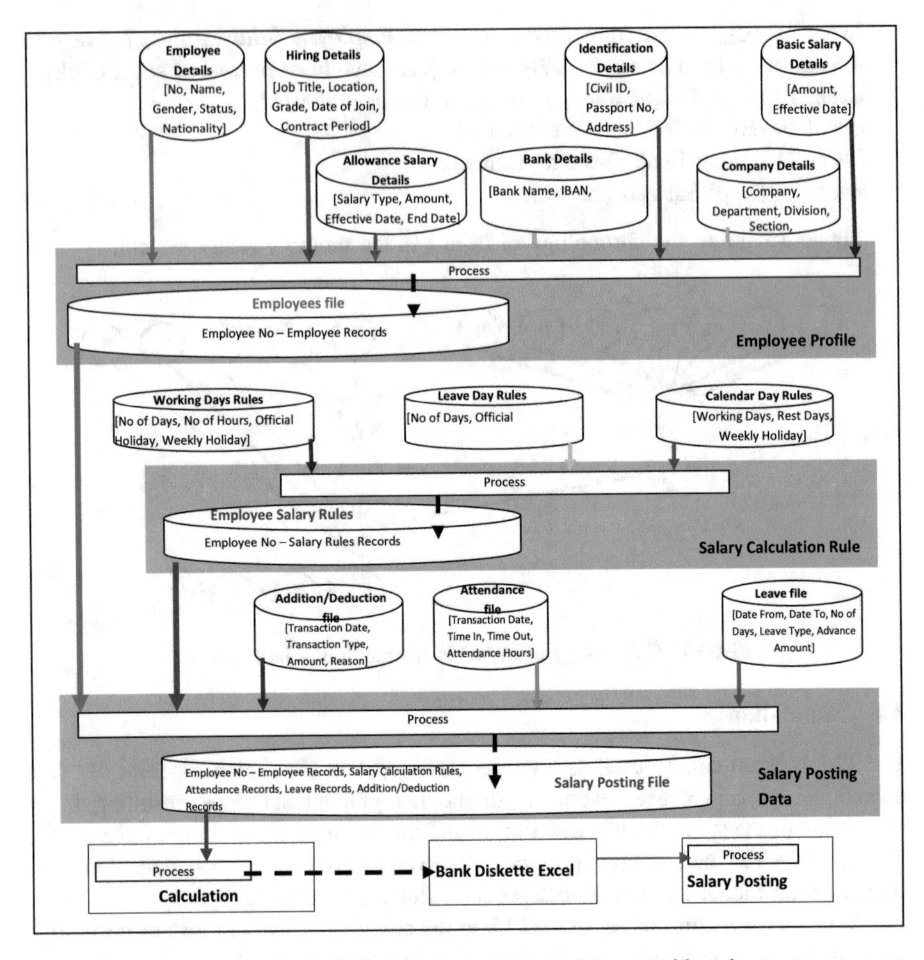

Fig. 16. Simplification of the TM representation of the monthly salary system.

- The release, transfer, and receive stages were removed, based on the assumption that the arrow's direction is enough to indicate the flow of things. Note that, in general, it may not be possible to eliminate these stages due to constraints.
- Assume that triggering indicates creation of a new thing. In general, it may also be impossible to omit the create stage, which may include decisions.

5 Conclusion

This paper has demonstrated that a new model—a TM—can serve as a conceptual framework that imposes uniformity across the task of describing software systems architecture. A TM diagram can be used at various levels of granularity and complexity, as in the case of nontechnical use. The viability of the model was demonstrated by applying it to an actual monthly salary system. The resultant model seems to serve as a foundation for planning and understanding different stakeholders.

References

1. Clements, P., Bachmann, F., Bass, L., Garlan, D., Ivers, J., Little, R., Merson, P., Nord, R., Stafford, J.: Documenting Software Architectures: Views and Beyond, 2nd edn. Addison-Wesley, Boston (2010). ISBN 978-0-321-55268-6
2. Ganesan, D.: Software Architecture, Software, IN SlidesShare, 9 October 2017. https://www.slideshare.net/dganesan11/software-architecture-80589682
3. Perry, D.E., Wolf, A.L.: Foundations for the study of software architecture (PDF). ACM SIGSOFT Softw. Eng. Not. **17**(4), 40 (1992). CiteSeerX 10.1.1.40.5174. https://doi.org/10.1145/141874.141884
4. Rozanski, N., Woods, E.: Viewpoints, Software Systems Architecture, Web site 2005–2019. https://www.viewpoints-and-perspectives.info/home/viewpoints/
5. Simon Brown Visualising Software Architecture with the C4 Model, Presentation, ReactiveConf, Las Vegas, 24–26 October 2018. https://www.youtube.com/watch?v=1zYK615kepE
6. Al-Fedaghi, S., Atiyah, Y.: Modeling with thinging for intelligent monitoring system. In: IEEE 89th Vehicular Technology Conference: VTC2019-Spring, Kuala Lumpur, Malaysia, 28 April–1 May 2019 (2019)
7. Al-Fedaghi, S., Hassouneh, A.: Modeling the engineering process as a thinging machine: a case study of chip manufacturing. In: The 8th Computer Science On-line Conference (CSOC 2019). To appear in Springer Advances in Intelligent Systems and Computing - ISSN 2194-5357
8. Al-Fedaghi, S.: Privacy as a base for confidentiality. Presented in the Fourth Workshop on the Economics of Information Security, Harvard University, Cambridge, MA (2005)
9. Al-Fedaghi, S., Al-Azmi, A.A.R.: Experimentation with personal identifiable information. Intell. Inf. Manag. (IIM) **4**(4), 123–133 (2012)
10. Al-Fedaghi, S.: Pure conceptualization of computer programming instructions. Int. J. Adv. Comput. Technol. **3**(9), 302–313 (2011)
11. Al-Fedaghi, S.: Modeling communication: one more piece falling into place. In: The 26th ACM International Conference on Design of Communication (SIGDOC 2008), Lisboa, Portugal, 22–24 September 2008
12. Al-Fedaghi, S., Alkhaldi, A.A.: Thinging for computational thinking. Int. J. Adv. Comput. Sci. Appl. (IJACSA) **10**(2), 620–629 (2019)
13. Al-Fedaghi, S.: Thinging for software engineers. Int. J. Comput. Sci. Inf. Secur. (IJCSIS) **16** (7), 21–29 (2018)
14. Al-Fedaghi, S., Al-Huwais, N.: Conceptual modeling of inventory management processes as a thinging machine. Int. J. Adv. Comput. Sci. Appl. (IJACSA) **9**(11), 434–443 (2018)
15. Hassan, N.R., Mingers, J., Stahl, B.: Philosophy and information systems: where are we and where should we go? J. Eur. J. Inf. Syst. **27**(3), 263–277 (2018). Special issue on Philosophy
16. Heidegger, M.: The thing. In: Poetry, Language, Thought, pp. 161–184 (trans: Hofstadter, A.). Harper & Row, New York (1975)

GPA-ES Algorithm Modification for Large Data

Tomas Brandejsky(✉)

University of Pardubice, Studentská 95, 532 10 Pardubice, Czech Republic
tomas.brandejsky@upce.cz

Abstract. This paper discusses improvement of Genetic Programming Algorithm to large data sets with respect to future extension to big data applications. On the beginning it summarizes requirements on evolutionary system to be applicable in the area of big data and ways of their satisfaction. Then GPAs and especially their improvements by solution constant optimization (so called hierarchical and hybrid genetic programming algorithms) are discussed in this paper. After a discussion of few experiment results of introduced novel evaluation scheme approach with floating data window is presented. Novel evaluation scheme applies floating data window to fitness function evaluation. After one evaluation step of GPA including tuning of parameters (solution constants) by embedded Evolutionary Strategy algorithm data window moves to new position. Presented results demonstrate that this strategy can be faster and more efficient than evolution of whole training data set in each evolutionary step of GPA algorithm. This modification can be starting point of future applications of GPA in the field of large and big data analytic.

Keywords: Big data · Fixed position data window · Floating data window · Genetic programming algorithm · Symbolic regression · Data analytic

1 Introduction

The development in sensors, intelligent and co-operating systems; and their networking allows to extract and store large amounts of data. Data volume produced by our civilization grows exponentially in time. By Cray company, 2.5 quintillion bytes of data are created every day. The main problem is now not how to produce and collect data but how to organize, extract and use them.

Nowadays, we meet frequently term "big data". Big data denotes situation when we need to process so large data sets to which standard techniques like relational databases and knowledge mining tools are inapplicable. But it is need to process and analyze them. Thus, the novel approaches known as "big data" are developed.

Till now, there is no commonly accepted definition of big data, but it possible to accept e.g. old definition of their main characteristic as the extreme volume of data, the wide variety of data types and the velocity at which the data must be processed [1]. These characteristics also explains why it is difficult to apply evolutionary techniques for processing of these data. Evolutionary techniques usually evaluates suitability of each individual in its population against whole fitness function, against whole data set.

R. Silhavy et al. (Eds.): CoMeSySo 2019, AISC 1046, pp. 98–106, 2019.
https://doi.org/10.1007/978-3-030-30329-7_9

It is extremely time consuming if the training data volume is large. Many evolutionary techniques like standard GPA are data type dependent, fitness function evaluation as well as mutation and crossover operators must be modified for used data types. Above mentioned arduous computations cause reduction of data processing velocity and makes applications depending on processing speed difficult.

In many situations it is impossible to apply natural solution of big data problem – their volume reduction. It is caused by requirements on simultaneous processing (analysis and modeling) of systems with both fast and slow dynamics or existence of so called stiff system. The data sampling period of such systems must be very small, thus the volume of processed data is big and cannot be reduced. Analogical problem is caused by data whose statistical parameters vary in time – they are non-stationary. In analysis and modeling of all these kinds of data the evolutionary algorithm application can be advantageous but the data volume reduction is impossible and algorithm efficiency must be reached alternative ways.

2 Genetic Programming

The application of Genetic Programming Algorithm (GPA) [2] to large and big data sets area requires elimination of basic GPA property – repeated evaluation of training data set. Especially, this property cases above discussed problems with data volume and concluding low velocity of their processing.

The first and most crucial question in investigating field of GPA application to big data field is if all data in the large collection are needed. The amount of needed training data depends on many factors, not only on parameters of GPA like population size, generation limit or chosen operator set but especially on system linearity and number of data dimensions. This dependency is also constrained by stationarity of training data set. Only if the used data set is stationary, it can be replaced by small subset to be efficiently computed. Previous work [3] points that such reduction is possible and in some situations the data set can be strongly reduced. On the opposite side, continuations of this research with deterministic chaos producing systems like Lorenz attractor [4] or Rössler [5] one bring opposite examples, which will be presented in the next chapter. Thus the application of this approach is limited and the research were focused to improvements of GP algorithm.

It is also needs to keep in mind that even historical Map-reduce approach [6] solves problem of data amount reduce in wider context. It solves not only selection of relevant data records, but also columns selection, data transformations and parallel implementation of this process with respect to constrained available memory volume and processing unit throughput. Because these significant problems are solved in many professional publications like e.g. [7–9], they will not be discussed herein.

The idea of GPA use to big data processing, especially to big data analytic, opens possibility to apply, as it is discussed in the work [3], GPA-ES hierarchical evolutionary algorithm or hybrid GP algorithm. McKay, Willis, and Barton [10] and Fröhlich and Hafner [11] presented a hybrid GPA including linear and non-linear optimization techniques to improve these numerical constants. Unfortunately, their research ended with conclusion that only a few steps of the time expensive non-linear optimization can be applied to each new solution to keep the total running time within limits. Such observation was limited by time of its origin – today available computation power is significantly greater than in the years 1995-6. This works are extended in [12]. The analysis of GPA-ES algorithm complexity [13] pointed to analogical constraint, but in many applications of this algorithm parameter optimization was typically constrained to 40 cycles. Parameter optimization prevents occurrence of bowing (creating of over complicated solutions) and can improve algorithm efficiency.

Non stationary data brings complication into application of evolutionary system. The non-stationarity can become significant when data sample is chosen small but the use of small data sets is efficient from the evolutionary algorithm viewpoint. The only solution of this contradiction is the preference of novel data against historical ones. This preference on the opposite side cannot be strict because old data can contain information about analyzed system behavior in such areas of state space, which are not visited in the nearest history. Solution preferring newest data against older ones is complicated by observed limited adaptability of many evolutionary algorithms to change of training data set [14], it is possible to categorize this problem as over-learning.

3 Experiments

The first experiments study influence of data window size to efficiency of evolutionary algorithm. In this context, the influence of data stationarity or data window movement is studied too. Four groups of experiments were performed. Two for Lorenz attractor system regression for fixed position and floating data windows of different sizes of 64, 128, 256 and 512 samples and analogical experiments with Rössler system regression for sets of 1024, 512 and 256 samples. Test data were precomputed from model Eqs. (1, 2) of Lorenz attractor and (3, 4) of Rössler one without additional noise containing only numerical errors caused limited precision of their computation.

$$\begin{aligned} x'(t) &= s(y-x) \\ y'(t) &= x(r-z)-y \\ z'(t) &= xy-bz \end{aligned} \tag{1}$$

$$s = 10,\ b = 8/3,\ r = 28 \tag{2}$$

$$\begin{aligned} x'(t) &= -y-z \\ y'(t) &= x+ay \\ z'(t) &= b+z(x-c) \end{aligned} \tag{3}$$

$$a = 0.2,\ b = 0.2,\ c = 5.7 \tag{4}$$

In herein presented experiments was used GPA-ES algorithm which optimizes capabilities of standard GPA to area of highly precise symbolic regression (symbolic regression producing results which are not over-complicated and are comparable to results produced by human analysts). This algorithm in each evolutionary step of GPA algorithm uses Evolutionary Strategy to optimize parameters of each individual in GPA population. This algorithm is suitable to small population operation (GPA populations of 3 to 20 individuals) which tends to very small total number of fitness function evaluations non looking to large number of GPA evolutionary cycles and thus can be suitable to big data applications.

4 Experiment Results

The first experiment sets studied influence of data window size to efficiency of evolutionary algorithm. In period of 512 to 128 samples in training data set the efficiency was increasing (number of evolutionary cycles was decreasing) and start of expected opposite behavior was observed since 64 samples (Fig. 1). This fact points to sensitivity of small data sets to local properties of training data, but also it concludes that even very small data sets can contain enough of data for model construction. Second experiment outlined at Fig. 2 presents the same study but with the rightmost data from samples vector. It points influence of data non-stationarity which it is easy to observe if the small data subsets are used for learning. Obtained models are equal, but the number of evolutionary steps need to create required quality models differs.

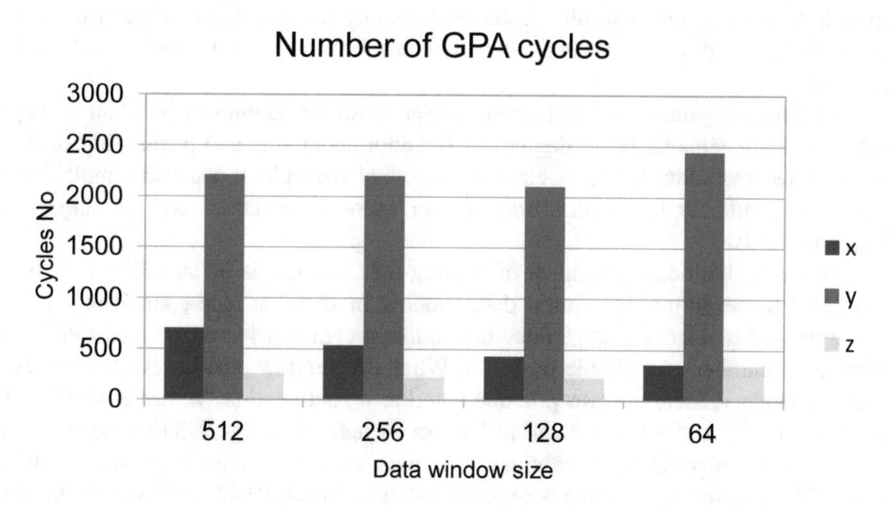

Fig. 1. Lorenz attractor – influence of window size for stationary window.

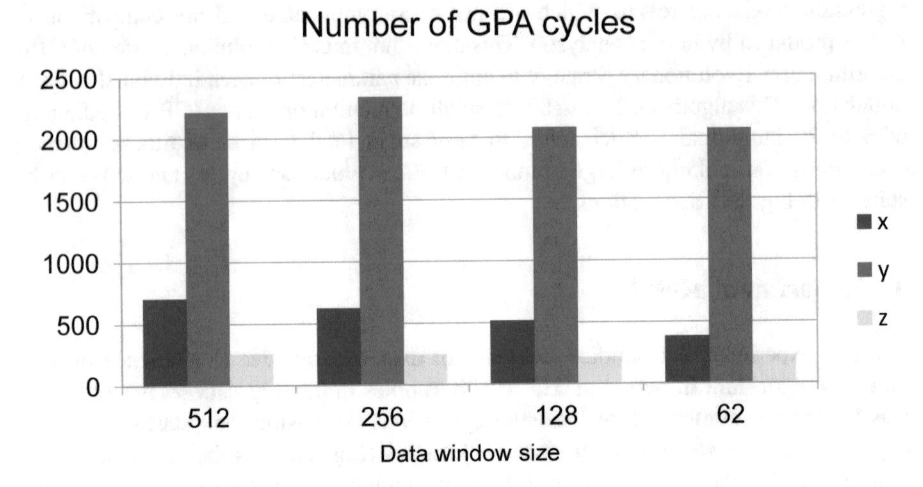

Fig. 2. Lorenz attractor – influence of window size for different position of data window.

5 Floating Window

Previous experiments point that it is not useful to constraint evaluation onto single position of data window. This is underlined uncertainty if any another position is not capable to produce better results, faster evolutionary process. Uncertainty is given by our inability to distinguish which data window position will be better without its evaluation.

The floating window is the logical answer to above mentioned problem of large data, but there remains many degrees of freedom about size and positioning of data window. Data window can move about fixed step of size at least one data sample, but it can move randomly too. Optimizing of data window movement will be subject of future research.

The most significant parameter of floating data window is its size. The following test case studies influence of used data window in situation where small data set is available. Using relatively large floating window on small data vector creates problem with small number of available positions. When the window position is reaching data vector boundary, there are two possible solution – continuation "across border" sketched on Fig. 3, or strict movement to left border and repeating. Solution sketched on Fig. 3 creates significant discontinuity in data and it is practically inapplicable. Used GPA was not able to find acceptable solution during 10000 iterations (with sum of error squares smaller than 10^{-7}). With static window this average number was less than 2500 in the worst case of y variable of Lorenz attractor system.

Problem of continuous evaluation border is probably more theoretical than real in the case of big data. Experiments points that floating window is capable to find correct solution even if data window is extremely small – e.g. 32 samples for Lorenz attractor system and thus we can expect that in real applications this problem will not be relevant.

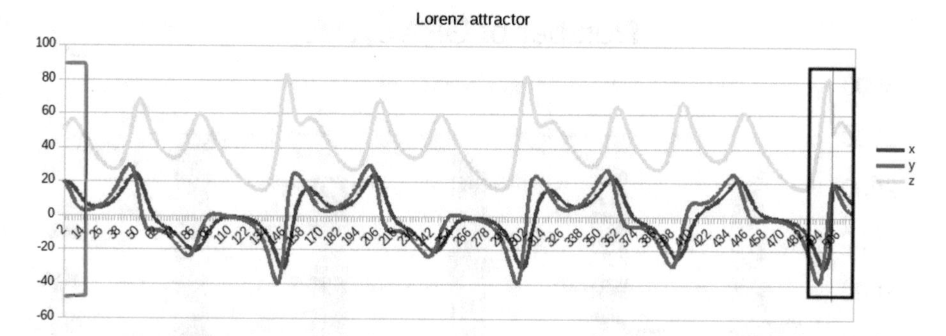

Fig. 3. Discontinuity problem in cyclic application of floating window.

To avoid such situation, the larger data vector evaluation 40000 samples vector of Lorenz attractor data was prepared. The lowest number of iteration gives large data set, where each window position is used once only and thus where no discontinuity on data set border can occur.

This result can be also explained by "stiff" character of deterministic chaos system data containing slow and fast dynamics. In such data set it is hard to observe whole dynamics from small sample. GPA in such situation produces a number of wrong hypotheses. On the opposite side, small but floating data window can be more efficient (Figs. 4, 5 and 6).

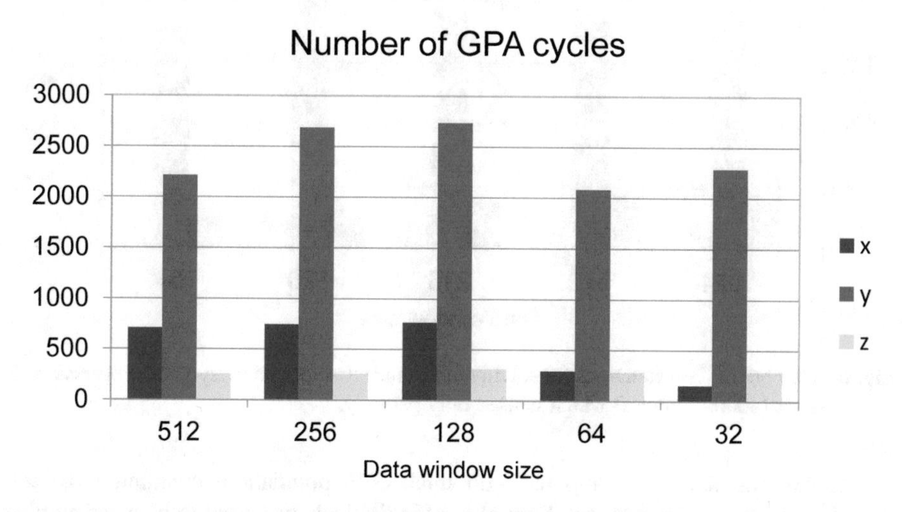

Fig. 4. Lorenz attractor symbolic regression dependency of iteration number on floating window size with discontinuity on data array border.

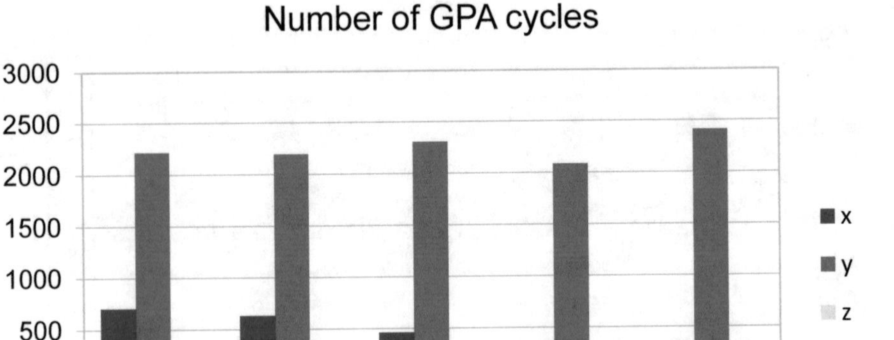

Fig. 5. Lorenz attractor symbolic regression dependency of number of GPA iterations on data window without discontinuity size for long data (40000 samples).

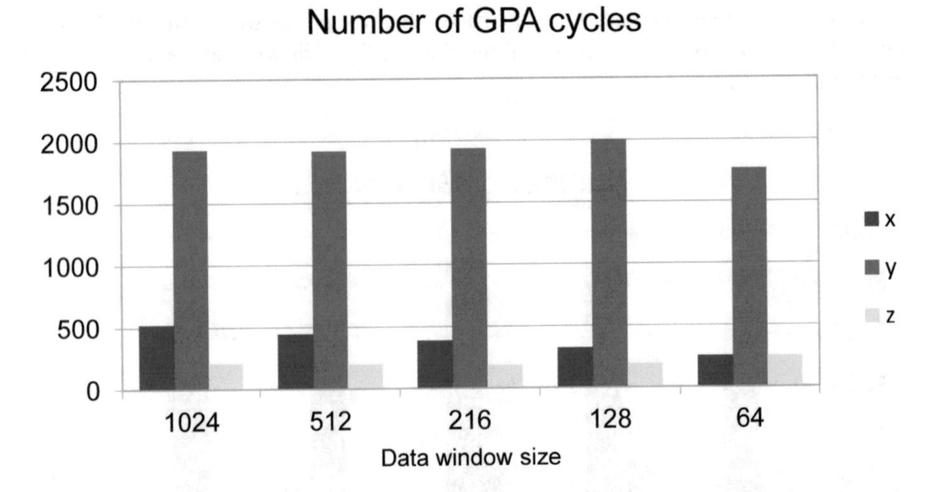

Fig. 6. Number of GPA iterations for solving of Lorenz attractor system symbolic regression for long data set (40000 samples) which is used only partially.

All experiments were computed with small GPA populations containing 10 individuals only. This atypically small number of individuals was used to increase number of populations and thus to increase sensitivity of this attribute used to measure quality of different approaches.

6 Conclusions

Presented paper brings idea of floating data window application to genetic programming. First experiments confirms not only its applicability but also demonstrates unexpected increase of GPA efficiency. Floating data window will be also suitable to big data analytic applications.

Acknowledgement. The work was supported from ERDF/ESF "Cooperation in Applied Research between the University of Pardubice and companies, in the Field of Positioning, Detection and Simulation Technology for Transport Systems (PosiTrans)" (No. CZ.02.1.01/0.0/0.0/17_049/0008394).

References

1. Laneym, D.: 3D Data Management: Controlling Data Volume, Velocity, and Variety. META Group Inc., file 949. https://blogs.gartner.com/doug-laney/files/2012/01/ad949-3D-Data-Management-Controlling-Data-Volume-Velocity-and-Variety.pdf. Assessed 19 Apr 2019
2. Poli, R., Langdon, W.B., McPhee, N.F., (with contributions by Koza, R.J.): A field guide to genetic programming (2008). Published via http://lulu.com and freely available at http://www.gp-field-guide.org.uk
3. Brandejsky, T.: Small populations in GPA-ES algorithm. In: Matousek, R. (ed.) MENDEL 2013, 19th International Conference on Soft Computing MENDEL 2013, Brno, 26–28 June 2013, pp. 31–36. Brno University of Technology, Faculty of Mechanical Engineering, Brno (2013). ISSN 1803-3814. ISBN 978-80-214-4755-4
4. Lorenz, E.N.: Deterministic nonperiodic flow. J. Atmos. Sci. **20**, 130–141 (1963)
5. Rössler, O.E.: An equation for continuous chaos. Phys. Lett. **57A**(5), 397–398 (1976). Bibcode: 1976PhLA…57..397R. https://doi.org/10.1016/0375-9601(76)90101-8
6. Miner, D., Shook, A.: MapReduce Design Patterns: Building Effective Algorithms and Analytics for Hadoop and Other Systems. O'Reilly Media, Inc. (2012). ISBN 1449341985
7. Abouzeid, A., Bajda-Pawlikowski, K., Abadi, D.J., Silberschatz, A., Rasin, A.: HadoopDB: an architectural hybrid of MapReduce and DBMS technologies for analytical workloads. In: VLDB, Lyon, France (2009)
8. Pavlo, A., Paulson, E., Rasin, A., Abadi, D.J., Dewitt, D.J., Madden, S., Stonebraker, M.: A comparison of approaches to large-scale data analysis. In: SIGMOD. ACM, June 2009
9. Yang, H.-C., Dasdan, A., Hsiao, R.-L., Parker, D.S.: Map-reduce-merge: simplified relational data processing on large clusters. In: SIGMOD (2007)
10. McKay, B., Willis, M.J., Barton, G.W.: Using a tree structured genetic algorithm to perform symbolic regression. In: Proceedings of the 1st International Conference on Genetic Algorithms in Engineering Systems: Innovations and Applications, UK, pp. 487–492 (1995)
11. Frohlich, J., Hafner, C.: Extended and generalized genetic programming for function analysis. J. Evol. Comput. (1996, submitted)
12. Raidl, G.R.: A hybrid GP approach for numerically robust symbolic regression. In: Koza, J. R., Banzhaf, W., Chhellapilla, K., Deb, K., Dorigo, M., Fogel, D.B., Garzon, M.H., Goldberg, D.E., Iba, H., Riolo, R. (eds.) Genetic Programming 1998: Proceedings of the Third Annual Conference, pp. 323–328. University of Wiconsin, Madison, Wisconsin, Morgan Kaufmann (1998)

13. Brandejsky, T.: Multi-layered evolutionary system suitable to symbolic model regression. In: Proceedings of the NAUN/IEEE.AM International Conferences, 2nd International Conference on Applied Informatics and Computing Theory, Praha, 26–28 September 2011, pp. 222–225. WSEAS Press, Athens (2011). ISBN 978-1-61804-038-1
14. Goldberg, D.E., Smith, R.E.: Nonstationary Function Optimization Using Genetic Algorithms with Dominance and Diploidy. ICGA (1987)

Periodic Formation Within a Large Group of Mobile Robots in Conditions of Limited Communications

Donat Ivanov(✉)

Southern Federal University, 2 Chehova Street, 347932 Taganrog, Russia
donat.ivanov@gmail.com

Abstract. This paper shows the prospects and areas of application of groups of mobile robots. The implementation of many practical tasks by groups of robots requires the formation of a specific configuration. Different approaches to solving the formation task in groups of mobile robots are considered. It is shown that the complexity of group's control increases with the number of robots. Moreover, in large groups of robots, the linear dimensions of the system may exceed the range of direct communication between robots using onboard communication devices. A formal statement of the formation task in a large group of mobile robots is given. The features of the periodic formation consisting of repeating geometric patterns are considered. A new method for the formation of a periodic system in a large group of mobile robots in conditions of limited communications has been proposed. The proposed method is based on the principles of decentralized control and swarm intelligence. The possibility of forming a complex periodic formations and multi-level formations in large groups of mobile robots is considered. The formation of such a system can be used in the tasks of monitoring, mapping, data collection and other applications.

Keywords: Swarm intelligence · Mobile robot · Formation task · Multi-robotics

1 Introduction

The possibilities of using single robots are limited by a small radius of action, the capacity of onboard energy resources and a small set of onboard actuators. In a non-deterministic environment, the likelihood of the task being performed by a single robot is noticeably reduced compared to a group of robots due to a possible failure. Currently, there are many projects for group use of mobile robots, both ground and air and underwater applications.

Groups of mobile robots can be used for video monitoring, the formation of phased antenna arrays [1], the deployment of mobile telecommunications networks and many other tasks.

The use of large groups of robots turns out to be effective in cases when the task can be done by using spatial separation, which will provide greater efficiency of work than is possible with a single robot. The problem arises of the formation of a specific system of mobile robots of the group in space.

R. Silhavy et al. (Eds.): CoMeSySo 2019, AISC 1046, pp. 107–112, 2019.
https://doi.org/10.1007/978-3-030-30329-7_10

With the increase in the number of mobile robots used in the group, the complexity of interaction within the group increases. Also, for a large group, the linear dimensions of the formed system can significantly exceed the range of the onboard telecommunications and sensor devices of individual mobile robots. In the framework of this work, groups of mobile robots are conventionally divided into large and small groups. In this case, a large group of robots will be considered as such in which at least one of the linear dimensions of the group in the initial or configured system exceeds the radius of the direct connection of the individual robot.

When using large groups of robots, greater efficiency is shown by flocking and swarm control methods. With swarm control, each robot of the group independently makes a decision about its further actions, based on independently obtained information about the environment and about some neighboring robots of the group.

The required (target) location of mobile robots in space is called a formation, and the task of forming a group of mobile robots of a given order is called a formation task. The formation task is one of the tasks of group robotics.

In this paper, an approach to solving the periodic formation task in a large group of robots, based on the principles of swarm intelligence, is considered. Under the periodic system we will understand such a system, which can be divided into identical component parts – patterns (see Fig. 1).

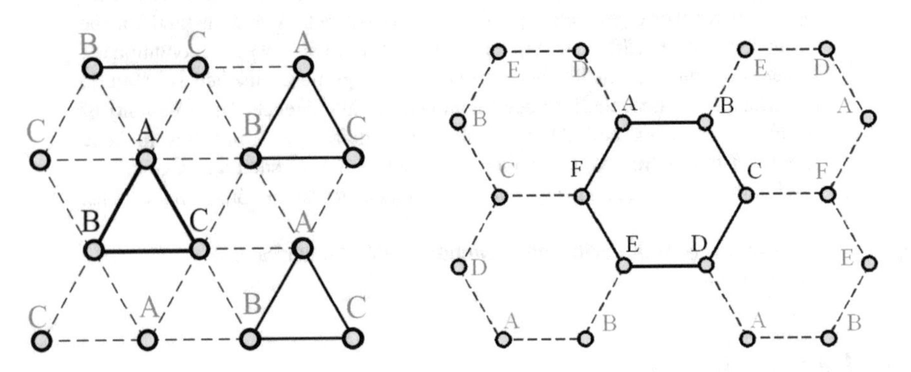

Fig. 1. An example of a periodic formation constituting a pattern: (a) a triangle; (b) a hexagon.

2 Problem Statement

Let there is a group $\mathbf{R}$ of N robots $r_i \in \mathbf{R}\left(i = \overline{1, N}\right)$. The state $\mathbf{s}_i(t)$ of each robot $r_i \in \mathbf{R}$ is described by a vector function $\mathbf{s}_i(t) = [s_{i,1}(t), s_{i,2}(t), \ldots, s_{i,h}(t)]^T$.

Under state variables $s_{i,h}(t)$ we should understand the coordinates $x_i(t), y_i(t), z_i(t)$ of the robot $r_i \in \mathbf{R}$ in space, its current speed, the remaining on-board supply of energy, etc.

The on-board telecommunications device of each robot allows it to maintain direct communication with some other robots of the group within the radius of direct communication. For each robot $r_i \in \mathbf{R}$, you can select a subgroup $\mathbf{R}_i$ of N_i robots, guaranteed to fall into the zone of direct communication of the robot $r_i \in \mathbf{R}$. Such a subgroup $\mathbf{R}_i$ will be called a local subgroup.

Each robot $r_i \in \mathbf{R}$ had information about its own state $\mathbf{s}_i(t)$, as well as information about the state $\mathbf{s}_j(t)$ (or at least the current position $x_j(t), y_j(t), z_j(t)$) of other robots of its local subgroup $\mathbf{R}_i$. Each robot $r_i \in \mathbf{R}$ has a control system, and can change the coordinates $\mathbf{x}_i(t) = x_i(t), y_i(t), z_i(t)$ of its position in space depending on the input controls $u_i(t)$.

In order to prevent collisions and mutual interference of robots, restrictions are imposed on the minimum allowable distance between robots:

$$\left|\mathbf{x}_i(t) - \mathbf{x}_j(t)\right| \geq \Delta_r, (i \neq j; i, j = \overline{1, N}), \tag{1}$$

where Δ_r is the minimum allowable distance between robots.

The initial position of the robots of the group is given by the description of the formation $\mathbf{D}_0$ (for example, using a matrix of distances between the robots, or a set of coordinates of the robots of the group), the current position $\mathbf{D}_t$ of the robots at the time t. The target formation is defined by the description $\mathbf{D}_f$ of the formation, consisting of some limited set of identical patterns, each of which consists of a certain number of target positions $v_\mu \in \mathbf{V}(\mu = \overline{1, N})$ for individual robots. At the same time, the description of the target system contains a table of distances between the target positions in the pattern and between the patterns in the target system.

The formation task in the group of robots is to determine such a sequence of controls (a vector-function of controls) $\mathbf{u}(t) = [u_1(t), u_2(t), \ldots u_N(t)]^T$ which, with restrictions on the position of robots in space (1), led the group of robots from the initial position $\mathbf{D}_0$ to the target formation $\mathbf{D}_f$ in an acceptable time.

3 Overview of Existing Approaches

In [2–5], methods for solving a formation task for small groups of mobile robots are proposed, in which the size of the formation does not exceed the zone of direct communication by the onboard communication devices of each group member.

There are various methods for solving a drill task in groups of mobile robots, based on the behavioral approach [3, 6], the master-slave approach [4, 7–9], the "virtual structure/virtual leader" approach [10], game theory [5] and others. However, some of these methods allow the formation of a system of only a certain shape. Others, for their implementation, require significant computational resources that are not possessed by the onboard computing systems of small mobile robots.

Moreover, in practice, in monitoring and forming phased array antennas, the observance of the required distances between mobile robots is more important than the positioning accuracy of mobile robots in absolute coordinates.

There is a need for a computationally simple method for solving a construction problem, which in real time allows forming a system of any form using weak computing devices and ensuring exact observance of the distances between mobile robots.

4 Proposed Method

To solve the drilling problem in large groups of mobile robots in this paper the proposed pattern method. This method is based on the principles of swarm interaction [11].

The proposed method for solving the drill problem is as follows. Each vertex of the pattern is assigned an identifier (for example, the letter of the Latin alphabet) and the following requirements are met: the orientation $\alpha_{f,\mu}$ is known either to the side of the world or to another vertex of the pattern, the orientation $\alpha_{f,h}$ and distance $d_{f,h}$ to the vertex with the same identifier of the neighboring pattern is known.

Each mobile robot $r_i \in \mathbf{R}$ has the following requirements:

- the presence of a unique identifier number;
- communication with neighboring mobile robots of the group;
- the ability to determine the direction of the world and the azimuth to the neighbor;
- ability to measure distance to neighboring robots;
- mobility;
- ability to avoid collisions with other robots of the group due to small deviations from the course.

The target formation in the group is formed as follows:

Step 1. Each identifier of the pattern is assigned an identifier-letter (see Fig. 2a);

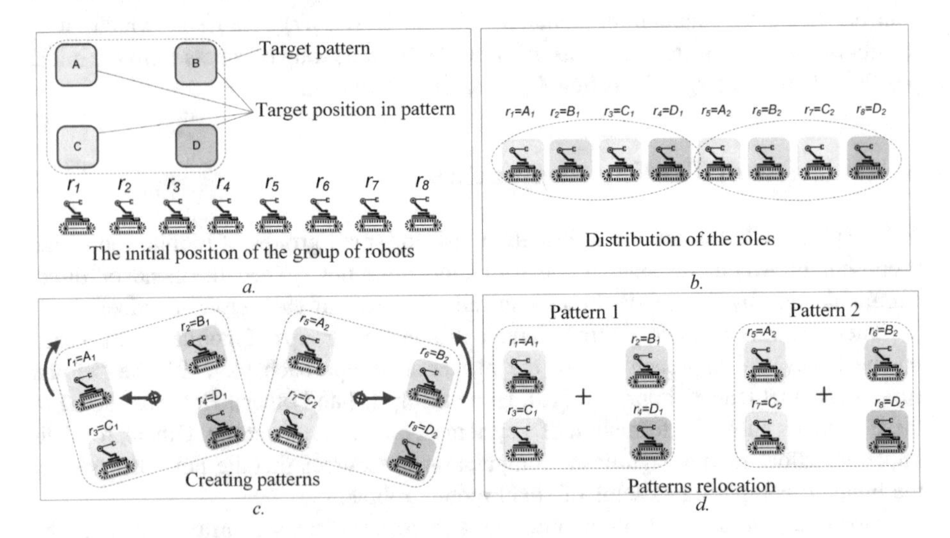

Fig. 2. Example of periodic formation by proposed method

Step 2. The distribution of the roles of the mobile robot r_i of the group $\mathbf{R}$ in a certain proportion (for example: equally A, B, C and D in each local area). The algorithm for distributing tasks in a large group of robots, proposed in [12] (see Fig. 2b), is suitable for this;

Step 3. In each local area $\mathbf{R}_i$, mobile robots with the role "A" are positioned in accordance with the requirements for the top "A" of the pattern $\mathbf{d}_{f,k}$, mobile robots with the role "B" are looking for an unoccupied pair "B" and lining up in relation to it the desired distance $d_{f,h}$ and azimuth $\alpha_{f,h}$, then mobile robots with the role of "B" are looking for an unoccupied pair of "A" and "B", and so on until the required number of patterns is formed. In each pattern, a local drill task is solved for the formation of a non-large order, taking into account the constraints (1). For this purpose it is proposed to use the method of circles, considered in [2, 13] or method of spheres [14] for 3D-formations. Some mobile robots may be left without a place in the pattern and will play a backup role.

Step 4. Moving patterns relative to each other with collision avoidance: If the two patterns intersect with each other, then they "push out" each other from themselves. For this, a search is made for the geometric centers of the patterns and for each of the patterns a vector is constructed with the beginning in the geometric center and the direction opposite to the geometric center of the second pattern. The movement of patterns along these vectors allows us to separate patterns (see Fig. 2c);

If the two patterns do not intersect with each other, then they approach each other so that the distances between A_i and A_j $(i < j)$ are at the required distance specified in the description of the target formation.

Step 4 is repeated until the current position $\mathbf{D}_t$ of the robots $r_i \in \mathbf{R}$ satisfies the requirements for the target formation $\mathbf{D}_f$ (see Fig. 2d).

5 Multi-level Formations and Pattern Nesting

In some cases, you can apply the method sequentially: at the first stage, patterns are formed from robots, and at the second stage, patterns are formed from already obtained patterns, and so on. This will make it possible to obtain such structures as, for example, rectangles, at the vertices of which there are triangular patterns (Fig. 3).

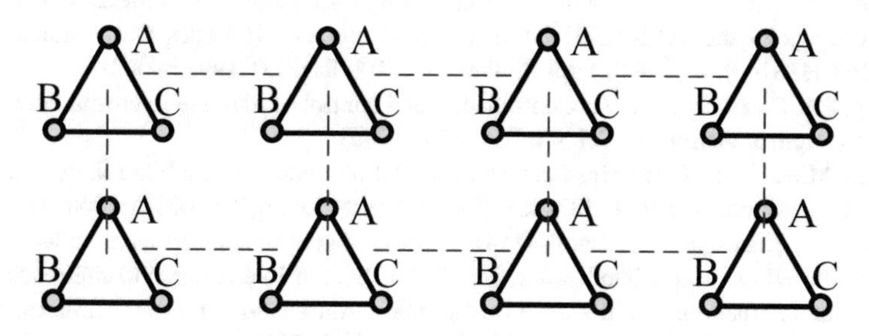

Fig. 3. An example of a periodic pattern with rectangular patterns, triangular patterns are located at the vertices of the rectangles.

6 Conclusions

The proposed method allows you to form a system in groups of tens or even hundreds of mobile robots. At the same time, the computational complexity of the method and the volumes of data transmissions required for its implementation in a telecommunications network are relatively small and allow the method to be used for groups of small mobile robots equipped with weak onboard computing devices.

Acknowledgement. The reported study was funded by RFBR according to the research projects №17-29-07054, №19-07-00907.

References

1. Tonetti, S., Hehn, M., Lupashin, S., D'Andrea, R.: Distributed control of antenna array with formation of UAVs. In: World Congress, pp. 7848–7853 (2011)
2. Ivanov, D., Kalyaev, I., Kapustyan, S.: Formation task in a group of quadrotors. In: Robot Intelligence Technology and Applications 3, pp. 183–191. Springer (2015)
3. Lawton, J.R.T., Beard, R.W., Young, B.J.: A decentralized approach to formation maneuvers. IEEE Trans. Robot. Autom. **19**, 933–941 (2003). https://doi.org/10.1109/TRA.2003.819598
4. Mesbahi, M., Hadaegh, F.Y.: Formation flying control of multiple spacecraft via graphs, matrixinequalities, and switching. In: Proceedings of the 1999 IEEE International Conference on Control Application (Cat. No. 99CH36328), 2 (1999). https://doi.org/10.1109/CCA.1999.801145
5. Erdoğan, M.E., Innocenti, M., Pollini, L.: Obstacle avoidance for a game theoretically controlled formation of unmanned vehicles. In: 18th IFAC (2011)
6. Balch, T., Arkin, R.C.: Behavior-based formation control for multirobot teams. IEEE Trans. Robot. Autom. **14**, 926–939 (1998). https://doi.org/10.1109/70.736776
7. Wang, P.K.C.: Navigation strategies for multiple autonomous mobile robots moving in formation. J. Robot. Syst. **8**, 177–195 (1991)
8. Desai, J.P., Ostrowski, J., Kumar, V.: Controlling formations of multiple mobile robots. In: Proceedings of the 1998 IEEE International Conference Robotics and Automation (Cat. No. 98CH36146), 4 (1998). https://doi.org/10.1109/ROBOT.1998.680621
9. Wang, P.K.C., Hadaegh, F.Y.: Coordination and control of multiple microspacecraft moving in formation. J. Astronaut. Sci. **44**, 315–355 (1996)
10. Lewis, M.A., Tan, K.-H.: High precision formation control of mobile robots using virtual structures. Auton. Robot. **4**, 387–403 (1997). https://doi.org/10.1023/A:1008814708459
11. Kaliaev, I., Kapustjan, S., Ivanov, D.: Decentralized control strategy within a large group of objects based on swarm intelligence. In: 2011 IEEE 5th International Conferrence on 2011 IEEE 5th International Conferrence on Robotics, Automation and Mechatronics, pp. 299–303 (2011). https://doi.org/10.1109/RAMECH.2011.6070500
12. Ivanov, D.Y.: Distribution of roles in groups of robots with limited communications based on the swarm interaction. Procedia Comput. Sci. **150**, 518–523 (2019)
13. Ivanov, D., Kalyaev, I., Kapustyan, S.: Method of circles for solving formation task in a group of quadrotor UAVs. In: 2014 2nd International Conference on Systems and Informatics (ICSAI), pp. 236–240 (2014)
14. Ivanov, D., Kapustyan, S., Kalyaev, I.: Method of spheres for solving 3D formation task in a group of quadrotors. In: Interactive Collaborative Robotics. LNCS, First International Conference ICR 2016, Budapest, Hungary, 24–26 August 2016, vol. 9812 (2016)

Aligning Decision Analytics with Business Goals: The Case of Incident Management

Vanja Bevanda(✉)

Faculty of Economics and Tourism "Dr. Mijo Mirkovic",
The Juraj Dobrila University of Pula, Preradoviceva 1/1, Pula, Croatia
vbevanda@unipu.hr

Abstract. Solving the problem of modelling decision performance can be facilitated with developing a more comprehensive framework based on an application of the requirement engineering principles to business alignments of decision analytics. The authors applied these principles in existing PPINOT framework based on incident management's case. With the proposed framework based on managers' requirements, it can be anticipated specific Decision Model and Notation Standard (DMN) extensions for decision performance measure purposes or for development of new analytics/monitoring tools. A traditional data-mining algorithm was applied to a decision (DPI) and process (PPI) performance measures in order to fulfill some of the decision analytic requirements identified within the framework.

Keywords: Decision management · Decision model and notation standard-DMN · Performance measurement · Decision analytics · Decision mining

1 Introduction

The Decision Model and Notation standard- DMN [14] has been released with the aim of enabling modelling routine and operational business decision, next to business processes and their data. Each of these decisions come in large numbers and represent value for organizations; for example; in granting a loan, simple diagnostics, customer call center, incident management etc. They need to be executed less costly and more efficiently especially in online and mobile applications. Only a few types of research concern DMN extension for enabling decision performance measurement [7–10]. The main conclusions of these previous researches are that "decision management can be significantly enriched by considering performance indicators and their relationships with decisions in three different perspectives: the impact of decisions on process performance, the measurement of the performance of decisions and the use of process performance indicators (PPIs) and performance measures as an input for decisions." [7]. Last mentioned paper considers introducing DMN extension in form of Decision Performance Indicators (DPIs) into process performance context using Process Performance Indicators (PPI) described by PPINOT's metamodeling. We used the case of incident management described in that previous work for identification of DMN performance measurement problems.

R. Silhavy et al. (Eds.): CoMeSySo 2019, AISC 1046, pp. 113–124, 2019.
https://doi.org/10.1007/978-3-030-30329-7_11

This paper deals with a "softer" but important issue of decision performance considering neglecting business goals and business requirements to be achieved with decisions modelling. The aim is to support decision performance measurement development efforts with business alignments using a structured approach. We propose that decision measurement cannot be separated from their implementation in business process with their performance indicators. Top-down approach with an adaptation of the framework borrowed from the area of requirement engineering [17] and also used for SNA business alignment [12, 13], can improve further development in the area of decision mining and performance management. Contributions made in this paper relating to introducing requirements engineering into the field of decision modelling and supporting companies' alignment actual DM effort with their business' goals using a structured approach.

The structure of the paper is organized as follows. Section 2 shortly introduces the architecture of the decision management system. Section 3 overviews DMN standard with an analysis of the relationships between decision and performance indicators in a real case from [7]. Section 4 describes the framework adaptation of business alignments [12] with their extension to decision management. In Subsect. 4.1, a traditional data mining algorithm (C4.5) [18] was applied on DPI and PPI in order to try to fulfill some of the decision analytic requirements identified within a new framework. Section 5 concludes the paper.

2 Decision Management System

Decision management appears as a systematic approach to automating and improving operational business decisions toward increasing the efficiency and effectiveness of decision making [19]. It is an approach for developing a decision management system (DMS) through modelling decisions, building and deploy IT components that combine advanced business analytics with business rules and optimization models. DMS tries to improve the effectiveness of a decision flow observed as a set of alternatives/choices associated with an operational task in a specific workflow. At this point it is necessary to emphasize the word "specific" decision, because these systems try to automate certain repeatable operational and micro decisions mostly about: customers, their churn, direct marketing, retail assortment planning, incident management, fraud detection, customer service, credit risk, price optimization, human capital management etc. For such a decision, DMS adds elements of different business analytics: descriptive, predictive or decision models integrated with explicit business rules or/and optimization models.

The core of the DMS system is Decision service component (Fig. 1) that runs on the standard enterprise platforms in use today.

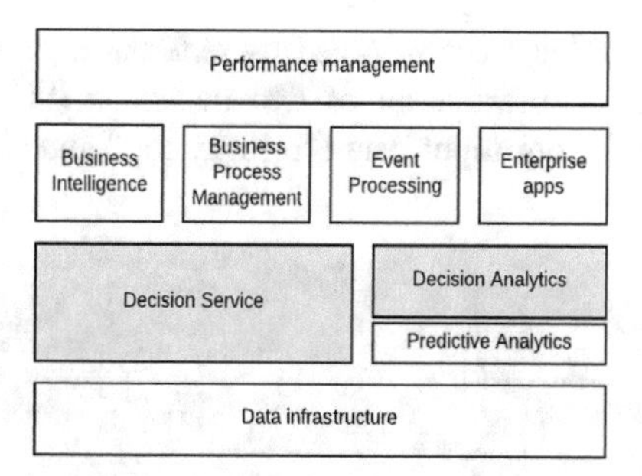

Fig. 1. Decision management system architectural context

The data infrastructure supplies operational data to the Decision Services and can also provide in-database analytics. Business intelligence system typically uses the same data infrastructure provided insight into human decision makers and often complements the function of decision service by handling the exceptions. With a link to performance management, it is possible to track and analyze the effect of some decision to key performance indicators (KPI). Decision service can be integrated with existing Business Process Management Systems (BPMS) where it is triggered with the task in processes; or with Event Processing System, where triggering some action on detected patterns of data through a variety of external connections. The decision itself, in the decision service, is made by executing generated code on the underlying platform, business rules on a deployed business rule engine, optimization models on a solver and predictive analytic models on a model execution engine. Decision Services also need to be able to log what happened each time a decision was made-which rules were fired, what model scores were calculated and which outcomes were selected by the optimization model. Decision analytics integrates all techniques of data and decision (log) mining techniques necessary to monitor the success of the decision in a particular process of interest.

3 DMN Standard and Performance Measurement

DMN is a standard for describing and modelling routine decisions within organizations, next to business and business data. It is an application-independent modelling language for business decisions. It consists of two levels: Decision Requirement Diagram (DRD) and Decision logic. DRD is a first level for describing the requirements of decisions with a set of elements and their connection using specific notation. At the second level of DMN, the decision logic captures the logic to make each decision in DRD. The expression (S-FEEL) language is provided by the standard for the description of decision logic in decision tables.

Figure 2 shows a DMN model based on real decisions made as a part of the IT incident management process of a public organization described in work [7]. All related data taken from events log and presented along this paper were taken from the aforementioned paper. The IT Department classifies incidents in different priorities (P1, P2 and P3), according to their impact (major, high, medium and low) and urgency (high, medium and low) that were resolved by external agents. The decision named "Priority setting" requires information of two decisions ("Urgency resolution" and "Impact resolution") and from input data ("priority log" and "IT incident log").

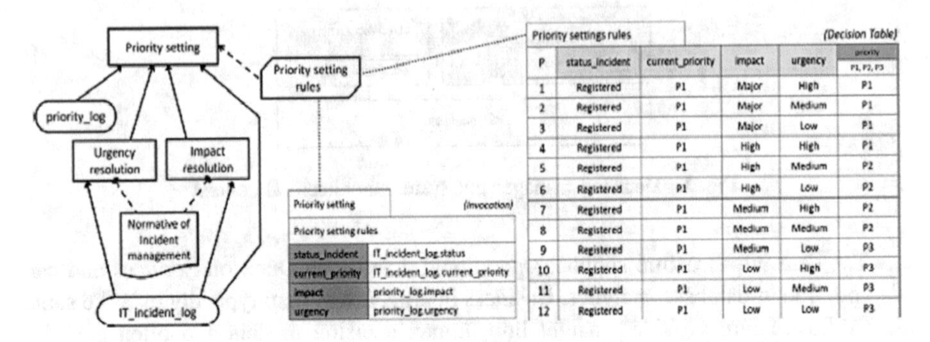

P	status_incident	current_priority	impact	urgency	priority P1, P2, P3
1	Registered	P1	Major	High	P1
2	Registered	P1	Major	Medium	P1
3	Registered	P1	Major	Low	P1
4	Registered	P1	High	High	P1
5	Registered	P1	High	Medium	P2
6	Registered	P1	High	Low	P2
7	Registered	P1	Medium	High	P2
8	Registered	P1	Medium	Medium	P2
9	Registered	P1	Medium	Low	P3
10	Registered	P1	Low	High	P3
11	Registered	P1	Low	Medium	P3
12	Registered	P1	Low	Low	P3

Fig. 2. DRD with decision logic for IT incident management [7]

Since their introduction, the DMN standard has an intention to model decision outside the process. From the other side, Business Process Management (BPM) puts the focus on modelling the control flow with tracking the activities, their data and resources, often neglecting to model the decisions. The focus on actual DMN's researches is the exigency for networking decision models into a dynamic environment, where a chain of decisions drives the business process. Current research shows [8–11] that modelling a decision logic is not the same as modelling execution steps in processes. Decision model can be executed in different ways depending on preferred criteria/goal with different implications to specific processes (or more them) and their business performance.

3.1 The Excerpt of DMN Extension in PPINOT

The decision as a part of the business process can have an impact on their performance and can be observed through their impact on their process performance indicators (PPI) [7]. If the PPI changes his values depending on output in a selected decision, it can be said that decision can impact on a PPI. PPINOT is a metamodel for the definition and modelling of PPI, enabling traceability of process elements and their performance [4]. It consists of five attributes: goals, target values to be achieved, a scope of measure (a subset of instances to be included in the calculation), a set of human resources involved and a measure definition (concrete specification of computation). In their work, Estrada- Torres et al. [7], built PPINOT metamodel extending PPI definition with DMN elements (Fig. 3). They created Decision Performance Indicator (DPI) for the case of Incident management, as quantifiable metrics that share the same attributes with a PPI. In that way, it was possible to measure three dimensions of decisions: how many times each output occurs, the time a decision takes and information involved in the decision. In this specific case, DPI could be defined as "the average time spent to assign an IT incident priority, whose scope is comprised by all instances and its target is set to less than 4 h. It is calculated as the duration between the time instant when Activity "IT incident registration" becomes active and Decision "Priority Setting" output becomes assigned" [7]. Further, the authors experimented and concluded that, with these extensions, it is possible to measure process performance and use PPI information as an input to make decisions and vice versa: to evaluate PPI values as a result of DPIs' changes. Shown work complement the previous work from [6]. The authors derived, without using DMN notation, decision criteria based on past process executions and proposed a formal framework to derive decision models from events logs using DMN and BPMN, but they were not concerned understanding the output of each decision.

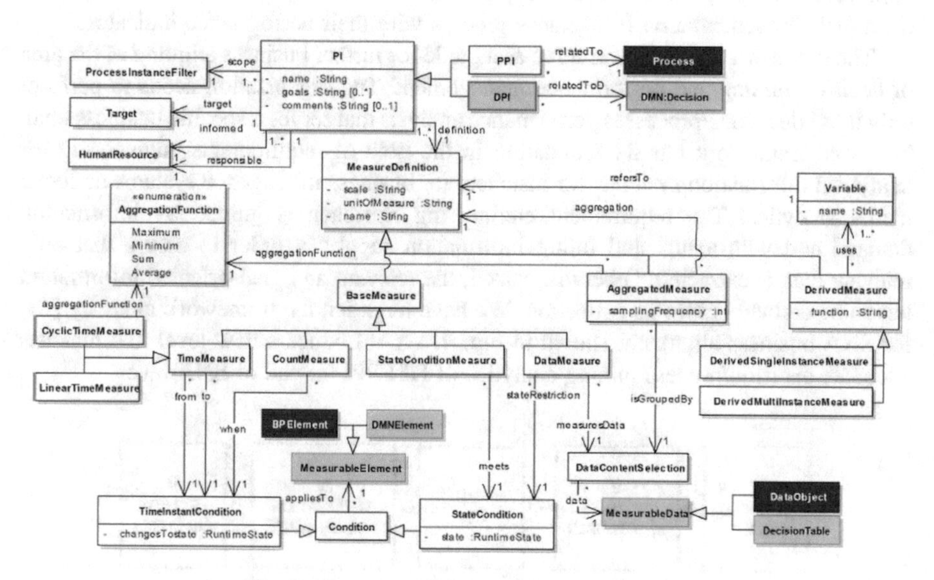

Fig. 3. Excerpt of the PPINOT metamodel [7]

With the novel concept of DPI and PPI included in PPINOT "it is possible to track decisions' performance based on the data generated along with the execution of the process" [7]. New problems arise in the form of supporting these tasks with machine learning techniques and by aggregating information from different sources of data with expert's experience needed for creating decision output. We think that the solving of these problems can be facilitated with developing a more comprehensive framework in the efforts to decision measurement development.

4 The Framework for Business Alignment

Described attempt to support decision performance measure deals with the details and it seems congruent with past development of customer relationship management systems (CRM), business analytics and social network analytics (SNA) where IT developers focus on topics like how to collect and store data, rather than on the more important aspect of what to do with it and why to do it. None of the researchers, to the best of our knowledge, are engaged in analyzing the managers' requirements or considering the business objectives that managers need to improve, assembly with indicators to be monitored or analysis to be done. Hasić et al. [8, 9]. mentioned that most existing literature approaches the matter of simplistic point of view by dealing with decisions embedded in a single place in the process, hence making the decision process- dependent. The same authors highlight that "the framework to support the decision representation must provide a holistic approach to decision modelling and mining that is control flow agnostic, i.e. a decision-first rather than a process-first approach." We propose that decision performance measurement cannot be separated from their implementation in business process with their performance indicators.

The top-down approach can serve as a guide for further business aligning of the area of decision mining and performance management. The organization needs to perform only those decisions/processes performance analysis that serves a specific business goal. The used framework has its foundation in the field of requirements engineering for analytical information systems, for instance, for business intelligence systems or social media analytics. The requirement engineering approach is guided by information demand and with actual and future information supply in order to assure that only relevant data is extracted. Otherwise stated, the relevant and goal-oriented information has to be defined before its collection. We have modified the framework made by [12] for SNA business alignment shown in Fig. 4. We introduce a new level that matches adequate decision/process mining analytics of DPI/PPI instead of SNA layer.

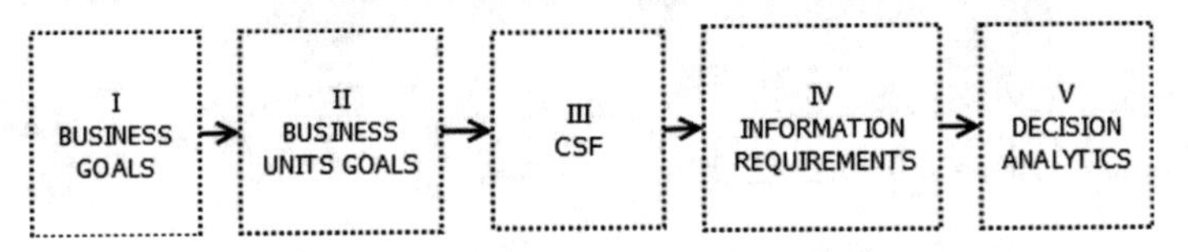

Fig. 4. Framework for the business alignments of decision analytics (modified from [12])

The starting point of the business alignment process is high-level business goals (I) which followed by the identification of business units and their individual objectives (II) that should possibly benefit from decision analytics. The relationships between I and II can be many-to-many type meaning that multiple units can follow the same goals. The third element (III) is related to critical success factors of the respective business units that need to be measured and monitored for performance and serve as a basis for determination of information requirements (IV) that need to be extracted from a suitable decision analytic. In the sequel of the article, we used previously described incident management example for developing business alignments framework for decision management. Business goals (I) were established like in most organizations through their strategy map and they need to be derived to the business goals of specific units (II). In our example, managing the incidents (as reducing time in resolving them) and optimal allocation of IT resources in their resolving as a business goal concerns IT service desk as a business unit in the organization.

To assist companies in implementing their strategies and avoiding costly failures, researchers and consultants have suggested a range of critical success factors (CSFs). It can be defined as 'those few things that must go well to ensure success for a manager or an organization, and, therefore, they represent those managerial or enterprise areas that must be given special and continual attention to bring about high performance' [4]. Numerous lists of the CSFs for different specific projects are available in literature and practice. For CSFs in Incident management, and for the purpose of this paper, we accept CFSs' list from IT Infrastructure Library [15]. According to ITIL, the main CSFs' metrics for incident management are: Total number of incidents, Number of incidents at each stage (logged, in progress, closed), Percent of incidents resolved within an agreed-upon (SLA) time, Average resolution time and cost, Percent of incident resolved already at service desk, Percent of incorrectly assigned incidents, Percent of incorrectly categorized incidents, Percent of incidents that have been re-opened, percent of incident that require on-site support and Dispersion of incidents per time of day. According to an available excerpt of data about DPIs and PPIs for incident management shown in [7], it was possible to consider only two related CSFs mentioned above: Number of incidents received and number (without percentage) of incorrectly categorized incidents (incidents that change priority). As can be seen, the focus was only on those CSFs that can be supported with decision analytics and their DPIs and PPS. With identified relevant CSFs, executives were able to determine their information requirements.

For the identification of information required for decision analytics in incident management, it can be listed relevant functional requirements needed for calculating CSFs (without specification of their quality requirements (in contexts of their granularity or standardization) and constraints (in context of data privacy issues) mentioned in from [12]. The identified functional requirements are identification of processes which have high number incident in time, identification of specific incident with changing priority, analysis of workload/time alerts in processes, calculation of the execution time and an average time for resolving changing priorities, analysis of time/workload alert relations on priorities changes. Bringing together all identified elements resulted in the framework for alignments of DMN performance measure efforts to business analytics' requirements (Fig. 5). This proposal provides transparency in a holistic approach to modeling decision performance measures by breaking up the whole process into manageable parts where the specific design of DMN extensions will serve specific goals and will be relevant to organizations' success. A common business goal in incident management is efficient management (level I in Fig. 5) as restoring normal service operation as quickly as possible and minimize the adverse impact on business operations, ensuring that agreed levels of service quality are maintained [15]. Specific incident management objectives (at level II, Fig. 5) can be: the management of the incidents should be aligned with the priorities of the particular business (Appropriate priorities identification), IT resources allocation according to priority and reducing time and cost of resolving incidents. At level III, some of ITIL' main CSFs' metrics for incident management were used: Total number of incidents, Number of Average resolution time and cost, Percent of incorrectly assigned incidents, Percent of incorrectly categorized incidents. Functional information requirements, in this case, include the detail specification of the relevant business information for executives in managing the incidents in organizations.

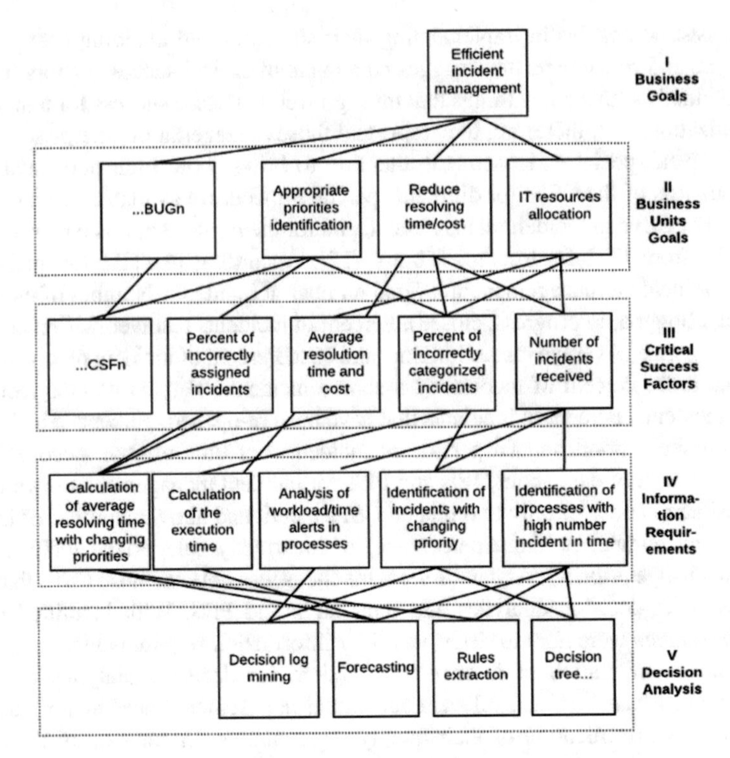

Fig. 5. Example of framework for the business goal of Incident management

Level IV, at Fig. 5. shows some of the possible information requirements without intent to make the list exhaustive. With defined information required as a foundation, adequate decision analysis concept need to be found in order to extract the required data from the specified business processes platforms.

4.1 Traditional Data Mining Algorithm in Decision Analysis

Induction can be based on C4.5. the algorithm introduced by [18]. Cases can be anything that we can describe from important aspects. In symbolic logic, it is done with attributes and their values. One value of every attribute is assigned to each of the cases that form logic rules. Cases described with the same rule are considered similar. Therefore, the case base is used to extract rules from experience. As it goes from the particular experiences towards the general rules, it is called induction. All cases form a disordered set, where the order is defined as homogeneity by benchmark values (value of outcome attribute), which means that cases in one subset have the same benchmark value. The attribute is searched, which contributes the most to the order. Their strength in making an order is measured by an entropy-gain calculating algorithm. Determining informativity (I_b) of attribute b is as follows: Let C be the set of cases in node, *a* the benchmark, $a_1 \ldots a_n$ its values, and $w_{a1} \ldots w_{an}$ ($\sum i\ w_{ai} = 1$) their rates in set C. Then entropy of benchmark in set C can be written: $E_c = -\sum w_{ai} \log_n w_{ai}$. Let $b_1 \ldots b_n$ be the values of attribute b, β is a set of them. Disjoint β into not empty subsets $\beta_1 \ldots \beta_m$. Then $U_i\ \beta_i = \beta$. Disjoint C into subsets $C_1 \ldots C_m$ attributing b of all elements of C_i in β_i for each i. Let w_i be the weight of C_i in C ($\sum i\ w_i = 1$). Then $I_b = EC - \sum i\ w_i EC_i$, or informativity is an increment of entropy resulted from disjoining $\beta_1 \ldots\ \beta_m$. The real output of computing is I_{bmax} of optimal selection. The most informative attributes are chosen and this first level of subset values are further divided using the same algorithm until all subsets are homogenous by benchmark values. The result of the classification of this case appears in the form of a decision tree that can be converted to production rules.

For exploring possibilities of supporting decision analytics with traditional data mining techniques we use available descriptions of DPI in incident management example. Case base consists of five attributes with values:

- Alert (Time and Workload), describes possible outcomes of alerting events' characteristics from the business process;
- "Priority from" (P1, P2, P3) describes starting values of incident's priority assigned before the decision in the specific process;
- "Priority to" (P1, P2, P3) describes a target values of the incident's priority assigned after the decision in the specific process;
- Execution time (<20, >=20, >=45, >=70, >=120) represents PPI value in average hours spent in solving an incident that changed priority; for those cases in which the output of the decision (priority from) is changed to another one (priority to);
- A number of incidents received (0–249; 250–499; 500–799; 800–1000; > 1000) describe aggregates measures of PPI as a number of instances registered in the scenario event log.

A small pilot knowledge base consists of eleven examples (cases) modified from [7], describing the use of decision performance measures in the incident management process (Table 1).

Table 1. Cases with attributes and their values in Incident management DPI

Alert	Priority from	Priority changes to	Execution time	A number of incidents
Time alert	P1	P1	>=20	250–499
Time alert	P1	P2	>=45	0–249
*	P1	P3	<20	0–249
Time alert	P2	P1	>=45	0–249
Workload alert	P2	P2	>=20	800–1000
Time alert	P2	P3	>=20	0–249
Time alert	P3	P1	>=70	0–249
Time alert	P3	P2	>=120	250–499
Workload alert	P3	P2	>=120	250–499
Time alert	P3	P3	>=120	>1000
Workload alert	P3	P3	>=120	>1000

After applying the induction algorithm on those values, we got the induction tree (Fig. 6).

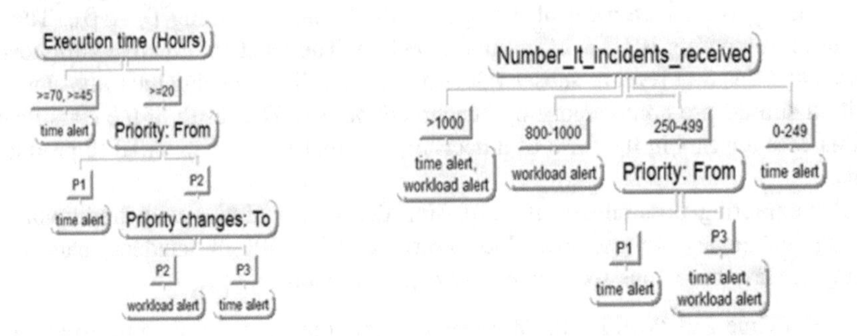

Fig. 6. Resulting decision trees for different views of DPI relationships in case of incident management

Resulting induction trees can be interpreted by "If-Then" rules. The rules may be read from the root of the graph towards its leaves, where the values of the outcome are shown. For example (from the right side on the graph in the left side of Fig. 6): If attribute "Execution time in hours" has a value " >=20" then if attribute "Priority: From" has a value "P1" than value of attribute "Alert" has a value "time alert". If attribute "Priority" does not change value "P2" alert is workload alert". If Priority changes from P2 to P3, the value of alert also changes to "time alert". Cases addition, deletion or some kind of modification will be described by different induction tree. Once the expert accepted the induction trees, a new knowledge base can be created, which contains only the informative attributes. This new reasoning reduces an existing model and their rules with positioning each new case according to its features by the informative attributes. The result of reduction (extraction of rules) means that if the informative attribute was known it is possible to previously evaluate future decisions. Table 2 depicts the extract rules. The star (*) representing "Don't care" value.

Table 2. Extraction of rules

Priority from:	Priority changes to:	Alert
P1	*	Time alert
P3	*	Time alert
P2	P1	Time alert
P2	P3	Time alert
P2	P2	Workload alert

For example, the first row at Table 2. means: If Priority: From changes to Priority any other (* don't care) priority, Alert will remain as "time alert". If during the fine-tuning of the reduced knowledge base implied changes of attributes and/or values, these changes should be applied to decision graph and using it as feedback some elements of tacit knowledge can be acquired in the initial set of cases.

5 Conclusion

Parasuraman & Sheridan [16] point out that automation does not only replace or supplement human activity but also changes it, in ways that are not firstly anticipated by designers. With the proposed framework based on managers' requirements for performance measurement, as a result of this paper, it can be anticipated specific DMN extensions and development of new analytics/monitoring tools. It can also serve as a trace for planning indispensable executives' competency profile for a new kind of analytics job's family as a prerequisite for the adoption of self- service analytics in organizations. Such an approach improves decision modelling in ways that executives do not need DMN notation knowledge in order to apply specific analysis and enable developers' anticipation of these requirements in modelling specific DMN extension. At least two limitations of this work can be identified. The first one refers to the example used in this paper as a snippet of published data in incident management in the real case of DMN extension published in the scientific literature. Second is related to listed decision analytics that was not finite. The list of decision analytics tools identifies a new problem of an exigency developing new decision analytic techniques that will be capable to dynamically manipulate with the feedback of decision outputs from processes.

References

1. Alavi, M., Carlson, P.: A review of MIS research and disciplinary development. J. Manag. Inf. Syst. **8**, 45–62 (1992)
2. Bazhenova, E., Weske, M.: Deriving decision models from process models by enhanced decision mining. In: Reichert, M., Reijers, H.A. (eds.) BPM 2015. LNBIP, vol. 256, pp. 444–457. Springer, Cham (2016). https://doi.org/10.1007/978-3-319-42887-136
3. Bevanda, V.: Decision engineering: settling a lean decision modeling approach. In: Studzieniecki, T., Kozina, M., Skalamera Alilovic, D. (ur.) Economic and Social Development 33rd International Scientific Conference on Economic and Social Development – "Managerial Issues in Modern Business". Varaždin, pp. 102–111(2018)
4. Boynton, A.C., Zmud, R.W.: An Assessment of Critical Success Factors. Sloan Manag. Rev. **25**(4), 17–27 (1984)
5. del Rio-Ortega, A., et al.: On the definition and design-time analysis of process performance indicators. Inf. Syst. **38**(4), 470–490 (2013)
6. del Rio-Ortega, A., et al.: Using templates and linguistic patterns to define process performance indicators. Enterp. Inf. Syst. **10**(2), 159–192 (2016)
7. Estrada-Torres, B., del-R´ıo-Ortega, A., Resinas, M., Ruiz-Cort´es, A.: On the relationships between decision management and performance measurement. In: Krogstie, J., Reijers, H.A. (eds.) CAiSE 2018. LNCS, vol. 10816, pp. 311–326. Springer, Cham (2018). https://doi.org/10.1007/978-3-319-91563-019
8. Hasić, F., De Smedt, J., Vanthienen, J.: Developing a modelling and mining framework for integrated processes and decisions. In: On the Move to Meaningful Internet Systems. OTM 2017 Workshops, Lecture Notes in Computer Science, vol. 10697. Springer, Heidelberg (2017)

9. Hasić, F., De Smedt, J., Vanthienen, J.: Augmenting processes with decision intelligence: principles for integrated modelling. Decis. Support Syst. **107**, 1–12 (2018)
10. Hasić, F., Devadder, L., Dochez, M., Hanot, J., De Smedt, J., Vanthienen, J.: Challenges in refactoring processes to include decision modelling. In: Teniente, E., Weidlich, M. (eds.) BPM 2017. LNBIP, vol. 308, pp. 529–541. Springer, Cham (2018). https://doi.org/10.1007/978-3-319-74030-042
11. De Smedt, J., Hasić, F., Vanthienen, J.: Towards a holistic discovery of decisions in process-aware information systems. In: Business Process Management, Lecture Notes in Computer Science, vol. 10445, pp. 183–199. Springer, Heidelberg (2017)
12. Kleindienst, D., et al.: The business alignment of social media analytics. In: Proceedings of the 23rd European Conference on Information Systems (ECIS), Münster, Germany, May 2015. https://www.fim-rc.de/Paperbibliothek/Veroeffentlicht/486/wi-486.pdf. Accessed 15 Jan 2019
13. Munro, M.C., Wheeler, B.R.: Planning, critical success factors, and management's information requirements. MIS Q. **4**(4), 27–38 (1980)
14. Object Management Group: Decision Model and Notation (DMN) V1.1 (2016)
15. Office of Government Commerce (OGC): ITIL Service Operation. London, UK: TSO (The Stationery Office) (2007)
16. Parasuraman, R., Sheridan, B.T.: A model for types and levels of human interaction with automation. IEEE Transact. Syst. Man Cybern. Part A Syst. Hum. **30**(3), 286–297 (2000)
17. Pohl, K., Rupp, C.: Requirements Engineering Fundamentals: A Study Guide for the Certified Professional for Requirements Engineering Exam. Rocky Nook, Massachusetts (2011)
18. Quinlan, J.R.: C4.5. Programs for Machine Learning. Morgan Kauffman, Burlington (1992)
19. Taylor, J.: Decision Management Systems Platform Technologies Report. Version 7, Update 4, March 25 (2016). http://www.decisionmanagementsolutions.com/wp-content/uploads/2014/11/The-Decision-Management-Systems-Platform-Report-V7U4.pdf. Accessed 19 Aug 2018

Influencing Factors of Mobile Technology Adoption by Small Holder Farmers in Zimbabwe

Fine Masimba, Martin Appiah, and Tranos Zuva(✉)

Department of ICT, Vaal University of Technology, Andries, Potgieter, Vanderbijlpark 1911, South Africa
fine.masimba@gmail.com, zuvat@vut.ac.za

Abstract. Agriculture is the primary occupation of the majority of Zimbabweans living in rural areas which constitutes of about 70% of Zimbabwe's population. In the present times of technological development, mobile technology particularly mobile phones have become the most important tools of communication which can be accessed by farmers for agricultural related information and knowledge, hence understanding the adoption of this technology is important. This paper contributes to the existing literature by comprehensively reviewing the role of mobile technology in agriculture and the factors that influence the adoption of this technology among small holder farmers in Zimbabwe. The study informs policy makers and technology startup companies to adopt appropriate strategies to encourage the adoption of mobile technology among small holder farmers having different demographic characteristics and other influencing factors highlighted in the study.

Keywords: Mobile technology · Mobile phones · Technology adoption · Small holder farmers

1 Introduction

Agriculture plays a crucial role in the economy of developing countries, and provides the main source of food, income and employment to their rural and poor urban populations [1]. Agriculture is also believed to be the world's largest industry as it employs more than one billion people and generates over $1.3 trillion dollars, worth of food annually. In many developing countries, agriculture is being viewed as a major contributor to social and economic development given that it is the major contributor to economic growth and stability [2]. Agriculture is also viewed as a great contributor towards reduction of poverty and hunger by 2030 in many developing countries as stated in the Sustainable development Goals of the United Nations.

In Zimbabwe around 70% of the 14.7 million people reside in the rural areas and most of these people are smallholder farmers whose livelihood is heavily depended on agriculture. Furthermore, in Zimbabwe, the agricultural sector is the back borne of the economy, as a result it is the sector which underpins the country's economic growth, food security and poverty eradication. The agricultural sector is pivotal to the economy

R. Silhavy et al. (Eds.): CoMeSySo 2019, AISC 1046, pp. 125–134, 2019.
https://doi.org/10.1007/978-3-030-30329-7_12

of Zimbabwe, providing 14–18% of the Gross Domestic Product (GDP), 40% of export earnings and 60% of raw materials for industry [3].

On the other hand, mobile technology penetration has continued to expand in Zimbabwe. The third Quarter Telecommunications report by the Postal and Telecommunications Regulatory Authority of Zimbabwe (Potraz) reported an increase in Zimbabwe's mobile penetration rate from 97% in Second Quarter to 100% in Third Quarter of 2017 (Potraz Telecoms Report 2017). With a high mobile penetration rate of 100% and a strong agricultural system, mobile farming platforms in Zimbabwe have a great potential to transform the agricultural sector in Zimbabwe. This is supported by [4] who highlighted that the strategic application of ICTs to the agricultural industry, the largest economic sector in most African countries, offers the best opportunity for economic growth and poverty alleviation on the continent. Among other ICTs, mobile telephony has emerged as the technology of choice of the majority of the urban and even the rural masses and mobile phones have been reported to be widely possessed ICT tool among farmers [5]. The possession of mobile phones particularly has become a necessity in the contemporary society irrespective of age, status, profession, income groups or place of residence. As such, mobile phones have been regarded as the widely accessed tool among the farmers for communication and also accessing agriculture-related information particularly for the marketing of produce [6].

Anecdotal evidence of research findings pointed that mobile phones are the most important tools of communication which can be accessed by farmers for agricultural related information and knowledge [6]. In this context, mobile technologies can offer the means for development in developing countries [7]. The rapid growth of mobile technology and the introduction of mobile-enabled information services provide ways to improve information dissemination to the knowledge intensive agriculture sector and also helps to overcome information asymmetry existing among the group of farmer [8].

2 Role of Mobile Technology in Agriculture

Mobile technology is now being used by farmers as a platform where they receive important and relevant information for their different farming activities. In a survey that was carried out on farmers in 2011 in states of India, it was noted that with access to information through mobile technology, farmers are now better connected to the markets and mobile technology have helped them to get better prices too [8]. [9] found that the introduction of mobile technology decreased price dispersion and wastage by facilitating the spread of information for fishermen in Kerala. Mobile phone usage by farmers can reduce information search costs, thereby dramatically lowering transaction costs and enabling greater farmer participation in commercial agriculture [10]. [11] states that the use of mobile phones for information search by farmers result in increased income among farmers. They further argued that farmers' incomes are expected to increase with the farmers access to real time information on prevailing prices and available markets made available by the mobile platforms.

[12] highlighted that mobile technology contribute to the sharing of agricultural information on farming decisions such as which crop to grow at a particular time of the season, when to grow, what to grow, which agrochemical to use and when to harvest as

well as information on weather reports. Agricultural information on timely soil preparation and planting, irrigation methods, weeding methods, cultivation methods, harvesting and storage methods can be sent to farmers through texts (SMS), and this information will help farmers to improve productivity. Agriculture extension officers can utilize mobile phones to make direct calls to farmers and avoid the need for travelling down to where these farmers are to deliver face-to face teaching on good agricultural practices. [13] noted that mobile technology can help farmers with information to decide where to sell and what price to sell their produce. With just a simple Short Message Service (SMS) to producers and suppliers, farmers can obtain their supplies such as herbicides and fertilizer at their door steps. This will in addition reduce transportation costs, accidents, theft, perishability as well as frustration among farmers.

In Uganda, [14] also conclude that farmers used their mobile phones for a range of farming activities, especially to coordinate access to agricultural inputs (such as training, seeds or pesticides) (87% of farmers), accessing market information (70%), requesting agricultural emergency assistance (57%), monitoring financial transactions (54%) and consulting with expert advice (52%). Mobile services are at work in the field of agriculture, mostly for sharing and obtaining information to increase productivity [15]. Mobile phones have the potential to provide solution to the existing information asymmetry in the sectors like agriculture [16].

3 Mobile Farming Platforms in Zimbabwe

Due to the potential role of mobile technology to reduce information asymmetry, cut transaction costs, improve market access, increase agricultural productivity and to improve income amongst small holder farmers, different technology start-up companies in Zimbabwe have managed to develop different mobile based farming platforms as a solution to the problem of information asymmetry with the aim of improving overall performance among smallholder farmers. [17] states that currently in Zimbabwe, there are four main mobile farming platforms/applications that are being used in the field on agriculture and these are Ecofarmer, Esoko, eMkambo and eHurudza. According to [22] the EcoFarmer platform, a farming insurance initiative being offered by Zimbabwe foremost mobile network provider Econet Wireless started operations in 2013. It is a revolutionary way of farming using mobile technology.

The EcoFarmer platform offers Zimbabwe's first Micro Insurance product designed to insure inputs and crops against drought or excessive rainfall. With Ecofarmer, a farmer registers first and then pays daily subscription. When fully registered and paid daily subscription, the farmer will get daily weather data from a weather station linked to your field, farming and market tips, free daily rainfall advice, free weekly best farming prices, free weekly crop data, free monthly market pricing requests, crop information, credit rating, free adverts and marketing links and financial linkages.

E-Hurudza is also another new mobile farming platform, which is an electronic farm management software solution developed by Hurudza Africa, a local agriculture technology company. This mobile application was launched in 2016 and it gives small holder farmers agricultural information for all regions. It gives tutorials on how to grow crops specific to your region including land preparation. It also gives information on input requirements for example seeds, fertilizers, insecticide/chemicals, manpower and expected yield per hectare. The information will enable the smallholder farmer to make right farming decisions in terms of farming the right crops at the right time and on the right soil.

Another application is eMkambo, [17] further states that this is a mobile farming application system that was launched in 2012 and it started as a call centre at Mbare Musika where they receive constant updates from farmers on the various produce ready for the market and relay that information to the relevant mix of traders in their database. The traders will then follow up with the farmers and bargain on prices agreeing on logistics of delivery and payments. This method leads to traders bidding for the produce in a typical auction style resulting in smallholder farmers settling for the best prices. It has been developed into a mobile app. The owners of the system visit the market place and talk to farmers who bring their produce to the market. If a farmer expresses interest, s/he submits details about their farming activities and place where they come from.

Esoko is another mobile platform that is also being used in the agricultural sector. According to [18] the Esoko mobile platform started its operations in Zimbabwe in 2012. They currently service 17 fresh produce markets scattered across Zimbabwe and they cover 33 commodities. They provide services to over 170 000 small holder farmers. Esoko uses a number of NGOs and contractors to push out their services to farmers. Services offered by the Esoko platform include Sms Push and Pull which allows interactive communication with recipients, bids and offers to facilitate trading of agricultural commodities between farmers and buyers as well as profiling of sms recipients using such criteria as age, gender, ward and crop.

All these findings point to the various areas of relevance of mobile technology services to agricultural operations, so literature has proven beyond any reasonable doubt that mobile technology can enhance agricultural activities but however the issue of the actual adoption of these mobile technologies remains a challenge. In spite of the presence of these several mobile farming platforms in Zimbabwe's agricultural sector (i.e. Eco-farmer, E-Hurudza, E-Soko and E-Mkambo), the majority of these platforms are not being utilized, this alone is a challenge as highlighted by [19] who indicated that technology is of little value, unless it is accepted and used hence the need to carry out an in-depth examination of the factors that affect adoption of technology particularly mobile technology.

4 Factors Influencing the Adoption of Mobile Technology Among Small Holder Farmers

Previous studies have been focusing much on the role of mobile technology in agriculture and their use by farmers and other agricultural agents [12]. However, little attention has been paid to the study of the real factors that affect the actual adoption or acceptance of mobile technology for agricultural activities among small holder farmers. This is supported by [20], who highlighted that among mobile innovations deployed for agriculture in developing countries, only 16% make it to widespread adoption. This is further noted [21] who highlighted that although the technology is advancing, its utilization is lacking. ICT uptake in agriculture is continuing to be less than expected worldwide [22].

Technology is of little value, unless it is accepted and used [19, 22]. [22] further argues that although technology has become essential to effective agricultural production, the adoption of these e-innovations in agriculture is still limited especially among smallholder farmers. Therefore, given the importance of mobile technology to modern agriculture, this section sought to identify the factors affecting use of mobile technology for agricultural production by small holder farmers in Zimbabwe. This is done by undertaking in-depth qualitative examination of these factors. The findings from this study contribute to a larger picture of the technology adoption process among small holder farmers specifically in Zimbabwe.

4.1 Demographic Factors

There is a good number of studies describing the importance of the demographic context in use and adoption of new technology [23]. [24] argue that demographic variables have a substantial impact on the adoption of modern technologies. According to those studies, demographic factors include variables such as age [25, 26], income and household [27], education [27]. Demographic characteristics like gender, age, occupation, income and education level needs to be explored in order to find out the behavior pertaining to adoption of mobile technology by farmer [28]. A study of demographic characteristics enhances an understanding of how demographic features of farmers influence their perception of mobile technology acceptance and their adoption decisions.

In technology adoption literature, age is one the most discussed demographic factors. [25] conducted a study amongst rural farmers in India, the study revealed that (62%) of mobile technology users are people within the age group of 20 to 40. [23], further argued that people within the age group of 20 to 30 are more likely to be receptive to a wider range of mobile technology services. [22] conducted a study among small holder farmers in Hungary and concluded that the majority of younger farmers use smart phones for agriculture than elderly farmers. [29] argues that majority of technology users tend to be young adults aged between 20–45 years old. [30] in their study highlighted that the majority of farmers who belong to the age category of above 25 are using mobile technology to gather agricultural information for increasing efficiency in farm operations. Several studies through regression and correlation analysis, have confirmed that there is a relationship between age and usage of mobile technology.

Education is another demographic factor that influences the adoption of mobile technology by farmers. [12] indicated that education has a positive effect on mobile technology adoption because more educated people are more aware of mobile technology, how to operate it, how to obtain it and how to use it. [31], revealed that the results of many studies suggest that as education level increases, farmers shift to modern ICT technologies for gathering agricultural information. [32] further argued that more educated people are better able to learn and use new technology and hence they are more likely to be innovative. [22] highlighted that farmers who are more educated adopt smart phones for agriculture than less educated farmers. [33], argues that diffusion of new technology among communities of less educated people is relatively slow due to their low education level. [34] in his study found that people with tertiary level of education have higher access and use of mobile devices than those of lower levels.

Many previous studies revealed that usage of technologies and income are significantly and positively associated which results in rich large farmers being able to get more information as compared to poor farmers [35]. There is a positive correlation between the level of income and the adoption of new technology [32]. There is also need to pay for initial purchase of the mobile device and thereafter to pay for the subscription fees if the user is to get some services which will be offered through the mobile device and because of this, only small holder farmers with much income are likely to afford. [36] in their study revealed that farmers' average daily income has a high statistical significant influence on mobile phone use in communicating agricultural information. Their study results further revealed that the use of mobile technology to communicate agricultural information was highly influenced by income levels.

Farmers with high income are comparatively resourceful and have personal electronic devices like computer, internet connection, television and mobile technology which can be used to grab and gather agricultural information [31]. [27] argues that the adoption of a certain technology (for example post paid vs pre-paid services and fixed vs mobile connection) depends on the level of household income over time. [37] in his study highlighted that the majority of mobile technology users are youngsters who are more educated with higher incomes, he also validates that there is a significant positive correlation between mobile technology adoption and income of the farmers.

Gender level has a significant effect on adoption of mobile technologies. Ownership of mobile phones among women is less than 21% as compared to their male counterparts in spite of skyrocketing increase in the penetration levels of mobile diffusion across the world. [23] conducted a study on gender ratio of ownership of mobile phone adoption by farmers in Bangladesh, the results of the study indicated that the majority of farmers who own mobile phones were males. In that same study, results also indicated that men make 70% of the mobile calls as compared to women, which indicate a distinct gender difference. In the context of mobile banking, mobile technology usage is more popular among males, who are more educated and with high incomes [38]. [30] argues that males and females both are using mobile technology to gather agricultural information but males perceive mobile technology to be more useful in comparison to females.

4.2 Facilitating Conditions

Facilitating Conditions is the degree to which an individual believes that an organizational and technical infrastructure exist to support the use of a system [21]. [39] describe Facilitating Conditions (FC) as the support given to the users while interacting with the technologies, like learning the technology from a friend. [25] highlighted that the choice of service provider is affected by facilitating factors such as network coverage, easy availability of subscriptions, bill payment centers and service quality. In mobile technology facilitating conditions include variables such as rural connectivity and access time [43], technological infrastructure [27] quality and availability of support services [25] market structure and mechanism [40].

4.3 Tech-Service Attributes

[23] refer to the properties or characteristics of a certain technology, system, or service that distinguish it from other technologies, systems or services. Farmers' perceptions of technology characteristics significantly affect their adoption decisions [41]. Tech service attributes include variables such as service characteristics of the technology [42], cost of handsets and services [43], technology characteristics for example network characteristics and interface characteristics [44]. [45] argued that cost is a significant barrier to the adoption of mobile based services.

4.4 Tech-Service Promotion

Tech-service promotion is a process of informing people about the product or initiative. Any technological innovation has to be promoted so as to create awareness which is a pre requisite for the adoption of any new technology. [46] highlighted that lack of awareness is one the main reasons for farmers not adopting the new technology. Awareness can be characterized as one of the steps towards adoption of any initiative. [47] therefore suggests that suppliers or technology startup companies must promote their initiatives in order to create awareness among users.

4.5 Perceived Usefulness (PU) and Perceived Ease of Use (PEU)

Previous studies suggest that these two factors are the most significant in mobile service usages and they are also the most cited factors that influence the attitude and behavioral intentions of a perso [48]. People will use a system or technology if they believe it is useful in their work. Perceived ease of use is equally important because even if a technology is useful, users may forego its performance benefits if it is hard to use. [49] found that perceived ease of use and perceived usefulness accounted for 88% of the variance in behavioral intention. Perceived usefulness is related to factors such as new possibilities brought by the new technology [45], productivity that is with the new technology you can save money and make more money, convenient/time saver and indispensable for business [50].

4.6 Social Influence

[51] find that the influence of social norms on individuals' behavioral intentions in some cases is stronger than the influence of attitudes. The perception of societal norms may prevent a person's behavior in accordance to his or her personal attitudes. [42] argue that in addition to neighbors, there are some other sources of influence which affect the adoption of technology and these include relatives, friends and influential persons in the community. [25] in their study on Indian rural farmers noted that farmer's adoption of technology was more influenced by the neighbors' usage of that particular technology as well as media. Farmers want to share their farming experience and to do this, they rely mainly on educated family members, friends and neighbors who are either early adopters or have knowledge about the products and services.

5 Conclusion

The study has revealed that mobile technology adoption provides a basis for improvement in the agriculture performance as it plays a very pivotal role in enhancing the operations and activities of farmers. However, demographic factors, facilitating conditions, tech-service attributes, tech-service promotion, perceived usefulness and perceived ease of use and social influence have a strong influence on the adoption of this technology by farmers. These factors have to be addressed if this technology is to adopted and full utilized by farmers. The factors which affect technology adoption which have been discussed in this study can be useful for policy makers, technology start-up companies, service and technology designers, marketers and researchers having particular interest in technology acceptance.

6 Future Work

In future, other different non-demographic variables like risk orientation, individual characteristics and demographic variables like culture and ethnic background, social class and household composition can be considered for similar studies, but in this study, only a limited number of demographic and non-demographic variables were considered.

References

1. Alhassan, H., Kwakwa, P.A.: The use of mobile phones by small scale farmers in Northen Ghana: benefits and challenges. J. Entrepreneurship Manag. **1**, 40 (2012)
2. Aker, J.C.: Dial a for agriculture: using information and communication technologies for agricultural extension in developing countries. In: Conference Agriculture for Development-Revisited, pp. 1–2. Berkeley University of California (2010)
3. Amid: Zimbabwe Agriculture Sector Policy Draft 1, Harare: Ministry of Agriculture, Mechanisation and Irrigation Development, Government of Zimbabwe. May 2012a
4. Zyl, V.O.: The use of ICTs for Agriculture in Africa (2012)
5. Hassan, M.S., Hassan, M.A., Samah, B.A., Ismail, N., Shafrill, H.A.M.: Use of information and communication technologies among agri-based entrepreneurs in Malaysia (2008)
6. Chhachhar, A.R., Querestic, B., Khushk, G.M., Ahmed, S.: Impact of ICTs in agriculture development. J. Basic Appl. Sci. Res. **4**(1), 281–288 (2014)
7. Rashid, A.T., Elder, L.: Mobile phones and development: an analysis of IDRC- supported projects. Electron. J. Inf. Syst. Dev. Ctries. **36**(2), 1–16 (2009)
8. Mittal, S., Mehar, M.: How mobile phones contribute to growth of small farmers? Evid. India **51**(3), 227–244 (2012)
9. Jensen, R.: The digital provide: information (technology), market performance and welfare in the south indian fisheries sector. Q. J. Econ. **122**(3), 879–924 (2007)
10. De Silva, H., Ratnadiwakara. D.: Using ICT to reduce transaction costs in agriculture through better communication: a case-study from Sri Lanka (2008)
11. De Silva, H., Ratnadiwakara, D.: Using ICT to improve agriculture a case-study from Sri Lanka (2008)

12. Falolaand, A.: Constraints to use of mobile telephony for agricultural production in Ondo State, Nigeria (2017)
13. Abraham, R.: Mobile phones and economic development: evidence from the fishing industry in India. In: Information Technologies and International Development (2007)
14. Martin, B.L., Abbott, E.: Mobile phones and rural livelihoods: diffusion, uses, and perceived impacts among farmers in rural Uganda. Inf. Technol. Int. Dev. **7**(4), 17–34 (2011)
15. Chisita, C.T., Malapela, T.: Towards mobile agricultural information services in Zimbabwean libraries: challenges and opportunities for small scale farmers in utilizing ICTs for sustainable food production (2014)
16. Jehan, N., Aujla, K.M., Shahzad, M.: Use of mobile phones by farming community and its impact on vegetable productivity. Pak. J. Agric. Res. **27**(1), 58–63 (2014)
17. Musungwini, S.: Harnessing mobile technologies for sustainable development of smallholder agriculture in Zimbabwe: challenges and lessons learned (2016)
18. Odunze, D., Hove, M.T.: An Analysis of the impact of the use of mobile communication technologies by farmers in Zimbabwe. In: A Case Study of Esoko and EcoFarmer Platforms (2015)
19. Qiang, C.: Barriers to adoption of mobile technology in Agriculture (2012)
20. Qiang, C., Kuek, S., Dymond, A., Esselaar, S.: Mobile applications for Agriculture and Rural Development, ICT Sector Unit World bank (2012)
21. Venkatesh, N., Morris, M.G., Davis, G.B., Davis, F.D.: User acceptance of information technology: toward a unified view. MIS Q. **27**(3), 425–478 (2003)
22. Csoto, M.: Mobile devices in agriculture: attracting new audiences or serving the tech-savvy? (2015)
23. Islam, S.M., Gronlund, A.: Factors influencing the adoption of mobile phones among the farmers in Bangladesh: theories and practices (2011)
24. Crabbe, M., Standing, C., Standing, S., Karjaluoto, H.: An adoption model for mobile banking in Ghana. Int. J. Mobile Commun. **7**, 515–543 (2009)
25. Jain, A., Hundal, B.S.: Factors influencing mobile services adoption in rural India. Asia Pac. J. Rural Dev. **17**(1), 17–28 (2007)
26. Hultberg, L.: Women Empowerment in Bangladesh: A Study of the Village Pay Phone Program. Thesis: Media and Communication Studies, School of Education and Communication (HLK) Jönköping University, Spring Term (2008)
27. Kalba, K.: The adoption of mobile phones in emerging markets: global diffusion and the rural challenge. Int. J. Commun. **2**, 631–661 (2008)
28. Mattila, M.: Factors affecting the adoption of mobile banking services. J. Internet Bank. Commer. **8**, 0306-04 (2003)
29. Frimpong, G.: Comparison of ICT knowledge and usage among female distance learners in endowed and deprived communities of a developing country. J. E-Learning **6**, 167–174 (2009)
30. Jain, P.R.: Impact of demographic factors: technology adoption in agriculture. SCMS J. Indian Manag. **14**(3), 93–102 (2017)
31. Mittal, S., Mehar, M.: Socio-economic factors affecting adoption of modern information and communication technology by farmers in india: analysis using multivariate probit model. J. Agric. Educ. Ext. **22**(2), 199–212 (2015)
32. DiMaggio, P., Cohen, J.: Information inequality and network externalities: a comparative study of the diffusion of television and the internet. The Economic Sociology of Capitalism, Working paper, no. 31 (2003)
33. Fuglie, K.O., Kascak, C.A.: Adoption and diffusion of natural-resource-conserving agricultural technology. Rev. Agric. Econ. **23**(2), 386–403 (2001)
34. Alampay, P.: What determines ICT access in Philippines? (2003)

35. Mittal, S., Gandhi, S., Tripathi. G.: Socio-economic impact of mobile phone on indian agriculture. In: ICRIER Working Paper no. 246. International Council for Research on International Economic Relations, New Delhi (2010)
36. Nyamba, S.Y., Mlozi, M.R.S.: Factors Influencing the Use of Mobile Phones in Communicating Agricultural Information: A Case of Kilolo District, Iringa, Tanzania. Int. J. Inf. Commun. Technol. Res. **2**(7), 558–563 (2012)
37. Cheong, W.H.: Internet adoption in Macao. J. Comput. Mediated Commun. **7**(2), 11–18 (2002)
38. Sulaiman, A., Jaafar, N.I., Mohezar, S.: An overview of mobile banking adoption among the urban community. Int. J. Mobile Commun. **5**, 157–168 (2006)
39. Seneler, C.O., Basoglu, N., Daim, T.U.: A Taxonomy for Technology Adoption: A Human Computer Interaction Perspective. In: PICMET 2008 Proceedings, South Africa (2008)
40. Hobijn, B., Comin, D.: Cross-Country Technology Adoption: Making the Theories Face the Facts. FRB NY Staff Report, no. 169, June (2003)
41. Adesina, A.A., Baidu-Forson, J.: Farmers perceptions and adoption of new agricultural technology: evidence from analysis in Burkina Faso and Guinea. West Africa Agric. Econ. **13**, 1–9 (1995)
42. Kargin, B., Basoglu, N.: Factors affecting the adoption of mobile services. In: PICMET Proceedings, Portland, USA (2007)
43. Lu, N., Swatman, P.M.C.: The MobiCert project: integrating Australian organic primary producers into the grocery supply chain. J. Manuf. Technol. Manag. **20**, 887–905 (2009)
44. Sarker, S., Wells, J.D.: Understanding mobile handheld device use and adoption. Commun. ACM **46**(12), 35–40 (2003)
45. Carlsson, C., Hyvönen, K., Repo, P., Walden, P.: Adoption of mobile services across different technologies. In: 18th Bled eConference, Slovenia, 6–8 June (2005)
46. Doss, C.R.: Understanding farm level technology adoption: lessons learned from CIMMYT's micro surveys in Eastern Africa. In: CIMMYT Economics Working Paper, 03–07. Mexico, D.F.: CIMMYT (2003)
47. Cook, R.: Awareness and Influence in Health and Social Care: How You Can Really Make a Difference, p. 256. Radcliffe Publishing Ltd., Milton Keynes (2006)
48. Davis, F.D.: Perceived usefulness, perceived ease of use, and user acceptance of information technology. MIS Q. **13**(3), 319–340 (1989)
49. Agarwal, R., Karahanna, E.: Time flies when you're having fun: Cognitive absorption and beliefs about information technology usage. MIS Q. **24**, 665–694 (2000)
50. Donner, J.: The social and economic implications of mobile telephony in Rwanda: an ownership/access typology. In: Glotz, P., Bertschi, S., Locke, C. (eds.), Thumb culture: The Meaning of Mobile Phones for Society, pp. 37–52 (2005)
51. Stiff, J.B., Mongeau, P.A.: Persuasive Communication, 2nd edn. Guilford Press, New York (2003)

Context-Sensitive Case-Based Software Security Management System

Mamdouh Alenezi and Faraz Idris Khan(✉)

Security Engineering Lab, Prince Sultan University, Riyadh, Saudi Arabia
malenezi@psu.edu.sa, fikhan00@gmail.com

Abstract. An increasing number of security attacks on software have motivated the need for including secure development practices within the software development life cycle. With this urgent need, software security management system has received considerable attention and there are various efforts that can be found in this direction. In this paper, we highlighted the need for including application context-sensitive modeling within case-based software security management system proposed by the authors in [3]. Therefore, in this paper, we extend the previous work [3] to include application context modeling. The proposed idea constructs software security models using an application context.

Keywords: Software security · Context-aware security · Ontology-based security · Secure software development

1 Introduction

Recently, we have witnessed a huge number of businesses being automated and brought online by using state of the art web development technologies. E-commerce based applications are transforming business processes and along with that, security attacks on such applications have increased tremendously. Raising a number of security attacks has incurred a huge loss for the organizations, which are dependent on their e-commerce based applications for generating revenue. Security attacks affect the functionality of the application that usually leads to unavailability of the service on the internet that has a direct impact on customer satisfaction. These security attacks are experienced because of software flaws or vulnerabilities left during the software development process. Software development process usually lacks attention towards ensuring security within the product. Hence, most recently ensuring security within the software has received considerable attention from the research community. Moreover, the development team involved in developing software usually lacks security expertise. In order to tackle these security practices within the development, processes have gained popularity and much advancement in this domain is observed nowadays. Software security is a term used to describe security during the whole software development process.

In order to increase the security of the software, it is required that the software engineers are equipped with the necessary knowledge and required skills for secure software development lifecycle [1]. With this knowledge, software engineers can tackle security attacks and deal with security errors in a correct manner. The process is

R. Silhavy et al. (Eds.): CoMeSySo 2019, AISC 1046, pp. 135–141, 2019.
https://doi.org/10.1007/978-3-030-30329-7_13

complemented by security artifacts, which assist in understanding the security of the software. In order to help the software engineers to gain insight into the security of the software, there is a need for an automated system that manages the security knowledge and depending on the cases present recommendations to the software engineer.

As an example, SHIELDS project targets constructing secure software engineering environment that is assisted by the repository of the software security knowledge [2]. With the help of the repository security models can be shared and stored representing the expertise of the experts. The project provides a modeling tool but lacks the relationship between artifacts and software security knowledge. Hence, the authors in [3] proposed a management system that manages software security knowledge and security artifacts created during the development process and relate them as a case. The system assists software engineers who are not expert in security to model and analyze heterogeneous cases of software security. The work lacks application context related cases. Modeling software security knowledge in a context-sensitive manner using ontologies can be found in [4] where software security related knowledge is extracted by assessing the application context at hand.

Context awareness is essential for interactive applications. Human to the human conversation is more successful in conveying ideas due to the factors like the richness of language, common understanding of how things work in the world and implicit knowledge of everyday situations. Hence, humans are able to utilize implicit situational information that is often referred to as context. The ability to transfer ideas well is hampered when humans interact with computers due to the lack of inclusion of context. Human-computer interaction is unable to take full advantage of the context. With the improvement of computer's access to context, richness in communication is increased with which it is possible to develop beneficial computational services. Hence, efforts in including context within the application have received considerable attention from the research community and industry. This has led to a new domain of context-aware computing. Application developers include context within computing machines and develop applications that are aware and responsive to the context [5].

Context-aware security techniques have already received attention from the research community, there various efforts that can be found in literature such as in [6, 7]. The authors in [6] proposed a context-aware authentication technique for the internet of things. In [7] context-aware security management system for pervasive computing is proposed. The security management system consists of a context-aware role-based access control model. Context-aware software engineering artifacts can be found in literature such as in [8]. In order to improve software security processes application context awareness needs to be included in the software engineering artifacts used for modeling software security.

We anticipate the need for inclusion of application context sensitivity within the case-based management systems proposed in [3]. In this paper, we propose a context-sensitive case-based software security management system. The paper is organized as follows in Sect. 2 we discuss related work. In Sect. 3, we discuss our proposed idea. Section 4 concludes the paper.

2 Related Work

There are various efforts in the area of ontology modeling and using semantic technologies to assist software security. Ontology is being used to provide application context related to security information such as in [4]. Moreover, efforts to model security requirements using ontology can be found in literature such as in [9]. In another work, the authors in [10] proposed an ontology-based technique called SecWAO for securing the web application. In some of the research work, ontology is being used to design security and risk assessment. One such work can be found in [11]. An ontology-based approach to model security vulnerabilities can be found in [12]. The authors in [13] presented ontology of attacks which improves the detection capability of web attacks on web applications.

Lately, efforts have been made to include security-relevant context within software applications. Security relevant context provisioning access control and policy enforcement in emerging applications have already received attention from the researchers. One such work can be found in [14]. Context-aware security can also be found in distributed health care applications such as in [15]. The authors proposed a dynamic context-aware security infrastructure that has on-demand authentication, extensible context-aware access control, and dynamic policy enforcement. In another work in [16], the authors proposed context-aware authorization architecture. With the help of context-aware security, the hosts within the network are allowed or denied access to resources by assessing the perceived security of the host. Contextual policies deployment in policy enforcement points i.e. firewall, intrusion detection system, to enforce context-aware access control can be found in [17]. As a result of including context awareness changes of the network, devices configuration is automated and applied by assessing the context. In [18], the authors stressed new security mechanisms and techniques for context-aware applications. Hence, the context-aware role-based access control mechanism for context-aware applications is proposed. In literature, efforts can be found where contextual information is used to secure ad-hoc communication [19]. With the help of such contextual information, secure groups of mobile devices can be created which can initiate secure communication within the group. It is believed by the authors that this work can be the foundation for building secure infrastructures that use contextual information.

Software security is ensured by knowledge management systems, which usually present security recommendations on the basis of software security knowledge. Efforts in this direction can be found in [20]. Portals managing software security knowledge can also be found in literature such as in [21]. The portal assists software engineers to extract related software security knowledge useful in ensuring security within the software.

3 Proposed Idea

In this paper, we extend the work of the authors in [3] inspired by the work done in [4] to include application context related cases for software security produced by case based management systems proposed in [3].

Context-sensitive software security information can be modeled using ontology [4]. The proposed idea has three main steps. The first step is giving the ontology as input to the module, which selects the security pattern [4]. The second step is using the security pattern by Core Security meta-model module from Secure Engineering Process [22] with a set of functions for case-based management system as proposed in [3]. The third step is to produce context sensitive domain based model (Domain security metamodel) which will be used to produce case based security artifacts.

Elaboration of ontology-based security pattern component is shown in Figs. 1 and 2. The authors in [23] proposed an ontology-based approach for security pat-tern selection that can be used with domain-specific pattern representation as pro-posed in [24] to produce domain-specific security pattern. This domain-specific pat-tern is associated with a case produced by the case-based management system in Fig. 1.

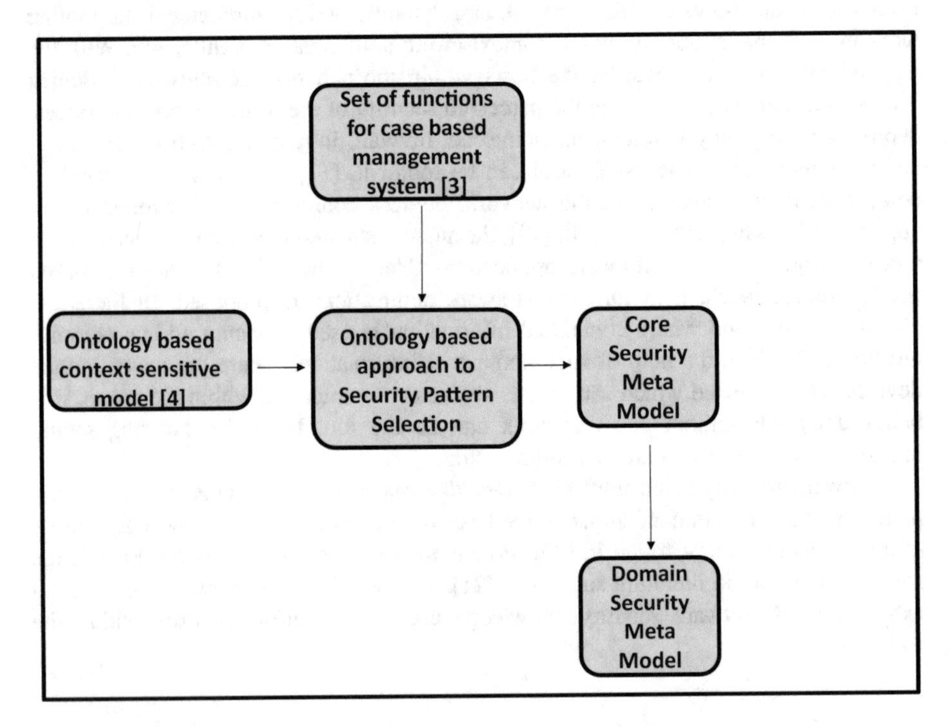

Fig. 1. Context sensitive case-based software security management system

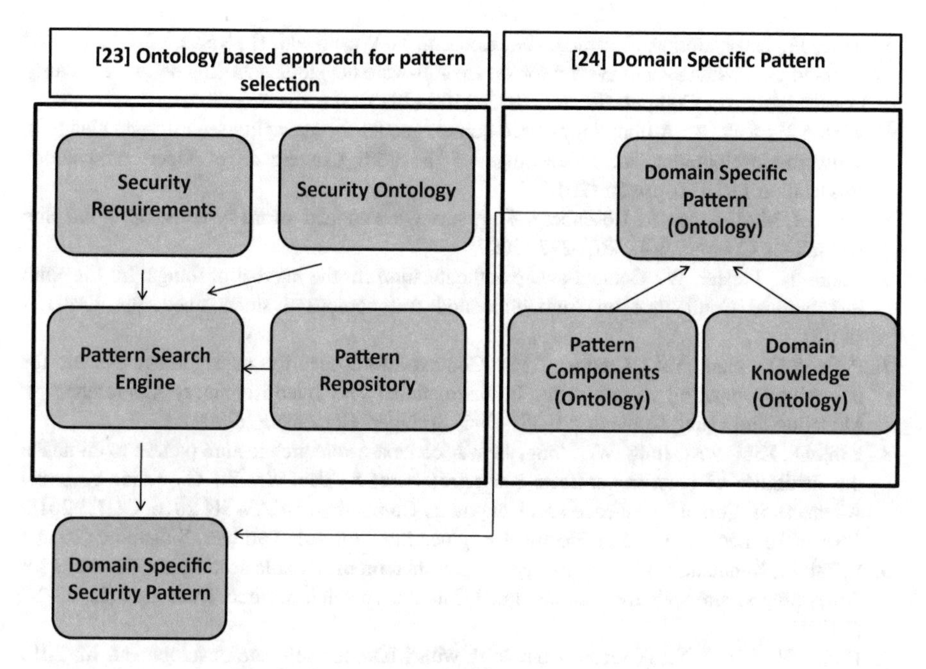

Fig. 2. Ontology-based security pattern selection

4 Conclusion

In this paper, we propose a context-sensitive case-based software security management system. Software developers and engineers lack software security expertise. Hence, there is a desire need to model software security that assists software engineers to analyze the security of the software. Lately, an increasing number of security attacks on web-based applications has raised concerns for the enterprises developing a commercial software. We identified a lack of context-sensitive information in the current software security management system. Therefore, we extend the previous work done by the authors in [3] inspired by the authors in [4] to include application context based software security modeling. In the future, we intend to improve the proposed idea and implement it in order to test its effectiveness.

References

1. Abunadi, I., Alenezi, M.: An empirical investigation of security vulnerabilities within web applications. J. UCS **22**(4), 537–551 (2016)
2. Hakon, P., Ardi, M.S., Jensen, J., Rios, E., Sanchez, T., Shahmehri, N., Tondel, I.A.: An architectural foundation for security model sharing and reuse. In: Proceedings of ARES 2009, pp. 823–828 (2009)

3. Saito, M., Hazeyama, A., Yoshioka, N., Kobashi, T., Washi-zaki, H., Kaiya, H., Ohkubo, T.: A case-based management system for secure soft-ware development using software security knowledge. Proc. Comput. Sci. **60**, 1092–1100 (2015)
4. Wen, S.F., Katt, B.: An ontology-based context model for managing security knowledge in software development. In: Proceedings of the 23rd Conference of Open Innovations Association FRUCT, pp. 56 (2018)
5. Baldauf, M., Dustdar, S., Rosenberg, F.: A survey on context-aware systems. Int. J. Ad Hoc Ubiquitous Comput. **2**(4), 263–277 (2007)
6. Habib, K., Leister, W.: Context-aware authentication for the internet of things. In: Eleventh International Conference on Autonomic and Autonomous Systems fined, pp. 134–139 (2015)
7. Park, S.H., Han, Y.J., Chung, T.M.: Context-aware security management system for pervasive computing environment. In International and Interdisciplinary Conference on Modeling and Using Context, pp. 384–396, Springer, Heidelberg, (2007)
8. Kim, D., Kim, S.K., Jung, W., Hong, J.E.: A context-aware architecture pattern to en-hance the flexibility of software artifacts reuse. In: Park, J., Pan, Y., Yi, G., Loia, V. (eds.) Advances in Computer Science and Ubiquitous Computing. UCAWSN 2016, CUTE 2016, CSA 2016. Lecture Notes in Electrical Engineering, vol. 421, Springer, Singapore (2016)
9. Salini, P., Kanmani, S.: Ontology-based representation of reusable security requirements for developing secure web applications. Int. J. Internet Technol. Secured Transact. **5**(1), 63–83 (2013)
10. Kang, W., Liang, Y.: A security ontology with MDA for software development. In: 2013 International Conference on Cyber-Enabled Distributed Computing and Knowledge Discovery, pp. 67–74. IEEE (2013)
11. Guo, M., Wang, J.A.: An ontology-based approach to model common vulnerabilities and exposures in information security. In: ASEE Southest Section Conference (2009)
12. Marques, M., Ralha, C.G.: An ontological approach to mitigate risk in web applications. In: Proceedings of SBSeg (2014)
13. Razzaq, A.: Ontology for attack detection: an intelligent approach to web application security. Comput. Secur. **45**, 124–146 (2014)
14. Covington, M.J., Fogla, P., Zhan, Z., Ahamad, M.: A context-aware security architecture for emerging applications. In: 18th Annual Computer Security Applications Conference 2002, pp. 249–258. IEEE (2002)
15. Hu, J., Weaver, A.C.: A dynamic, context-aware security infrastructure for distributed healthcare applications. In: Proceedings of the First Workshop on Pervasive Privacy Security, Privacy, and Trust, pp. 1–8 (2004)
16. Park, S.H., Han, Y.J., Chung, T.M.: Context-role based access control for context-aware application. In: International Conference on High Performance Computing and Communications, pp. 572–580. Springer, Berlin (2006)
17. Wullems, C., Looi, M., Clark, A.: Towards context-aware security: an authorization architecture for intranet environments. In: IEEE Annual Conference on Pervasive Computing and Communications Workshops, pp. 132–137 (2004)
18. Preda, S., Cuppens, F., Cuppens-Boulahia, N., Alfaro, J.G., Toutain, L., Elrakaiby, Y.: Semantic context aware security policy deployment. In: Proceedings of the 4th International Symposium on Information, Computer, and Communications Security, pp. 251–261. ACM (2009)
19. Shankar, N., Balfanz, D.: Enabling secure ad-hoc communication using con-text-aware security services. In: Workshop on Security in Ubiquitous Computing, in Proceedings of the Ubicomp, vol. 2002 (2002)

20. Barnum, S., McGraw, G.: Knowledge for software security. IEEE Secur. Priv. **3**, 74–78 (2005)
21. Mead, N.R., McGraw, G.: A portal for software security. IEEE Secur. Priv. **3**, 75–79 (2005)
22. Ruiz, J.F., Rudolph, C., Maña, A., Arjona, M.: A security engineering process for systems of systems using security patterns. In: 2014 IEEE International Systems Conference Proceedings, pp. 8–11. IEEE (2014)
23. Guan, H., Yang, H., Wang, J.: An ontology-based approach to security pattern selection. Int. J. Autom. Comput. **13**(2), 168–182 (2016)
24. Montero, S., Díaz, P., Aedo, I.: A semantic representation for domain-specific patterns. In: International Symposium on Metainformatics, pp. 129–140. Springer, Heidelberg (2004)

Performance Evaluation of Hardware Unit for Fast IP Packet Header Parsing

Danijela Efnusheva(✉)

Faculty of Electrical Engineering and Information Technologies,
Computer Science and Engineering Department,
Skopje, Republic of North Macedonia
danijela@feit.ukim.edu.mk

Abstract. Modern multi-gigabit computer networks are faced with enormous increase of network traffic and constant growth of number of users, servers, connections and demands for new applications, services, and protocols. Assuming that networking devices remain the bottleneck for communication in such networks, the design of fast network processing hardware represents an attractive field of research. Generally, most hardware devices that provide network processing spend a significant part of processor cycles to perform IP packet header field access by means of general-purpose processing. Therefore, this paper proposes a dedicated IP packet header parsing unit that allows direct and single-cycle access to different-sized IP packet header fields with the aim to provide faster network packet processing. The proposed unit is applied to a general-purpose MIPS processor and a memory-centric network processor core and their network processing performances are compared and evaluated. It is shown that the proposed IP header parsing unit speeds-up IP packet headers parsing when applied to both processor cores, leading to multi-gigabit network processing throughput.

Keywords: Header parser · IP packet processing · Memory-centric computing · Multi-gigabit networks · Network processor

1 Introduction

The rapid expansion of Internet has resulted with increased number of users, network devices, connections, and novel applications, services, and protocols in the modern multi-gigabit computer networks [1]. As technology has been advancing, the network connection links were gaining higher capacities, (especially with the development of fiber-optic communications) [2], and consequently the networking devices were experiencing many difficulties to cope with the increased network traffic and to timely satisfy the novel imposed requirements of high throughput and speed, and low delays [3].

Network processors (NPs) have become the most popular solution to the bottleneck problem for constructing high speed gigabit networks. Therefore, they

R. Silhavy et al. (Eds.): CoMeSySo 2019, AISC 1046, pp. 142–154, 2019.
https://doi.org/10.1007/978-3-030-30329-7_14

are included in various network equipment devices, such as: routers, switches, firewalls or IDS (Intrusion Detection Systems). In general, NPs are defined as chip programmable devices that are particularly tailored to provide network packet processing at multi-gigabit speeds [2–4]. They are usually implemented as application specific instruction processors (ASIP), with customized instruction set that is based on RISC, CISC, VLIW or some other instruction set architecture [3]. Over the last few years many vendors have developed their own NPs (Intel, Agere, IBM etc.), which resulted with many NP architectures existing on the market [3]. Although there is no standard NP architecture, most NP designs in general include: many processing engines (PE), dedicated hardware accelerators (coprocessors or functional units), adjusted memory architectures, interconnection mechanisms, hardware parallelization techniques (ex. pipelining), and software support [4,5]. NPs architecture design is an ongoing field of research, expecting that the NPU market will achieve strong growth in the near future. What is more, many new ideas, such as the NetFPGA architecture [6], or software routers [7] are constantly emerging.

The most popular NPs that are in use today include one or many homo- or heterogeneous processing cores that operate in parallel. For example, Intel's IXP2800 processor [8], consists of 16 identical multi-threaded general-purpose RISC processors organized as a pool of parallel homogenous processing cores that can be easily programmed with great flexibility towards ever-changing services and protocols. Furthermore, EZChip has introduced the first network processor with 100 ARM cache-coherent programmable processor cores [9], that is by far the largest 64-bit ARM processor yet announced. Along with the general-purpose ARM cores, this novel chip also include a mesh core interconnect architecture that provides a lot of bandwidth, low latency and high linear scalability.

The discussed NPs confirm that most of the network processing is basically performed by general-purpose RISC-based processing core (as a cheaper but slower solution) combined with custom-tailored hardware units (as more expensive but also more energy-efficient and faster solution) for executing some complex tasks like traffic management, fast table look-up etc. Therefore, if network packet processing is analyzed on general-purpose processing cores then it can be easily concluded that a significant part of processor cycles is spent on packet's headers fields access, especially when the packet's headers fields are non word-aligned. In such case, some bit-wise logical and arithmetic operations are needed in order to extract the field's value from the packet header, (i.e. parsing of the header) that should be further processed.

Assuming that network processing usually begins by copying the packet header into a memory buffer that is available for further processing by the processor, this paper proposes a specialized IP header parsing unit that performs field extraction operations directly on the memory buffer output, before forwarding the IP header to be processed by the processor. This way, the bit-wise logical and arithmetic operations for extraction of IP header fields that are non word-aligned, are avoided, and the packet header fields are directly sent to the processor's ALU in order to be further evaluated and inspected by the processor.

The proposed IP header parsing unit is applied to a general purpose MIPS processor [10] and a memory centric processor that operates with on-chip memory [11], and then the performance gain of IP header parsing speed for the both processors is compared, discussed and evaluated.

The rest of this paper is organized as follows: Sect. 2 gives an overview of different approaches for improving network packet processing speed. Section 3 describes the proposed IP header parsing unit and provides details about its design and way of operation. Section 4 presents and evaluates the simulations results of IP header parsing speed attained when the proposed parser is applied to a general purpose processor architectures (ex. MIPS and RISC-based memory centric processor). Section 5 concludes the paper, outlining the performance gain that is achieved with the proposed IP header parsing unit.

2 Approaches of Improving Network Processing Speed

NPs development starts in the late 1990-ies when network devices were insufficient to handle complex network processing requirements [2]. Generally, NPs are used to perform fast packet forwarding in the data plane, while the slow packet processing (control, traffic shaping, traffic classes, routing protocols) is usually handled in software by the control plane. NP operation begins with the receipt of an input stream of data packets from the physical interface. After that, the IP headers of the received packets are inspected and its content is analyzed, parsed and modified [12], so some IP header fields are validated, checksum is calculated, Time to Live - TTL is decremented etc. During the packet processing some specialized hardware units may be used to perform tasks like: classification of packets, lookup and pattern matching, forwarding, queue management and traffic control [3]. For example, the forwarding engine selects the next hop in the network where the packet is to be sent, while the routing processor deals with routing table updates. Once the IP address look up in the forwarding table is finished, the outport is selected and the packet is ready to be forwarded. Actually, the packet scheduler decides which packets should be sent out to the switching fabric and also deals with flow management.

According to the research that is presented in [13], the packet processing can be accelerated if some most time consuming network processing operations are simplified, and appropriate choices of routing protocol functionalities are made. So far many different approaches have been proposed, including: label concept used to accelerate the look-up operations, dedicated hardware units intended to perform complex and slow operations (ex. header checksum calculation), or several algorithms for faster routing table lookup. For example [14] proposes a route lookup mechanism that when implemented in a pipelined fashion in hardware can achieve one route lookup every memory access, while [15] presents a lookup scheme called Tree Bitmap that performs fast Hardware/Software IP Lookups. Moreover, the authors of the research given in [13] suggest an approach that avoids the slow routing table lookup operations, by use of the IP protocol source routing option. Other similar approach that also makes use of source routing is given in [16].

In general, network processing software is getting closer to the network processing hardware, such as in [17], where part of the packet processing is offloaded to application-specific coprocessors, which are used and controlled by the software. This way, the hardware handles the larger part of the packet processing, at the same time leaving the more complex and specific network traffic analyses to the general-purpose processor. As follows, a flexible network processing system with high throughput can be build. Some researchers also try to unify the view on the various network hardware systems, as well as their network offloading coprocessors, by developing a common abstraction layer for network software development [18].

When it comes to packet parsing, many proposed approaches make big use of FPGA technology, as it is very suitable for implementation of pipeline architectures and thus it is ideal for achieving high-speed network stream processing [19]. According to that, the reconfigurable FPGA boards can be used to design flexible multi-processing systems that adjust themselves to the current packet traffic protocols and characteristics. This approach is given in [20], where the authors propose use of PP as a simple high-level language for describing packet parsing algorithms in an implementation-independent manner. Similarly, in the research given in [21], a special descriptive language PX is used to describe the kind of network processing that is needed in a system, and then a special tool generates the whole multi-processor system as an RTL description. Afterwards, this system may be mapped to an FPGA platform, and may be dynamically reconfigured.

3 Design of Hardware Unit for IP Header Parsing

The basic idea for designing a dedicated hardware unit for IP packet header parsing is to provide direct and single-cycle access to each field of the IP packet header. When accompanied with a general purpose processor such IP header parsing unit should allow same access time for an IP packet header field as the access to any random memory word, even when the field is not word-aligned. It is expected that this approach would have huge impact on the IP packet processing speed, and thus would provide increased data throughput of the appropriate NP device.

In order to achieve single-cycle access, the proposed IP packet header parsing unit uses part of the memory address space to directly address various IP packet header fields. This technique is known as memory aliasing, and allows each IP header field to be accessed with a separate memory address value (field address). When the field address is sent to the IP header parsing unit, it is used to select the corresponding word from memory (headers buffer), where the given field is placed. Afterwards, the word is processed and depending on the field address, the value of the field is extracted. This process may include some operations such as: shifting of the word and/or modification of its bits. A schematic of the IP header parsing unit, used for performing read access to a single IP header filed, is given on Fig. 1.

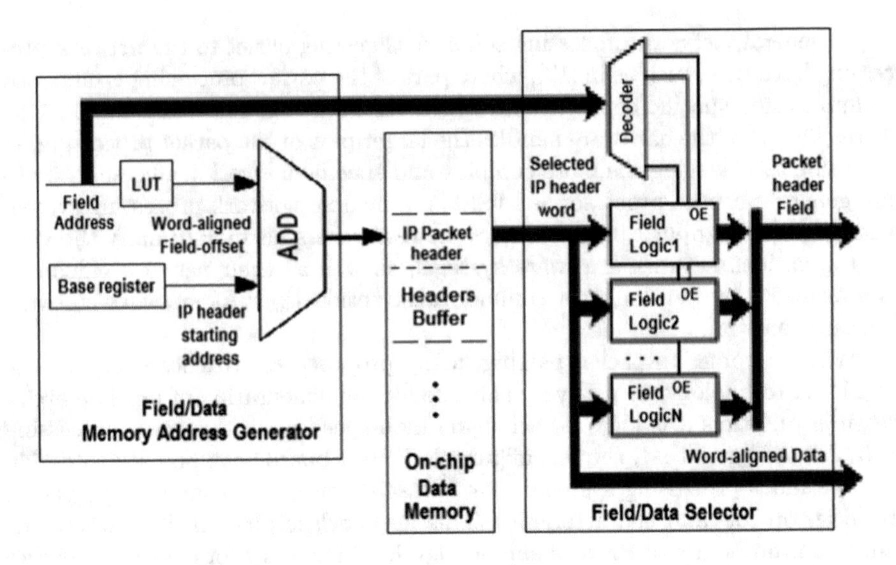

Fig. 1. Read access to an IPv4/IPv6 header field with the IP header parsing hardware unit

The IP header parsing unit is designed so that it assumes that the IPv4 or IPv6 packet headers are placed in a fixed area (headers buffer) of the memory, before they are being processed. The descriptions of the IPv4 and IPv6 packet headers that are supported by the IP header parsing unit include type of the IP header and its location in memory as first line, while each following line contains the definition of a single field. For each IP header field, the name and its size in bits are specified, whereas the fields are defined in the order as they appear in the IP header.

The IP packet header starting address that is specified in the IP header description is used to set the base register value inside the field/data memory address generator of the IP header parsing unit. This address generator module also receives a field address that is translated into a field offset by the lookup table (LUT). Actually, the field offset is a word-aligned offset to the starting IP header packet address, thus it points to the location where the given IP packet header field is placed in the headers buffer. This means that if the length of some field is smaller than the memory word length, then the closest word-aligned offset is placed in the LUT table for the given IP header field. For example, Field Offset for IPv4 fields placed in the first word of an IPv4 header (Version, Header Length, Type of Service and Total Length) is 0000h, while for the second word of an IPv4 header (Identifier, Flags and Fragment Offset) is 0001h etc.

According to Fig. 1, the selected field offset from the LUT table is added to the IP header starting address and the address of the memory word that holds the required IP header field is generated. This address is applied to the memory (headers buffer) and then the read word is forwarded to the field/data selector. This selector module consists of separated field logic (FL) blocks purposed to

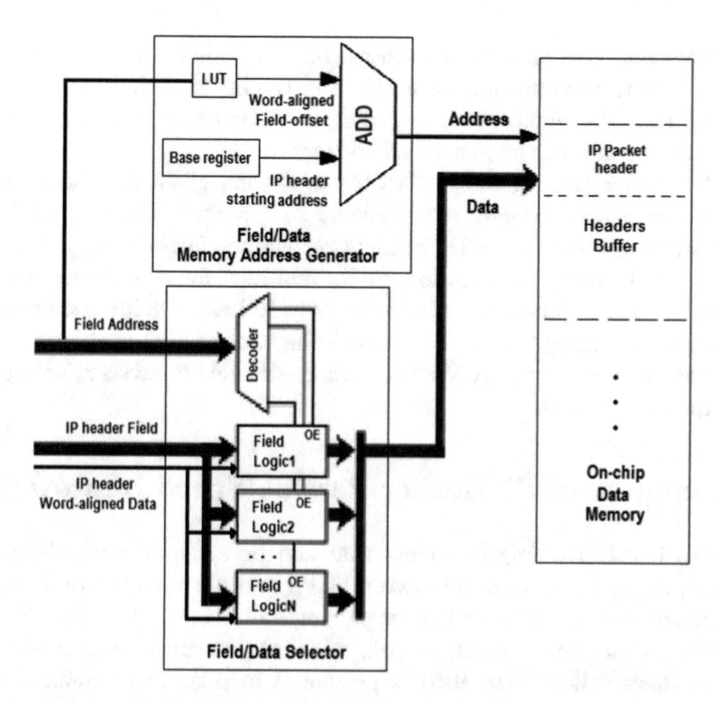

Fig. 2. Write access to an IPv4/IPv6 header field with the IP header parsing hardware unit

extract the value of the various IP header fields (FieldLogic1... N). In fact, each field is extracted with a separate field logic (FL) block that is activated by the output enable (OE) signal connected to a decoder output. The given decoder is driven by the field address, which causes only one of the FL blocks to be selected at a given moment. Afterwards, the selected FL block performs some bit-wise and/or shifting operations in order to extract and then zero-extend the appropriate IP header field. In the case when the IP header field is word-aligned, then its FL block is empty and the word is directly forwarded from the memory to the output of the field/data selector module.

The IP header parsing unit, presented on Fig. 1 shows the hardware that is used to read out a single IP header field from the headers buffer. The same concept is used for writing directly to an IP header field in the headers buffer, as shown on Fig. 2. According to Figs. 1 and 2 it can be noticed that the both modules use the same field/data memory address generator logic to generate the address of the memory word that holds the required IP packet header field that should be accessed.

The only difference between the two modules given on Figs. 1 and 2 is in the field/data selector logic, since the FL blocks of the parsing unit receive two inputs during writing: the IP header word-aligned data that was read from the memory and the IP header field that will be written to the memory. In order to provide write access to the required field, the decoder that is driven by the field address activates only one of the FL blocks. This FL block sets the input IP header field to the appropriate position in the input IP header word-aligned data. After that the whole word, including the written IP header field is stored in the headers buffer at the generated address.

The IP header parsing unit is flexible to design, given that there are well-defined packet header formats that should be supported. The proposed parsing unit currently operates with IP headers, providing further support for other packet header formats. In addition to the abilities for flexible extension, the presented hardware approach of direct access to IP header fields also brings much faster packet processing in comparison with the bare general-purpose processing, used by nearly all network processors. A more detailed analysis referring to this is given in the next section.

4 Estimation of IP Header Parsing Speed Improvement

In order to justify the improvements that can be achieved with the proposed IP header parser, a comparison between MIPS and memory-centric RISC-based processors without and with IP header parsing unit is made. Detailed description of the MIPS processor is given in [10], while the memory-centric RISC-based processor, (here called MIMOPS) is presented in [11]. The extended versions of the MIPS and MIMOPS processors include IP header parsing unit that is added next to the on-chip cache and data memory, accordingly. The following analysis compares the IP header parsing speed (for IPv4 and IPv6), achieved by the both processors when they operate without and with the IP header parsing unit, appropriately.

Figure 3 shows MIPS and MIMOPS assembly programs that parse an IPv4 header, without and with the IP header parsing unit. The given programs extract the IPv4 fields: version, header length, type of service, total length, identifier, flags, fragment offset, TTL, protocol, header checksum, source and destination IP addresses; and some fields that are not standard IPv4 fields (ex. first word first half, sec. word sec. half, etc.), but are used to simplify IPv4 header checksum calculation. The fields extraction basically involves bit-wise logical and shifting operations. The MIPS processor supports right and left arithmetical or logical shifts, specified as a separate instruction, where the shifting amount is given as a 5-bit constant. On the other hand, the MIMOPS processor provides shifting of the second flexible source operand and then execution of an ALU operation, using just a single instruction. Besides the shifting dissimilarities, the main difference between these two processors is that MIPS has to load the IP header from the cache memory into the GPRs in order to process it, while MIMOPS can operate directly on the IP header, placed in the headers buffer of its on-chip data memory.

The first program, shown on Fig. 3 implements IPv4 header fields access in the pure MIPS processor. This program uses the r0 register as a zero-value register, and the r1 register as a pointer for the header words. Additionally, the register r2 is purposed to hold the header words that are read from memory and further used to extract every IPv4 header field into registers r3–r21, appropriately. In

MIPS	MIPS with IP header parser	MIMOPS	MIMOPS with IP header parser
mov r0, #0	--Set Base Register	--Set Base Registers	--Set Base Register
mov r1, #0	la t0, Header	SetBRs Fields, Header, Header	SetBR1, Header
lw r2, Header[r1]			
--Version	--Version	--Version	--Version
and r3, r2, 15	lw r3, h0	and Fields[0], Header[0], 15, 110b	//h0
--Header Length	--Header Length	--Header Length	--Header Length
srl r4, r2, 4	lw r4, h1	lw t1, 15, 000b	//h1
and r4, r4, 15		and Fields[4], t1, Header[0] srl 4, 101b	
--Type of Service	--Type of Service	--Type of Service	--Type of Service
srl r5, r2, 8	lw r5, h2	lw t2, 255, 000b	//h2
and r5, r5, 255		and Fields[8], t2, Header[0] srl 8, 101b	
--First Word First Half	--First Word First Half	--First Word First Half	--First Word First Half
and r6, r2, 65535	lw r6, h3	and Fields[12], Header[0], 65535, 110b	//h3
--Total Length	--Total Length	--Total Length	--Total Length
srl r7, r2, 16	lw r7, h4	srl Fields[16], Header[0], 16, 110b	//h4
mov r1, #4			
lw r2, Header[r1]			
--Identifier	--Identifier	--Identifier	--Identifier
and r8, r2, 65535	lw r8, h5	and Fields[20], Header[4], 65535, 110b	//h5
--Flags	--Flags	--Flags	--Flags
srl r9, r2, 16	lw r9, h6	lw t3, 7, 000b	//h6
and r9, r9, 7		and Fields[24], t3, Header[4] srl 16, 101b	
--Fragment Offset	--Fragment Offset	--Fragment Offset	--Fragment Offset
srl r10, r2, 19	lw r10, h7	srl Fields[28], Header[4], 19, 110b	//h7
--Sec.Word Second Half	--Sec.Word Second Half	--Sec.Word Second Half	--Sec.Word Second Half
srl r11, r2, 16	lw r11, h8	srl Fields[32], Header[4], 16, 110b	//h8
mov r1, #8			
lw r2, Header[r1]			
--Time to Live	--Time to Live	--Time to Live	--Time to Live
and r12, r2, 255	lw r12, h9	and Fields[36], Header[8], 255, 110b	//h9
--Protocol	--Protocol	--Protocol	--Protocol
srl r13, r2, 8	lw r13, h10	and Fields[40], t2, Header[8] srl 8, 101b	//h10
and r13, r13, 255			
--Third Word First Half	--Third Word First Half	--Third Word First Half	--Third Word First Half
and r14, r2, 65535	lw r14, h11	and Fields[44], Header[8], 65535, 110b	//h11
--Header Checksum	--Header Checksum	--Header Checksum	--Header Checksum
srl r15, r2, 16	lw r15, h12	srl Fields[48], Header[8], 16, 110b	//h12
mov r1, #12			
--source IP address	--source IP address	--source IP address	--source IP address
lw r16, Header[r1]	lw r16, h13	add Fields[52], Header[12], 0, 110b	//h13
--Fourth Word First Half	--Fourth Word First Half	--Fourth Word First Half	--Fourth Word First Half
and r17, r16, 65535	lw r17, h14	and Fields[56], Header[12], 65535, 110b	//h14
--Fourth Word Sec. Half	--Fourth Word Sec. Half	--Fourth Word Sec. Half	--Fourth Word Sec. Half
srl r18, r16, 16	lw r18, h15	srl Fields[60], Header[12], 16, 110b	//h15
mov r1, #16			
--dest. IP address	--dest. IP address	--dest. IP address	--dest. IP address
lw r19, Header[r1]	lw r19, h16	add Fields[64], Header[16], 0, 110b	//h16
--Fifth Word First Half	--Fifth Word First Half	--Fifth Word First Half	--Fifth Word First Half
and r20, r19, 65535	lw r20, h17	and Fields[68], Header[16], 65535, 110b	//h17
--Fifth Word Sec. Half	--Fifth Word Sec. Half	--Fifth Word Sec. Half	--Fifth Word Sec. Half
srl r21, r19, 16	lw r21, h18	srl Fields[72], Header[16], 16, 110b	//h18

Fig. 3. Assembly programs that perform IPv4 header parsing in MIPS and MIMOPS processors, without and with IP header parsing hardware unit

the case of the Version field, only logical AND operation is needed to set all bits to zero, except the last 4 that hold the field's value. The Header length field on the other hand needs a shifting first. After that an AND instruction is used to select the last 4 bits of the shifted word, which hold the field's value. All the other fields are also retrieved by shifting and logical operations. The second program is an equivalent to the previous, except that it refers to a MIPS processor that operates with the IP header parsing unit. This program directly addresses the fields, by using mnemonics starting with the letter 'h' followed by the number of the field, as specified in the header description. Accordingly, only one instruction is needed to read out a header field to a register.

The third and the fourth program, shown on Fig. 3, implement IPv4 header fields access in the pure MIMOPS processor, and the MIMOPS processor that includes IP header parsing unit, accordingly. Referring to that, it can be noticed that the third program has many similarities with the first one, while the fourth program is similar to the second one. Although the third program (which is purposed for the MIMOPS processor) operates directly with the IP header words that are placed in the on-chip memory, it still has to extract the fields that are not word-aligned. The instructions that perform field's extraction can address up to three operands, where a 3-bit immediate value signifies which of the operands implement base addressing. The extracted fields are placed on continuous memory locations in the Fields array, so afterwards they can be directly accessed and processed by the ALU unit. On the other hand, the fourth program (which is purposed for the MIMOPS processor that includes IP header parsing unit) only has to set the base register to point to the starting address of the IP header, in order to provide direct access to the IP header fields (specified with mnemonics, as in the second program). According to that, an instruction that decrements the TTL (h9) field could be simply given as SUB h9, h9, 1, allowing the ALU unit to instantly process the extracted TTL field. This simplification could significantly speed-up the complete network processing of an IP packet.

Figure 4 shows MIPS and MIMOPS assembly programs that parse an IPv6 packet header, without and with the IP header parsing unit. The given programs perform parsing of an IPv6 packet header, by extracting its fields: version, traffic class, flow label, payload length, next header, hop limit, source IP address and destination IP address. These programs are very similar to the ones related to IPv4 header parsing and also consist of many bit-wise logical and shifting operations.

The comparative analysis between MIPS and MIMOPS processors that operate without or with IP header parsing unit is shown on Fig. 5. This analysis verifies that the proposed parsing unit improves the IPv4/IPv6 header parsing speed, providing impressing speed-up for the MIMOPS processor.

The results of the comparative analysis are given on Fig. 5, where Fig. 5a/b shows the execution time of an IPv4/IPv6 header parsing program, while Fig. 5c/d illustrates the IPv4/IPv6 parsing speed improvement that is achieved by the use of IP header parsing unit in MIPS and MIMOPS processors, accordingly. Referring to these results, it can be noticed that the MIPS processor that

MIPS	MIPS with IP header parser	MIMOPS	MIMOPS with IP header parser
mov r0, #0	*--Set Base Register*	*--Set Base Registers*	*--Set Base Reg.*
mov r1, #0	la t0, Header	SetBRs Fields, Header, Header	SetBR1, Header
lw r2, Header[r1]			
--Version	*--Version*	*--Version*	*--Version*
and r3, r2, 15	lw r3, h0	and Fields[0], Header[0], 15, 110b	//h0
--Traffic Class	*--Traffic Class*	*--Traffic Class*	*--Traffic Class*
srl r4, r2, 4	lw r4, h1	lw t1, 255, 000b	//h1
and r4, r4, 255		and Fields[4], t1, Header[0] srl 4, 101b	
--Flow Label	*--Flow Label*	*--Flow Label*	*--Flow Label*
srl r5, r2, 12	lw r5, h2	srl Fields[8], Header[0], 12, 110b	//h2
mov r1, #4			
lw r2, Header[r1]			
--Payload Length	*--Payload Length*	*--Payload Length*	*--Payload Length*
and r6, r2, 65535	lw r6, h3	and Fields[12], Header[4], 65535, 110b	//h3
--Next Header	*--Next Header*	*--Next Header*	*--Next Header*
srl r7, r2, 16	lw r7, h4	and Fields[16], t1, Header[4] srl 16, 101b	//h4
and r7, r7, 255			
--Hop Limit	*--Hop Limit*	*--Hop Limit*	*--Hop Limit*
srl r8, r2, 24	lw r8, h5	srl Fields[20], Header[4], 24, 110b	//h5
--source IP address	*--source IP address*	*--source IP address*	*--source IP address*
mov r1, #8	lw r9, h6	add Fields[24], Header[8], 0, 110b	//h6
lw r9, Header[r1]			
mov r1, #12	lw r10, h7	add Fields[28], Header[12], 0, 110b	//h7
lw r10, Header[r1]			
mov r1, #16	lw r11, h8	add Fields[32], Header[16], 0, 110b	//h8
lw r11, Header[r1]			
mov r1, #20	lw r12, h9	add Fields[36], Header[20], 0, 110b	//h9
lw r12, Header[r1]			
--dest. IP address	*--dest. IP address*	*--dest. IP address*	*--dest. IP address*
mov r1, #24	lw r13, h10	add Fields[40], Header[24], 0, 110b	//h10
lw r13, Header[r1]			
mov r1, #28	lw r14, h11	add Fields[44], Header[28], 0, 110b	//h11
lw r14, Header[r1]			
mov r1, #32	lw r15, h12	add Fields[48], Header[32], 0, 110b	//h12
lw r15, Header[r1]			
mov r1, #36	lw r16, h13	add Fields[52], Header[36], 0, 110b	//h13
lw r16, Header[r1]			

Fig. 4. Assembly programs that perform IPv6 header parsing in MIPS and MIMOPS processors, without and with IP header parsing hardware unit

implements IP header parsing unit provides a parsing speed improvement of 37.5/51.7% in comparison with a pure MIPS processor, while the MIMOPS processor that implements IP header parsing unit provides a parsing speed improvement of 95.6/93.7% in comparison with a pure MIMOPS processor, for IPv4/IPv6 header parsing, appropriately. The complete improvement of MIMOPS processor that includes IP header parsing unit, given on Fig. 5e/f shows that this processor achieves 96.8/96.5%, 95/92.8%, 95.6/93.7% better IPv4/IPv6 header parsing speed results, in comparison with the processors: MIPS, MIPS with IP header parser and MIMOPS, accordingly. This analysis verifies that the

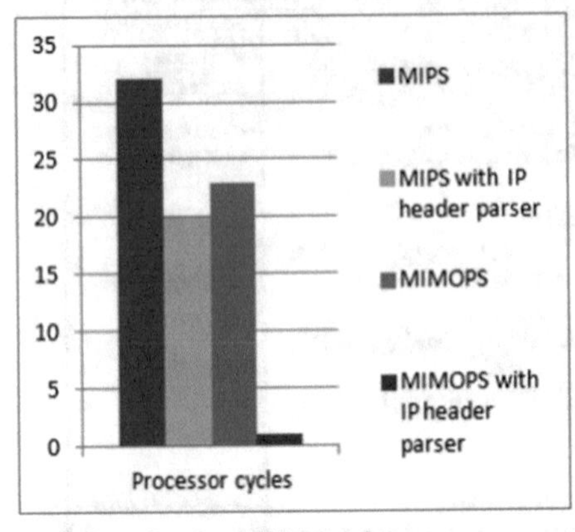

a) Execution time of IPv4 parsing

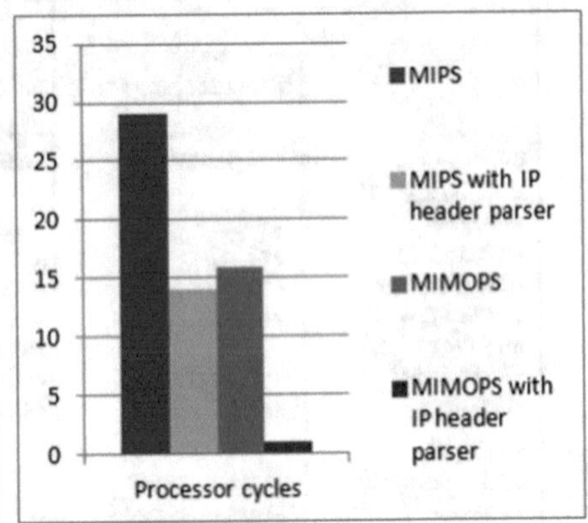

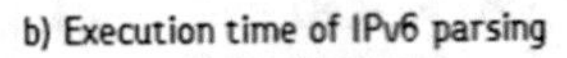
b) Execution time of IPv6 parsing

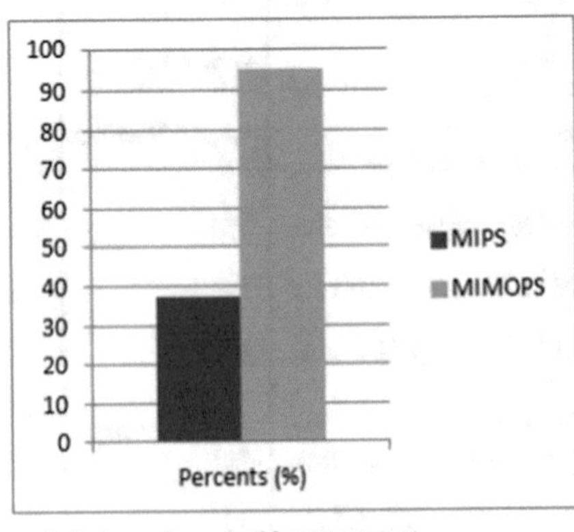

c) IPv4 parsing speed improvement

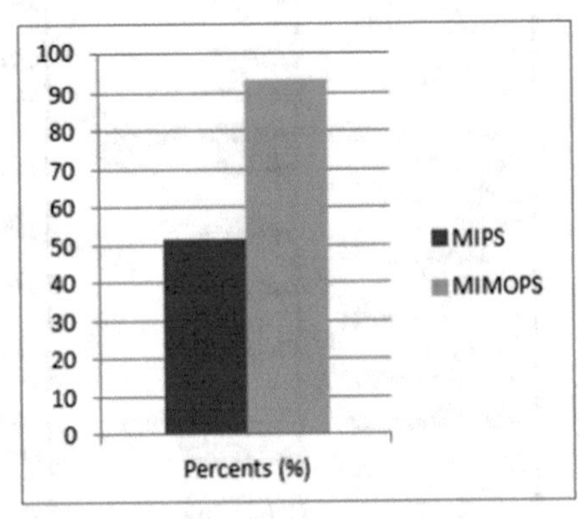

d) IPv6 parsing speed improvement

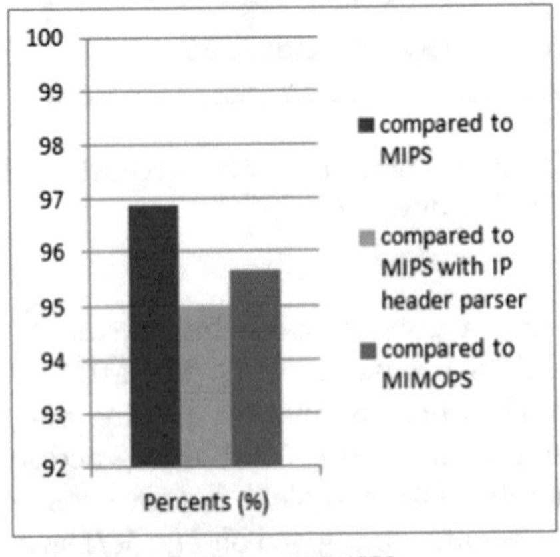

e) Overall Improvement of MIMOPS with IP header parser for IPv4 parsing

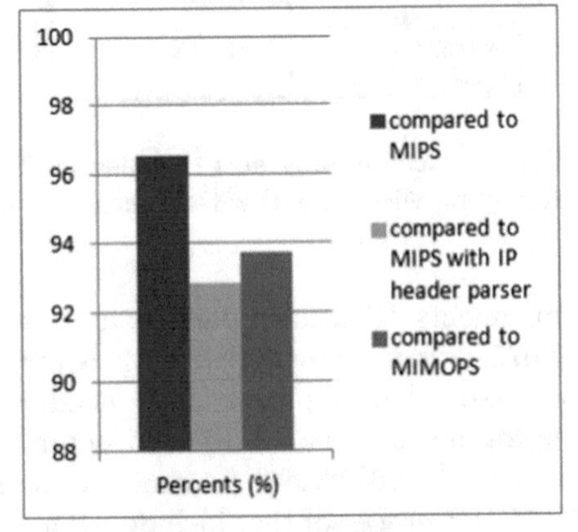

f) Overall Improvement of MIMOPS with IP header parser for IPv6 parsing

Fig. 5. Analysis of IPv4/IPv6 header parsing speed on MIPS and MIMOPS processors, without and with IP header parsing unit

proposed IP header parsing unit can generally improve the IPv4/IPv6 header parsing speed, on various general-purpose processor architectures.

5 Conclusion

This paper proposes an extension of MIPS and MIMOPS processors with specific hardware unit for IP header parsing, which adjusts them for network processing application. The proposed IP header parser is added next to the processor's on-chip memory (i.e. headers buffer) in order to perform field extraction operations directly on the memory output. Additionally, the IP header parsing unit uses an addressing scheme, which implements a memory aliasing technique to provide direct and single-cycle access to different-sized IP header fields. This approach allows the processor to avoid execution of logical or arithmetical operations, needed to extract the fields that are not word-aligned. Consequently, a MIPS processor that implements IP header parser achieves speed improvement of 37.5/51.7%, compared to a pure MIPS processor, while parsing IPv4/IPv6 headers. Additionally, a MIMOPS processor that implements IP header parser achieves speed improvement of 95.6/93.7%, compared to a pure MIMOPS processor, while parsing IPv4/IPv6 headers. Generally, it can be concluded that the proposed IP header parser provides speed-up of IP header parsing when applied in various processor architectures, showing impressive results for the memory-centric MIMOPS processor. Further extension of the parsing unit in order to provide support for other packet formats (not only IP) would require extension of its look up table and definition of novel FL blocks in the field/data selector. Therefore, one of the possible directions for future work is to make the proposed parser reconfigurable (for various packet header formats), by means of FPGA technology, which has proven as the best match for high speed and low price.

References

1. Giladi, R.: Network Processors - Architecture, Programming and Implementation. Elsevier, USA (2008)
2. Chao, H.J., Liu, B.: High Performance Switches and Routers. Wiley-IEEE Press, Canada (2007)
3. Ahmadi, M., Wong, S.: Network processors: challenges and trends. In: 17th Annual Workshop on Circuits, Systems and Signal Processing, pp. 222–232 (2006)
4. Lekkas, P.C.: Network Processors: Architectures, Protocols and Platforms. McGraw-Hill Professional, USA (2003)
5. Shorfuzzaman, M., Eskicioglu, R., Graham, P.: Architectures for network processors: key features, evaluation, and trends. In: Communications in Computing, pp. 141–146 (2004)
6. Naous, J., Gibb, G., Bolouki, S., McKeown, N.: NetFPGA: reusable router architecture for experimental research. In: SIGCOMM PRESTO Workshop. USA (2008)
7. Petracca, M., Birkea, R., Bianco, A.: HERO: high speed enhanced routing operation in software routers NICs. In: IEEE Telecommunication Networking Workshop on QoS in Multiservice IP Networks. Italy (2008)

8. Intel Corporation: IXP2800 and IXP2850 network processors. Product Brief (2005)
9. Doud, B.: Accelerating the data plane with the Tile-mx manycore processor. In: Linley Data Center Conference. USA (2015)
10. Patterson, D.A., Hennessy, J.L.: Computer Organization and Design: The Hardware/software interface. Elsevier, USA (2014)
11. Efnusheva, D., Tentov, A.: Design of processor in memory with RISC-modifed memory-centric architecture. In: Advances in Intelligent Systems and Computing. AISC, vol. 574. Springer, Heidelberg (2017)
12. Moestedt, A., Sjödin, P., Köhler, T.: Header Processing Requirements and Implementation Complexity for IPv4 Routers. White paper, UK (1998)
13. Hauger, S., Wild, T., Mutter, A., Kirstädter, A., Karras, K., Ohlendorf, R., Feller, F., Scharf, J.: Packet processing at 100 Gbps and beyond—challenges and perspectives. In: 10. ITG Symposium on Photonic Networks. Germany (2009)
14. Gupta, P., Lin, S., McKeown, N.: Routing lookups in hardware at memory access speeds. In: IEEE Conference on Computer Communications. pp. 1240–1247 (1998)
15. Eatherton, W., Varghese, G., Dittia, Z.: Tree bitmap: hardware/software IP lookups with incremental updates. Sigcomm Comput. Commun. Rev. **34**(2), 97–122 (2004)
16. Jakimovska, D., Dokoski, G., Kalendar, M, Tentov, A.: Network processor architecture design for multi-gigabit networks. In: VIII International Symposium on Industrial Electronics. BiH, pp. 357–362 (2010)
17. Kekely, L., Push, V., Kořenek, J.: Software defined monitoring of application protocols. In: IEEE Conference on Computer Communications. pp. 1725–1733 (2014)
18. Bolla, B., Bruschi, R., Lombardo, C., Podda, F.: OpenFlow in the small: a flexible and efficient network acceleration framework for multi-core system. IEEE Transact. Netw. Serv. Manag. **11**(3), 390–404 (2014)
19. Push, V., Kekely, L., Kořenek, J.: Design methodology of configurable high performance packet parser for FPGAs. In: 17th International Symposium on Design and Diagnostics of Electronic Circuits Systems. pp. 189–194 (2014)
20. Attig, M., Brebner, G.: 400 Gb/s programmable packet parsing on a single FPGA. In: 7th ACM/IEEE Symposium on Architectures for Networking and Communications Systems. pp. 12–23 (2011)
21. Brebner, G., Jiang, W.: High-speed packet processing using reconfigurable computing. IEEE Micro **34**(1), 8–18 (2014)

Classification and Reduction of Hyperspectral Images Based on Motley Method

Maria Merzouqi(✉), Elkebir Sarhrouni, and Ahmed Hammouch

Electronic Systems Sensors and Nanobiotechnologies (E2SN) Research Team, ENSET, Mohammed V University, B.P. 6207, Rabat, Morocco
merzouqimaria@hotmail.com, sarhrouni436@yahoo.fr, hammouch_a@yahoo.com

Abstract. The Hyperspectral image is a substitution of more than one hundred images of the same region called bands. To exploit the richness of this image it is necessary to reduce its dimensionality. The question is how to reduce it? This high dimensionality is also a source of confusion and adversely affects the accuracy of the classification bands and increases the compute time. Many methods are introduced to solve this confusion. Our proposed approaches are based on three-step. As a first principle, we start by scheduling according to a Mutual information criterion. Then the selection according to the filter in which we retried two filters MidMax for the first algorithm and MIFSU for the second. In the last step we added a wrapper strategy which is based on the probability of error. The classifier we had chosen is the Support Vector Machine (SVM). This algorithm gets high performance of classification accuracy. The study is established on HSI AVIRIS 92AV3C.

Keywords: Hyperspectral images · Classification · Feature selection · Mutual information · Error probability · Quadratic error redundancy

1 Introduction

The hyperspectral image represented as a cube of images where its third dimension in the spectral domain is defined by hundreds of spectral waves of length. It involves the principle that the choice of bands does not depend on the optimization; but it is necessary to differentiate between the selected bands. It contains hundreds of relevant and irrelevant bands. For the irrelevant bands, there are the redundant ones "that make the learning system difficult and produce a wrong prediction [11], and those affected by the atmospheric virus, which have a negative effect on the accuracy of the classification".

The reduction of the dimensionality is an important step to make use of this kind of data. There are, in fact, different methods of selection. One is selection by picking up the relevant bands only (selection from all or just the subsets). The other method is the extraction of new bands, from the original bands, that have a large amount of information about the classes, with the use of functions, either logical or numerical (Characteristic Selection Method) [8, 9]. The focal point of this work is the reduction of dimensionality of the hyperspectral images by the feature selection; in this regard we proposed a new algorithm based on three important parts. The first part deals with the

R. Silhavy et al. (Eds.): CoMeSySo 2019, AISC 1046, pp. 155–164, 2019.
https://doi.org/10.1007/978-3-030-30329-7_15

principle of ranking based on the mutual information gained. The second is on the selection of bands done by filters. The wrapper part is devised to ensure if the selected bands decrease the error probability or the quadratic error in order to test the attributes picked by the SVM classifier.

In this paper, we use the Hyperspectral image AVIRIS 92AV3C (Airborne Visible Infrared Imaging Spectrometer) [2]. It contains 220 images taken on the region of "Indiana Pine" in the "north-western Indiana", USA [3]. The 220 bands are taken between 0:4 μm and 2:5 μm. Each band has 145 lines and 145 columns. The ground truth map is also provided, but only 10366 pixels are labeled from 1 to 16. Each label indicates one from 16 classes. The zeros indicate pixels that are not classified yet (Fig. 1).

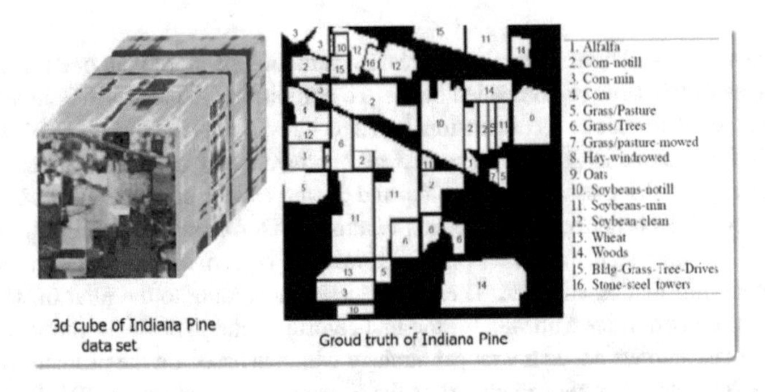

Fig. 1. 3D cube of Indiana pines data set (in the right), three-band

2 Methodology

This effort brings a new method that can be called motley method since it is based on several stages that are not of the same nature. The first step in this method is the ranking according to the Mutual Information Gain in order to ease the task of the selection to the filter, then the strategy wrapper for the precision. This motley method aims to retain the benefit of each step in terms of the precision of the classifications and the reduction of the calculation time.

The rest of this article will be presented as follows: The following is the part of the body on which we will detail the proposed method, then the interpretation of the results and a conclusion as the end of the work.

2.1 Mutual Information

The mutual information is generally applied to select bands for dimensionality reduction of HSI [13]. Sarhrouni [5] exploited this information to develop several schemes to maintain the integrity of the specifications in order to reduce the dimensionality applied to the HSI. Guo [4] used the mutual information in filters to select and classify HSI AVIRIS 92AV3C [3]. To quantify the amount of information contained by a random variable A Shannon entropy, denoted by H(A) [8], where P(A) is the probability density function of A.

$$H(A) = \sum\nolimits_{A} P(A) \log_2 P(A) \tag{1}$$

The mutual information is used to find the relation between two random variables [1]. In our case the taking bands are noted as (A) and the ground truth are noted as (B). When the value is high the bands are strongly related while the zero MI shows that they are independent

$$I(A,B) = \sum_{A,B} P(A,B) \log_2 \frac{P(A,B)}{P(A)P(B)} \tag{2}$$

2.2 Second Stage: Filter Method

Mid-Max Filter Based on PSNR

Simple mid-Maximum [15] Method: In this method, the resultant built image B_f is obtained by taking the band that has the maximum value of the mutual information and the ground truth while applying the following formula:

$$B_f = \sum\nolimits_{i=1}^{N} \sum\nolimits_{j=1}^{N} \max[A(i,j)B(i,j) \times 3] \div 4 \tag{3}$$

Where A and B are source images and B_f is a merged image that should minimize the PSNR.

Peak Signal-to-Noise Ratio (PSNR)

The PSNR is a report often used as a peak signal-to-noise ratio metric for verifying fused image quality. When PSNR has a significant value, it implies that the merged image is better. This report is defined in decibels, according to the following formula:

$$PSNR = 20 \log_{10} \frac{1}{REQM} \tag{4}$$

The Quadratic Error

(EQM) is a well-known parameter to evaluate the quality of the merged image. This parameter is calculated between B_f image and the ground truth and it is defined as:

$$REQM = \sqrt{EQM}$$

With:

$$EQM = \frac{(C - B_f)^2}{(m * n)} \tag{5}$$

Where C is the input image, B_f is the fused band to evaluate, m is the row of numbers and n is the number of columns. When the mean squared error becomes smaller, it indicates better fusion results.

MIFSU Filter

Kwak and Choi proposed an improvement of MIFS e, which they called MIFSU [14]. It is a filter method founded on statistical measures, the mutual information.

$$MIFSU = (C, A) - \beta \sum\nolimits_{A \in S} \frac{I(A, B)}{H(B)} I(B, C) \tag{6}$$

The principle of this filter is to select the bands which have the greatest value of mutual information and reduce the redundancy with the help of the formula that is already stated; the latter consists of two parts. The first defines the dependence between the chosen band and the ground truth, whereas the second part represents the redundancy information on the band A and the band B being already selected in S "by reference to the ground truth". The factor β is a factor controlling the penalty of the term of redundancy; the latter directly influences the selection of the bands.

2.3 The Last Stage: Performance Measures Using SVM-RBF

The control of the redundancy is a crucial step to deal with this type of image. Fano [13] confirmed in his Fano wrapper approach that the relationship between the mutual information and error probability is proportional, according to the following formula:

$$\frac{H(C / \mathrm{B_c}) - 1}{\log_2 N_C} \leq P_e \leq \frac{H(C / \mathrm{B_c})}{\log_2 N_C} \tag{7}$$

With:

$$\frac{H(C / \mathrm{B_c}) - 1}{\log_2 N_C} = \frac{H(C) - I(C; B_c) - 1}{\log_2 N_C} \tag{8}$$

The conditional entropy expression $H(C / \mathrm{B_c})$ is computed between the ground truth (that is, the C classes) and the sample band B_c. $\mathrm{N_c}$ is the number of classes. The value of mutual information of B_c and the probability of error varies inversely. This explains the decrease in the probability of error. When the B_c characteristics have a higher value of the MI with the ground truth, it implies that the selected band is closer to the truth on the ground.

3 Principle of Selection Based on the Filter Approach

Among the methods currently used to overcome the noticeable hyperspectral image constraint is the selection and elimination of redundant bands to reduce dimensionality and facilitate its deployment. This selection is made in particular by several methods such as the filters reproduced in our work, based on the measurement of the mutual information used as a metric to rank the band in descending order; then the control of the redundancy by the MidMax formula OR the MIFSU equation. It consists of introducing selection procedures independent of the learning algorithm that make this approach fast in terms of calculation time. Their physical form lies in their speed, with a high rate, but they also represent a number of limitations.

4 Principle of Selection According to the Motley Method

To face the challenge of high dimensionality of the HIS, several solutions are introduced. In our case the solution is always framed by selection of bands: There are two types of methods: the filter approach noted beforehand in our previous work which does not require feedback from the classifier and which estimates the performance of classification indirectly. Conversely, the wrapper approach regularly gives good results of subsets chosen directly according to the accuracy of the classification. Their reliance on the SVM [11] proved that it can produce better performances, although having the disadvantage of the high computational cost.

For this reason, we had the right to combine both approaches to benefit from their advantages. The selection will be specified on a particular filter based on the classification of MI as a common departure and on other criteria previously defined for each filter. Therefore, to increase the precision of classification we added wrapper Fano to improve the result obtained. We will then test the subset selected in the previous stage by the classifier SVM-RBF [4]. Therefore, we can conclude that they complete each other (Fig. 2).

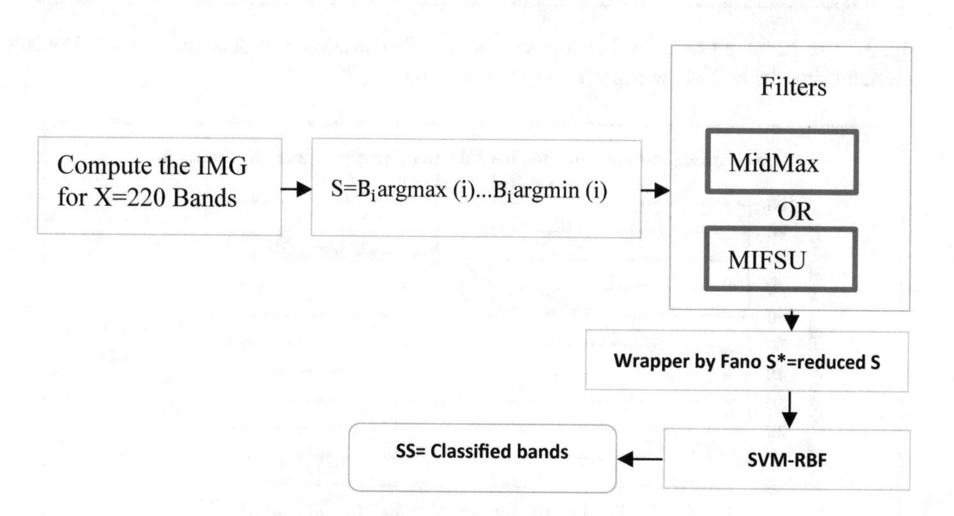

Fig. 2. The flowchart above illustrates the proposed method

These algorithms were applied on the Hyperspectral image AVIRIS 92AV3C [3], 50% of the labeled pixels are randomly chosen and used in training; and the other 50% are used for testing classification [4, 5, 7]. The classifier used is the SVM [10–13].

5 Findings and Data Analysis

The curves below illustrate the evolution of the results of the classification of the bands selected according to several thresholds. We can note the efficiency in the accurate selection of bands as well as the elimination of redundant bands using the methods proposed in the first two tables (Figs. 3, 4 and 5).

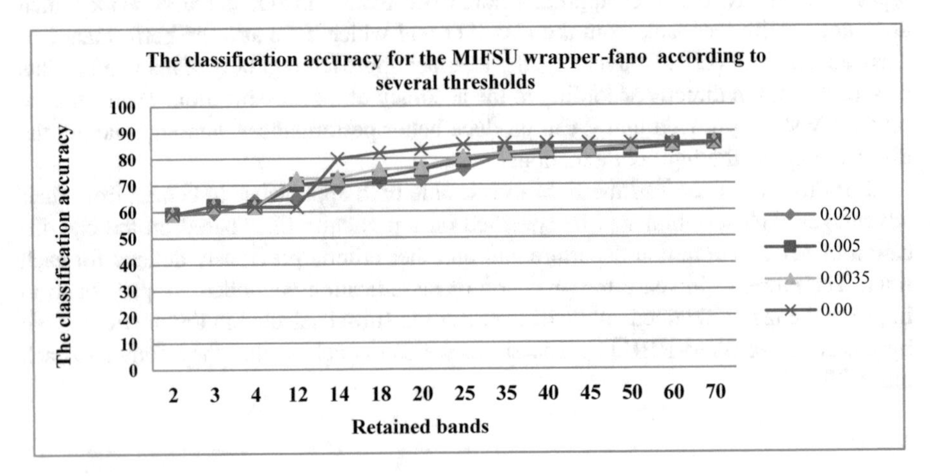

Fig. 3. The curve present the development of the classification rate according to the bands selected using the MIFSU wrapper Fano SVM method.

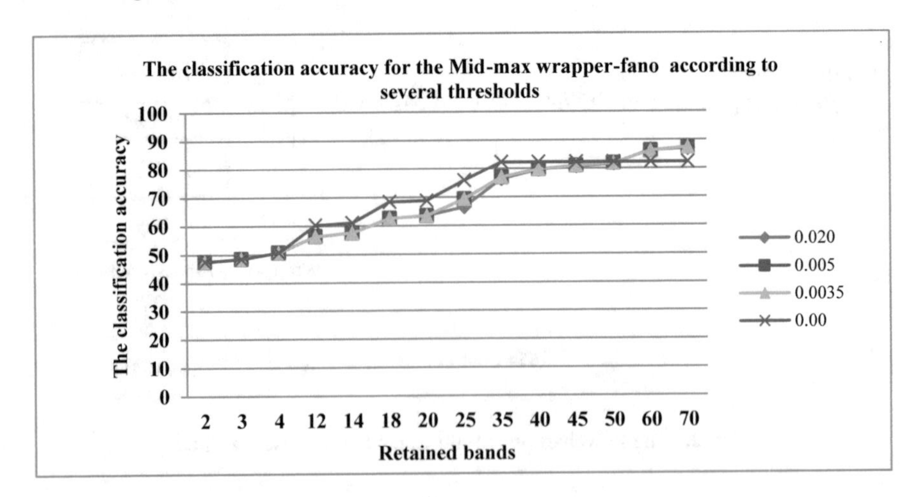

Fig. 4. The curve present the development of the classification rate according to the bands selected using the Mid-Max wrapper Fano SVM method.

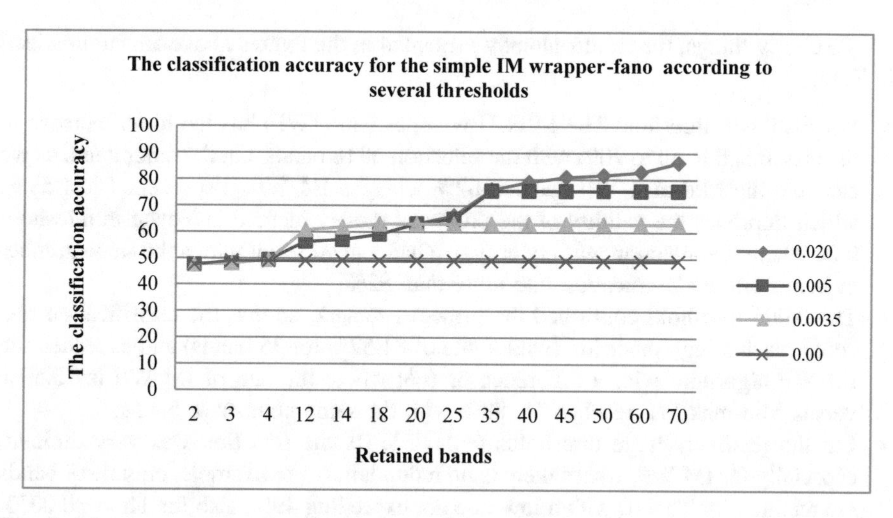

Fig. 5. The curve present the development of the classification rate according to the bands selected using the simple IM wrapper Fano SVM method.

5.1 Data Analyse

For the aim of the reduction of dimensionality of the hyperspectral images, we used almost three methods. To confirm the reliability of the methods, the classification rate achieved by the proposed algorithms was also compared with simple wrapper Fano [13].

The poor choice of bands affects the competing of these algorithms since it can result in misleading classification outputs. In this respect, a motley method is well implemented based on MIFSU and Mid-max filters after the ranking using the MI Gain and, as a third stage the wrapper Fano so that we take advantage of its precision on the results as well as manual thresholding to control redundant bands.

Table 1. The classification accuracy for each method separately according to several thresholds.

Numbers of bands	The classification accuracy for the IMG/Mid-max/MIFSU wrapper-Fano according to several thresholds											
	0.020			0.005			0.0035			0.00		
	IM-FW	Mid-max-WF	MIFSU-WF	IM-WF	Mid-max-WF	MIFSU-WF	IM-WF	Mid-max-WF	MIFSU-WF	IM-WF	Mid-max-WF	MIFSU-WF
2	47,44	47,52	58,93	47,44	47,52	58,93	47,44	47,52	58,93	47,44	47,52	58,93
3	47,87	48,49	59,52	47,87	48,49	62,09	47,87	48,49	62,09	48,92	48,49	62,09
4	49,31	50,67	63,8	49,31	50,67	61,72	49,31	50,67	61,72	*	50,67	61,72
12	56,3	56,3	64,85	56,3	56,3	70,23	60,76	56,3	72,37	*	60,26	79,84
14	57,00	57,53	69,16	57,00	57,53	71,48	61,8	57,53	72,59	*	61,04	80,30
18	**59,09**	**62,56**	**71,15**	**59,09**	**62,56**	**72,88**	**63**	**62,56**	**75,88**	*	**68,38**	**82,14**
20	63,08	63,45	71,85	63,08	63,45	75,59	*	63,45	76,49	*	68,75	83,44
25	66,12	66,51	75,67	64,89	69,24	79,08	*	69,24	80,6	*	75,84	85,25
35	76,06	76,27	81,2	75,59	76,88	81,57	*	76,88	81,96	*	82,23	85,6
40	78,96	79,56	81,9	*	79,66	82,21	*	79,66	83,27	*	*	*
45	80,85	80,65	82,47	*	80,89	83,01	*	80,89	83,42	*	*	*
50	81,63	81,57	83,21	*	81,78	83,17	*	81,78	84,75	*	*	*
60	82,74	86,34	84,38	*	86,15	84,88	*	86,15	85,47	*	*	*
70	86,28	86,87	85,1	*	87,32	85,64	*	87,32	86,09	*	*	*

To clarify things, the results already presented in the figures above are summarized in Table.

- For the 0.020 threshold The MIFSU wrapper Fano (WF) has led to an increase in the classification up to 70% with the selection of 18 bands. On the other hand, as we can see, the Mid-max WF reached 62% whereas IM WF [13] reached just 59%, which improves the validity of the proposed motley method and even in threshold 0.00 where the selection was not easy, we had a good result with a the same number of bands and a classification rate more than 82%.
- The 0.005 threshold confirmed the previous remark, so that the classification rate could reach a high precision (with a rate of 81.57% for 35 bands) and exceeded the IM WF algorithm with a difference of 6%; where the rate of IM WF is 75.59% versus Mid-max WF reaches 76. 88% with the same number of bands.
- For the relatively large thresholds (−0.0035, 0), the selection was very difficult, especially for IM WF, where there is no redundancy. For example, only three bands are retained for Th = 0 with a low rate not exceeding 49%, also for Th = −0.0035, where it cannot select more than 18 bands at an average rate 63%. On the other hand, the second proposed algorithms Mid-max WF increase the classification rate up to 82% or more with a conserved band of 35 for Th = 0.
- The proposed motley feature selection method enables to optimize feature subsets and reduce the computation cost while improving the classification accuracy. The number of bands used for classification was reduced from 220 to 18, while the classification accuracy increased from 59.96 to 82.14%.
- We demonstrate that the robustness of the proposed algorithm lies on its capacity to choose the relevant and control the redundant ones in a short time, which explains its positive impact on classification accuracy compared to the IM WF as shown in the Table 1.

5.2 Discussion

The goal we are trying to achieve is to reach a higher classification rate with a minimum of selected bands. Hence, as already presented in the curves, our method has confirmed its ability to accomplish the purpose for which it was introduced. Seeing that the Mid-max WF and MIFSU WF algorithms confirm their ability to find the global optimum in a short time with a high classification rate; and this reveals their superiority over the reproduced algorithm IM-WF [13]. The MIFSU WF also proves its effectiveness compared to the first proposed algorithm.

6 Conclusion

In this paper, we discuss the problems of Hyperspectral image which is represented on its high dimensionality. In brief, we have presented above an algorithms that reduces the number of bands in order to classify this HIS. This method is made by several principles. The first is the ranking based on the MI value, to facilitate the choice of bands in the next step which is the filtering stage, according to the two filters reproduced MIFSU or MidMax, which serves to select the noisy bands and make the set of attribute smaller than it was. Another criterion was introduced as last stage in our approach (the inequality of Fano). To put it differently, these measures the probability of error to differentiate between bands that can reproduce an estimated ground truth, and reduce the probability error by adding manually thresholds for the aim of removing the redundant bands before going to the final step in this stage, that involves the validation test by the SVM-RBF. The bands which can decrease the P_e and build the ground truth estimated close to the actual will be selected. We keep the useful redundancy by adjusting the threshold manually.

In conclusion, the result of the algorithms shows the credibility of motley method as a rigorous bands selection tool. It makes the advantage of both approaches. It's not as fast as a pure filter, but it can give better results than the wrapper. In addition, the computation time and complexity have been reduced by the filters used first, which provide a narrow subset of bands.

References

1. Huges, G.: On the mean accuracy of statistical pattern recognizers. IEEE Trans. Inf. Thaory **14**(1), 55–63 (1968)
2. ftp://ftp.ecn.purdue.edu/biehl/MultiSpec/
3. Guo, B., Gunn, S.R., Damper, R.I., Nelson, J.D.: Band selection for hyperspectral image classification using mutual information. IEEE Geosci. Remote Sens. Lett. **3**(4), 522–526 (2006)
4. Guo, B., Gunn, S.R., Damper, R.I., Nelson, J.D.: Customizing kernel functions for SVM-based hyperspectral image classification. IEEE Trans. Image Process. **17**(4), 622–629 (2008)
5. Sarhrouni, E., Hammouch, A., Aboutajdine, D.: Dimensionality reduction and classification feature using mutual information applied to hyperspectral images: a filter strategy based algorithm. Appl. Math. Sci. **6**(101–104), 5085–5095 (2012)
6. Gorretta-Monteiro, N.: Proposition d'une approche de'segmentation dimages hyperspectrales. Ph.D. thesis, Universite Montpellier II, Fvrier (2009)
7. Denton, E.R.E., Holden, M., Christ, E., Jarosz, J.M., Russell-Jones, D., Goodey, J., Cox, T. C.S., Hill, D.L.G.: The identification of cerebral volume changes in treated growth hormone-deficient adults using serial 3D MR image processing. J. Comput. Assist. Tomogr. **24**(1), 139–145 (2000)
8. Holden, M., Denton, E.R.E., Jarosz, J.M., Cox, T.C.S., Studholme, C., Hawkes, D.J., Hill, D.L.: Detecting small anatomical changes with 3D serial MR subtraction images. In: Hanson K.M. (ed.) Medical Imaging: Image Processing. Proceedings of SPIE, vol. 3661, pp. 44–55. SPIE Press, Bellingham (1999)
9. Otte, M.: Elastic registration of fMRI data using Bezier-spline transformations. IEEE Trans. Med. Imaging **20**(3), 193–206 (2001)
10. Chang, C.-C., Lin, C.-J.: LIBSVM: a library for support vector machines. ACM Trans. Intell. Syst. Technol. **2**, 27:1–27:27 (2011). Software available at http://www.csie.ntu.edu.tw/cjlin/libsvm
11. Hsu, C.-W., Lin, C.-J.: Comparaison des méthodes pour les machines à vecteurs supportant plusieurs classes. Transactions IEEE sur les réseaux de neurones **13**(2), 415–425 (2002)

12. Liu, H., Liu, L., Zhang, H.: Sélection des fonctionnalités à l'aide d'informations mutuelles: une étude expérimentale. In: Conférence internationale sur l'intelligence artificielle du Pacifique. Springer, Heidelberg (2008)
13. Sarhrouni, E., Hammouch, A., Aboutajdine, D.: Dimensionality reduction and classification feature using mutual information applied to hyperspectral images: a wrapper strategy algorithm based on minimizing the error probability using the inequality of Fano. Appl. Math. Sci. **6**(101–104), 5073–5084 (2012). (Cit#e en page 104). arXiv preprint arXiv:1211.0055
14. Kwak, N., Choi, C.H.: Input feature selection by mutual information based on Parzen window. IEEE Trans. Pattern Anal. Mach. Intell. **24**(12), 1667–1671 (2002)
15. Morris, C., Rajesh, R.S.: Techniques de fusion d'images primitives modifiées pour le domaine spatial. Informatologia **48** (2015)

Algorithm and Improved Methodology for Clustering Data with Self-learning Elements for Intrusion Detection Systems

Valeriy Lakhno[1], Timur Kartbaev[2], Aliya Doszhanova[2], Feruza Malikova[2], Zhuldyz Alimseitova[2(✉)], Sharapatdin Tolybayev[2], and Madina Sydybaeva[3]

[1] National University of Life and Environmental Sciences of Ukraine, Kiev, Ukraine

[2] Almaty University of Energy and Communications, Almaty, Kazakhstan
zhuldyz_al@mail.ru

[3] Kazakh Academy of Transport and Communications named after M. Tynyshpaev, Almaty, Kazakhstan

Abstract. The article proposes an algorithm with elements of self-learning for intrusion detection systems, as well as an improved methodology of data clustering recorded by the system, which relate to information security events. The proposed approaches differ from the known ones by the use of an entropy approach, which makes it possible to present data as homogeneous groups, and each such group (or cluster) can correspond to predetermined parameters. The proposed solutions concern the possibilities of estimating dynamic dependencies between clusters characterizing the analyzed classes of intrusions. During the research it was found that in case of a new sign of information security events, the corresponding scale describing the distances between the clusters also changes. In order to test the performance and adequacy of the proposed solutions there was conducted a computational experiment, the results of which confirmed the sufficient reliability of the results obtained in the work.

Keywords: Cybersecurity · Decision making support systems · Expert systems · Clusters

1 Introduction

The growing role of information and cybersecurity factors (hereinafter IS and CS) in the overall national security system (NS) of countries implementing the policy of digitalization of economies is fixed in the relevant strategies and doctrines of many countries, for example, the USA, the European Union, etc. [1–3]. The growing amount of cyber threats, as well as the amount and complexity of destructive, including targeted effects on information and telecommunications tools and information systems (hereinafter referred to as informatization objects – IO), makes the trend to strengthen IS and CS policies as a priority in the development of modern digital society. It should be noted that the policy of counteraction to the use of the potential of cyber-attacks in

R. Silhavy et al. (Eds.): CoMeSySo 2019, AISC 1046, pp. 165–173, 2019.
https://doi.org/10.1007/978-3-030-30329-7_16

the modern high-tech world is becoming one of the priority tasks that need to be solved for many states that focus on the development of information technology (IT).

Guided by the paradigm of providing IS and CS of IO, one of the key tasks is the task of resisting to destructive computer attacks. The last one can be achieved by using technical means, including intrusion detection tools (IDT).

In this context, one of the subtasks solved by IDT for the protected IO is the analysis of events related to IS in IO (hereinafter referred to as EIS). As a result of this analysis there should be formed recommendations and adjusted the information security policy. In this case, a natural contradiction arises between the large amount of registered network events in the IDT and the need promptly to correct the local IS policy for a particular IO. Currently used models, methods and tools for EIS analysis using database control systems (DBCS) do not always give the desired result [4]. This is especially obvious with the increase in IO scale [5].

The works conducted by many companies [2, 3] and independent researchers [4, 5] in the IS and CS of IO segments is connected with the study of the possibility of resolving the above mentioned contradiction. And as one of the solution is a variant of the application of data mining methods (DMM), clustering and machine learning methods in the tasks of large data arrays processing in IDT.

Therefore, we believe that the relevance and subjects of our study, caused by the need for additional study of the capabilities of DMM in IDT, are obvious. This becomes especially obvious if we consider the tasks of IS and CS of IO in the context of development of highly efficient tools for EIS network analysis.

2 Literature Review

Researches related to the use of DMM in IDT according to a number of authors is quite promising. Especially this task is relevant in order to reduce the amount of false alarms in IDT.

In [6] there was described a method based on the traditional Bayesian approach in combination with boosting [7]. But according to the authors, they did not overcome the problem of boosting, associated with the need for retraining.

In [8] there was considered the possibility of combining DMM with the methods of fuzzy logic (FL), and associative rules. The researches in this direction have not been completed by the authors.

In [9] it was proposed to combine the advantages of genetic algorithms (GA), FL, and cellular automata. In [9] the authors noted a reduction in the time spent on IDT learning. However, the problem of false alarms was not resolved.

The intrusion detection task for rare classes of attacks was considered in [10]. The authors note that the priority in their research is the ability to apply the developed algorithm in real time.

In [11] it was proposed to train IDT to be implemented on the basis of training rules and clustering methods. Work in this direction is underway.

In [12] there were considered the principles of using FL and GA in IDT. Work in this direction continues.

In [13] there was investigated the possibility of joint use of support vector and decision tree methods in attack detection problems. The authors came to the following conclusions: (1) the decision tree method is more preferable for attack detection of the classes Probe, U2R, R2L, however, the support vector method works better in situations of DoS/DDoS attacks detection.

However, as the analysis of the reviewed researches has shown, in general, the development of the problems of the use of DMM for IDE remains rather low [14, 15]. The above mentioned determines the relevance of chosen research topic, aimed at the development of methodological and algorithmic support at the study of EIS statistics.

3 Purpose of the Study

The purpose is to improve the methods and models for analytical data processing for intrusion detection systems at informatization objects.

In order to achieve the goal of the research it is necessary to solve the following tasks:

To develop an algorithm and to improve the method of data clustering with self-learning elements and taking into account the monitoring of the consistency of specific IS events recording time.

To develop a methodology that makes it possible to assess the levels of cyber threats for the informatization objects, guided by the expertise and retrospective data analysis associated with EIS.

4 Models and Methods

In order to analyze data related to EIS for large-scale IO, these data need to be divided into homogeneous groups – clusters [16]. During the process of clustering, as was shown in our previous researches [16, 17], the Shannon entropy measure and the Kullback-Leibler information-distance criterion can be used as a criterion for the effectiveness of training the decision support system in anomalies or attacks detection tasks.

The following criteria are used as criteria for processing events related to IS for IO: the amount of elements in a homogeneous (of the same type) data group (cluster); the value of the proximity of the elements in the group; the value of the dynamic dependence, which describes the consistency according to the EIR recording time for various clusters.

Each cluster corresponds to a derivative of a statistical entity. Network activity can be represented as a series of events recorded in time. Therefore, it is possible to compare the obtained clusters in pairs to determine the degree of consistency according to EIS time, which are included in different clusters. At the next stage of assessing the threats for IS and CS of various IOs, there is performed an examination and analysis of the retrospective data related to the EIS. And at the final stage, there is performed a projection of previous experience and accumulated information in knowledge bases, for example, an expert system or DSS [18, 19] to synthesize the current assessment, which characterizes the level of analyzed threats.

As an example, let consider the following basic features of cyber threats for IO: (1) the signature number; (2) and (3) IP addresses of the source and target, respectively; (4) and (5) port number of source and target, respectively.

Clustering was performed in several stages:

Stage 1. Determination of the most informative feature (hereinafter – IFt: *IS*) for the studied set of elements:

$$IS_i = \frac{IS_i^1}{IS_i^2}, \tag{1}$$

where

$$IS_i^1 = \sum\nolimits_{j \neq i} \left(\begin{array}{c} \sum_{n=1}^{s_i} p_{i(n)} \log_2 p_{i(n)} + \sum_{m=1}^{s_j} p_{j(m)} \log_2 p_{j(m)} - \\ \sum_{n=1}^{s_i} \sum_{m=1}^{s_j} \left(p_{i(n),j(m)} \log_2 p_{i(n),j(m)} \right) \end{array} \right);$$

$$IS_i^2 = \sum\nolimits_{j \neq i} \left(\sum\nolimits_{n=1}^{s_i} \sum\nolimits_{m=1}^{s_j} \left(p_{i(n),j(m)} \log_2 p_{i(n),j(m)} \right) \right);$$

i,j – feature numbers; $n = 1, \ldots, s_i$, $m = 1, \ldots, s_j$ – number of possible values for features i,j, respectively; $p_{i1}, \ldots, p_{i(s_i)}$, $p_{j1}, \ldots, p_{j(s_j)}$ – probabilities of occurrence of corresponding values in features i,j, respectively; $p_{i(n),j(m)}$ – probability of simultaneous occurrence in the element of values with numbers n, m for features i,j, respectively; $1 \leq n \leq s_i$; $1 \leq m \leq s_j$.

Then we will assume that the feature that has the highest value *IS* will contain the most information. Accordingly, we will consider this feature of threat, anomaly or attack as the most informative.

Stage 2. Determination of the most informative value (hereinafter – IVl) among the set of IFt:

$$IS_{i(n)} = \frac{IS_{i(n)}^1}{IS_{i(n)}^2}, \tag{2}$$

where

$$IS_{i(n)}^1 = \sum\nolimits_{j \neq i} (A - B);$$

$$A = \left(p_{i(n)} \log_2 p_{i(n)} + p_{\overline{i(n)}} \log_2 p_{\overline{i(n)}} + p_{i(n),j} \log_2 p_{i(n),j} + p_{\overline{i(n)},j} \log_2 p_{\overline{i(n)},j} \right);$$

$$B = \left(\begin{array}{c} \sum_{m=1}^{s_j^{i(n)}} \left(p_{i(n),j(m)} \log_2 p_{i(n),j(m)} \right) - \\ \sum_{m=1}^{s_j^{i(n)}} \left(p_{\overline{i(n)},j(m)} \log_2 p_{\overline{i(n)},j(m)} \right) \end{array} \right) - p_{\overline{i(n)},j} \log_2 p_{\overline{i(n)},j};$$

$$IS^2_{i(in)} = \sum\nolimits_{j \neq i} \left(\begin{array}{c} \sum_{m=1}^{s_j^{i(n)}} \left(p_{i(n),j(m)} \log_2 p_{i(n),j(m)}\right) + \\ \sum_{m=1}^{s_j^{i(n)}} \left(p_{\overline{i(n)},j(m)} \log_2 p_{\overline{i(n)},j(m)}\right) \end{array} \right) + p_{\overline{i(n)},j} \log_2 p_{\overline{i(n)},j};$$

$IS_{i(n)}$ – a parameter that determines whether the feature i is informative and corresponds to the value n; $p_{i(n)}$ – probability of occurrence of an element with a value n in the feature i; $p_{i(n),j}$ – probability of occurrence of an element with an arbitrary value for the feature i, (we assume that this occurrence of the value n has already been recorded at least once before for i); $p_{\overline{i(n)},j}$ – probability of occurrence of an element with an arbitrary value for the feature i, (we assume that this occurrence of the value n was not been recorded before for i); $p_{\overline{i(n)},j(m)}$ – probability of occurrence of an element with a value m for j, but different from the value m for i; $s_j^{i(n)}$ – amount of combinations for the values of n features with numbers i, j.

Stage 3. Sampling of the elements of the initial set, for which the value of IFt corresponds to the IVl [19–21].

Step 4. Determination of the homogeneity or uniformity parameter for the obtained sets:

$$ho = \frac{\sum_{j=1}^{k} z_{\max(j)}}{z \cdot k},$$

where z – amount of elements for the studied set; k – amount of features for a particular element; $z_{\max(j)}$ – maximum amount of elements with identical values of the feature with the number j.

Stage 5. The uniformity parameter is checked. If this parameter is less than the threshold value, then the obtained set is placed in the list of initial sets and again repeat steps 1–5.

Stage 6. A check of the amount of elements in the set is implemented. If their amount is less than the threshold value, then these elements are considered statistically insignificant. Unimportant elements are placed in a set of data that are nonclustered.

Stage 7. Those groups of elements that satisfy both threshold values, and correspond to the final clusters for which the template is formed. This template is stored in a database (DB) or in the KBof DSS or ES.

Further, there is defined a parameter that characterizes the consistency of information about the time of specific IS events recording. Moreover, this parameter is calculated in pairs for all clusters. The proposed algorithm is shown on Fig. 1.

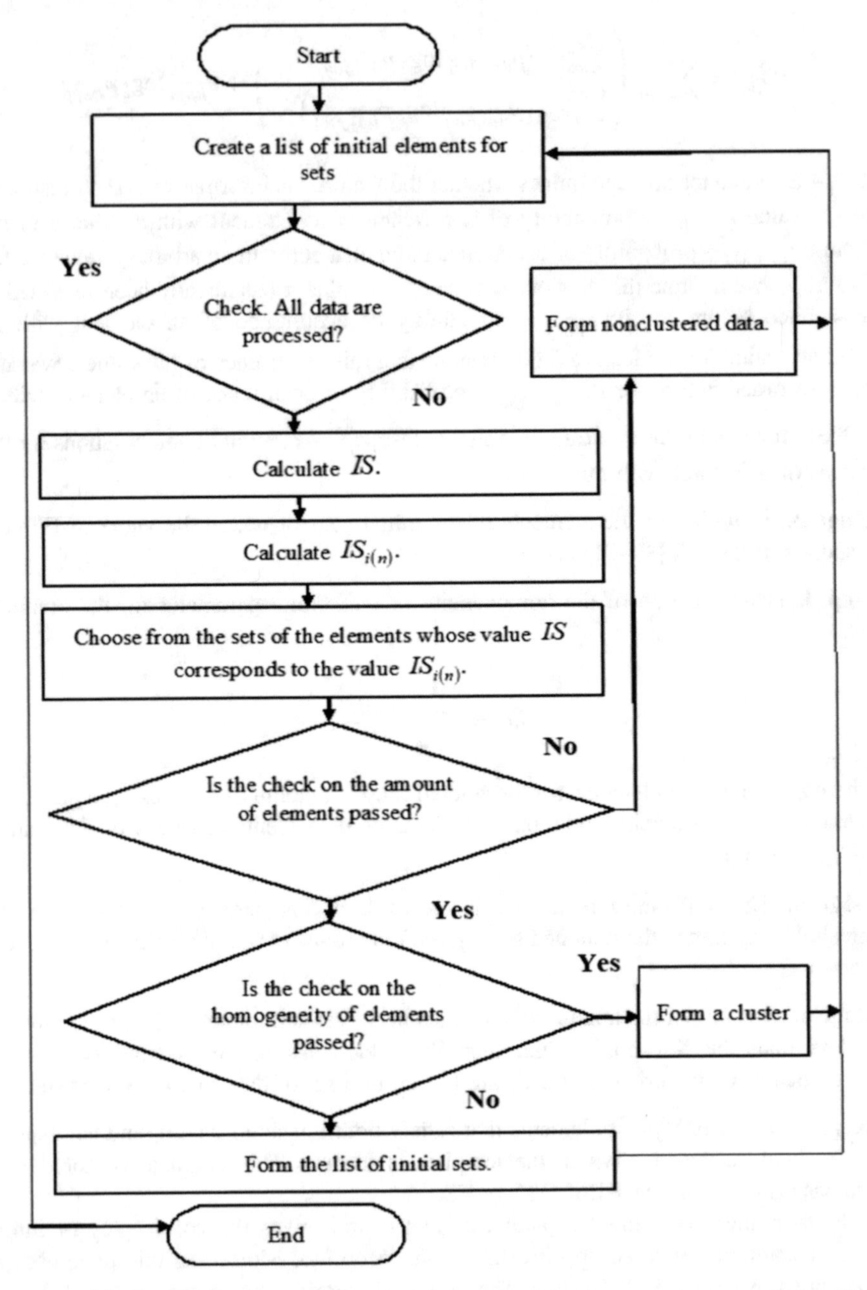

Fig. 1. The block diagram of the algorithm for the formation of clusters for the analysis of EIS

In order to confirm the correctness of the proposed algorithms, a computational experiment was performed.

5 Computational Experiment

The amount of initial data – 295 elements. As the initial data there were taken network EIS that were recorded using the Snort IDT [17, 18]. IP addresses were generated as random data. The results were determined for the following parameters: Z_{min} – min amount of elements in the analyzed class; HO_{min} – min threshold value of homogeneity of elements in the analyzed class. During the course of the computational experiment, it was assumed: $Z_{min} = 10$; $HO_{min} = 80$.

The results of the experiment are presented in Table 1.

Table 1. Clustering of EIS features

Template number	Analyzed features					Amount of elements
	Signature number	Source IP-address	Target IP-address	Source port number	Target port number	
	1	2	3	4	5	
1.	101	36.254.254.80	–	65536	443	32
2.	152	–	195.213.95.121	65536	25	13
3.	195	205.102.45.132	–	53	65536	16
4.	106	202.205.12.28	–	–	65536	34
5.	117	207.219.38.34	180.209.188.171	80	65536	28
6.	115	–	200.231.16.154	80	65536	16
7.	115	–	200.231.16.179	80	65536	11
8.	148	–	40.101.36.112	80	65536	13
9.	162	–	197.67.214.141	80	65536	13
10.	175	201.23.211.208	–	80	65536	22
11.	186	–	198.81.213.31	8080	65536	29
12.	186	–	198.81.213.31	80	65536	68

6 Discussion and Prospects for Further Research

Analysis of the computational experiment results allows to formulate the following conclusions:

The step-by-step calculation of the parameters of informative features and values (IFt, IVl), as well as the homogeneity of the sets (stages 1–4 of the clustering procedure) for the control sample made it possible to form sufficiently informative cluster data structures with characteristic attributes;

The calculated dynamic dependencies between clusters allow reliably to determine the sets of information security events that can become initial data for the subsequent automatic assessment of the current threats level recorded by the EIS for IO.

At the current stage of the research, we have limited ourselves with a relatively small sample of initial data for a computational experiment, which is a definite disadvantage of the work. Therefore, a logical continuation of the research in this direction will be an increase in the amount of initial data for the experiment and the development of software for automated analysis of EIS based on the proposed models and algorithms.

7 Conclusion

There was developed an algorithm with self-learning elements of IDT and was improved the clustering method of data recorded by the system. These methods are related to information security events and differ from the known ones by using an entropy approach that allows data to be presented as homogeneous groups, each such group (or cluster) can correspond to the predetermined parameters.

There was developed an algorithm and method for estimating dynamic dependencies between clusters characterizing the analyzed classes of intrusions and differing from the known ones by a model that characterizes the degree of consistency in time of events related to IS. It was found that in case of a new feature of the EIS, the corresponding scale describing the distance between the clusters also changes.

Acknowledgments. The work was carried out within the framework of the grant financing of the project AP05132723 "Development of adaptive expert systems in the field of cybersecurity of critically important informatization objects" (Republic of Kazakhstan).

References

1. Johanson, D.: The evolving US cybersecurity doctrine. Russ. J. Int. Secur. **19**(4), 37–50 (2013)
2. Harknett, R.J., Stever, J.A.: The new policy world of cybersecurity. Public Adm. Rev. **71**(3), 455–460 (2011)
3. Newmeyer, K.P.: Elements of national cybersecurity strategy for developing nations. Nat. Cybersecur. Inst. J. **1**(3), 9–19 (2015)
4. Bass, T.: Intrusion detection systems and multisensor data fusion. Commun. ACM **43**(4), 99–105 (2000). https://doi.org/10.1145/332051.332079
5. Lakhno, V., Kazmirchuk, S., Kovalenko, Y., Myrutenko, L., Zhmurko, T.: Design of adaptive system of detection of cyber-attacks, based on the model of logical procedures and the coverage matrices of features. Eastern-Eur. J. Enterpr. Technol. **3**(9), 30–38 (2016). https://doi.org/10.15587/1729-4061.2016.71769
6. Rahman, C.M., Farid, D.M., Rahman, M.Z.: Adaptive intrusion detection based on boosting and naïve Bayesian classifier. Int. J. Comput. Appl. **24**(3), 12–19 (2011). https://doi.org/10.5120/2932-3883
7. Jyothsna, V.V.R.P.V., Prasad, V.R., Prasad, K.M.: A review of anomaly based intrusion detection systems. Int. J. Comput. Appl. **28**(7), 26–35 (2011)
8. Harshna, N.K.: Fuzzy data mining based intrusion detection system using genetic algorithm. Int. J. Adv. Res. Comput. Commun. Eng. **3**(1), 5021–5028 (2014)
9. Sree, P.K., Babu, I.R.: Investigating cellular automata based network intrusion detection system for fixed networks (NIDWCA). In: International Conference on Advanced Computer Theory and Engineering, ICACTE 2008, pp. 153–156 (2008)
10. Dokas, P., Ertoz, L., Kumar, V., Lazarevic, A., Srivastava, J., Tan, P.N.: Data mining for network intrusion detection. In: Proceedings of the NSF Workshop on Next Generation Data Mining, pp. 21–30 (2002)
11. Chan, P.K., Mahoney, M.V., Arshad, M.H.: Learning rules and clusters for anomaly detection in network traffic. In: Managing Cyber Threats, pp. 81–99. Springer, Boston (2005)

12. Borgohain, R.: FuGeIDS: fuzzy genetic paradigms in intrusion detection systems. arXiv preprint arXiv:1204.6416 (2012)
13. Peddabachigari, S., Abraham, A., Thomas, J.: Intrusion detection systems using decision trees and support vector machines. Int. J. Appl. Sci. Comput. **11**(3), 118–134 (2004)
14. Pan, S., Morris, T., Adhikari, U.: Developing a hybrid intrusion detection system using data mining for power systems. IEEE Trans. Smart Grid **6**(6), 3104–3113 (2015). https://doi.org/10.1109/TSG.2015.2409775
15. Buczak, A.L., Guven, E.: A survey of data mining and machine learning methods for cyber security intrusion detection. IEEE Commun. Surv. Tutorials **18**(2), 1153–1176 (2016)
16. Lakhno, V., Tkach, Y., Petrenko, T., Zaitsev, S., Bazylevych, V.: Development of adaptive expert system of information security using a procedure of clustering the attributes of anomalies and cyber attacks. Eastern-Eur. J. Enterpr. Technol. **6**(9), 32–44 (2016). https://doi.org/10.15587/1729-4061.2016.85600
17. Akhmetov, B., Lakhno, V., Akhmetov, B., Alimseitova, Z.: Development of sectoral intellectualized expert systems and decision making support systems in cybersecurity. In: Intelligent Systems in Cybernetics and Automation Control Theory, CoMeSySo 2018. Advances in Intelligent Systems and Computing, vol. 860, pp. 162–171. Springer, Cham (2019)
18. Akhmetov, B., Lakhno, V., Akhmetov, B., Myakuhin, Y., Adranova, A., Kydyralina, L.: Models and algorithms of vector optimization in selecting security measures for higher education institution's information learning environment. In: Intelligent Systems in Cybernetics and Automation Control Theory, CoMeSySo 2018. Advances in Intelligent Systems and Computing, vol. 860, pp. 135–142. Springer, Cham (2019)
19. Lakhno, V., Akhmetov, B., Korchenko, A., Alimseitova, Z., Grebenuk, V.: Development of a decision support system based on expert evaluation for the situation center of transport cybersecurity. J. Theor. Appl. Inf. Technol. **96**(14), 4530–4540 (2018)
20. Akhmetov, B., Lakhno, V.: System of decision support in weakly formalized problems of transport cybersecurity ensuring. J. Theor. Appl. Inf. Technol. **96**(8), 2184–2196 (2018)
21. Al Hadidi, M., Ibrahim, Y.K., et al.: Intelligent systems for monitoring and recognition of cyber attacks on information and communication systems of transport. Int. Rev. Comput. Softw. **11**(12), 1167–1177 (2016)

Obtaining Generic Petri Net Models of Railway Signaling Equipment

İlker Üstoğlu[1], Daniel Töpel[2], Mustafa Seckin Durmus[3], Roman Yurievich Tsarev[4(✉)], and Kirill Yurievich Zhigalov[5,6]

[1] Control and Automation Engineering Department, Yildiz Technical University, Istanbul, Turkey
ustoglu@yildiz.edu.tr
[2] Sheffield Hallam University, Sheffield, UK
[3] Sadenco (Safe, Dependable Engineering and Consultancy), Antalya, Turkey
[4] Department of Informatics, Siberian Federal University, Krasnoyarsk, Russia
tsarev.sfu@mail.ru
[5] V.A. Trapeznikov Institute of Control Sciences of Russian Academy of Sciences, Moscow, Russia
[6] Moscow Technological Institute, Moscow, Russia

Abstract. This paper describes a model for railway signals in the context of interlocking. Since the design and construction of the software models shall be obtained in a formal way, the models were created using Petri nets to allow automatic verification and validation. A generic model for railway signals was designed and further utilized for the German railway signaling systems (The Ks signaling system). Simplified Petri net models for points, track segments and signals are given as a frame. The obtained models are also applied to a real station layout.

Keywords: Railway signaling · Ks-signaling system · Timed arc Petri nets

1 Introduction

Railways are said to be a safe transport system when compared to other means of transport over the land. The interlocking is a safety-critical system that controls the movement of trains in a station and between neighboring stations, and it needs to ensure that the system prevents dangerous situations.

In the railway engineering domain, Petri nets are used in some applications like supervisory control approach [1], scheduling [2], signaling systems [3], modeling and simulation [4] and developing interlocking [5]. Various more attempts have been made to model railway interlocking with Petri nets [6]. Doing this has two main reasons. The first is to ease verification and validation of the model. The second is to get one step closer to generating interlocking software, as done in [7]. This paper will concentrate on the elaboration of those parts of the Petri net which represent the railway signals. A basic interlocking model will also be included as a frame.

A Petri net of a railway signal is shown in [8–10] uses a simple signal model. In these nets, two points are missing: The first point is blown bulbs. For the case a missing

R. Silhavy et al. (Eds.): CoMeSySo 2019, AISC 1046, pp. 174–188, 2019.
https://doi.org/10.1007/978-3-030-30329-7_17

light would upgrade the signal aspect, measurements must be taken to prevent this situation. This could be done by designing signal aspect which is immune to such failures or by using a diagnose mechanism. The second point missing is the full functionality for distant signaling. In the cases mentioned above a signal changes its aspect according to the next signal's aspect when the route is opened, but after that no longer listens the evolution at the following signal. Whereas the signals above know only a small number of aspects, German railway signals know a much wider variety of signal aspects. Speed restrictions are specified in increments of 10 km/h and route letters can indicate a direction. Additionally three major signaling systems are currently in use. Further complexity is added by the fact that not all signals can show the same aspects. Therefore, the Petri net for every signal looks different.

This paper tries to find patterns to determine the characteristic attributes of a signal and to generate a Petri net from this information. Section 2 will give an overview on Petri Nets. The model is described in Sect. 3 and a concrete example is presented in Sect. 4. Finally, a conclusion is given in Sect. 5.

2 Preliminaries

2.1 Abbreviations and Notations

Ks system: Kombinationssignale (combination signal)
Hp system: Hauptsignale (main signal, West Germany signaling system)
Hl system: East Germany signaling system
PN: Petri net
TAPAAL: Tool for Verification of Timed-Arc Petri Nets
TAPN: Timed Arc Petri net
PNML: Petri net Markup Language
XML: Extensible Markup Language

2.2 Petri Nets and Timed-Arc Petri Nets

A Petri net is a kind of bipartite directed graphs consisting of four types of objects. These objects are transitions, places, directed arcs, and tokens. Transitions (the active components) symbolize the actions (signified by bars or rectangles), and they model the activities which can occur thus changing the marking of the Petri net. A marking can be considered as an assignment of tokens to the places of a Petri net.

The places, signified by circles, symbolize the conditions (=states) that need to be met before an action can be carried out. Directed arcs connect places to transitions or transitions to places. The places from which an arc runs to a transition are known as the input places of the transition and the places to which arcs run from a transition are called the output places of the transition. Places may contain tokens that may move to other places by executing firing actions.

Transitions are only allowed to fire if all the preconditions for the activity are fulfilled, i.e., there are enough tokens available in the input places. Recall that the tokens can represent the jobs performed, the flow control, resource availability or synchronization conditions. When the transition fires, it removes tokens from its input places and adds some at all of its output places. The number of tokens added or removed depends on the weight of each arc.

Let $\mathbb{N}$ denote the set of natural numbers and let $\mathbb{N} = \{0\} \cup \mathbb{N}$. The Petri net *PN* as a mathematical discrete event simulation modeling tool can be defined by a quintuple as follows:

$$PN = (P, T, Pre, Post, M_0) \tag{1}$$

- $P = \{p_1, p_2, \ldots, p_k\}$ is the finite set of places,
- $T = \{t_1, t_2, \ldots, t_z\}$ is the finite set of transitions,
- $Pre : (P \times T) \rightarrow \mathbb{N}$ represents the directed ordinary arcs from places to transitions,
- $Post : (T \times P) \rightarrow \mathbb{N}$ represents the directed ordinary arcs from transitions to places,
- $M_0 : P \rightarrow \mathbb{N}\ 0$ is the initial marking,
- $P \cap T = \emptyset$ and $P \cup T \neq \emptyset$.

Timed arc Petri nets as defined in [11, 12] are an extension to standard PNs which has invariants, inhibitor arcs (circle arrow tip, ⊸), enabling arcs (→) and transport arcs (diamond arrow tip, ◆), and are characterized by a tuple as follows:

$$TAPN = (P, T, F, c, F_{tarc}, c_{tarc}, F_{inhib}, c_{inhib}, t) \tag{2}$$

- $F \subseteq (P \times T) \cup (T \times P)$ is the set of normal arcs called the flow relation,
- $c : F|_{P \times T} \rightarrow I$ is the function assigning time intervals to arcs from places to transitions,
- $F_{tarc} \subseteq (P \times T \times P)$ is the set of transport arcs,
- $F_{inhib} \subseteq (P \times T \times P)$ is the set of inhibitor arcs,
- $c_{tarc} : F_{tarc} \rightarrow I\ \mathbb{N}\ 0$ is the function assigning time intervals to transport arcs,
- $t : P \rightarrow I$ is the function assigning invariants to places.

As mentioned previously, the places in the model may contain tokens, which are associated with a real numbers giving the age of the token. Invariants can be assigned to places that restrict the ages of tokens in that place. The normal arcs from places to transitions will consume tokens whereas normal arcs from transitions to places will produce tokens. The only difference of a transport arc is that any token produced will have the same age as the one consumed.

Inhibitor arcs are used to test for the absence of tokens of particular ages in places. In other words, they allow a transition to fire only when the inhibiting place is empty. Therefore it is a read-only arc. Recall that it is only used from places to transitions. The enabling arc is a bidirectional transport arc, giving back the used tokens unaltered, which makes it also a read-only arc.

As mentioned previously, the places in the model may contain tokens, which are associated with a real numbers giving the age of the token. Invariants can be assigned to places that restrict the ages of tokens in that place. The normal arcs from places to transitions will consume tokens whereas normal arcs from transitions to places will produce tokens. The only difference of a transport arc is that any token produced will have the same age as the one consumed. Inhibitor arcs are used to test for the absence of tokens of particular ages in places. In other words, they allow a transition to fire only when the inhibiting place is empty. Therefore it is a read-only arc. Recall that it is only used from places to transitions. The enabling arc is a bidirectional transport arc, giving back the used tokens unaltered, which makes it also a read-only arc.

2.3 PNML

The files read and written by TAPAAL are written in PNML. The general format of PNML is a labeled graph with two kinds of nodes: places and transitions [13]. PNML is based on the XML [14]. The Petri net, the objects, and the labels are represented as XML elements. A Petri Net file contains (in this case) shared places and nets. A net contains places, transitions, and arc [15].

2.4 The Ks Signaling System

A railway yard is made up of the following components: Train lines are split up into ***track segments***, and each segment is associated with an axle counter or a track circuit which can detect if a train is in the segment.

A ***point*** is a mechanical tool used to merge two lines into one line and lets the trains to be guided from one segment to another at a railway intersection. The two possible positions of a point are called normal and reverse. These two positions are determined at the mounting stage and cannot be changed furthermore. A train can drive over a point if it has been locked into a position physically.

Routes are sequentially-connected track segments that begin and end at signals. They are defined on the interlocking control tables at the beginning of the design phase. A ***signal*** is a device placed next to the track which uses color lights or other means to give information to the train driver. A signal aspect encodes the information provided to the train driver, for example *one green light above a yellow light*. A signal book describes the appearance and meaning of a signal aspect. A *signal information* is an actual information given to the driver, for example *proceed with* 40 km/h.

There are three types of signals: Main, distant and combined signals. Main signals show stop or an allowed speed and optionally a direction indicator and a left track indicator. A distant signal announces the information from the main signal. If two signals follow within breaking distance, a combined signal may show the distant information for the next signal. The signaling system presented in this paper is the Ks signal system used in Germany. This system was designed to replace the West German `Hp` system and the East German `Hl` system with a single new one and like the `Hl` signals, they combine the functions of main and distant signals in one single head by indicating the speed after the signal (`Zs3`, a white number above the signal) as well as the speed limit from the next signal (`Zs3v`, a yellow number below the signal). If such a yellow number is displayed then the green light will flash. Note that the `Ks` system has three lights; green, yellow and red. It uses one colored light which shows the number of open blocks. Here, red means no block is open, whereas yellow and green mean one and two blocks are free, respectively. A green light can also mean that information is only available for one block. At distant signals, the track section following the signal is treated like a block which will always be open, therefore never allowing a red light. For each block a speed and direction information can be given. Alphanumerical indicators are used to display that information. The following aspects defined in the signal-book [16] will be used here (See Fig. 1):

- `Hp0`: One red light, Stop,
- `Ks1`: One green light, Proceed,
- `Ks2`: One yellow light, Proceed, expect to stop,
- `Zs2`: A white letter, Route is set to a direction specified by the letter,
- `Ks2v`: A yellow letter, Route at the next signal is set to a direction specified by the letter,
- `Zs3`: A white number, Proceed with speed indicated by the number multiplied by 10 km/h,
- `Zs3v`: A yellow number, Expect speed restriction at the next signal with speed indicated by the number multiplied by 10 km/h,
- `Zs6`: A white symbol shaped like a backslash, Proceed on the left track.

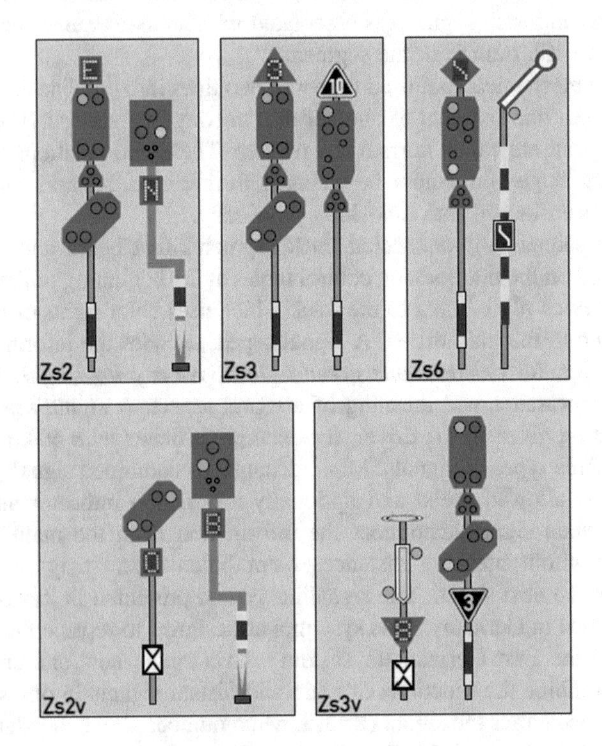

Fig. 1. Example of Ks signals.

3 Modeling

3.1 Signal Model

A signal net has subnets for each of the lamps. These subnets have a locally shared place (*indicated with dashed outer line*) for a request for lighting and turning off the lamp as well as for lit and unlit lamp. It also has a place for being in a malfunctioning state (See Fig. 2 and Table 1).

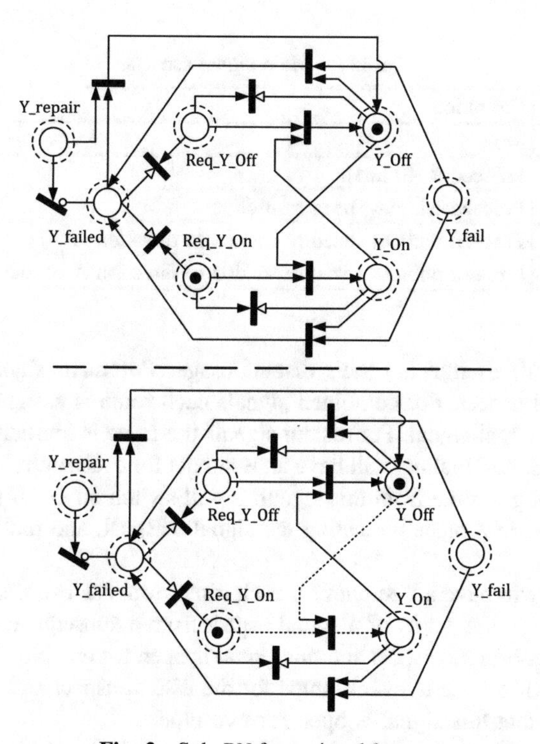

Fig. 2. Sub-*PN* for a signal lamp.

Table 1. Yellow lamp states.

States	Definition
Y_Off	Yellow lamp is unlit
Y_On	Yellow lamp is lit
Y_Failed	Yellow lamp is failed (and unlit)
Req_Y_Off	Yellow lamp is requested to be switched off
Req_Y_On	Yellow lamp is asked to be switched on
Y_Fail	Yellow lamp fails
Y_Repair	Yellow lamp is repaired

The signal aspect to be shown will be divided into main and distant aspect. All possible main aspects and all possible distant aspects will be listed. Each route requires the signal to show a specific main aspect. The net will be designed in a way that on the (granted) request of a route the lamps will be switched on or off as required by the main aspect. Therefore the main signal aspects have to be defined. They are prefixed H (=Hauptsignal). Next is the speed allowed in km/h, assuming a general limit of 160 km/h, followed if applicable by a direction letter and a "\" for the left track indicator. Table 2 shows some examples.

Table 2. Main signal aspects.

States	Definition
H0	Stop
H40	Proceed at 40 km/h
H160	Proceed without speed restriction
H160A	Proceed without speed restriction to direction A
H160A\	Proceed without speed restriction to direction A on the left track

V (=Vorsignal) to indicate their distant usage. For main signals, each route is assigned a signal aspect. For combined signals each route is assigned a signal aspect and optionally a target signal. For distant signals the route is implicit and only a target signal is assigned. A signal net will have an H-subnet for each route. This subnet will be accessed by setting a route from this signal. It will switch on or off the relevant lamps and leave a token in a place indicating the signal's aspect, and that it is open for one block.

A signal net will have a V-subnet for each route from each subsequent signal. Each subnet will observe the status of a signal aspect from a subsequent signal. When that signal shows the observed aspect and this signal is open for one block to that signal, the net switches on or off the relevant lamps for the distant aspect and leave a token in a place indicating that this signal is open for two blocks.

A signal net will have one 0-subnet. This subnet will be accessed by closing this signal. It will switch on or off the relevant lamps and leave a token in a place indicating the signal's stop aspect, and that it is closed. An example for a full signal net for a given signal will be shown in Sect. 4.

3.2 Point Model

Figure 3 shows a Petri Net similar to the yellow lamp model from Fig. 2. The model represents an item, which can be in one of two states. It can be requested to change to each of the states without checking before if it already is in this state. The net does not check if the change of status is desired. Additionally, it has states which represent the time delay while the point is moving.

It is assumed that the point is in Normal position initially. By an incoming position request (Req_W1_R) the token in W1_Normal moves to place N_to_R. Later, the token in place N_to_R moves to place W1_Reversed by an incoming position sensor information from the railway field. Requesting both position at the same time is prohibited by using the enabling arcs.

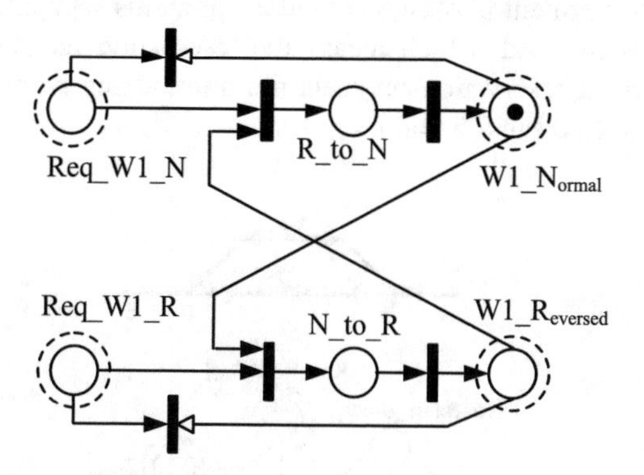

Fig. 3. Point model.

3.3 Track Model

A track segment will maintain information about a physical occupation and a logical reservation (see Fig. 4). An occupation is caused by the train entering or leaving the segment. The occupancy detector type is unspecified. The only information gathered from the detector is if a train is in the track segment. A reservation is caused by the interlocking. Each track segment can only be used by one route at a time. A free track segment can be reserved by a route. After it has been occupied and unoccupied again, the reservation will be released.

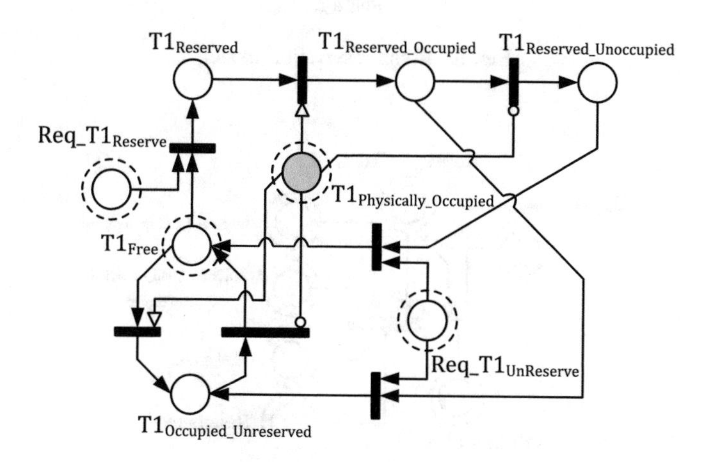

Fig. 4. Track model.

3.4 Route Reservation and Route Release Model

The interlocking checks if the required track segments are free, i.e., neither reserved nor occupied. If that is the case, the segments are reserved, and the points are set. After the reservations and points are confirmed, the signal is opened and the route marked as open. A Place `Dispatcher_busy` prevents other routes from setting at the same time. When the last segment is occupied, and all previous segments have been unoccupied, the route is released, which means the reservation on the segments will be removed. The segment reservation prevents the interlocking from setting conflicting routes automatically (see Figs. 5 and 6).

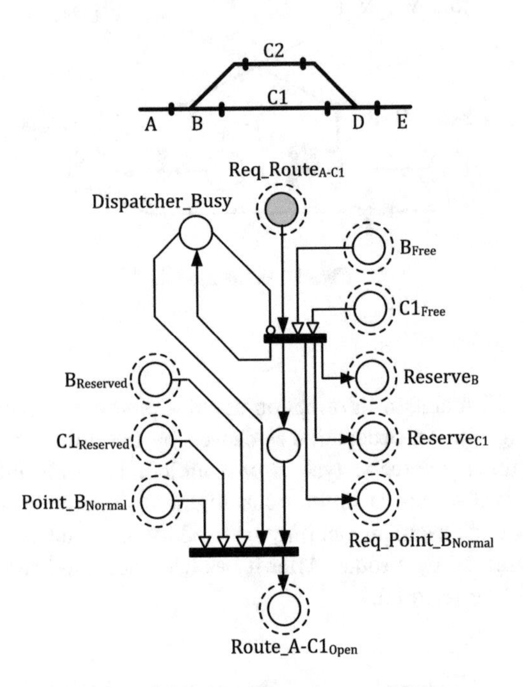

Fig. 5. Route reservation model.

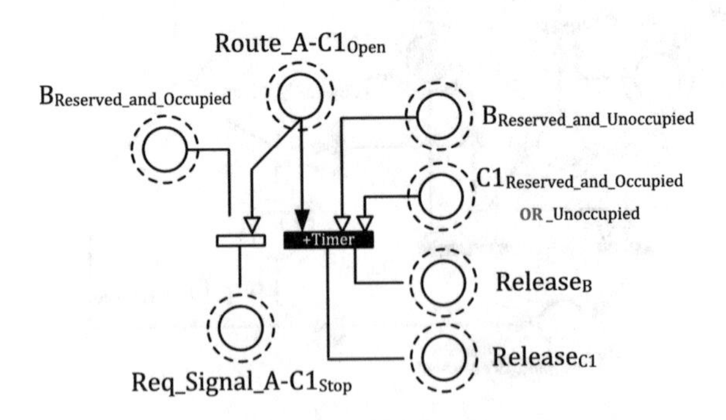

Fig. 6. Route release model.

4 Case Study

The diagram shows the southern part of the station *Obervellmar* where the line splits into the lines to *Kassel Hauptbahnhof* and to junction Berg where the line further splits to R*angierbahnhof* and W*ilhelmshöhe*. The letters "R" and "W" are shown as direction indicators (Zs2). The *PN* model of signal A will be created (Fig. 7).

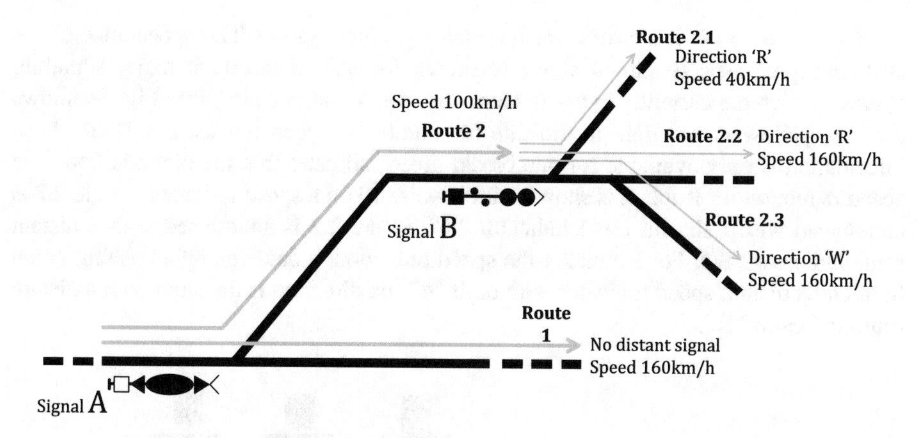

Fig. 7. Railway layout for the case study.

From signal A (Ks signal) there are two possible routes. Route 1 allows unreduced speed. The next signal on this route has its distant signal, so signal A does not show distant aspects for this route. Route 2 allows 100 km/h and will end at signal B (Hp signal). From there three subsequent routes are possible, influencing the main aspect of signal B. Signal A can show the following main aspects:

- H0: Stop (no route set),
- H160: Route 1 (no speed restriction),
- H100: Route 2 (100 km/h).

Signal B has a direction indicator, which will show "R" for Routes 2.1 and 2.2 and "W" for Route 2.3. Route 2.1 has a speed limit of 40 km/h. Therefore Signal B can show the following main aspects:

- H0: Stop (no route set),
- H40R: Route 2.1 (expect speed restriction to 40 km/h, direction indicator "R"),
- H160R: Route 2.2 (no speed restriction, direction indicator "R"),
- H160W: Route 2.3 (no speed restriction, direction indicator "W").

Additionally, if the route 2 is set, then the main (H-Hauptsignal) aspects of signal B becomes distant (V-Vorsignal) aspects for signal A. This allows the following combinations for Signal A, as seen in Fig. 8:

- H0: Stop (no route set),
- H160: Route 1 (no speed restriction),

- `H100 + V0`: Route 2 (100 km/h and expect stop),
- `H100 + V40R`: Route 2 + 2.1 (100 km/h and expect 40 km/h, direction indicator “R”),
- `H100 + V160R`: Route 2 + 2.2 (100 km/h and expect no speed restriction, direction indicator “R”),
- `H100 + V160W`: Route 2 + 2.3 (100 km/h and expect no speed restriction, direction indicator “W”),

The lamps to be lit are different for each signaling system. The appearance of the abstract signal aspects shown above is shown for one of the three major signaling systems which are currently in use in Germany. The `Ks` signal [16], (see Fig. 8), shows a green, yellow or red light to indicate the number of open blocks. For Route 1, as information is only available for one block, green indicates that the block is free. The speed reduction for Route 2 is shown with a white 10 on a speed indicator. Route 2.2 is announced with a distant route indicator “R”. Route 2.3 is announced with a distant route indicator “W”. For Route 2.1 the speed reduction is declared by a flashing green light and a distant speed indicator with digit “4”. Its direction is declared with a distant route indicator “R”.

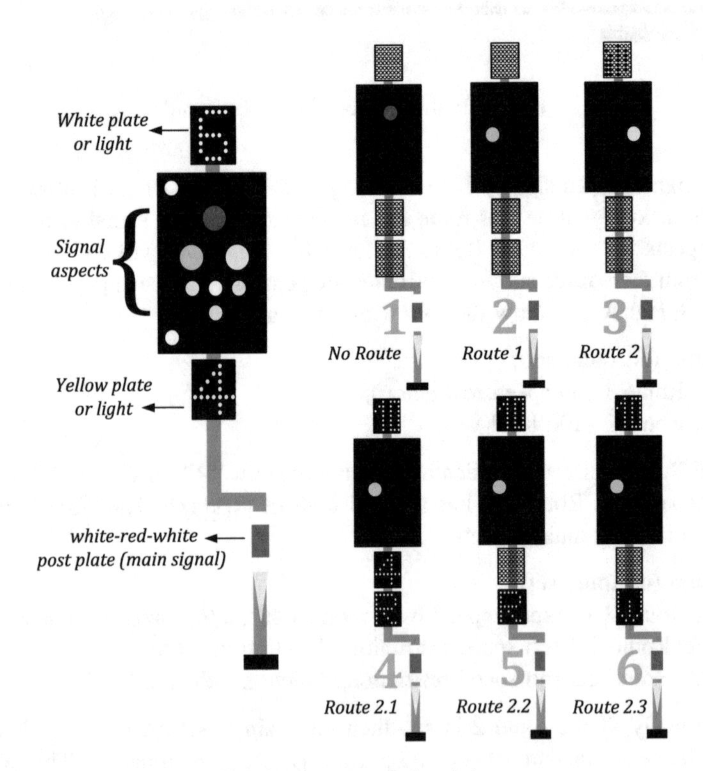

Fig. 8. Aspects of signal A.

Figure 9 shows the whole *PN* for this signal. The boxes `H`, `V` and `0` have to be created depending on the abstract aspect required by the route and the signaling system requiring a specific appearance of the aspect. The net is accessed by the six public places on the left, which are:

- Request route 1,
- Request route 2,
- Signal B shows man aspect `H160W`,
- Signal B shows man aspect `H160R`,
- Signal B shows man aspect `H40W`,
- Close signal.

The boxes `H1` and `H2` are magnified in Fig. 10. From the `V`-boxes only box `V2.3` is shown in Fig. 11. Box `V2.2` would look similar as only the `Zs2v` indicator shows a different letter. Box `V2.1` would additionally require a `Zs3v` indicator and a flashing green light. The current lamp net is not able to handle a flashing lamp, but could be easily extended to do so. The box `0` is not shown at all. It would switch the red light on and all other lights off. As a request to switch off a lamp does not require information if the light is on, this box can be created for all routes equally.

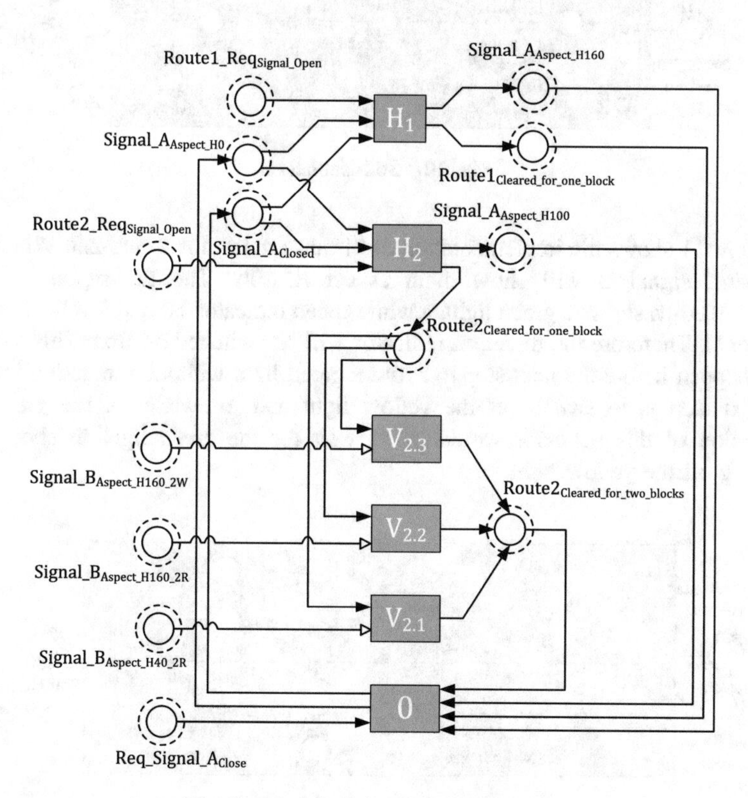

Fig. 9. Model of signal A (the whole).

Figure 10 shows the implementation of main aspects for routes 1 and 2 for the Ks signaling system. For Route 1 only the green light is switched on and the red light is switched off. For Route 2, a distant signal information is available. Therefore the most restrictive distant signal aspect V0 is assumed. After switching on the speed indicator and the yellow light, the red light is switched off. The verification later should make sure, that if route two is opened, the signal never shows a yellow light alone.

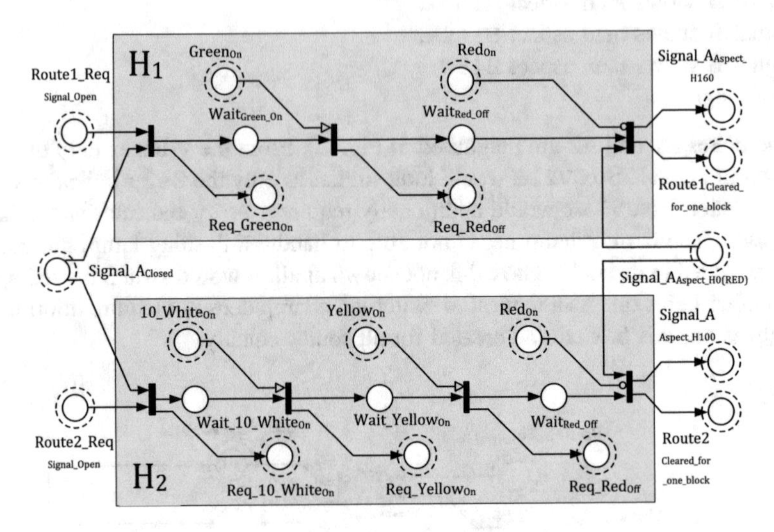

Fig. 10. Both subnets H.

Figure 11 shows the implementation of distant aspects for Route 2.3. When Route 2.3 opens, signal B will show main aspect H160W. The Ks signal aspect for H100 + V160W shows a green light, a white speed indicator 10 and a yellow direction indicator W. Therefore the direction indicator will be switched on first. This switching has to happen before the next step to avoid a green light without a direction indicator. The next step is to switch off the yellow light and to switch on the green light. A variation of this behavior would be to wait for the green light to show before switching off the yellow light.

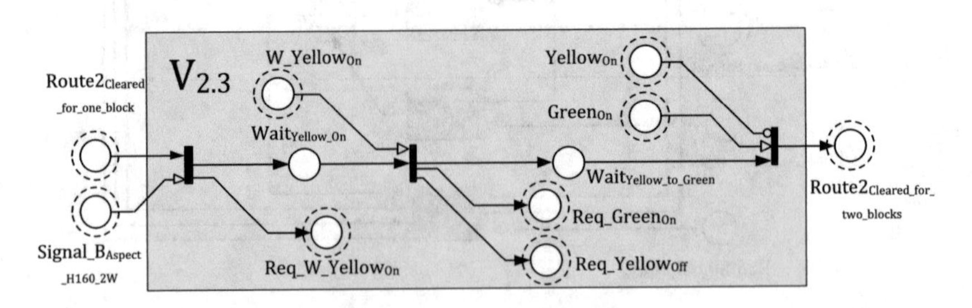

Fig. 11. Subnet V for route 2.3.

5 Conclusion and Future Works

In this study, a generic signal model was created for a railway layout. The model of the point was derived from the bulb sub-model. An example entrance signal model is constructed. The example was specified in PNML for use in TAPAAL for future studies. The signal and point models created in this paper will be used to create an interlocking model, including track segment models and route models. That model will be applied to a simple station layout and extended by a train model and a dispatcher model. The models will be generated automatically from a given layout.

References

1. Giua, A., Seatzu, C.: Modeling and supervisory control of railway networks using Petri nets. IEEE Trans. Autom. Sci. Eng. **5**(3), 431–445 (2008)
2. Ren, X., Zhou, M.C.: Tactical scheduling of rail operations: a Petri net approach. In: International Conference on Systems, Man and Cybernetics, Vancouver BC, Canada, pp. 3087–3092 (1995)
3. Hei, X., Takahashi, S., Nakamura, H.: Toward developing a decentralized railway signalling system using Petri nets. In: Conference on Robotics, Automation and Mechatronics, Chengdu, China, pp. 851–855 (2008)
4. Hörste, M.M., Schneider, E.: Modeling and simulation of train control systems using Petri nets. In: Wing, J.M., Woodcock, J., Davies, J. (eds.) World Congress on Formal Methods in the Development of Computing Systems. LNCS, vol. 1709, pp. 1867–1883. Springer, Heidelberg (1999)
5. Hei, X., Takahashi, S., Nakamura, H.: Distributed interlocking system and its safety verifications. In: The 6th World Congress on Intelligent Control and Automation, Dalian, China, pp. 8612–8615 (2006)
6. Anunchai, S.V.: Verification of interlocking tables using coloured Petri nets. In: 10th Workshop and Tutorial on Practical Use of Coloured Petri Nets and the CPN Tools, Aarhus, Denmark, pp. 139–158 (2009)
7. Yıldırım, U., Durmuş, M.S., Söylemez, M.T.: Application of functional safety on railways part II: software development. In: The 8th Asian Control Conference, Kaohsiung, Taiwan, pp. 1096–1101 (2011)
8. Yıldırım, U., Durmuş, M.S., Söylemez, MT.: Fail-safe signalization and interlocking design for a railway yard: an automation Petri net approach. In: The 7th International Symposium on Intelligent and Manufacturing Systems, Sarajevo, Bosnia Herzegovina, pp. 461–470 (2010)
9. Durmuş, M.S., Yıldırım, U., Söylemez, M.T.: Automatic generation of Petri net supervisors for railway interlocking design. In: The 2nd Australian Control Conference, Sydney, Australia, pp. 180–185 (2012)
10. Durmuş, M.S., Üstoglu, İ.: An application of discrete event systems based fault diagnosis to German railway signaling system. Int. J. Syst. Appl. Eng. Dev. **11**, 70–77 (2017)
11. Jacobsen, L., Jacobsen, M., Møller, M.H.: Modeling and verification of extended timed-arc Petri nets. MSc. thesis. Aalborg University, Denmark (2010)
12. Byg, J., Jorgensen, K.Y., Srba, J.: An efficient translation of timed-arc Petri nets to networks of timed automata. In: Breitman, K., Cavalcanti, A. (eds.) Formal Methods and Software Engineering. LNCS, vol. 5885, pp. 698–716. Springer, Heidelberg (2009)

13. Weber, M., Kindler, E.: The Petri net mark up language. In: Ehrig, H., Reisig, W., Rozenberg, G., Weber, H. (eds.) Petri Net Technology for Communication-Based Systems. LNCS, vol. 2472, pp. 124–144. Springer, Heidelberg (2003)
14. World Wide Web Consortium (W3C): Extensible Markup Language (XML), 22 February 2019. http://www.w3.org/XML/
15. ISO/IEC 15909-1: Systems and software engineering - High-level Petri nets - Part 1: Concepts, definitions and graphical notation (2004)
16. DB Netz AG, Richtlinie 301 – Signalbuch, Aktualisierung 9, 22 February 2019. https://fahrweg.dbnetze.com/resource/blob/1355720/e8f119eef14f0b915191c78dbbd11c51/rw_301_aktualisierung_10-data.pdf

Use of Multiple Data Sources in Collaborative Data Mining

Carmen Anton(✉), Oliviu Matei, and Anca Avram

Electric, Electronic and Computer Engineering Department,
Technical University of Cluj-Napoca, North University Center Baia Mare,
Dr. Victor Babes 62A, Baia-Mare 430083, Romania
{carmen.anton,oliviu.matei}@cunbm.utcluj.ro, anca.avram@ieee.org
https://inginerie.utcluj.ro/

Abstract. Agriculture is one of the domains that depend on weather forecasts and which would improve its performance if some features could be predicted. Because of that, we try to define a concept that is capable of generating predictive results for temperature, based on multiple sources and use of virtual machine learning. We define collaborative approaches in data mining that use multiple sources and we analyze the results from a comparative point of view with the independent process. For each approach, we use a machine learning algorithm which is applied to the combination of data sources from the weather stations. The research proposes a model of the data mining process for a collaborative variant with multiple sources. This leads to the conclusion that a collaborative approach generates better results than a standalone one.

Keywords: Data mining · Collaborative · Virtual machine learning

1 Introduction

Decisions making in agriculture depends on many factors, one system based on technology advancement of the day is a guarantee that can make improves the results in the domain. Use of the predictive analysis can influence the results of the agriculture operations. By predicting some data that can influence the process, decision results can be optimized and can have a significant impact on efficiency. Predictive analyzes require complete data to be successful, but some times this data is incomplete or incorrect and the process of prediction offer not usable results. A solution for this case may be the multiply the sources of the data usable in the process.

The rapid evolution of IoT has generated various solutions for agricultural predictions. Lai et al. in [2] describe a methodology for temperature and rainfall forecasting, based on some necessary data preprocessing technique and the dynamically weighted time-delay neural networks. The results confirm that the relationship between factors contributing to certain weather conditions can be estimated to some extent.

R. Silhavy et al. (Eds.): CoMeSySo 2019, AISC 1046, pp. 189–198, 2019.
https://doi.org/10.1007/978-3-030-30329-7_18

Many factors underlie decisions in agriculture, so some predictive solutions consider combining them. Putjaika et al. in [6] take into account for the decision-making process the data detected by a sensor system and the weather information, using a decision tree model that generates the weather condition and develops a set of rules that automatically devise the operation of an irrigation system. Mahmood et al. in [11] proposed the Cumulative Distribution Function (CDF) for the modeling and analysis of complex data for weather forecasting, which gives them better accuracy in the event of climate change in the near future. In [12], Gadekallu et al. used data capture techniques to predict maximum temperature, precipitation, evaporation and wind speed. They use different algorithms, such as an artificial decision tree, naive Bayes, random forest, K-close neighbors algorithms; and concludes that if case data is sufficient then data extraction techniques can be used successfully in meteorological predictions analysis.

Actual predictive analytic puts the focus for many types of applications which involves in the prediction of future outcomes. Different approaches are encountered in [13] by Poornima et al., where predicted forecast value is compared with real-time data, or by Kale et al. in [14], where they use a combination between the fuzzy logic and machine learning, for obtaining a good decision for the farmers to cultivate crops.

With regard to the concept of data mining collaborative, Moyle in [4] establishes three interpretations of the concept, namely: "(1) multiple software agents applying Data Mining algorithms to solve the same problem; (2) using modern collaboration techniques to apply Data Mining to a Single, Defined Problem; (3) Data Mining to the Artifacts of Human Collaboration." The notion of human collaboration can also be applied to the data required in a prediction process, in order to achieve an increase in the accuracy of the results.

Collaborative filtering is often found in applications, trying to determine the user's options based on other users' data, so by Ambulgekar et al. in [15] is presented a series of collaborative filtering applications in several areas of study. Domingos and Richardson in [1] uses a combined data set. The set of independent entities is seen as a social network and modeled as a random Markov field, the study utilizing a collaborative database and gaining advantages in this approach.

A combination of data used in the data mining is also encountered at Jin et al. in [3] where experiments on real data sets containing data usage and semantic data show that this approach can get better recommendations, compared to systems that use only usage information.

Matei et al. in [7] obtained positive results, for data mining collaborative, study in which collaborative data exploitation techniques are applied in combination with machine learning algorithms in order to discover and extract useful knowledge about users behavior and the devices used. Because the data available in any field is growing, it is useful to structure the sources and levels of them, Matei et al. in [9] presented a structure, to highlight the advantages and provide a starting point in a data mining process.

Janeja et al. in [10] provides a way to validate results by using them collaboratively, and using techniques to exploit them in a collaborative manner, making them more useful and accurate than existing techniques.

The research presented by Anton et al. [16] has established that a collaborative data mining process can deliver better results than an independent approach, and has attempted to determine the algorithm that generates better results in a collaborative process. There was testing four algorithms: k-Nearest Neighbor model, Generate Weight, Neural Net model and Support Vector Machine. The conclusion was that the k-NN (k-Nearest Neighbor model) algorithm is the one with the highest percentage for the collaborative model. A future development proposed by the study is the collaborative data mining process with three sources. This article gives a continuation of research from the point of view of collaborative mining with multiple sources and realizes a analyze of that.

2 Methods

2.1 Subject of Research on the Use of Multiple Sources

In this paper, we proposed for the process of predicting the temperature in a certain area of the Transylvania Plateau, a system that allow to use the data from the nearby meteorological station. The purpose was to demonstrate that the data from a similar station can be useful and can increase the accuracy of the results for the station predicted.

Temperature forecasting for a source was performed in three different systems, using data from the standalone station, using data from two stations for predicting one (collaborative with 2 sources) and using data from three stations (collaborative with 3 sources). The main idea of this research is to determine that a collaborative approach can lead to better outcomes than an independent one, and can generate a usable process for predictions. The relationships between the data used and the results obtained represent a second objective of the research and can lead to a generalized predictive process based on collaborative mining data.

2.2 Data Acquisition

In addition to the previously published results, we evaluate the performance of a collaborative data mining process over a standalone process and demonstrate the feasibility and applicability of this concept in a real scenario.

The data used for the collaborative data mining process was downloaded from Reliable Prognosis[1]. This website provides weather forecasts and observational data from 10400 stations. Weather information is provided in a tabular form indicating the exact point in time. We used data from 16 weather stations in the central region of Transylvania. The structure of a file downloaded contain values for air temperature (degrees Celsius) and the relative humidity (%) at 2 m

[1] https://rp5.ru/Weather_in_the_world.

height above the earth's surface, atmospheric pressure, wind direction, amount of precipitations etc. The time period selected for the process was February 1, 2010 - April 30, 2010 because for 12 weather stations, the data acquired was complete in this interval. The excel files contained data recorded daily starting from 5.00 a.m. until 23.00 p.m. with a 3-hour distance between values. The stations selected for the trial were 12, namely Baisoara, Bistrita, Campeni, Campia Turzii, Dej, Dumbraveni, Gherla, Targu Mures, Tarnaveni, Turda, Lakauti, Sarmasu. From each meteorological station we used temperature and humidity as data, the process referring to them as data sources marked with S1, S2, ... , S12.

2.3 Data Preparation

The data for the temperature and humidity from the files, for the selected stations, were reduced to values that represented a daily average of them.

The following Table 1 gives the structure of the final file of data after the steps of the preprocessing and reducing the initial data.

Table 1. Structure of the final data

Date	S1_TEMP	S1_UMID	S2_TEMP	S2_UMID
1.02.2010	−8.686	94.000	−2.843	84.143
2.02.2010	−11.350	71.250	−6.200	81.250
3.02.2010	−7.200	62.500	−2.733	71.333
4.02.2010	−5.938	54.250	−3.075	81.250
5.02.2010	−3.362	24.000	−5.662	73.375
6.02.2010	−3.650	57.875	−0.350	70.875
7.02.2010	−5.038	90.875	−0.375	88.375

The columns prefixed with "Date" refer to the registered day, columns prefixed with "S1_TEMP" and "S2_TEMP" represents average value of temperature of the air (degrees Celsius) at 2 m height above the earth's surface, and the columns prefixed with "S1_UMID" and "S2_UMID" refers to the relative humidity (%) at 2 m height above the earth's surface. The file contains daily data from February 1, 2010 to April 30, 2010 for all 12 sources, and therefore has 90 lines and 25 columns.

2.4 Experimental Methods - Collaborative with 2 Source

In this article, we propose a new virtual machine learning model that uses data from multiple sources and generates a better prediction for collaboration technique than without it. Processes have been applied to predict air temperature (degrees Celsius) for the next day. The processes applied to the data from the source file have been addressed from three different points of view. The three approaches assumed are:

- Standalone variant (predictions based on data from the observed station).
- Collaborative with 2 sources (predictions based on data from 2 stations).
- Collaborative with 3 sources (predictions based on data from 3 stations).

The algorithm used was k-Nearest Neighbor model (k-NN), because starting from a previous study presented in [16] by Anton et al. was the algorithm that provided the best results in collaborative data mining. The processes were performed using the RapidMiner Studio application (version 7.6).

In the standalone approach, where p_t_a stands for prediction_trend_accuracy, the values obtained were: S1: p_t_a = 0.482; S2: p_t_a = 0.459; S3: p_t_a = 0.494; S4: p_t_a = 0.482; S5: p_t_a = 0.518; S6: p_t_a = 0.529; S7: p_t_a = 0.518; S8: p_t_a = 0.494; S9: p_t_a = 0.482; S10: p_t_a = 0.482; S11: p_t_a = 0.518; S12: p_t_a = 0.565.

The values obtained were in the range [0.482, 0.565]. From this point we can see that the sources could be divided into two categories: those that obtained prediction values less than 0.500 (7 sources) and those that obtained prediction values greater than 0.500 (5 sources).

Table 2. Predictive values obtained in the collaborative process with 2 sources

Sta	0.482	0.459	0.494	0.482	0.518	0.529	0.518	0.494	0.482	0.482	0.518	0.565
	S1	S2	S3	S4	S5	S6	S7	S8	S9	S10	S11	S12
S1		0.576	0.647	0.529	0.506	0.553	0.506	0.541	0.541	0.529	0.529	0.494
S2	0.506		0.647	0.447	0.553	0.541	0.553	0.459	0.518	0.447	0.565	0.447
S3	0.494	0.529		0.576	0.482	0.482	0.482	0.529	0.482	0.576	0.506	0.518
S4	0.435	0.541	0.659		0.435	0.494	0.435	0.518	0.494	0.482	0.553	0.424
S5	0.553	0.518	0.576	0.553		0.553	0.518	0.576	0.518	0.553	0.553	0.506
S6	0.553	0.588	0.635	0.541	0.541		0.541	0.529	0.541	0.541	0.6	0.553
S7	0.553	0.518	0.576	0.553	0.518	0.553		0.576	0.518	0.553	0.553	0.506
S8	0.471	0.541	0.635	0.576	0.541	0.494	0.541		0.506	0.576	0.553	0.506
S9	0.447	0.553	0.600	0.494	0.506	0.459	0.506	0.529		0.494	0.518	0.494
S10	0.435	0.541	0.659	0.482	0.435	0.494	0.435	0.518	0.494		0.553	0.424
S11	0.482	0.541	0.581	0.576	0.506	0.518	0.506	0.506	0.494	0.576		0.529
S12	0.541	0.588	0.659	0.529	0.518	0.529	0.518	0.518	0.565	0.529	0.565	
nvc	**7**	**11**	**11**	**10**	**5**	**5**	**5**	**11**	**10**	**10**	**10**	**0**

In the second approach, collaborative with 2 sources, the results is visible in Table 2, where the abbreviations are: S1 to S12 for the 12 sources in the process, Sta = predictive value for standalone process, nvc = number of collaborative values higher than the standalone. Following the application of the collaborative process with two sources and based on the obtained results, the following aspects were observed:

- Sources that obtained, in a standalone process, values below 0.500 have obtained in the collaborative process, in the majority of cases, a higher predictive value and number of them exceed half of the possible (11 possible values). Examples are the sources: S1, S2, S3, S4, S8, S9, S10.
- The sources that obtained over 0.500 in the standalone process, have obtained lower predictive values in the collaborative process, the number of them is smallest than half of the possible values. Examples are the sources: S5, S6, S7, S12.

A special situation is the S11 source, which has a prediction value obtained in the standalone process greater than 0.500, but it is nevertheless noticed that the number of predictive values obtained in collaboration with 2 sources is high, ie equal to ten.

Pairs S7, S5 and S10, S4 have obtained identical results in both processes because the temperatures recorded in the two sources are identical given the geographic proximity of the weather stations from which source data originated. (Dej-Gherla 12.78 km respectively Turda-Campia Turzii 8.17 km). Given this observation, the S7 and S10 sources have been removed from the process with three collaborative sources. The second approach, based on the collaborative model with two sources, has succeeded in reinforcing the idea that the 12 sources have different behaviors and the third approach would deserve to follow the behavior according to the two distinct groups.

2.5 Experimental Methods - Collaborative with 3 Source

Given the results obtained in the collaborative process with 2 sources and the observations made on them, two categories of sources were created for the next stage of work. Thus, the first category was composed of the sources: S1, S2, S3, S4, S8, S9, S10, S11, called Lower500. The second category consisting of the sources: S5, S6, S12, named Higher500.

Since the S11 source had different results from the two categories, before it was included in particular one, the correlations between the stations were generated and are presented in the Fig. 1, where can be seen that this source has lower values for correlation with other stations and all were under 0.625.

Table 3. Results for a collaborative process with 3 sources for Lower500 group

Sources	S1	S2	S3	S4	S8	S9	S11
Number of CDM values greater than SDM values	9	15	15	14	12	10	13
Percentage of CDM values greater than SDM values	60.0%	100%	100%	93.3%	80.0%	66.6%	86.6%

As a result, it was included in the Lower500 study category for the collaborative process with 3 sources.

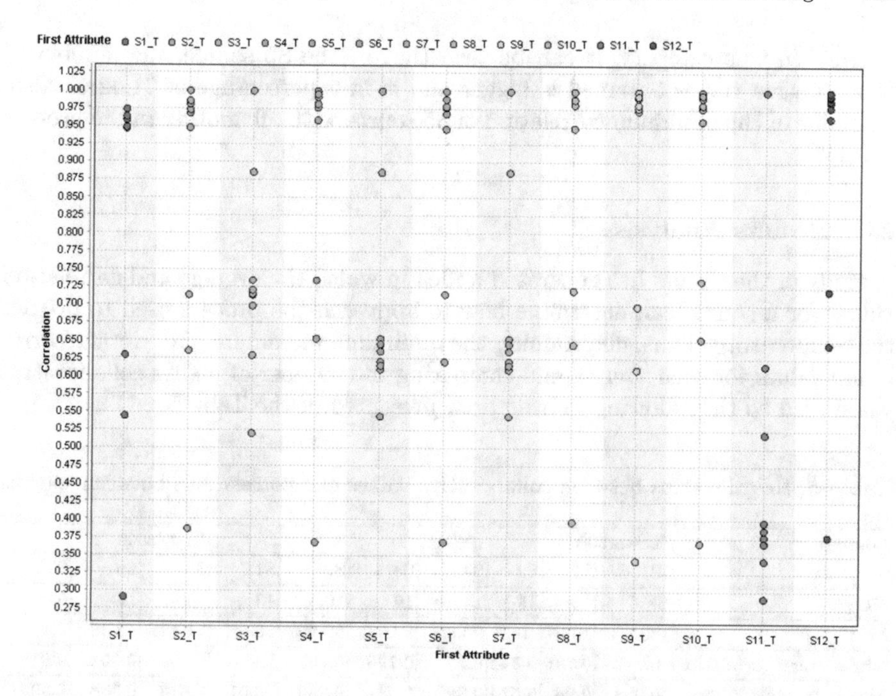

Fig. 1. Correlations between sources.

The collaborative process with 3 sources was tested for the Lower500 group for all possible combinations of stations, ie each station was predicted by combining all the other stations in pair. The number of values obtained for each station was 15. These were compared to the value obtained in the standalone process, the results being visible in the Table 3.

It is noticeable that S2 and S3 sources recorded the increase of all the collaborative values. The minimum value of the correlations with the other sources for S3 is 0.521, and in S2 is 0.388.

The Higher500 group was tested using the sources from Lower500 group. For source 1 it was selected each station from Higher500 and for sources 2 and 3, all possible pairs of the Lower500 group. It was obtained 21 collaborative predictive values for sources: S5, S6, S12. The results can be seen for each source in the Table 4.

Table 4. Results for a collaborative process with 3 sources for Higher500 group

Sources	S5	S6	S12
Number of CDM values greater than SDM values	15	11	0
Percentage of CDM values greater than SDM values	100%	52.38%	0%

Also in this category, it can be seen that at the S5 source, the number of collaborative values obtained is bigger and is in a percentage of 71.42%. Also, the value of the minimum correlation of S5 source with all another in the process is 0.545.

2.6 Results Analysis

A study of the results has generated a idea in which the average and correlation values for a source can determine how to approach the process used to predict the temperature. Thus, determining the minimum, maximum and average correlation values for each source, and comparing the number of higher collaborative values, led to the following information, presented in the Table 5.

Table 5. Relationships between collaborative values and correlations between sources

Sources	Lower500							Higher500		
	S1	S2	S3	S4	S8	S9	S11	S5	S6	S12
NCV	9	15	15	14	12	10	13	15	11	0
PCV	60.0%	100%	100%	93.3%	80.0%	66.6%	86.6%	71.42%	52.38%	0%
The average correlations	0.772	0.795	0.544	0.797	0.796	0.787	0.298	0.580	0.901	0.918
Maximum correlation	0.988	0.982	0.575	0.993	0.981	0.993	0.501	0.878	0.988	0.996
Minimal correlation	0.227	0.299	0.501	0.274	0.311	0.249	0.229	0.467	0.294	0.284
MinDV	−0.047	0.047	0.035	−0.035	−0.035	−0.023	−0.047	−0.083	−0.070	−0.165
MaxDV	0.047	0.153	0.200	0.106	0.071	0.094	0.058	0.129	0.071	−0.024

where: NCV if number of collaborative values (from process with 3 sources) higher than the standalone value, PCV is percentage of collaborative values higher than the standalone value, MinDV is minimum difference value (collaborative minus standalone) and MaxDV means maximum difference value (collaborative minus standalone).

By comparing the S5 and S3 sources in the two study groups, it is observed that at an average correlation between 0.400 and 0.750 the collaborative data mining process has a higher than average rate and produces significantly better results than the standalone.

By comparing the values obtained using a Collaborative Data Mining (CDM) process, with a Standalone Data Mining (SDM), certain relationships between the station correlations and the final values were achieved.

Thus, on a more detailed analysis, it is observed that 11 sources out of 12, they obtained CDM values greater than SDM values, which represent 91.66%. The numbers of values CDM greater than SDM for each source is situated between 9 and 11, a percentage ranging from 50% to 100%.

In this research, the result obtained by comparing the CDM values with the SDM values led to the results that reveal a good percentage of sources that increase the accuracy of prediction for k-NN algorithm applied in collaborative manner. An exception is source S12, but if we analyze the correlations graphic from Fig. 1, we see that the source has the many values of correlations near 1.00 and very few below 0.950.

The results reveal that a collaborative approach with 2 or 3 sources can have better outcomes if we are using the right combination of sources. It can be noticed that an average of correlations between 0.500 and 0.600 leads to higher results of the collaborative process, with values ranging from 0.035 to 0.200.

3 Discussions

In this technological era, most things are interconnected and adapted to the constantly changing technology. According to Khan et al. in [5] the Internet of Things (IoT) is evolving to make machine-to-machine learning (M2M). In this case, IoT offers connectivity and solutions for embedding some intelligence in connected the objects in order to make decisions based on innovative services. Del Giudice in [8] said that the use of IoT technologies is on a progressive scale, being essential in the evolution of the learning and application process supported by standardization efforts.

In this article, the conclusions that emerged from collaborative processes with multiple sources are in favor of it, and so we have:

- If the predictive values obtained in the standalone process are below 0.500 then the collaborative process with multiple sources produces higher values in most cases.
- If the predictive values obtained in the standalone process are above 0.500 then the collaborative process with multiple sources produces higher values in cases where the predicted source has correlations situated around 0.500.
- Correlation sources ranging from 0.500 to 0.600 record a significant increase in values in collaborative multi-source mode compared to the standalone approach.
- The values obtained in collaborative data mining have registered an increase situated between 0.035 and 0.200 range.
- The percentage of sources that recorded increases in predictive values in collaborative mode compared to standalone is 91.66%.

Given the results and conclusions that emerged from this study, a future approach could be to treat the collaborative process with multiple sources for cases where one of the sources does not present study data. In this hypothesis, the predictive value for the missing data source could be positively influenced by sources correlated with it and able to deliver results with greater accuracy.

References

1. Domingos, P., Richardson, M.: Mining the network value of customers. In: Proceedings of the Seventh ACM SIGKDD International Conference on Knowledge Discovery and Data Mining, vol. 7, pp. 57–66. ACM, August 2001

2. Lai, L., Braun, H., Zhang, Q., Wu, Q., Ma, Y., Sun, W., Yang, L.: Intelligent weather forecast. In: Proceedings of 2004 International Conference on Machine Learning and Cybernetics (IEEE Cat. No. 04EX826), vol. 7, pp. 4216–4221, August 2004
3. Jin, X., Zhou, Y., Mobasher, B.: A maximum entropy web recommendation system: combining collaborative and content features. In: Proceedings of the Eleventh ACM SIGKDD International Conference on Knowledge Discovery in Data Mining, pp. 612–617. ACM, August 2005
4. Moyle, S.: Collaborative data mining. In: Data Mining and Knowledge Discovery Handbook, pp. 1043–1056. Springer, Boston, MA (2005)
5. Khan, R., Khan, S.U., Zaheer, R., Khan, S.: Future internet: the internet of things architecture, possible applications and key challenges. In: 2012 10th International Conference on Frontiers of Information Technology, pp. 257–260. IEEE, December 2012
6. Putjaika, N., Phusae, S., Chen-Im, A., Phunchongha, P., Akkarajitsakul, K.: A control system in an intelligent farming by using arduino technology. In: Fifth ICT International Student Project Conference (ICT-ISPC) IEEE, pp. 53–56, May 2016
7. Matei, O., Di Orio, G., Jassbi, J., Barata, J., Cenedese, C.: Collaborative data mining for intelligent home appliances. In: Working Conference on Virtual Enterprises, pp. 313–323. Springer, Cham (2016)
8. Del Giudice, M.: Discovering the Internet of Things (IoT) within the business process management: a literature review on technological revitalization. Bus. Process Manag. J. **22**(2), 263–270 (2016)
9. Matei, O., Rusu, T., Bozga, A., Pop-Sitar, P., Anton, C.: Context-aware data mining: embedding external data sources in a machine learning process. In: International Conference on Hybrid Artificial Intelligence Systems, pp. 415–426. Springer, June 2017
10. Janeja, V.P., Gholap, J., Walkikar, P., Yesha, Y., Rishe, N., Grasso, M.A.: Collaborative data mining for clinical trial analytics. Intell. Data Anal. **22**(3), 491–513 (2018)
11. Mahmood, M., Patra, R., Raja, R., Sinha, G.: A novel approach for weather prediction using forecasting analysis and data mining techniques. In: Innovations in Electronics and Communication Engineering, pp. 479–489. Springer (2019)
12. Gadekallu, T., Kidwai, B., Sharma, S., Pareek, R.: Application of data mining techniques in weather forecasting. In: Sentiment Analysis and Knowledge Discovery in Contemporary Business, pp. 162–174. IGI Global (2019)
13. Poornima, S., Pushpalatha, M., Shankar, J.: Analysis of weather data using forecasting algorithms. In: Computational Intelligence: Theories, Applications and Future Direction, vol. I, pp. 3–11. Springer (2019)
14. Kale, S., Patil, P.: Data mining technology with fuzzy logic, neural networks and machine learning for agriculture. In: Data Management, Analytics and Innovation, pp. 79–87. Springer (2019)
15. Ambulgekar, H., Pathak, M., Kokare, M.: A survey on collaborative filtering: tasks, approaches and applications. In: Proceedings of International Ethical Hacking Conference, pp. 289–300. Springer, Singapore (2019)
16. Anton, C., Matei, O., Avram, A.: Collaborative data mining in agriculture for prediction of soil moisture and temperature. In: Advances in Intelligent Systems and Computing (to appear)

Framework for Controlling Interference and Power Consumption on Femto-Cells In-Wireless System

P. T. Sowmya Naik[1(✉)] and K. N. Narasimha Murthy[2]

[1] Department of Computer Science and Engineering, City Engineering College, Bengaluru, India
sowmya.vturesearch@gmail.com
[2] Faculty of Engineering, Christ (Deemed to be University), Bengaluru, India

Abstract. Utilization of femto-cells is one of the cost effective solution to increase the internal network connectivity and coverage. However, there are various impediment in achieving so which has caused a consistent research work evolving out with solution. Review of existing literature shows that maximum focus was given for energy problems in cellular network and not much on problems that roots out from interference. Therefore, the proposed system has presented a very simple and novel approach where the problems associated with interference and energy in using large groups of femto-cells are addressed. Adopting analytical research methodology, the proposed model offers on-demand utilization of the selective femto-cells on the basis of the traffic demands. The study outcome shows that proposed system offers better performance in contrast to existing approach.

Keywords: Femto-cells · Cellular network · Energy efficiency · Power consumption · Coverage

1 Introduction

There has been significant improvement in the area of cellular network in the last decade with inclusion of more advancement in communication system [1]. However, the increasing demands of bandwidth and superior quality-of-services, it is not enough at present in its current state [2]. The potential challenges in this aspect are coverage of network, deployment-supportability of modern communication devices, and Long Term Evolution (LTE) based network. All these problems doesn't offer any solution when it comes to developing wireless system for in-building infrastructure and femto-cell is one effective solution in this regards [3]. Femto-cells offer multiple base-station, which is essentially made for improving the coverage and solve any form of connectivity issues. It offers beneficial prospects e.g. allowing the internet to be connected to core network reducing cost of communication, compensation of low coverage, secured, supports concurrent data transmission, compatible with existing connection of network, etc. [4]. However, there are certain challenges too viz. no control over usage bandwidth, not cost effective in complex and hierarchical connections, dependent on higher

R. Silhavy et al. (Eds.): CoMeSySo 2019, AISC 1046, pp. 199–208, 2019.
https://doi.org/10.1007/978-3-030-30329-7_19

network resources and location information etc. [5]. At present, there are many literatures that have been concentrating on various problems on femto-cells [6–10]. However, one common problem associated with the femto-cell deployment is energy consumption with respect to transmit power of base station. Apart from this, interference is also another significant problem which is also not much studied in past.

Therefore, this paper hypothesizes that there is a close relationship of interference and power. By controlling interference, both transmission qualities as well as power efficiency can be improved. The organization of the paper is as follows: Briefing of existing research contribution is carried out in Sect. 2 followed by problem's highlights in Sect. 3. Adopted research methodology is briefed in Sect. 4 while discussion of constructed algorithm is carried out in Sect. 5. Analysis of results is done in Sect. 6 while conclusion is written in Sect. 7.

2 Related Work

This section discusses about the existing research work being carried out towards addressing energy problems in cellular network using femto-cells as an extension of our prior review work [10]. Most recently, problems associated with the power controlling factor has been addressed by Brahmi et al. [11] focusing on quality-of-service. Joint problem of load balancing and energy efficiency has been presented by Huang et al. [12] using Markov process to offer better quality-of-service. The work of Thakur et al. [13, 14] have considered the case study of remote networks of femto-cells and developed a framework connecting cost with energy factor. Existing system is also found to use sensing of spectrum method in order to control energy consumption as seen in work of Zuo and Nie [15]. The works of Zhang et al. [16] have addressed an optimization problem towards capacity and energy efficiency for femto-cells in indoor applications. Adoption of clustering approach was reported by Zhang et al. [17] towards energy and better throughput performance. Wang et al. [18] has briefed that clustering the femto-cells can improve the power control aspects in femto-cells. An investigation carried out by Sheikh et al. [19] has showed that energy efficiency can be retained upon the using femto-cells irrespective of any degree of fading effect. Bouras and Diles [20] have presented a unique access strategy using diversified sleep pattern to prove that it is feasible to save power in advance networks. Energy efficiency can be also increased using macro-femto network in order to support heterogeneity in the nodes (Tayade and Gulhane [21]). Adoption of next generation multiple access technology towards minimizing the transmit power is seen in the work of Mili et al. [22]. Rao et al. [23] have presented a statistical modeling for energy saving with more insights on its spectrum and energy behavior. Adoption of game theory towards power control in distributed manner has been carried out by Mao et al. [24]. Apart from this, existing research was also carried out towards sleep scheduling for base station (Kim et al. [25]), learning-based approach for energy efficiency (Gao et al. [26]), investigational study using dense network of femto-cell (Kim et al. [27]), selective power control (Lin et al. [28]), allocation of power (Zinali et al. [29]), and large-scale deployment (Al Haddad and Bayoumi [30]). The next section discusses about the research problem.

3 Problem Description

After reviewing the existing approaches, it can be seen that adopted methods are more focused on energy efficiency and offers less emphasis towards considering the networking artifacts. The presence of noise and interference modeling is highly essential while working on large number of femto-cells, which is seen missing in the existing system. Apart from this, there is a closer relationship of radio scheduling being connected with the energy factor as well as the interference aspect, which is also significantly missing in the existing approaches. Therefore, the problem statement is "*To develop a computational framework that can maintain balance between energy consumption as well as interference while optimizing the deployment of number of femto-cells in cellular network.*"

4 Proposed Methodology

The idea of the proposed system is continued after our prior model [31] where traffic and energy problems over femto-cells are addressed. By adopting an analytical research methodology, the proposed system targets to jointly address the interference and energy issues associated with cellular network using femto-cells. The adopted architecture of proposed system is as follows (Fig. 1):

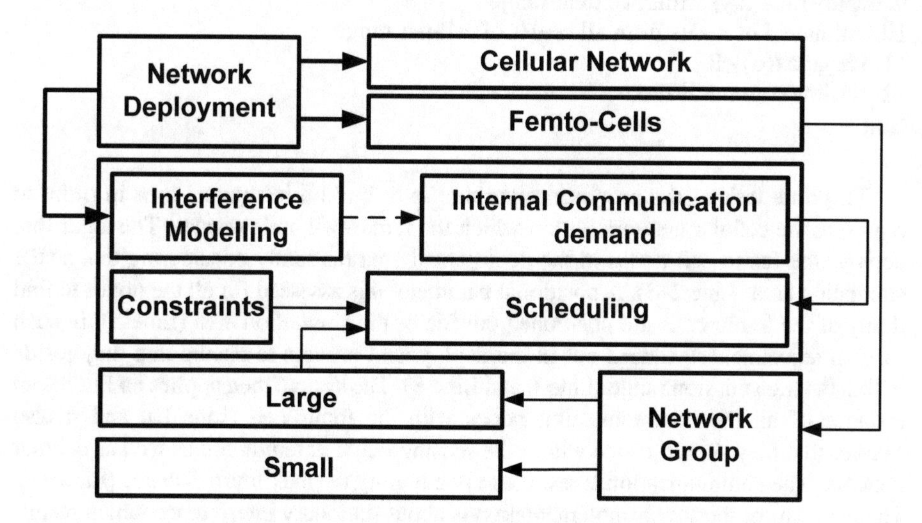

Fig. 1. Proposed notion of methodology

The complete operation of proposed system can be briefed in three stages. The first stage is about network deployment while the second stage is focused on modeling the interference model on the basis of internal communication (uplink transmission). The third stage of implementation is focused on scheduling the transmission on the basis of the network of the femto-cells. The adopted strategy is mainly targeting to process the incoming traffic request by femto-cells on the basis of the classified demands of large and small scale. The next section elaborates the algorithm design.

5 Algorithm Implementation

The design strategy of the proposed algorithm is constructed for balancing the minimization of adverse interference and maximizing the energy efficiency by extending the coverage. The primary task of the algorithm is to eliminate any form of interference existing between any overlapping and interfering signals among the neighboring femto-cell. Hence, an algorithm is initially constructed in order to deploy the cellular network with femto-cells. The steps of the algorithm are as follows:

Algorithm for Network Deployment
Input:t_r (transmission range of femto-cell)
Output: x/y (positioning of nodes in cellular area)
Start
1. init tr
2. **For** i=1: tr
3. deploy random nodes
4. **End**
5. **If** $\delta \geq 0$
6. re-position $(x,y)^{++}$
7. **Else**
8. re-position $(x,y)^{--}$
9. deploy rand(x,y) within cellular range
10. estimate d of nodes from all center of cellular range
11. **If**argmax(d) ≥R
12. Allow communication

End

Th prime task of the proposed algorithm is to build a hexagonal area in order to represent the cellular network within which the femto-cell node resides. The algorithm deploys the femto-cell nodes in the deployment area randomly considering that as the simulation area (Line 2–3). A positional parameter δ is assessed for all the nodes to find if any of the femto-cells are positioned outside of the simulation area (Line-5). In such case an adjustment is carried out in the deployment process to ensure that they reside within the transmission range (Line-6 and Line-8). The system then applies an Euclidean distance of all the communicating nodes with the femto-cell (Line-10) and it also ensures that they should reside within the sensing range of femto-cell as well as cellular network. The communication is resumed upon finding the maximum distance (Line-11). The next part of the study implementation is about stationary interference which results in signal weakening too. This formulates the basis of the noise modeling as follows:

$$g(z) \rightarrow 1 - e^{a}$$

In the above expression, g is a function representing the data distribution of noise where z represents natural number and the exponent a empirically represents $a = -z/$ average signal-to-noise ratio. This expression assists in evaluating the actual distribution of the noise for the assigned femto-cells. The proposed study considers that all the communication channel of the user are highly different from each other (in order to represent heterogeneous network) and all the information associated with the adjoining networks could be accessible from the respective base station. The communication is initiated at variable transmission rate. The system then computes the probability that a user under coverage of femto-cell forwarding encoded data to its destination node.

$$\alpha(p,q) = \int_{\theta_p}^{\theta_{p+1}} dg_q(z) = \Delta \mathrm{a}$$

In the above expression, the variables p and q represents index equivalent to noise value while q is specific user. The proposed system also uses a cut-off attribute θ of the threshold in while Δa represents difference of two exponential parameter with power of cut-off attribute θ and mean signal-to-noise ratio. This is an essential consideration of the proposed system which attempts to find the noise and compensates the noise using a cut-off value. The choice of cut-off value can be carried out either using probability theory or using some standard application. The study implementation is focused on selection of best number of femto-cells in order to resist noise and interference. The overhead is highly controlled as only one femto-cell is selected (out of many) on the basis of their best coverage and thereby saving the energy of others. In order to perform this form of classification of femto-cells, the concept looks for best user which can be defined as a user that has better condition of communication channel. The base station performs provisioning of the best user and not all the n number of users. It will also mean that base station retains the right to schedule communication of the user with the respective femto-cell. The empirical modeling to do this is as below:

$$\Phi_{\mathrm{n}} \rightarrow \mathrm{arg}_{\mathrm{max}}(\mathrm{rand},\ \mathrm{j})$$

In the above formulation, the variable Φ_{n} can be said to be the interference associated with the set of users n, while *rand* is a random variables and j is the next node. This representation actually means that without any inclusion of the any external parameters or sophisticated scheme, the proposed system offer simple optimization of the network variables to ensure that communication always happens through interference-free channel. Even in presence of interference, the proposed system computes the degree of compensation and allows the communication to be continued using the channel with best compensation score. An assessment of this part of the study will be distinctly carried out to check how it affects the data forwarding performance along with energy efficiency. The next part of the study is associated with energy efficiency of it where more emphasis is offered on the signal-to-interference-noise ratio. The proposed study splits the network of femto-cells into two parts viz. N_1 and N_2. The first type i.e. N_1 represents femto-cells with smaller coverage area while N_2 will represent femto-cell with larger coverage area. The same expressions were used towards controlling the interference, overhead, and energy consumption even in smaller and larger coverage with a difference of more value of cut-off parameter in larger coverage femto-cell and smaller cut-off for smaller coverage of femto-cells. The next section briefs of the outcome obtained by implementing proposed system.

6 Results Discussion

From the prior section, it can be seen that proposed system is focused on two goals viz. (i) solution for communication in presence of interference and (ii) energy conservation. Hence, these two are considered as the prominent performance parameter for assessing the proposed outcomes. Scripted in MATLAB, a simulation model is constructed where the femto-cell network is assumed to be working on 30 MHz channel capacity while the mean size of packet is considered to be 2000 bytes while transmission rate is considered as 25 MBPS (Min). Apart from this, the outcome of the proposed system is compared with the work carried out by Wang et al. [32, 33] as it is the only research work with higher number of citation and relevant with the solution for addressing interference and energy efficiency.

The outcome in Fig. 2 shows that proposed system offers better control over the interference in comparison to existing system. The prime reason behind this is existing system is specific to a network model where degree of constraint is more and hence it can control the interference and noise if the traffic exceeds its peak limit. However, proposed system doesn't have any specification while it allocates many numbers of femto-cells in entire cellular network area; however, not all are used. It identifies the communication demands and uses only the femto-cells that have higher coverage area in order to increases it level of connectivity.

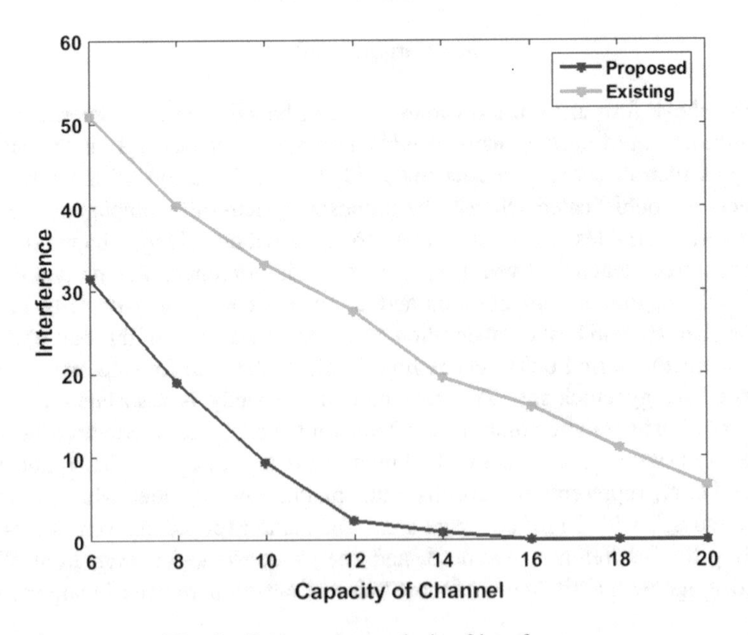

Fig. 2. Comparative analysis of interference

The second part of analysis deals with energy saving which is amount of energy being not used over increasing channel capacity. Figure 3 clearly indicates that proposed system offers better energy saving performance in contrast to existing system. The prime reasons behind this are many factors. The proposed system works on the basis of group formation on the basis of smaller and larger coverage femto-cells, which makes faster classification towards better decision making of forming transmission. As there are two types of femto-cell allocation that is carried out dynamically; therefore, specific demands of the traffic on the basis of the load-level in it can be scheduled and implemented. Irrespective of type of networks or resources used, proposed system can actually take a decision by using the femto-cells wisely on the basis of the demands. An objective function is constructed to ensure that any form of load of traffic has to be identified and necessary scheduling operation has to be carried out. Hence, this save significant amount of resources of the nodes as well as for the entire network. Hence, proposed system offer better energy saving in contrast to existing system. The existing system on the other hand uses a single femto-cell network in one base station making the case too much narrowed. Hence, existing system exhibits energy consumption even if the increasing channel capacity is allocated which also means increasing traffic load. It can save the energy of the node but the rate of energy saving trend is too slow and below the proposed system. Therefore, the proposed system offers a good balance between interference reduction and energy conservation.

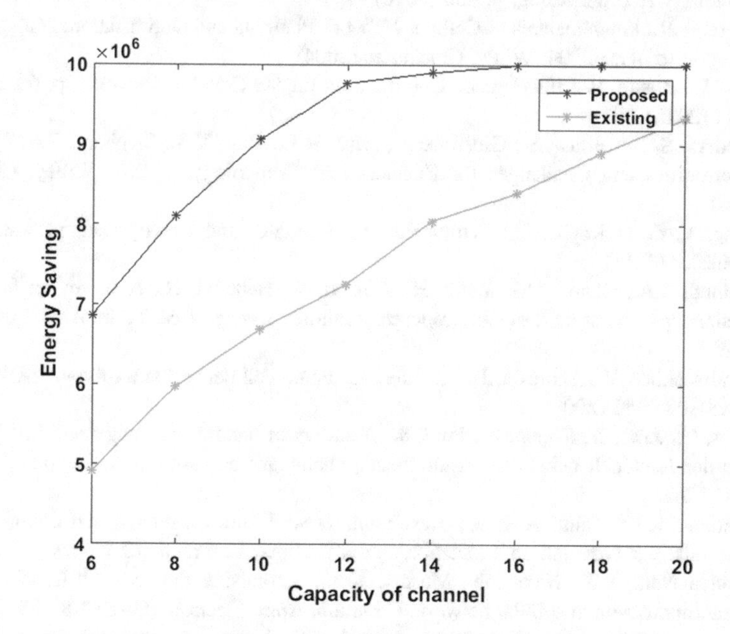

Fig. 3. Comparative analysis of energy saving

7 Conclusion

The current mechanisms of deployment of the femto-cells are associated with the problems as it is a disruptive technology. It has been seen that there are already good approaches towards controlling power dissipation; however, there was no standard model where power consumption is controlled by addressing interference problems. Therefore, the proposed system constructs a novel interference model, where an unique deployment model is created. According to the model, the cellular network is designed on the basis of femto-cell that can cover both large and small scale networks discretely. The model constructs an empirical expression which can compute the transmission possibility in case of any level of signal to interference. A cut-off value is used for narrowing down the decision to transmit data using femto-cell that has higher coverage without involving all of them. The simulated outcome of the proposed study shows that proposed system offer better interference mitigation and maximal energy conservation in contrast to existing system.

References

1. de Alencar, M.S., de Melo Carvalho Filho, D.: Cellular Network Planning. River Publishers, Technology & Engineering, p. 200 (2016)
2. Mishra, A.R.: Fundamentals of Cellular Network Planning and Optimisation: 2G/2.5G/3G… Evolution to 4G, p. 304. Wiley, Chichester (2004)
3. Yang, L., Zhang, W.: Interference Coordination for 5G Cellular Networks, p. 66. Springer, Berlin (2015)
4. Saunders, S.R., Carlaw, S., Giustina, A., Bhat, R.R., Rao, V.S., Siegberg, R.: Femtocells: Opportunities and Challenges for Business and Technology, p. 252. Willey, Chichester (2009)
5. Zhang, J., de la Roche, G.: Femtocells: Technologies and Deployment, p. 328. Willey, Chichester (2011)
6. Mahmud, S.A., Khan, G.M., Zafar, H., Ahmad, K., Behttani, N.: A survey on femtocells: benefits deployment models and proposed solutions. J. Appl. Res. Technol. **11**(5), 733–754 (2013)
7. Chandrasekhar, V., Andrews, J., Gatherer, A.: Femtocell networks: a survey. arXiv preprint arXiv:0803.0952 (2008)
8. Gódor, G., Jakó, Z., Knapp, Á., Imre, S.: A survey of handover management in LTE-based multi-tier femtocell networks: requirements, challenges and solutions. Comput. Netw. **76**, 17–41 (2015)
9. Suleiman, K.E., Taha, A.-E.M., Hassanein, H.S.: Handover-related self-optimization in femtocells: a survey and an interaction study. Comput. Commun. **73**, 82–98 (2016)
10. Sowmya Naik, P.T., Narasimha Murthy, K.N.: Appraising the research trend of energy utilization problem in cellular network. Commun. Appl. Electron. (CAE) **3**(8), 28–36 (2015)
11. Brahmi, I., Mhiri, F., Zarai, F.: Power control method based on users and applications QoS priorities (UAQP) in femtocell network. In: 2018 IEEE/ACS 15th International Conference on Computer Systems and Applications (AICCSA), pp. 1–5. IEEE (2018)
12. Huang, X., Tang, S., Zheng, Q., Zhang, D., Chen, Q.: Dynamic femtocell gNB on/off strategies and seamless dual connectivity in 5G heterogeneous cellular networks. IEEE Access **6**, 21359–21368 (2018)

13. Thakur, R., Mishra, S., Murthy, C.S.R.: An energy and cost aware framework for cell selection and energy cooperation in rural and remote femtocell networks. IEEE Trans. Green Commun. Netw. **1**(4), 423–433 (2017)
14. Thakur, R., Swain, S.N., Murthy, C.S.R.: An energy efficient cell selection framework for femtocell networks with limited backhaul link capacity. IEEE Syst. J. **12**(2), 1969–1980 (2018)
15. Zuo, X., Nie, H.: A spectrum sensing scheme with dynamic sensing period adjustment for femtocells in LTE systems. In: 2017 IEEE International Conference on Electro Information Technology (EIT), pp. 240–244. IEEE (2017)
16. Zhang, W., Zhang, G., Zheng, Y., Xie, L., Yeo, C.K.: Energy efficiency consideration for indoor femtocell networks in TV white spaces. IEEE Access **6**, 1565–1576 (2018)
17. Zhang, J., Labiod, H., Hammami, S.E., Afifi, H.: Scalable energy efficient routing in multi-layer femtocell networks. In: 2017 13th International Wireless Communications and Mobile Computing Conference (IWCMC), pp. 1540–1545. IEEE (2017)
18. Wang, L.-C., Cheng, S.-H., Tsai, A.-H.: Data-driven power control of ultra-dense femtocells: a clustering based approach. In: 2017 26th Wireless and Optical Communication Conference (WOCC), pp. 1–6. IEEE (2017)
19. Sheikh, A.U.H., Khalifa, M.A., Zerguine, A.: Impact of fading on per-energy capacity in femto-macro environment. In: 2017 International Symposium on Wireless Systems and Networks (ISWSN), Lahore, pp. 1–5 (2017)
20. Bouras, C., Diles, G.: Energy efficiency in sleep mode for 5G femtocells. In: 2017 Wireless Days, pp. 143–145. IEEE (2017)
21. Tayade, S.N., Gulhane, V.A.: Designing of macro-femto heterogeneous network for improving energy efficiency of cellular system. In: 2016 International Conference on Wireless Communications, Signal Processing and Networking (WiSPNET), pp. 695–698. IEEE (2016)
22. Mili, M.R., Hamdi, K.A., Marvasti, F., Bennis, M.: Joint optimization for optimal power allocation in OFDMA femtocell networks. IEEE Commun. Lett. **20**(1), 133–136 (2016)
23. Rao, J.B., Fapojuwo, A.O.: An analytical framework for evaluating spectrum/energy efficiency of heterogeneous cellular networks. IEEE Trans. Veh. Technol. **65**(5), 3568–3584 (2016)
24. Mao, T., Feng, G., Liang, L., Qin, S., Bin, W.: Distributed energy-efficient power control for macro–femto networks. IEEE Trans. Veh. Technol. **65**(2), 718–731 (2016)
25. Kim, J., Jeon, W.S., Jeong, D.G.: Base-station sleep management in open-access femtocell networks. IEEE Trans. Veh. Technol. **65**(5), 3786–3791 (2016)
26. Gao, Z., Wen, B., Huang, L., Chen, C., Ziwen, S.: Q-learning-based power control for LTE enterprise femtocell networks. IEEE Syst. J. **11**(4), 2699–2707 (2017)
27. Kim, J., Jeon, W.S., Jeong, D.G.: Effect of base station-sleeping ratio on energy efficiency in densely deployed femtocell networks. IEEE Commun. Lett. **19**(4), 641–644 (2015)
28. Lin, M., Silvestri, S., Bartolini, N., La Porta, T.: Energy-efficient selective activation in femtocell networks. In: 2015 IEEE 12th International Conference on Mobile Ad Hoc and Sensor Systems, pp. 361–369. IEEE (2015)
29. Zinali, M., Mili, M.R., Khalili, M.: Increasing energy efficiency through maximizing channel capacity with a minimum power allocation in the Femtocell network. In: 2015 2nd International Conference on Knowledge-Based Engineering and Innovation (KBEI), pp. 1106–1112. IEEE (2015)
30. Al Haddad, M., Bayoumi, M.: Green energy solution for femtocell power control in massive deployments. In: 5th International Conference on Energy Aware Computing Systems & Applications, pp. 1–4. IEEE (2015)

31. Sowmya Naik, P.T., Narasimha Murthy, K.N.: Joint algorithm for traffic normalization and energy-efficiency in cellular network. In: Software Engineering Perspectives and Application in Intelligent Systems, pp. 47–57. Springer, Cham (2016)
32. Wang, Y., Dai, X., Wang, J.M., Bensaou, B.: Energy efficient medium access with interference mitigation in LTE femtocell networks. arXiv preprint arXiv:1508.01454 (2015)
33. Wang, Y., Dai, X., Wang, J.M., Bensaou, B.: Iterative greedy algorithms for energy efficient LTE small cell networks. In: 2016 IEEE Wireless Communications and Networking Conference, pp. 1–6. IEEE (2016)

Model-Based Prediction of the Size, the Language and the Quality of the Web Domains

Rexhep Shijaku[2] and Ercan Canhasi[1,2(✉)]

[1] Gjirafa, Inc., Rr. Ahmet Krasniqi, Veranda C2.7, Prishtinë, Kosovo
rexhepshijaku@gmail.com, ercan@gjriafa.com
[2] University of Prizren, Rruga e Shkronjave, nr. 1 20000, Prizren, Kosovo
ercan.canhasi@uni-prizren.com
http://www.gjirafa.com

Abstract. This article investigates possibilities for developing a model, which without completely downloading domains, is able to predict: (1) the size of domains, (2) the domains main language (as Albanian or not), (3) the multilingual domains, where at least one of the languages is Albanian, and (4) the quality of a domain. Proposed model excludes domains which are not written in Albanian from a given set of domains and ranks exclusively Albanian domains by their importance. Consequently, presented model offers higher flexibility and efficiency to a search engine which tends to index the Albanian web.

Keywords: Language specific web crawling · Multi-lingual domain detection · Domain quality estimation · Domain size prediction

1 Introduction

The World Wide Web is an ecosystem where different domains with different size, quality, language and importance live. These domains are large, medium and small, and by the language they mostly appear either as monolingual or multilingual. While small domains do not show tendency for frequent changes, large domains are considered the ones that constantly change.

The main goal of a search engine which aims to index the web based on a specific language (in our context, in Albanian) is to crawl and index only the pages which are written in its predefined language, to refresh its index efficiently and to show results with higher efficacy. Identifying Albanian and non-Albanian domains, big or small domains, multilingual or monolingual domains in other words relevant domains for a search engine which tends to index Albanian web is an important challenge since it results in a fewer crawling, storage and indexing resources, and a higher quality in its search results.

In this paper we present methods to predict: (1) the size of domains, (2) the domains main language (as Albanian or not), (3) the multilingual domains,

R. Silhavy et al. (Eds.): CoMeSySo 2019, AISC 1046, pp. 209–225, 2019.
https://doi.org/10.1007/978-3-030-30329-7_20

where at least one of its languages is Albanian, and (4) the quality of a domain. These are all combined into a single component which works periodically to generate a list of domains for crawlers next iterations, in which are included only Albanian domains ranked by their importance. The main criteria behind all of this process is to not download completely domains.

The paper is organized as follows: Sect. 2 presents related work. Section 3 presents our proposed model where in each subsection briefly are explained modules which constitute it. Subsequently, in Sect. 4, we report the models results and evaluations. Finally, Sect. 5 concludes the paper.

2 Related Work

There is a lot of work that has been done in topical crawling, document quality measuring and language identifying. Most of these works are performed in the context of a single document.

Topical crawling was first proposed by Chakrabati et al. in [1] as the way to find pages that are relevant to the predefined topics. Fish Search which was presented in [2] attempts to selectively download web pages that are relevant to a particular topic by using the common technique of keyword match. Another important method was introduced in [3] which presents language specific crawling strategies based on the assumption that there is a language locality on the Web. Language identification of a document is mostly implemented based on document content. The "common words" approach [4] is based on the observations that for each language, there is small class of words that carry little information but make up a large portion of any text. These are called function words or stop words and their presence is to be expected as word distribution follows Zipf's law. In [5] is presented a method which generates N-gram frequency profiles to represent each of the categories (languages). To classify a document the system first computes its N-gram frequency profile, compares this profile compared against each available language model then it identifies the documents language by the model having the smallest distance. There is URL based language identification approaches as well e.g. in [14] Baykan et al. proposed a method of determining the language of a web page using only its URL.

Finally, document quality estimation is another important topic that has been studied by many researchers, these contributions have resulted in development of highly successful graph algorithms such as PageRank [10], HITS [11] and SALSA [12], which estimate document quality by examining their neighborhood in the link graph. On the other hand, Bendersky et al. in [13] evaluate the quality of the document by taking into account its content.

To sum up, we propose a model which makes conclusions on domain context, instead of a single document, therefore, we experimented with a few of aforementioned works and adapted a couple of their methods to our model.

3 Proposed Model

We adapt the idea of topical crawling to the language specific web crawling. Our proposed model is based on our heuristic assumptions such as: (1) web domains have oriented graph structures where each page of them is represented by a vertex and each link between the pages by an edge where the root of the graph represents the index page of any domain, (2) there exists a notion as *website level* which is the distance of any page from the root of a domain, more concretely the number of links followed by a user in order to access a particular page, where the initial point of this is the index page of the domain. Breath-first search strategy is the best possible traversal method that we apply to our crawler, since we crawl the domains level by level. Figure 1 outlines the proposed model which has three main modules: crawler, analyzer and classifier. While crawler is responsible for downloading pages recursively until the third level crawl is done, analyzer is responsible to estimate and predict: the language, the size and the quality of the web domain, lastly classifier is authoritative to classify domains and to rank them based on their score. This crawler is a hybrid type of focused and incremental crawlers: (1) focused since it collects relevant (Albanian) documents and (2) incremental because it incrementally refreshes the existing collection of a domain list by visiting them periodically in the predefined time intervals (Fig. 2).

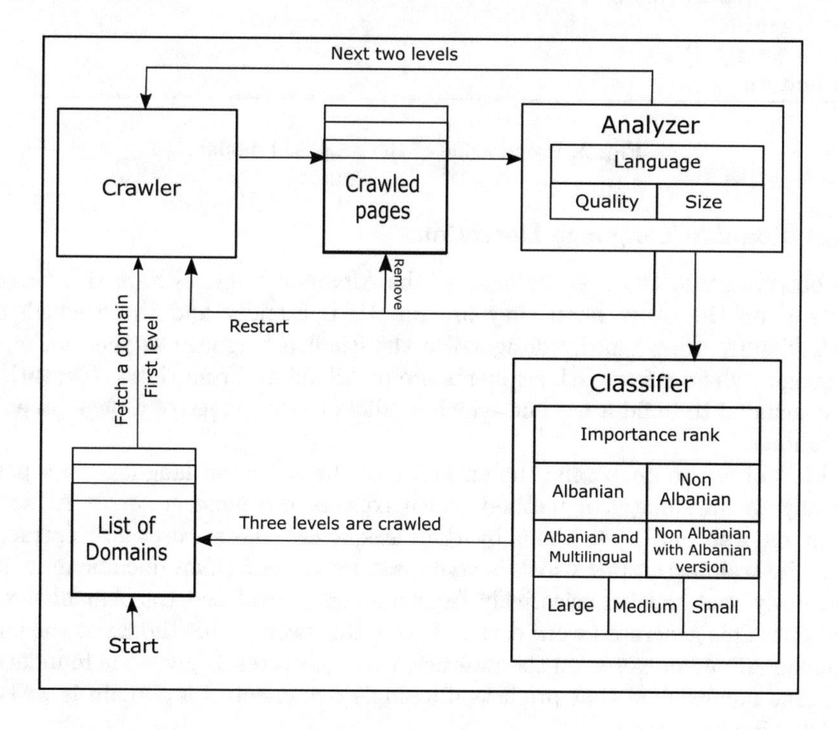

Fig. 1. Proposed model

Algorithm 1 Lotor

1: **for** each Domain *dname* **do**
2: $index_{page} \leftarrow download(dname)$
3: $l_1 \leftarrow extractpages(index_{page})$
4: $isalbanian \leftarrow isalbanian(index_{page}, l_1)$
5: $checkml \leftarrow true$
6: **if** *isalbanian* **then**
7: **if** *checkml and multilingual*(l_1) **then**
8: $preventnonalbaniancrawl()$
9: **end if**
10: $l_2, l_3 \leftarrow extractpages(l_1)$
11: $size \leftarrow size(l_1, l_2, l_3)$
12: $quality \leftarrow quality(l_1, l_2, l_3)$
13: $score \leftarrow score(albanian, size, quality)$
14: $classify()$
15: *continue*
16: **end if**
17: **if** $hasalbanian(l_1)$ **then**
18: $index_{page} \leftarrow getindexpage(l_1)$
19: $l_1 \leftarrow extractpages(index_{page})$
20: $checkml \leftarrow false$
21: $isalbanian \leftarrow true$
22: *repeat step* 6
23: **end if**
24: $classify()$
25: **end for**

Fig. 2. Pseudocode of the proposed model

3.1 Albanian Language Detection

We observed that the vast majority of the Albanian websites have rich textual content, on the other hand they are not URL friendly and those which are URL friendly are primarily designed in the English language as the "technical language" where a few URL segments are in Albanian. From these observations we concluded to build a method which predicts the language of a page based in its content.

First of all, to determine the existence of the Albanian language in a page, initially we formulated a method which collects the most common Albanian words on web and sorts them by their frequency, these words are extracted from the training corpus which is composed by the Albanian documents which previously was crawled selectively from manually identified top Albanian web domains. This generated word corpus D contains twenty nine thousand the most common Albanian words on the internet, which, as a result gives the foundation to define our method that predicts if a single document of a domain is written in Albanian or not:

$$D = \{\, x \mid x \text{ is an Albanian word} \,\} \tag{1}$$

Every web page which has a textual content is composed by a finite number of words:

$$A = \{\, x \mid x \text{ is a word found in page} \,\} \tag{2}$$

Inter-lingual homographs as well exits between Albanian and other languages (e.g Italian), our observations show that these words are usually short, also acronyms mostly appear as short words. Since we make term based analysis, these words may potentially affect the accuracy we tend to achieve. In order to attenuate this effect we eliminate all the words which are shorter than four letters:

$$B = \{\, x \mid \; x \in A \text{ and length of } x > 3 \,\} \tag{3}$$

and in the same way we filter the corpus words:

$$E = \{\, x \mid \; x \in D \text{ and length of } x > 3 \,\} \tag{4}$$

the cardinality of intersection between the derived sets gives all of the Albanian words in a single document which are composed minimum by four letters:

$$\alpha = \mid E \cap B \mid \tag{5}$$

and the cardinality of B will indicate the count of all the possible words in a page:

$$\omega = \mid B \mid \tag{6}$$

The ratio between α and ω gives Albanian rate of a document and it can be calculated as follows:

$$v(\alpha, \omega) = \begin{cases} \frac{\alpha}{\omega}, & \text{if } \omega \geq 1 \\ 0, & \text{otherwise} \end{cases} \tag{7}$$

We consider the Albanian language mean value of the first level documents as the second parameter in our method. If N is a vector which contains all the documents which are found in the first level of the domain $N = \{p_1, p_2, p_3 ... p_n\}$ this vector has two corresponding vectors: (1) Albanian word count vector $A = \{\alpha_1, \alpha_2, \alpha_3 ... \alpha_n\}$, and (2) all document words which are longer than three letters $\Omega = \{\omega_1, \omega_2, \omega_3 ... \omega_n\}$. By these vectors we can derive the language vector of the first level pages:

$$\Delta(A, \Omega) = \{v(A_1, \Omega_1), v(A_2, \Omega_2) ... v(A_n, \Omega_n)\} \tag{8}$$

and the Albanian rate of the first level pages μ is described by the following expression:

$$\mu(\Delta) = \begin{cases} \frac{\sum_{\delta \in \Delta} \delta}{|\Delta|}, & \text{if } |\Delta| \geq 1 \\ 0, & \text{otherwise} \end{cases} \tag{9}$$

where $|\Delta|$ is the total number of documents found in the first level of domain. Similarly, if p_{index} is the index page it should have two corresponding values α_{index} and ω_{index} which will result to the Albanian rate of the index page υ_{index}. The sum between the Albanian rates of the index page and the first level of a domain will result in the Albanian value of that domain:

$$\delta(\upsilon_{index}, \Delta) = \alpha\upsilon_{index} + \beta\mu(\Delta) \tag{11}$$

where α is a weight parameter for the index page of a domain and β is a weight parameter for the first level pages of a domain. Based on the experiments, we give a higher importance to the index page than to the first level pages $\alpha = 0.8$ and $\beta = 1 - \alpha$.

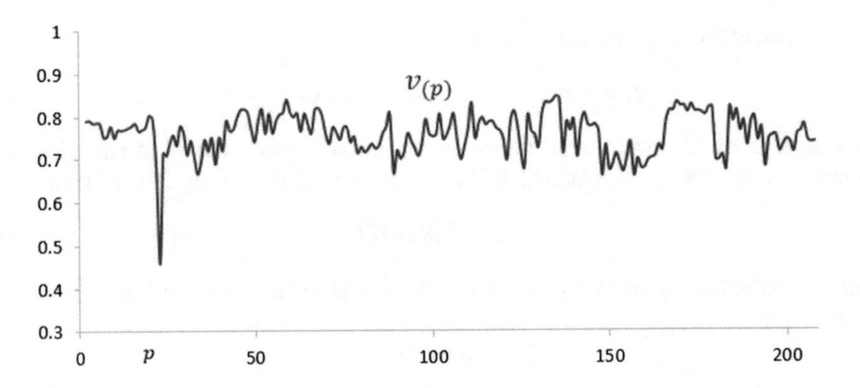

Fig. 3. Albanian Language detection on telegrafi.com

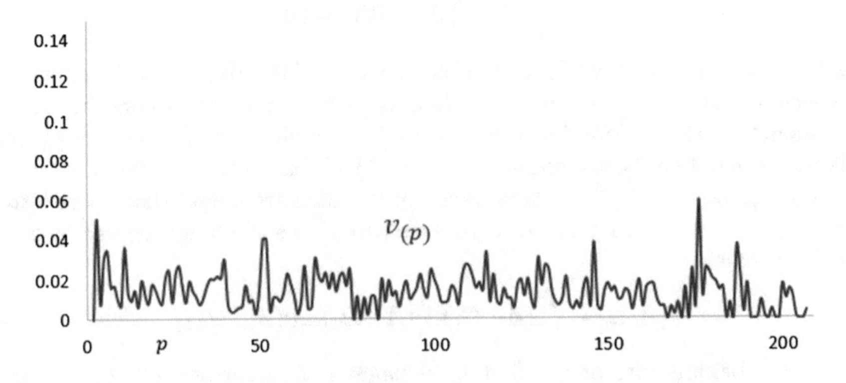

Fig. 4. Albanian Language detection on nytimes.com

Figures 3 and 4 show results of our model for given Albanian and non-Albanian domains. It is noticeable that when a domain is not Albanian function outputs very low values.

3.2 Multilingual Domain Detection

Many of nowadays websites are multilingual. A multilingual website is any website that offers content in more than one language. Examples of multilingual websites may include an Albanian newspaper with an English and an Italian version of its site.

While there is a plenty of literature in the language identification of multilingual documents, we are not aware of any prior work where a language specific web crawler learns to identify and classify the multilingual and monolingual web domains.

Multilingual domain detection for our language specific focused crawler is a crucial challenge, since with this we try to prevent it from collecting, processing and storing unimportant documents.

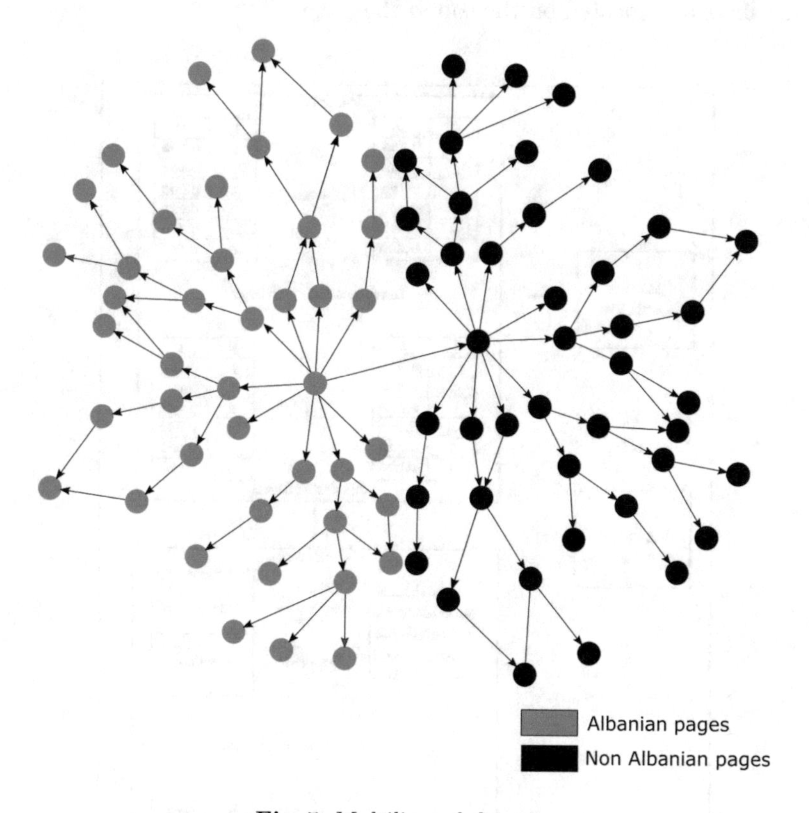

Fig. 5. Multilingual domain

In Fig. 5 is presented a multilingual domain which has an Albanian version and non-Albanian, it is apparent that the non-Albanian version of this domain is kinda a replica domain, which is wholly irrelevant for our crawler. In previous section we formulated Albanian detection function δ by which we get a binary answer for the primary language of a domain. However, identifying the main language of a domain is not the end of analyzers work, often there are two situations that crawlers have to deal with: (1) domains main language may be Albanian but it offers content in a different language also, or (2) the domains main language may not be Albanian but it has an alternative version in Albanian. If after the language determination process analyzer finds that the domain is not written in Albanian but it has an Albanian version, it informs crawler to stop crawling of the non-Albanian version of page and to start crawling only the Albanian part of the domain, otherwise if analyzer uncovers that the domain is written in Albanian it should find the non-Albanian versions of the domain and prevent the crawling of these parts.

By our observations, we concluded that navigation links between languages in Albanian web are present mostly in the index pages, they are shorter compared to other links and located on the top of the page.

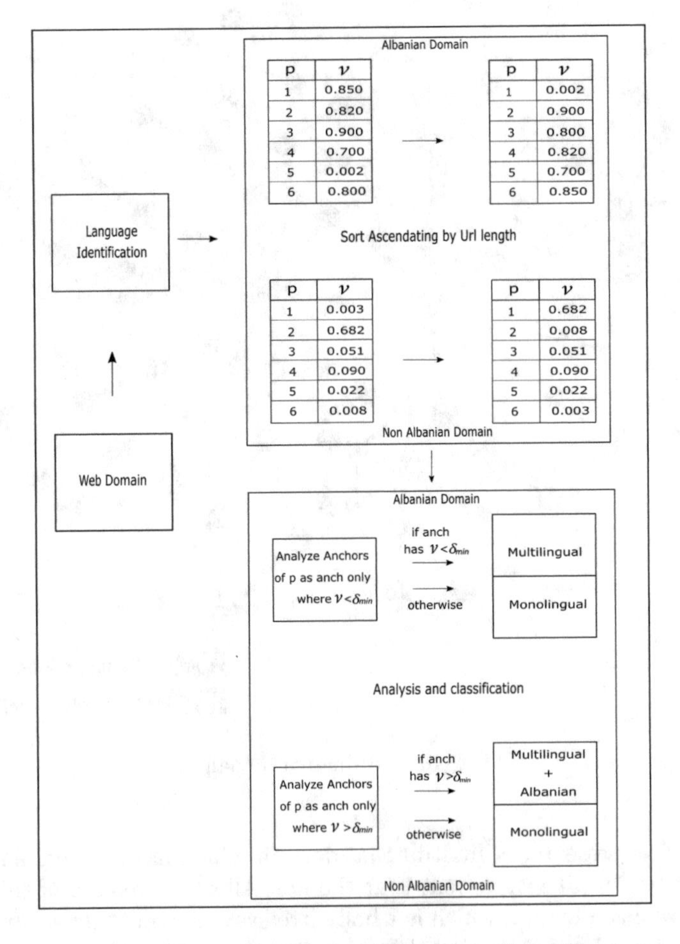

Fig. 6. Detecting if domain is Albanian and multilingual

(1) if a domains main language is Albanian most of its first level pages have high v, in this case we try to find pages where $v < \delta_{min}$ and analyze them:

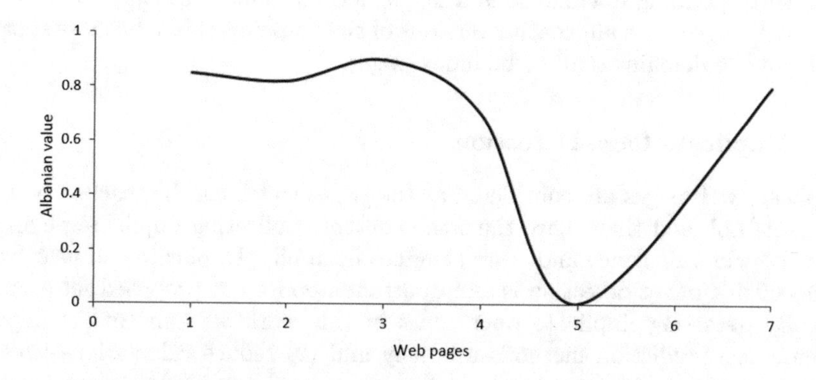

Fig. 7. Multilingual domain detection in an Albanian domain

In Fig. 7 are plotted first level language values of a domain whose main version is written in Albanian, but the fifth page seems as it may navigate to the part of the domain which is written in a different language.

(2) if language was detected as non-Albanian most of the pages in first level of domain will have low Albanian rates v, in this case we will try to find any page where $v > \delta_{min}$ and to analyze it.

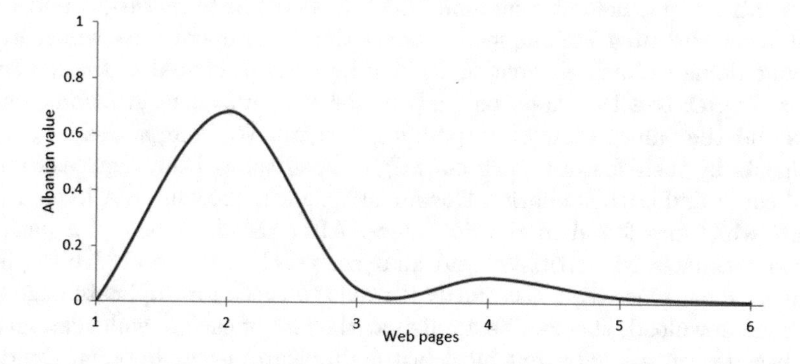

Fig. 8. Albanian detection in a non-Albanian multilingual domain

In Fig. 8 are plotted first level language values of a domain which is not written in Albanian, but the second page changes drastically and seems like it potentially may navigate to the part of the domain which might be written in

Albanian. We can draw a conclusion for both cases after we reorder the first level pages ascending by their URL length, and analyze each as the Fig. 6 delineates.

Furthermore, ordering the URLs ascending by their length before analyzing their corresponding documents is a faster and more accurate way to reach out the index page of the alternative versions of the domains, this is because shortest URLs in the domains tend to be index pages.

3.3 Duplicate Page Detection

Duplicate web pages are considered as the pages which can be reached by more than one link and they share the same content. Collecting duplicate pages can affect crawlers efficiency and search engines flexibility. Duplicate and near duplicate web document detection is an important step for our proposed set of methods. By detecting duplicate documents in this work we aim to: (1) improve domain size prediction methods accuracy and (2) reduce calculations time for estimating the quality and the language of a domain and (3) minimize storage resources for collected pages.

Detecting duplicate documents is an easier problem than detecting near duplicate documents. Many different techniques have been developed to identify pairs of documents that are duplicate or similar to each other. Brin et al. in [6] proposed a system for registering documents and then detecting complete copies or partial copies in which were described algorithms for such detection, and metrics required for evaluating detection mechanisms. Some other interesting and important papers are: Broder et al. in [7] proposed shingling technique for estimation degree of similarity between two documents and Manku et al. [9] who proposed near duplicate detection model based on Charikar's simhash technique in a multi-billion page repository [8].

We introduce a heuristic method which is based in term weight and traditional check-summing techniques to detect duplicate documents which live in the same domain. Each element of the word corpus presented in the beginning of this chapter is a key value pair where the keys represent Albanian unique words and the values their corresponding weights. We compare same domain documents by their features such as: weight, hash value, html tags (e.g. image, video) count and URL similarity. Document weight is the sum of Albanian word weights which are found in the document. After the download of a page our method estimates its attributes and analyzer checks for a probable duplicate document occurrence. Not necessarily it needs to check for duplicate page after each page download, this can be applied at the end of each downloaded level as well, because we suggests that most of the duplicates occur between the documents which live in the same domain level, hence, they are easier to detect in the context of level (Fig. 9).

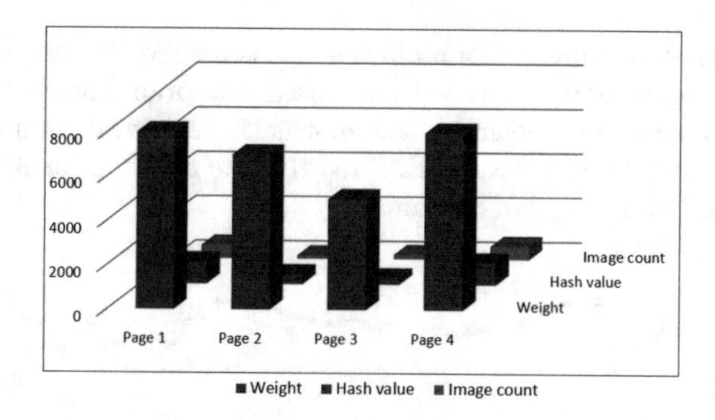

Fig. 9. Duplicate page detection

3.4 Evaluating the Quality of a Domain

Document quality estimation is a challenging topic. Previous works such as PageRank [10], HITS [11] and SALSA [12] do not take into account document content quality. On the other hand, Bendersky et al. in [13] proposed the quality-biased ranking method that promotes documents containing high-quality and penalizes low-quality documents. This method was formulated by document quality features such as: number of visible terms on the page (as rendered by a web browser), number of terms in the page title, field, average length of visible terms on the page, fraction of anchor text on the page, fraction of visible text on the page (as rendered by a web browser), entropy of the page content, stopword/non-stopword ratio, fraction of terms in the stopword list that appear on the page, the depth of the URL path (number of backslashes in the URL), fraction of table text on the page.

Since we aim to estimate the quality of a domain as a parameter for our ranking function, to attain this, firstly, we need a method which estimates the quality of a single document. For this purpose after certain analysis we decided to utilize and adapt the query-independent method which was proposed by Bendersky et al. in [13] and is based on a set of quality-based factors $\mathcal{L}(D)$ associated with the document node D:

$$\log \nu(D, \Lambda) = \sum_{L \in \mathcal{L}(D)} \lambda_L f_L(D) \tag{12}$$

where D is a document and Λ is a set of free parameters. As it is known that a domain incorporates a set of documents in it, we propose that the quality of a domain is represented by the quality of its documents. Based on the Eq. (12) we define the quality of a domain as follows:

$$q(D, \Lambda) = \frac{\sum_{d \in \mathrm{D}} \log \nu(d, \Lambda)}{|D|} \tag{13}$$

where D is a set of documents found in the first three levels of any given domain. In defining of some parameters we have used our prior knowledge. The used heuristics are commonly known by the information retrieval community.

In Fig. 10 are plotted values of Eq. (13) that we adapted for Albanian documents in two different type of domains.

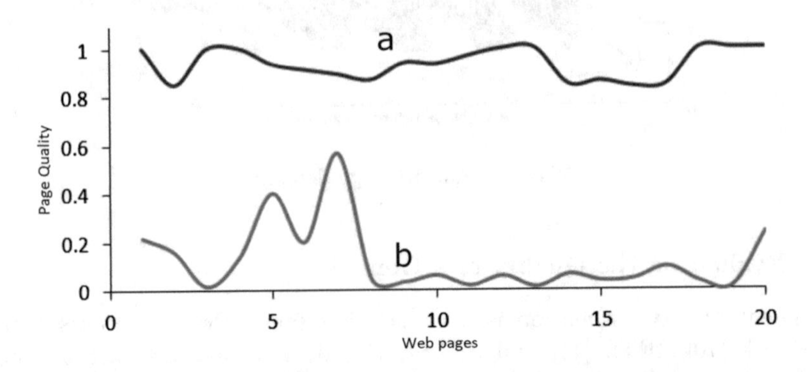

Fig. 10. Quality values of the pages in two different domains: (a) Qualitative (b) Non-Qualitative

3.5 Size Prediction and Ranking

Size prediction is another crucial parameter which has an impact in the evaluation of the importance of a web domain. Based on their sizes we classify web domains in three main groups: small domains, medium domains and large domains. We define the small domains as the domains whose graphs compress dramatically in the second eventually in the third level compared to the large domains whose graphs extend intensively through the first three levels (Figs. 11 and 12).

Since large domains frequently change by updating their content and increasing their document count we assume that they are more important to the web crawlers than small domains which offer static content and small amount of data. Before we make a prediction about the size of a domain, in the list of crawled pages of the domain we apply duplicate page detection method by which we try to remove all occurred duplicates. After the crawling process for a particular domain is finished, analyzer determines its size m and finally its score S.

We define the size of the domain as logarithmic function of the sum of document count found in the first three levels T and the gradient among first three levels γ:

$$m = \log(T + \gamma) \tag{14}$$

where:

$$T = \sum_{n \in N} n \tag{15}$$

and the gradient

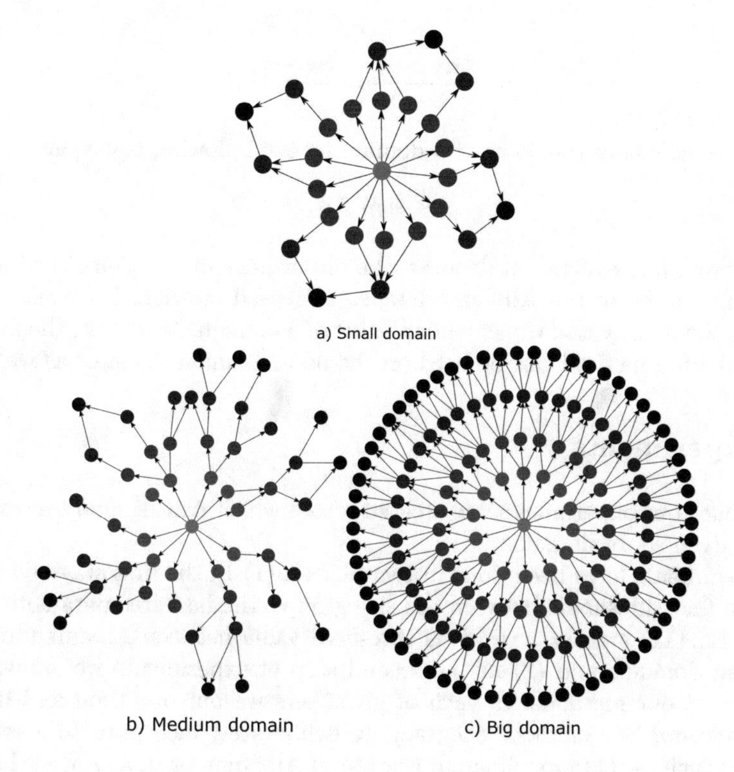

Fig. 11. Domain sizes presented by graphs

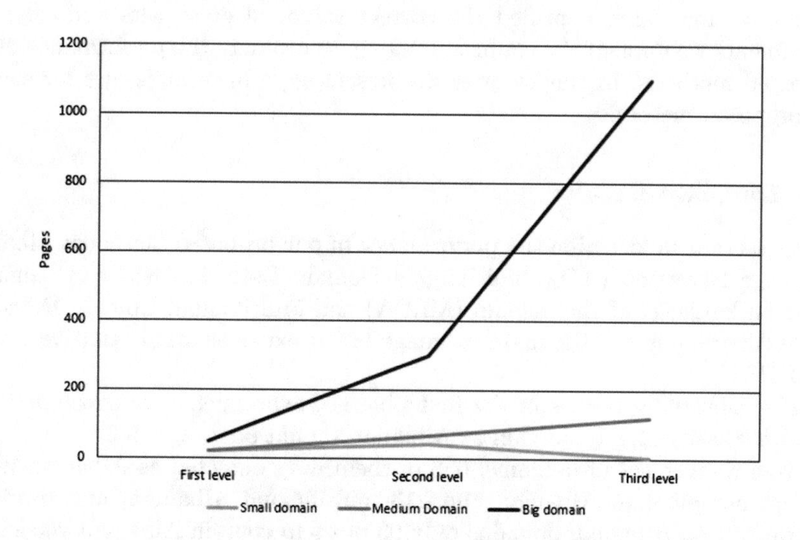

Fig. 12. Different domains sizes

$$\gamma = \frac{n_3 - n_1}{3 - 1} = \frac{n_3 - n_1}{2} \tag{16}$$

Finally we calculate the score of a domain by the following equation:

$$S = m\ \delta\ q \tag{17}$$

As the preceding expression denotes, the importance of any domain which is in the Frontiers list of the Albanian language focused crawlers is correlated with the size, the quality and the Albanian value of a domain. Moreover, the proposed size prediction method can be used on the non-Albanian domains as well.

4 Experiments

To conduct our experiments, we created a tool which crawls, analyzes and classifies a given set of domains.

Experiments have been done in two phases: (1) In the first stage we investigated to find thresholds between the size groups, the best free parameter values for the Eq. (11) and δ_{min} which is threshold value between Albanian and non-Albanian domains and (2) in the second step of experiments we analyzed the accuracy of our methods. In each of given sets we put one thousand domains and to observe our methods accuracy we deliberately incorporated a variety of domains such as: (1) monolingual: written in Albanian or in any other language and (2) multilingual: which provide the Albanian version of a domain and where the main version is either Albanian or not. Before the second phase of experiments, we manually identified the correct values of given sets and after the experiments we applied the confusion matrix to evaluate the performance of the presented methods. In this chapter we present experimental results for each of our proposed methods.

4.1 Language Results

In this section we describe the performance of our proposed methods: Albanian Language Detection (AD), Multilingual Domain Detection where Albanian is the main language of the domain (MLDA) and Multilingual Domain Detection where Albanian is not the main language but it exists as an alternative version (MLDNA).

After observing results in the first phase of experiments we concluded that δ_{min} for classifying any domain as Albanian should be $\delta_{min} = 0.2$.

From a given set of domains: 688 of them were detected as Albanian where 475 are monolingual Albanian and 213 multilingual Albanian, and from 332 discovered non-Albanian domains only 93 of them contain Albanian version.

Table 1 shows that proposed language identification methods are highly accurate. After manual analysis we observed that these methods have higher accuracy

Table 1. Albanian Language detection results

Method	Precision	Recall	Specification
AD	96.5%	99%	92.7%
MLDA	92.4%	83%	97%
MLDNA	92.4%	80.7%	96.9%

in the domains which are textually rich and have rational designs (especially on those which are search engine optimized). In the analysis of the misclassified domains we see that most of the errors in the Albanian language detection occurred in the domains which mostly have documents: (1) textually very poor, (2) Albanian and textually composed by internationalisms and (3) Albanian and textually composed by words which are Albanian but are not covered by our word corpus. On the other hand, multilingual domain detection methods show non real results mostly when: (1) the domains store their languages in the session variables, (2) the navigations between languages are processed by non server-side scripting languages and there is not information about language in URLs and (3) the domains where the anchors are monolingual.

4.2 Duplicate Page Detection Results

After the crawling of the first three levels in 688 domains which are detected as Albanian and only the first level of other non-Albanian domains, crawler found nearly 310 K web documents and analyzer marked nearly 66 K as duplicates. The results show how often a crawler deals with duplicate page situations. Dynamically generated links are the most common cause of this occurrence. These traps usually appear in the big domains which have dynamic content. An example of big domains are forums where thousands of comments are generated for a single topic and each of these comments have an anchor with identifier that references the same page with different URL. To evaluate the performance of the duplicate page detection method we analyzed 30 domains where were detected 440 duplicate pages from 1554 in total. Statistically expressed our duplicate page detection reaches up to Accuracy of 99.2%, Specification 91.1% and Precision 97.04%.

4.3 Size Results

First phase of experiments led us to identify thresholds between the presented size groups. Big domains are assigned to be in the interval of $[2.68, +\infty)$, mediums $(1.7, 2.68)$ and smalls $(-\infty, 1.7]$.

After the second phase of experiments the classifier shows a high accuracy in the specification and precision of the small and big domains. In certain situations it misclassified some big domains as medium and vice versa, but the most important point to emphasize is that there was not any big domain which was classified as small or conversely (Table 2).

Table 2. Size detection results

Method	Precision	Recall	Specification
LDC	93%	89.4%	96.4%
MDC	87%	87.8%	93.5%
SDC	95%	97.9%	97.5%

4.4 Quality Estimation Results

In the quality estimation experiments we compared high ranked domains with low ranked domains (Fig. 13). Documents of the domains which ranked as qualitative have higher textual content, fraction of stopwords and term length, which offer better readability and higher quality. In the first group are present educational websites, news websites and web forums. In the domains with low quality we observed spam pages, poor content, low fraction of stopwords and high anchor text usage.

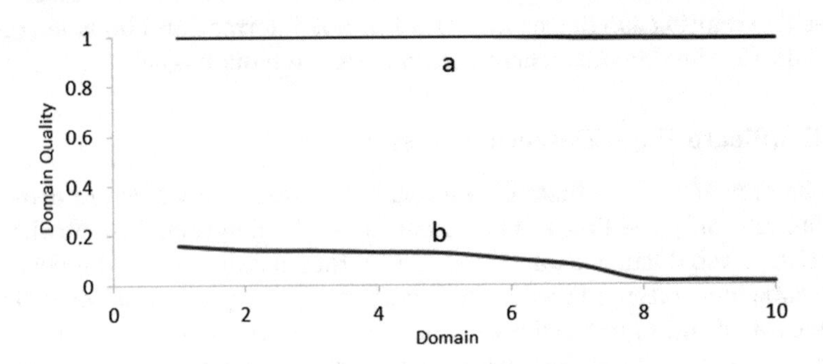

Fig. 13. The quality results of (a) 10 high ranked domains and (b) 10 low ranked domains

5 Conclusion

In this paper we proposed a set of methods which can be use to optimize the work of the language specific web crawler, more concretely for an Albanian language specific web crawler. These methods rank domains without downloading them completely.

We created a tool by these methods and crawled a thousand domains where experiments proved positive results. Ranking domains by their importance and determining their main attributes, is an important meta data for the language specific web crawlers, by which they can concentrate their work on important domains.

Acknowledgment. This work was completely supported by the Gjirafa, Inc. We also thanks Univesity of Prizren for allowing us to use their infrastructure.

References

1. Chakrabarti, S., van den Berg, M., Dom, B.: Focused crawling: A new approach to topic-specific web resource discovery. Comput. Netw. **31**(11–16), 1623–1640 (1999)
2. De Bra, P., Post, R.: Searching for arbitrary information in the WWW: The Fish-Search for Mosaic. In: WWW Conference (1994)
3. Somboonviwat, K., Tamura, T., Kitsuregawa, M.: Simulation study of language specific web crawling
4. Ingle, N.: A Language Identification Table. Technical Translation International (1980)
5. Cavnar, W.B., Trenkle, J.M.: N-gram-based text categorization, Ann Arbor, MI (1994)
6. Brin, S., Davis, J., Garcia-Molina, H.: Copy Detection Mechanisms for Digital Documents (1995)
7. Broder, A., Glassman, S.C., Manasse, M., Zweig, G.: Syntactic clustering of the web. Comput. Netw. **29**(8–13), 1157–1166 (1997)
8. Charikar, M.: Similarity estimation techniques from rounding algorithms. In: Proceedings of the 34th Annual Symposium on Theory of Computing (STOC 2002), pp. 380–388 (2002)
9. Manku, G.S., Jain, A., Sarma, A.D.: Detecting Near Duplicates for Web Crawling, May 8–12 (2007)
10. Brin, S., Page, L.: The anatomy of a large-scale hypertextual web search engine. Comput. Netw. ISDN Syst. **30**(1–7), 107–117 (1998)
11. Kleinberg, J.M.: Authoritative sources in a hyperlinked environment. J. ACM **46**(5), 604–632 (1999)
12. Najork, M.A.: Comparing the effectiveness of HITS and SALSA. In: Proceedings of CIKM, pp. 157–164 (2007)
13. Bendersky, M., Bruce Croft, W., Diao, Y.: Quality-Biased Ranking of Web Documents (2011)
14. Baykan, E., Henzinger, M., Weber, I.: Web Page Language Identification Based on URLs (2008)

Smart Injector Well Optimization for a Non-communicating Reservoir

D. A. Oladepo[1(✉)], C. T. Ako[1], O. D. Orodu[1], O. J. Adeleke[2], and A. S. Fadairo[1]

[1] Department of Petroleum Engineering, College of Engineering, Covenant University, Ota, Nigeria
david.oladepo@covenantuniversity.edu.ng
[2] Department of Mathematics, College of Science and Technology, Covenant University, Ota, Nigeria

Abstract. This study proposes a technique called intelligent well completions which uses water injection to improve sweep efficiency and recovery factor in oil production. The application of the technique in a water-flooding operation aims to optimize the outflow control valve (OCV) settings and injection rate of each segment of the oil production reservoir. A dynamic reservoir model was built using a reservoir flow performance simulator. It was observed that smart injection wells yield a better sweep efficiency with a favourable mobility ratio and improved pressure maintenance leading to an increase in field oil efficiency (FOE) when compared to the conventional wells and overall increase in net present value (NPV). In addition, the reservoir field oil efficiency was increased by 5% using the proposed technique.

Keywords: Field oil efficiency · Smart wells · Outflow Control Valve · Net present value

1 Introduction

Smart wells are non-conventional wells that make use of downhole sensors to monitor downhole temperature, pressure, flow saturation changes, phase composition. They are also completed with subsurface inflow control valves (ICVs) to control the flow of fluid from a segment of the reservoir to the well. They give real time information about the pressure, temperature, flow and phase composition at the subsurface (Fombad 2016). The smart well was originally used for shut off areas that were watered-out and as a measure taken against the early breakthrough during production. Improvements in smart well technologies and increasing amount of opportunities in more challenging assets, and the use of smart well technologies to improve recovery has caught significant attention in the oil and gas industry in the last decade. Several workflows have been developed and proposed in order to automate the whole process that integrates several sub processes focusing on specific parts of the surface or subsurface phenomena. Reservoir sweep is a crucial part of recovery efficiency, especially where significant investment is done by means of installing smart wells that feature ICVs,

R. Silhavy et al. (Eds.): CoMeSySo 2019, AISC 1046, pp. 226–232, 2019.
https://doi.org/10.1007/978-3-030-30329-7_21

which are remotely controllable. However, as it is a relatively newer concept, effective use of this technology has been a challenge (Ranjith et al. 2017).

Brouwer (2001) conducted a study to increase recovery through water flooding with smart well technology. In his work, water flooding was used to expand oil recovery after principal exhaustion. Zones with excessive permeability can affect the recovery of hydrocarbon, since they can bring about early water breakthrough and catching of by-passed oil. Smart well innovation provides the chance to balance these impacts by forcing a suitable pressure along the injection and production wells. In his research, water flooding is enhanced by changing the well profiles as indicated by some basic calculations that move stream ways far from the high porousness zone with a specific end goal to postpone water achievement. Aitokhuehi (2004) presented a study on the real-time optimization of smart by considering the use of both valve optimization and history matching. The author claimed that maximizing recovery in smart wells requires the estimation of the optimum settings of the control valves.

In another study, Almeida et al. (2007) used evolutionary optimization to control smart wells under technical uncertainties. Their work focused on an evolutionary algorithm that has the ability to optimize the mechanism for controlling the smart well technology currently used on the smart fields. El-Sayed et al. (2014) presented a study on the first intelligent well installation in UAE offshore field. The well was equipped with three inflow control valves and four permanent downhole gauges with remote control ability to control the flow from each zone monitoring the real time gauge data. The completion enabled commingled production from three reservoirs while balancing flow contribution from each reservoir and avoiding cross-flows from one reservoir into another. The intelligent well technology is one of the most important technologies with the capability of optimizing reservoir management, increasing recoverable reserve, enhancing oil recovery and speeding up the oil exploitation.

Goodyear et al. (1996) investigated on hot water flooding for high permeability viscous oil fields. In their investigated work, they discussed about reservoirs at the North Sea having oil of high viscosity present in permeable sands, they then stated that the technique they feel best to be recommended is injecting hot water along with the application of Horizontal wells of high length. Christensen et al. (2000) investigated the compositional simulation of water-alternating-gas processes. Their study incorporated the "compositional simulator" and "calculation process" to establish the relative permeability for the injection of WAG. The performance of the Hysteresis is executed for the water and gas stages (wetting and non-wetting stage). Oladepo et al. (2017) presented a study that shows how various WAG ratios are being varied with WAG injection program. From their study, various WAG ratios were used in optimizing the ultimate recovery and were then compared with two secondary injection schemes. Ekebafe (2012) presented a studied on smart well technology application in deep-water field development. The study puts more emphasis on integrated method and the implementations that aided it. Their study provided further insights into field developments.

The remaining parts of this paper are organized as follows. In Sect. 2, the research methodology is presented for the problem under study. The various results obtained through modeling and simulations are presented and discussed in Sect. 3. The concluding remarks are given in Sect. 4.

2 Methodology

2.1 Dynamic Model Description

A dynamic model was built using Eclipse E-100 black oil simulator. The history matching, predictions and sensitivity analysis were carried out. The individual injection rates for each segment from the intelligent injection well will be optimized for each of the layers to be considered in the reservoir by making use of the Frontsim application available in Eclipse. For this study, the optimization process will involve optimizing well allocation rates, OCV's, well location as control variables to improve the volumetric sweep efficiency of the Water flooding injection profile.

2.2 Numerical Reservoir Simulator

Schlumberger GeoQuest's Eclipse was used as the numerical reservoir simulator in this research. Eclipse can easily implement many types of production and economic constraints through its existing keywords. These constraints include economic production limit, production rate limit, and bottom hole pressure limit (BHP).

2.3 Optimizer and Links to Simulator

The optimizer, as indicated above, drives ECLIPSE for objective function evaluations. An interface establishes communication between the optimization routines and the simulator. Although the cases presented here are based on the maximization of the recovery factor and cumulative oil production, different objective functions, such as the net present value of the project, the minimization of water cut or the gas-oil ratio of individual wells, well groups or the entire field can easily be implemented. Multiple valves installed on different wells and their associated production rates/bottomhole pressures or injection rates/injection pressures can also be optimized along with the valve settings.

3 Results and Discussion

3.1 Case 1: FIELD X (Conventional Waterflood (Non-communicating))

In this case, the conventional water flood was performed, without downhole outflow control on injection. This case actuated on a field found to be towards total marginality due to rapid decline in pressure. There are three 40 ft thick producing units which are separated by 10 ft thick shale barriers. The reservoir dimensions are 2500 × 2500 × 50 ft3. The porosity across each of the fourteen layers was assumed to be homogenous. The layers are of inhomogeneous permeability from top to bottom. The shale barriers are impermeable. The ratio of the vertical to horizontal permeability in each layer was taken to be 0.1. The displacement is unfavourable, with an endpoint mobility ratio of 3.5. Water was injected at a bottomhole pressure target of 10000 psi and subject to the maximum injection rate of 6 MSTB/d. Production was specified to occur at a target bottomhole pressure of 1000 psi, subject to a maximum oil rate of 8 MSTB/d. In both the controlled and uncontrolled cases, we reduced the current liquid production rate by 60% whenever a sharp increase in water cut is noticed. A minimum oil production rate of 400 STB/d was also imposed as an economic constraint. The simulations were performed for 14620 days (40 years).

3.2 Case 2: FIELD X (Smart Waterflood (CHOKE CASE))

In this case, water was injected into the reservoir using downhole outflow control valve for optimal performance of the injector well. Three valves were used for the three reservoirs in the field. Sensitivity analysis was done on the Outflow Control Valves (OCV's) to achieve the optimum valve setting with a total of 30 runs. Water was injected at a bottomhole pressure target of 10000 psi and subject to the maximum injection rate of 6 MSTB/d. Production was specified to occur at a target bottomhole pressure of 1000 psi, subject to a maximum oil rate of 8 MSTB/d. In both the controlled and uncontrolled cases, the current liquid production rate was reduced by 60% whenever a sharp increase in water cut is noticed. A minimum oil production rate of 400 STB/d was also imposed as an economic constraint. The simulations were performed for 14620 days (40 years).

The results obtained from implementing the methods described above are presented in form of tables and figures. Each of the results is discussed appropriately. Note that COP denotes cumulative oil production and FOE stands for field oil efficiency.

From Table 1, Case 2 (choke case) gave a Field Oil efficiency of 36% compared the conventional case and base case which gave a FOE of 32% and 12.5% respectively. An increase in FOE for the choke case is as a result of the optimal settings of the OCV for the choke case.

Table 1. Field X performance overview for a non-communicating reservoir

	COP (MMSTB)	FOE (%)
Base case	18.75	12.5
Case 1	48	32
Case 2	54	36

In addition, Fig. 1 shows water production due to smart completion in comparison to the conventional waterflood process. An overall increase in oil recovery due to optimization was also observed.

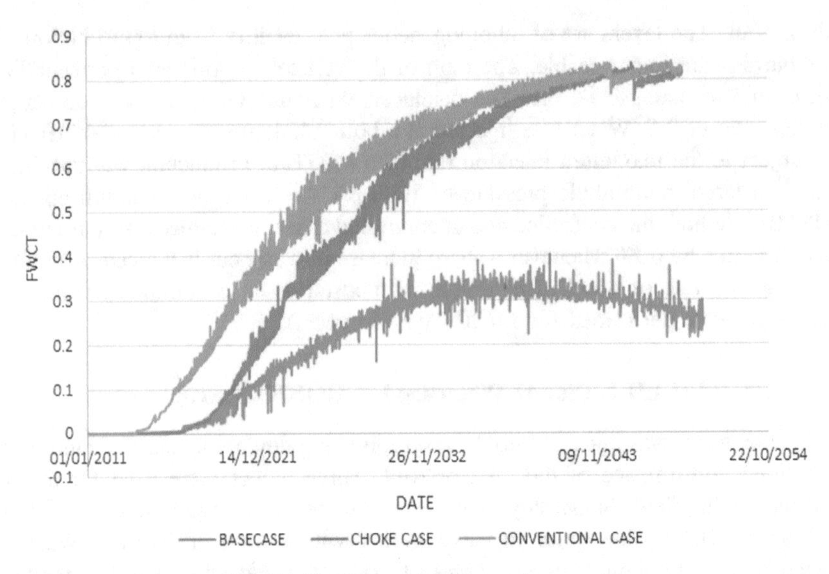

Fig. 1. Field water cut comparison case

3.3 Sensitivity of Valve Coefficient on Field Oil Efficiency Reservoir (Non Communicating)

This section describes the various results obtained from the sensitivity analysis performed on the valve coefficient. The analysis was conducted for the non-communicating well. VOC denotes valve opening coefficient.

Table 2 shows the sensitivity analysis done on different valve openings. Valve opening 0.3 gave us the optimal control setting for segment 1 with a corresponding FOE of 6.70% while at valve opening at 1 (fully opened valves) is at 6.50%.

Table 2. Segment 1 OCV size sensitivity analysis

VOC	0.1	0.2	0.3	0.4	0.5	0.6	0.7	0.8	0.9	1.0
FOE (%)	6.30	6.42	6.70	6.69	6.39	6.40	6.10	6.00	6.20	6.50

Table 3 shows the sensitivity analysis done for the OCV settings of Segment 2. It will be noticed that 0.5 optimal setting gave the highest FOE value and cumulative recovery compared with the OCV setting of 1 (fully opened), which gave an FOE of 16.2%.

Table 3. Segment 2 OCV settings sensitivity analysis

VOC	0.1	0.2	0.3	0.4	0.5	0.6	0.7	0.8	0.9	1.0
FOE (%)	16.50	16.42	16.70	16.69	16.40	16.10	16.10	16.00	16.40	16.20

3.4 Economic Results

From Fig. 2, it is easy to notice the improved economics of the field, a base case experiencing an increase in overall in oil revenue of about 140% for conventional waterflooding and 166.7% for smart development i.e. a net change of 26.7% from conventional waterflood to smart waterflood injection strategy, generally for non-communicating reservoir. Overall development with smart injection strategy yields best economic estimates as judged by the payout time (POT).

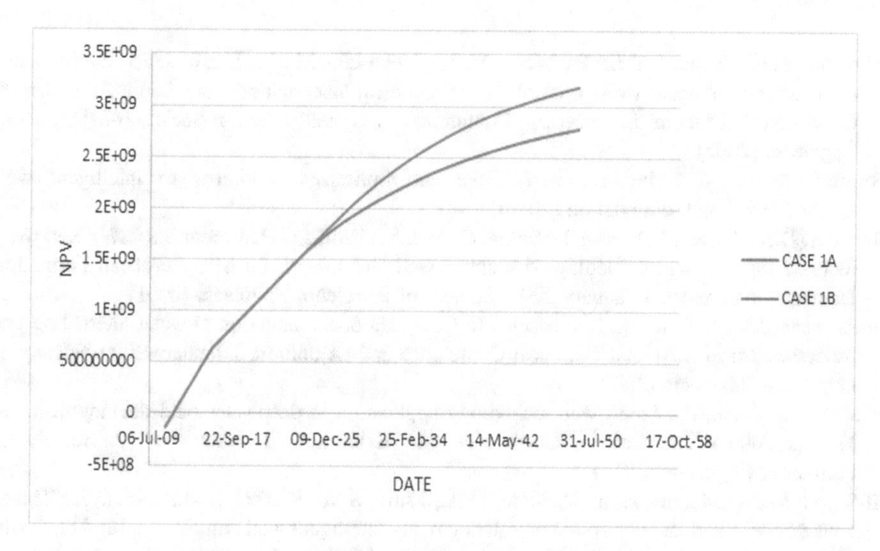

Fig. 2. Pay Out Time

It can be deduced that the POT for all case scenarios is averagely the same, but already established is the drawback of using the POT as an economic indicator, i.e. it does not take cognisance of the future project performance, hence not clearly representative, nonetheless smart well development strategy can be appreciated by virtue of the distinct differences in project performance. Also from the above indicator, it will be noticed that the performance of the NPV outputs increase with smart injection strategy.

4 Conclusion

The use of smart injector well in some oil producing regions of the world is relatively new and not much is known about its applications in terms of the control settings. Smart well technology aids better reservoir management and proper dynamics synthesis. This study has established a foundation for smart injector wells in terms of having an optimal setting for the outflow control valves. The major contributions of this project can be summarised as follows. The reactive control strategy used in this study for smart intelligent well resulted in a significant increase in oil recovery factor over the conventional well and also a cumulative improvement in production. Good sweep efficiency was also achieved in each of the segments and also shows a significant increase in the total field oil production as well as an improved field water cut curve.

Acknowledgement. The authors would like to appreciate the support of Covenant University Management and Covenant University Centre for Research, and Innovation (CUCRID) for their support and funding the publication of this research output.

References

Almeida, L.F., Tupac, Y.J., Pacheco, M.A., Vellasco, M.M., Lazo, J.G.: Evolutionary optimization of smart-wells control under technical uncertainties. In: Latin American & Caribbean Petroleum Engineering Conference, 1 January 2007. Society of Petroleum Engineers (2007)

Fombad, M.W.: A technology perspective and optimized workflow to intelligent well applications. Ph.D. dissertation (2016)

Brouwer, D.R., Jansen, J.D., Van der Starre, S., Van Kruijsdijk, C.P., Berentsen, C.W.: Recovery increase through water flooding with smart well technology. In: SPE European Formation Damage Conference, 1 January 2001. Society of Petroleum Engineers (2001)

Christensen, J.R., Larsen, M., Nicolaisen, H.: Compositional simulation of water-alternating-gas processes. In: SPE Annual Technical Conference and Exhibition, 1 January 2000. Society of Petroleum Engineers (2000)

Ekebafe, A., Ogan, A.: Smart well technology application in deepwater field development. In: Nigeria Annual International Conference and Exhibition, 1 January 2012. Society of Petroleum Engineers (2012)

El-Sayed, M., Al Mutairi, A.M., Hassane, M.A., Kutty, S.M., Karrani, S.M., Kurian, A.: Three-zone commingled and controlled production using intelligent well completion. In: Abu Dhabi International Petroleum Exhibition and Conference, 10 November 2014. Society of Petroleum Engineers (2014)

Goodyear, S.G., Reynolds, C.B., Townsley, P.H., Woods, C.L.: Hot water flooding for high permeability viscous oil fields. In: SPE/DOE Improved Oil Recovery Symposium, 1 January 1996. Society of Petroleum Engineers (1996)

Aitokhuehi, I.: Real-time optimization of smart wells. Master of Science dissertation. Stanford university, June 2004

Oladepo, D.A., Churchill, A., Fadairo, A., Ogunkunle, T.F.: Evaluation of different wag optimization and secondary recovery techniques in a stratified reservoir. Int. J. Appl. Eng. Res. **12**(20), 9259–9270 (2017)

Ranjith, R., Suhag, A., Balaji, K., Putra, D., Dhannoon, D., Saracoglu, O., Hendroyono, A., Temizel, C., Aminzadeh, F.: Production optimization through utilization of smart wells in intelligent fields. In: SPE Western Regional Meeting, 23 April 2017. Society of Petroleum Engineers (2017)

Towards Ontology Engineering Based on Transformation of Conceptual Models and Spreadsheet Data: A Case Study

Nikita O. Dorodnykh[1] and Aleksandr Yu. Yurin[1,2(✉)]

[1] Matrosov Institute for System Dynamics and Control Theory, Siberian Branch of the Russian Academy of Sciences, 134, Lermontov Street, Irkutsk 664033, Russia
iskander@icc.ru

[2] Irkutsk National Research Technical University, 83, Lermontov Street, Irkutsk 664074, Russia

Abstract. The ontology engineering is a complex and time-consuming process. In this regard, methods for automated formation of ontologies based on various information sources (e.g., databases, spreadsheets data, and text documents, etc.) are being actively developed. This paper presents a case study for the domain ontology engineering based on analysis and transformation of conceptual models and spreadsheet data. The analysis of conceptual models, which are serialized using XML, provides the opportunity to develop content ontology design patterns. The specific concepts for filling obtained ontology design patterns are resulted from the transformation of spreadsheet data in the CSV format. In this paper, we present statement of the problem and the approach for its solution. The illustrative example describes ontology engineering for the industrial safety inspection tasks.

Keywords: Ontology engineering · Ontology design patterns · OWL · Conceptual models · Spreadsheets · Transformations · Industrial safety inspection

1 Introduction

Ontologies [1] gained a wide popularity among knowledge base developers as powerful means for representing knowledge in the form of semantic nets with support for the description logic by the standard of the W3C consortium – OWL (Web Ontology Language). Ontology engineering is a complex and time consuming process. Generally, ontology engineering is performed in specialized ontological editors (e.g., Protégé, OntoStudio, WebOnto, Fluent Editor, etc.). However, they mainly target highly qualified specialists (programmers) and restrict access to any other sources of data and knowledge for ontology formation.

In this regard, methods for automated formation of ontologies based on various information sources (e.g., databases, spreadsheets data, and text documents, etc.) are being actively developed. One of such sources is conceptual models of subject domains (e.g., diagrams, schemes, flowcharting, concept maps, mind maps, etc.). Conceptual

R. Silhavy et al. (Eds.): CoMeSySo 2019, AISC 1046, pp. 233–247, 2019.
https://doi.org/10.1007/978-3-030-30329-7_22

modeling tools are more attractive for domain experts. Such tools make it possible to manipulate domain entities including domain-specific notations, for example, event trees, fault trees, Ishikawa diagrams, mind maps and others. The analysis of conceptual modeling tools including IHMC CmapTools, FreeMind, Coggle, Mindjet MindManager, TheBrain, XMind, IBM Rational Rose, Star UML revealed two main issues: the lack of ability of source code synthesis for ontologies or knowledge bases, but also the absence of a general document standard for representation of concepts and relationships. At the same time, most of such software uses incompatible XML-like formats. However, these formats can be used for automatic retrieving of information about concepts and relationships.

The use of conceptual models for ontology engineering provides reusing of a large amount of accumulated heterogeneous information and involvement of domain experts in the knowledge formalization process. Thus, the development of new methods and software for the ontology engineering based on the transformation of various conceptual domain models is relevant and confirmed by the following examples [2, 3].

It should be noted that automated analysis and transformation of conceptual models provides the ability to obtain ontology fragments at the abstract conceptual level (T-Box level).

At the same time, it is rather difficult to obtain specific knowledge (instances and their values) on the basis of conceptual models. For this purpose, it is necessary to use rather large sets of already existing previously accumulated information presented in the form of various structured data. In particular, such data can be spreadsheet data. Processing and extracting knowledge from them is also a relevant task today.

This paper describes the case study of ontology engineering based on automated analysis and transformation of conceptual models and spreadsheet data. Main results of transformation of conceptual models are ontology schemas in the form of content ontology design patterns (ODPs) presented in the OWL format. Main results of transformation of spreadsheet data are domain ontologies which are formed taking into account the ODPs obtained earlier.

Therefore, the automated ontology formation at the T-Box and A-Box levels involves a set of sequential transformations:

1. conceptual models to an ontology schema at the T-Box level;
2. arbitrary spreadsheet data to canonical form as an intermediate data representation;
3. canonical tables to ontology instances (domain ontology at the A-Box level).

Thus, such transformation can be formalized as follows:

$$T_1 : CM^{XML} \rightarrow ODP^{OWL},\ T_2 : SD^{CSV} \rightarrow CT^{CSV},$$
$$T_3(ODP^{OWL}) : CT^{CSV} \rightarrow Ont^{OWL}$$

where CM^{XML} is a source conceptual model in the XML format; ODP^{OWL} is ODP in the OWL format; SD^{CSV} is an arbitrary source spreadsheet data in the CSV format; CT^{CSV} is a target canonical (relational) table in the CSV format; Ont^{OWL} is a domain ontology with instances in the OWL format, T_1, T_2, T_3 are transformation operators which contain a set of transformation rules.

The approach proposed includes three steps and employs the following software: Knowledge Base Development System (KBDS) [4] for transformation of conceptual models to ODPs; TabbyXL [5] for extraction of canonical (relational) tables from arbitrary spreadsheet data in the CSV format; TabbyOWL prototype for OWL domain ontology generation based on transformation of canonical tables. Industrial safety inspection (ISI) tasks are selected as a subject domain.

The ISI is a procedure for assessing the technical condition of industrial facilities, in order to determine the residual life of the operated equipment and the degradation processes occurring on it.

Analysis of the ISI procedure stages [6] showed that the implementation of some stages (forming a map of initial data, developing the ISI program, analysis of diagnostic results including interpretation, etc.) requires the processing of a large volume of semi-formalized information, which can be enhanced by building and using ontologies.

2 Background

Ontology engineering for ISI tasks includes three stages of transformation, and each of them requires the use of specific methods and tools, as well as determining the type of ODPs created.

2.1 Ontology Design Pattern Engineering

Ontology Design Patterns (ODPs) [7] are a way for fixing typical solutions in ontology engineering that helps to avoid some frequently repeated mistakes. Currently, there are some ODP catalogues, e.g., Association for Ontology Design & Patterns (ODPA) [8], which contain different types of ODPs: structural, correspondence, content, reasoning, presentation, and lexico-syntactic. The content and structural ODPs are the most popular and frequently used ones.

Content ODPs encode conceptual design patterns and provide solutions to domain modeling problems. They affect only a specific region of an ontology dealing with such domain modeling problems. Logical ODPs are independent from a specific domain of interest (i.e. they are content-independent), but they depend on an expressivity of logical formalism used for representation. In other words, logical ODPs help to solve design problems where primitives of a representation language do not directly support certain logical constructs. Architectural ODPs affect the overall shape of ontology. Their goal is to constrain "how the ontology should look like". Thus, the use of common ontology fragments in the form of structural and content ODPs allows us to simplify and speed up further refinement (modification) of ontologies obtained through the use of standard solutions.

ODPs contain a set of concepts and relationships reflecting a semantic structure of a domain at a fairly high abstract level. In some cases this is not enough for solving practical tasks. In particular, structured data sets in the form of spreadsheet data can be used to fill ODPs.

In this paper, content ODPs will be created as the first step in the complex automation of ontology engineering.

2.2 Ontology Engineering Based on Model Transformations

Today, few researchers construct ODPs on the basis of conceptual model transformations. However, there are some examples with transformation support of different conceptual models for ontology engineering. A metamodel-driven model transformation approach for interchanging rules between OWL along with Semantic Web Rule Language (SWRL) and Object Constrained Language (OCL) along with Unified Modeling Language (UML) is proposed in [9]. REWERSE Rule Markup Language (R2ML) is used as an intermediate knowledge representation (rules). ATL is used to describe the transformation rules. In [10] researchers present an approach to transforming UML class diagrams into OWL ontologies using the Query/View/Transformation (QVT) standard.

Nowadays, the world accumulates a big volume of various data presented in the XML format. This format is a universal and most common way of integrating software and providing the exchange of information between applications. There are examples of transforming XML-like data into target ontologies [11–14]. Some solutions aim at generating linked data in the form of Resource Description Framework (RDF) triplets based on the transformation of various source XML data. In particular, the W3C consortium has proposed the Gleaning Resource Descriptions from Dialects of Languages (GRDDL) standard [15] for the support of the XML to RDF conversion. Lange in [16] proposes an extensible framework for extracting RDF in various notations from various XML languages, which can easily be extended to additional input languages. The implementation is done in eXtensible Stylesheet Language Transformations (XSLT) with a command-line frontend and a Java wrapper. A new query language called XSPARQL was introduced in [17], and it combines XQuery and SPARQL. It allows querying XML and RDF data, as well as converting data from one format to another. The implementation of XSPARQL is based on rewriting the XSPARQL query as a semantically equivalent XQuery query.

In this paper, we propose to use our Knowledge Base Development System (KBDS) [18] and a declarative domain-specific language for description of model transformations called Transformation Model Representation Language (TMRL) [19]. This tool is used to create converters for most XML formats and eliminates dependency on a specific type of conceptual models and tools. The tool and the language have been tested in the analysis of UML diagrams of classes represented in the XMI format.

2.3 Spreadsheet Data Analysis

Spreadsheet data can be used to fill ODPs. Over the past two decades, the methodological approaches for extracting and transforming data from spreadsheets have been actively developed. In particular, methods for the role and structural spreadsheet analysis (restoration of relations between cells) are proposed in [20–23]. The research in this area is also focused, in particular, on specialized tools for extraction and transformation arbitrary spreadsheet data into a structured form, including tabular data transformation systems [24–26], linked data extraction systems [27, 28], transformation systems of arbitrary tables to a relational form based on the search for critical cells [23], relational data extraction systems from spreadsheets with header hierarchies [22]. The tools considered have similar goals (to transform spreadsheet data from an arbitrary form to a relational form). However, they use specified models of source spreadsheets, in which their physical and logical composition (layout) are mixed. This fact limits the use of these tools for processing arbitrary spreadsheets presented in statistical reports.

Our review showed that the solutions considered above (each separately) fail to process the layouts found in ISI reports. The TabbyXL, which we developed in [5], looks promising in this aspect. TabbyXL is a command-line tool for spreadsheet data canonicalization. This tool is used to produce flat relational (normalized) spreadsheet tables from semi structured tabular data. The tool operates with CSV format documents. Data structures for representing table elements (cells, entries, labels, and categories). TabbyXL has the ability to customize certain layout types by using a domain-specific language – Cells Rule Language (CRL) [5].

2.4 RDF/OWL Ontology Generating from Spreadsheet Data

There are several recent studies that deal with spreadsheet data transformation. Such tools as RDF123, csv2rdf4lod, Datalift, Any2OWL, Excel2OWL, Spread2RDF, Any23, TopBraid Composer are used to solve issues of converting spreadsheet data to RDF or OWL formats. Some of the solutions also include own domain-specific languages: XLWrap, Mapping Master and RML.

Quite a few solutions are available in this area, especially for the generation of RDF documents. However, the generation of ontologies in the OWL format is usually poorly supported, which is why we develop our own OWL generator, focused on the format of canonical TabbyXL tables [5].

3 The Proposed Approach

The approach used in this study includes methodology and means for its implementation.

3.1 Methodology

The methodology based on the formal problem statement that we presented in introduction includes three main steps.

Step 1: Forming content ODPs based on the transformation of conceptual models. This step consists of the following main actions:

1.1 Forming a conceptual model. A domain expert represents their own knowledge in the form of a conceptual information model (perhaps, using domain-specific visual notations).

1.2 Serializing a conceptual model. The conceptual model is represented in the XML-based format for further transformation.

It should be noted that Actions 1.1 and 1.2 can be implemented using various software to support conceptual modeling and serializing models in the form of XML documents.

1.3 Analysing a XML document structure. This action involves extracting elements, their attributes and relationships from the XML tree.

1.4 Forming an ODP code in the OWL format. The main objective of this action is to obtain typical ontological fragments in the form of a set of classes and their relationships (object properties), which describe a certain domain and are based on the extracted XML elements.

A declarative domain-specific language called Transformation Model Representation Language (TMRL) [19] is used in Actions 1.3 and 1.4. This language is implemented in the Knowledge Base Development System (KBDS) [18] and provides a scenario (program) for model transformations. TMRL is designed for converting conceptual models into knowledge bases only.

The model transformation is one of the major concepts in the model-oriented approach (Model-Driven Engineering) [29]. It is an automatic generation of a target model from a source model according to a set of transformation rules. These rules together describe how a model in a source language can be transformed into a model in a target language. Consequently, a transformation rule is a description of how one or more constructs in a source language can be transformed into one or more constructs in a target language. At the same time, the use of meta-modeling is one of the main approaches to defining an abstract syntax of languages, including conceptual modeling and knowledge representation languages.

Thus, in 1.3 and 1.4 a user forms a transformation scenario in TMRL. This scenario is a set of rules for transformation of XML document elements of the conceptual model to ontological constructs. Each rule contains a correspondence between elements of the source and target metamodels with a certain priority (sequence) of this rule for an interpreter. The XML Schema is used as a metamodel for XML documents. The OWL language description is used as a metamodel for ODPs.

1.5 Editing ODP code. This action is additional and represents a refinement (modification) of the ODP obtained with the aid of various ontological modeling tools, for example, Protégé.

Thus, the main result of this step is a set of content ODPs, which define the ontology schema at the T-Box level.

Step 2: Transforming source spreadsheet data with an arbitrary layout to the canonical (relational) form. This step consists of the following main actions:

2.1 Analyzing the CSV file of spreadsheet data by using Cells Rule Language (CRL) rules [5]. CRL is a domain-specific language for expressing table analysis and interpretation rules. A set of the rules can be implemented for a specific task characterized by requirements for source and target data.

2.2 Canonicalization. The process of table canonicalization begins with loading tabular data from CSV files via Apache POI. Recovered semantics (entries, labels, and categories) allow transformation of the source spreadsheet data into the canonical form. The canonical form requires the topmost row to contain field (attribute) names. Each of the remaining rows is a record (tuple). It obligatorily includes the field named DATA that contains entries. Each extracted category constitutes a field that contains its labels. Each record presents recovered relationships between an entry and labels in each category.

Thus, let us formally define the canonical table for our case on the basis of [6]:

$$CT^{CSV} = \{DATA, RowHeading, ColumnHeading\},$$

where *DATA* is a data block that describes literal data values (named "entries") belonging to the same datatype (e.g., numerical, textual, etc.), *RowHeading* is a set of row labels of the category, *ColumnHeading* is a set of column labels of the category. The values in cells for heading blocks can be separated by the "|" symbol that is intended to divide categories into subcategories. Thus, the canonical table denotes hierarchical relationships between categories (headings). Detailed description of this process is presented in [5]. The result of this step is tables in the unified canonical form prepared for their further automated processing.

Step 3: Generating OWL domain ontology from canonical tables using ODPs. This step consists of the following main actions:

3.1 Semantic canonical table interpretation using obtained ODPs. This process is semi-automatic and aims to linking *RowHeading* and *ColumnHeading* cell values with entities of previously obtained ODPs (the ontology schema at the T-Box level). Each canonical table corresponds to a specific class of this ontology. In this case, a set of concept candidates (classes and data properties) for each *RowHeading* and *ColumnHeading* cell is determined basing on the obtained ontology schema. A cell value with separator ("|") is interpreted as a hierarchy of either classes ("subClassOf" construction) or data properties ("subPropertyOf" construction). Thus, ontology concepts, which are closest to each label, are determined. For this purpose, it is necessary to calculate an aggregated assessment (for example, a linear convolution or any other proximity measure) for all concept candidates and select a referent concept with the maximum rating. Note that context for canonical table and ontology schema should be equal. Otherwise, the aggregated assessments would be low due to the lack of correspondences between labels and ontology entities.

3.2 Forming a domain ontology at the A-Box level. The main objective of this action is to generate OWL axioms for representation of instances based on the obtained canonical table to supplement the existing ontology schema (the content ODP) with concrete instances. The obtained canonical table is divided into blocks (row sets) that correspond to a certain category group from *RowHeading* and *ColumnHeading*. At the same time, this group is annotated with a class from the ontology schema. Thus, each such row set is interpreted as an ontology instance and entities from *DATA* corresponding to data property values of instance.

3.2 Implementation

We implement our approach using KBDS, TabbyXL and TabbyOWL that provide transformations for Steps 1, 2, and 3, respectively.

KBDS is a tool for knowledge base engineering that performs transformation of various conceptual models using TMRL. KBDS is designed with the help of Model-View-Controller (MVC) design pattern, PHP, and Yii2 framework. jQuery and jsPlumb libraries are used to build graphic editors. KBDS has a client-server architecture and is described in detail in [18].

TabbyXL [5] is a command-line tool for spreadsheet data canonicalization. This tool is used to produce flat relational (normalized) spreadsheet data from semi structured tabular data. The tool operates with CSV format documents. Data structures for representing table elements (cells, entries, labels, and categories) are Java classes developed in accordance with naming conventions of JavaBeans 5 specification.

TabbyOWL is a tool (converter) for transformation of canonical tables into OWL ontology at the A-Box level. The input data for a converter is a set of transformation rules and a source canonical table. The output data is an ontology containing instances, which specifies the previously obtained content ODP. Thus, the main result is the domain ontology that reflects the abstract level of domain knowledge in terms of classes and their properties and contains specific axioms. These axioms can be useful for their further usage in intelligent systems for solving various types of domain issues.

3.3 Case Study

As an example, consider the ISI ontology engineering:

Step 1: Forming content ODPs based on transformation of conceptual models.

1.1 Forming conceptual models in the form of concept maps. At this step, conceptual models in the form of concept maps for ISI are used.

Conceptual models were created within the project for automation of the ISI procedure [30]. These models reflect different aspects and concepts of the ISI procedure, such as: technical object, geodesic measurements analysis, visual measurement control, nondestructive testing, hardness measurement, rapid diagnosis, technical conditions assessment, wall thickness test, technical documentation, testing, the ISI report generation and degradation processes that take place within the inspected object (e.g., hydrogen embrittlement, etc.) and are influenced by a combination of factors. The dataset of conceptual models [31] includes 26 models containing information about 235 concepts and 208 relationships. IBM Rational Rose and IHMC CmapTools were used when building models.

Figure 1 shows a fragment of a concept map describing storage tank elements.

1.2 Serializing concept maps to XTM (XML Topic Maps) format using IHMC CmapTools.

1.3 and 1.4 Analysing XTM document structures and forming the ODP code in the OWL format based on the transformation model in TMRL. This model is created by a user through a visual transformation model editor (a KBDS module) (see Fig. 2). The user defines correspondences between the source and target metamodel elements and assigns priority to each transformation rule.

Table 1 describes the obtained correspondences.

On the basis of the obtained correspondences, the transformation model is generated on TMRL. A fragment of the transformation model describing the transformation rules is given below:

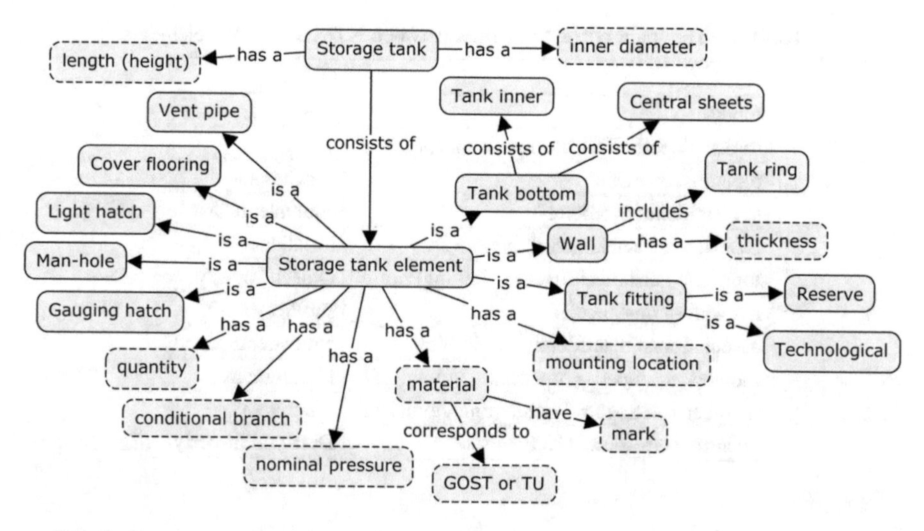

Fig. 1. A concept map fragment for a storage tank prepared with IHMC CmapTools.

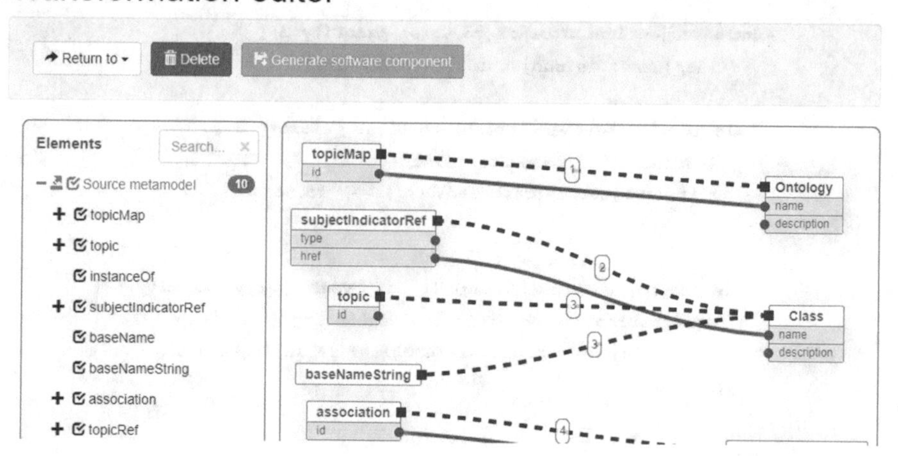

Fig. 2. GUI of the transformation model editor (a KBDS module).

Table 1. The main correspondences between XTM and OWL elements.

XTM	OWL
topicMap	Ontology
topic/subjectIndicatorRef = "concept"	Class
topic/instanceOf	subClassOf
subjectIdentity/topicRef	equivalentClass
baseName/baseNameString	rdf:about
topic/subjectIndicatorRef = "linkingPhrase"	ObjectProperty
association/instanceOf	subPropertyOf
association/subjectIdentity/topicRef	equivalentProperty
member roleSpec = "incoming"/topicRef	rdfs:domain
member roleSpec = "outcoming"/topicRef	rdfs:range
occurrence/resourceData	DatatypeProperty

```
Transformation XTM metamodel to Ontology metamodel {
  Rule topicMap to Ontology priority 1 [
    Ontology(name) is topicMap(id)
  ]
  Rule subjectIndicatorRef to Class priority 2 [
    Class(name) is subjectIndicatorRef(href)
  ]
  Rule (topic, baseNameString) to Class priority 3 [
    Class(name) is baseNameString [
      if (subjectIndicatorRef(href) == '#concept')
    ]
  ]
  Rule (topic, baseNameString) to DatatypeProperty priority 6 [
    DatatypeProperty(name) is baseNameString [
      if (subjectIndicatorRef(href) == '#linkingPhrase)
    ]
  ]
  ...
}
```

The KBDS transformation module (converter) uses this transformation model in TMRL and translates the XTM document to the OWL ODP code.

The OWL code (RDF/XML syntax) fragment of the obtained ODP is presented below:

```
<owl:Class rdf:about="StorageTank"/>
<owl:Class rdf:about="StorageTankElement">
  <rdfs:subClassOf rdf:resource="StorageTank" />
</owl:Class>
<owl:DatatypeProperty rdf:about="ConditionalBranch">
  <rdfs:domain rdf:resource="#StorageTankElement"/>
  <rdfs:range rdf:resource="xsd:positiveInteger"/>
</owl:DatatypeProperty>
```

Thus, concepts and relationships of concept maps were mapped into a set of ontological classes and their properties. The obtained ontology fragments at the T-Box level describe the patterns of various degradation processes, as well as the structure of petrochemical mechanical systems.

1.5 The OWL code obtained does not require any changes.

Step 2: Transforming spreadsheet data in the CSV format to the canonical (relational) form.

2.1 Analysis of CSV files is performed using TabbyXL. Six ISI reports were analyzed and 216 spreadsheets were extracted from these reports to fill the obtained ODPs by specific values (data) within the case study.

Figure 2(a) shows an arbitrary source table fragment that contains information about the storage tank.

2.2 The resulted canonical tables (see Fig. 2b) are generated based on analysis and transformation of the source arbitrary tables (see Fig. 2a).

The process of TabbyXL setting, layouts, discovered tables and rules for their processing are described in [32].

Step 3: Generating OWL domain ontology from the canonical tables.

3.1 and 3.2 Semantic interpretation of obtained canonical tables using generated ODPs and forming domain ontology at the A-Box level. These activities are made with the use of TabbyOWL prototype. In our case, the fragment presented in Fig. 3(b) corresponds to a single instance (individual) of "Technological" class ("TankFitting" subclass) from the resulting ontology schema at T-Box level. At the same time, labels in *RowHeading* cells describe the class, which the instance belongs. Labels in the *ColumnHeading* cells describe the data properties of this instance. Entities in *DATA* cells describe specific values for data properties of this instance.

The OWL code fragment in RDF/XML syntax of the obtained instance is presented below:

a) A source spreadsheet table

<table>
<tr><th rowspan="2">Name</th><th rowspan="2">Quantity, pcs.</th><th rowspan="2">Conditional branch, mm</th><th rowspan="2">Nominal pressure, MPa (kgf / cm2)</th><th rowspan="2">Mounting location</th><th colspan="2">Material</th></tr>
<tr><th>Mark</th><th>GOST or TU</th></tr>
<tr><td>Inlet-outlet</td><td>2</td><td>80</td><td rowspan="10">1,6 (16)</td><td>Wall</td><td>Steel 20</td><td>GOST1050</td></tr>
<tr><td>Man-hole</td><td>1</td><td>500</td><td>Wall</td><td>St3</td><td>GOST380</td></tr>
<tr><td>Reserve</td><td>1</td><td>200</td><td>Wall</td><td rowspan="4">Steel 20</td><td rowspan="4">GOST1050</td></tr>
<tr><td>Technological</td><td>1</td><td>20</td><td>Wall</td></tr>
<tr><td>Technological</td><td>1</td><td>50</td><td>Wall</td></tr>
<tr><td>Technological</td><td>1</td><td>250</td><td>Wall</td></tr>
<tr><td>Light hatch</td><td>2</td><td>500</td><td>Roof</td><td>St3</td><td>GOST380</td></tr>
<tr><td>Gauging hatch</td><td>1</td><td>150</td><td>Roof</td><td rowspan="3">Steel 20</td><td rowspan="3">GOST1050</td></tr>
<tr><td>Vent pipe</td><td>1</td><td>150</td><td>Roof</td></tr>
<tr><td>Technological</td><td>1</td><td>150</td><td>Roof</td></tr>
</table>

b) A canonicalized table

DATA	RowHeading	ColumnHeading
GOST1050	Technological	Material | GOST or TU
Steel 20	Technological	Material | Mark
Roof	Technological	Mounting location
1,6 (16)	Technological	Nominal pressure, MPa (kgf / cm2)
150	Technological	Conditional branch, mm
1	Technological	Quantity, pcs.
...		

Fig. 3. An example of a source spreadsheet data (a) and a canonical table (b) containing information about the storage tank elements.

```
<Technological rdf:about="Technological1">
  <MountingLocation rdf:datatype="xsd:string">Roof</MountingLocation>
  <Quantity rdf:datatype="xsd:positiveInteger">1</Quantity>
  <ConditionalBranch rdf:datatype="xsd:positiveInteger">150</ConditionalBranch>
  <NominalPressure rdf:datatype="xsd:string">1,6 (16)</NominalPressure>
  <Mark rdf:datatype="xsd:string">Steel 20</Mark>
  <GOSTorTU rdf:datatype="xsd:string">GOST1050</GOSTorTU>
</Technological>
```

In this example, 15 classes and 10 properties-values were obtained.

4 Discussion

In accordance with the paper objective we have made ontological engineering for ISI tasks. Methodologically, the chain of sequential transformations was made: conceptual models to ontology schema at the T-Box level; arbitrary spreadsheet data to canonical form as an intermediate data representation; canonical tables to ontology instances (domain ontology at the A-Box level).

In this case, the author's software tools were used: Knowledge Base Development System (KBDS) [18] for transformation of conceptual models to ODPs; TabbyXL [5] for extraction of canonical (relational) tables from arbitrary spreadsheet data in CSV format; the TabbyOWL prototype for the OWL domain ontology generation based on transformation of canonical tables.

The following estimates were obtained: recall – 85, precision – 89, when using KBDS for a set of 26 conceptual models containing 256 concepts and 208 relationships. The losses of recall and precision are mainly due to the complexity of transformation of the following types of correspondences for elements:

- *An ambiguous correspondence* (*synonymy*) – several XML Schema elements correspond to one of an OWL metamodel element,
- *An indistinguishable correspondence* (*homonymy*) – one of the XML Schema element corresponds to several OWL metamodel elements.

An indistinguishable correspondence can be illustrated as follows: it is necessary to map the element "*topic*" (which is a successor of the element "*concept*") into several ontological constructions: «*Class*» and «*DatatypeProperty*». To do this, a user has to configure the transformation model by adding a conditional statement to TMRL.

TabbyXL processed all tables from this data set. The TabbyXL experiment results on precision and recall for other data sets, in particular, TANGO [6] are presented in [5].

5 Conclusion

In this paper, we investigated the ability to automate the ontologies engineering on the basis of analysis and transformation of conceptual models and spreadsheet data. The methodology proposed enables reusing of a large amount of accumulated heterogeneous information: spreadsheet data and conceptual models for the automated formation of domain ontologies, which contain knowledge both at abstract (T-Box) and specific (A-Box) levels.

It should be noted that only 30% of tables from the processed ISI reports were used for filling ODPs because of the difficulties with the automatic interpretation of canonical table cell values (their semantics). This challenge requires a more detailed study of the table interpretation problem in future.

The ontologies obtained are used for prototyping rule-based expert system [33, 34].

Acknowledgement. The contribution of this work was supported by the Russian Science Foundation under Grant No. 18-71-10001.

References

1. Guarino, N.: Formal ontology in information systems. In: The First International Conference on Formal Ontology in Information Systems (FOIS 1998), vol. 46, pp. 3–15 (1998)
2. Starr, R.R., de Oliveira, J.M.P.: Concept maps as the first step in an ontology construction method. Inf. Syst. **38**, 771–783 (2013). https://doi.org/10.1109/EDOCW.2010.43
3. Herrero-Zazo, M., Segura-Bedmar, I., Martínez, P.: Conceptual models of drug-drug interactions: a summary of recent efforts. Knowl.-Based Syst. **114**, 99–107 (2016). https://doi.org/10.1016/j.knosys.2016.10.006
4. Berman, A.F., Dorodnykh, N.O., Nikolaychuk, O.A., Pavlov, N.Y., Yurin, A.Yu.: Fishbone diagrams for the development of knowledge bases. In: The 41st International Convention on Information and Communication Technology, Electronics and Microelectronics (MIPRO 2018), pp. 967–972 (2018). https://doi.org/10.23919/MIPRO.2018.8400177
5. Shigarov, A.O., Mikhailov, A.A.: Rule-based spreadsheet data transformation from arbitrary to relational tables. Inf. Syst. **71**, 123–136 (2017). https://doi.org/10.1016/j.is.2017.08.004
6. Tijerino, Y.A., Embley, D.W., Lonsdale, D.W., Ding, Y., Nagy, G.: Towards ontology generation from tables. World Wide Web **8**(3), 261–285 (2005). https://doi.org/10.1007/s11280-005-0360-8
7. Hitzler, P., Gangemi, A., Janowicz, K., Krisnadhi, A.A., Presutti, V.: Ontology Engineering with Ontology Design Patterns: Foundations and Applications. Studies on the Semantic Web. IOS Press/AKA, Amsterdam (2016)
8. Association for ontology design & patterns (ODPA). http://ontologydesignpatterns.org/wiki/ODPA. Accessed 18 Mar 2019
9. Milanović, M., Gašević, D., Giurca, A., Wagner, G., Devedžić, V.: Bridging concrete and abstract syntaxes in model-driven engineering: a case of rule languages. Soft. Pract. Exp. **39** (16), 1313–1346 (2009). https://doi.org/10.1002/spe.938
10. Zedlitz, J., Luttenberger, N.: Conceptual modelling in UML and OWL-2. Int. J. Adv. Softw. **7**(1), 182–196 (2014)
11. Bohring, H., Auer, S.: Mapping XML to OWL ontologies. In: Leipziger Informatik-Tage, vol. 72, pp. 147–156 (2005)
12. Rodrigues, T., Rosa, P., Cardoso, J.: Moving from syntactic to semantic organizations using JXML2OWL. Comput. Ind. **59**(8), 808–819 (2008). https://doi.org/10.1016/j.compind.2008.06.002
13. O'Connor, M.J., Das, A.K.: Acquiring OWL ontologies from XML documents. In: The 6th International Conference on Knowledge Capture (K-CAP 2011), pp. 17–24 (2011). https://doi.org/10.1145/1999676.1999681
14. Bedini, I., Matheus, C., Patel-Schneider, P.F., Boran, A., Nguyen, B.: Transforming XML schema to OWL using patterns. In: The 2011 IEEE Fifth International Conference on Semantic Computing, pp. 102–109 (2011). https://doi.org/10.1109/ICSC.2011.77
15. Gleaning resource descriptions from dialects of languages (GRDDL). https://www.w3.org/TR/grddl/. Accessed 18 Mar 2019
16. Lange, C.: Krextor - an extensible framework for contributing content math to the web of data. In: International Conference on Intelligent Computer Mathematics, pp. 304–306 (2011). https://doi.org/10.1007/978-3-642-22673-1_29
17. Bischof, S., Decker, S., Krennwallner, T., Lopes, N., Polleres, A.: Mapping between RDF and XML with XSPARQL. J. Data Semant. **1**(3), 147–185 (2012). https://doi.org/10.1007/s13740-012-0008-7

18. Dorodnykh, N.O.: Web-based software for automating development of knowledge bases on the basis of transformation of conceptual models. Open Semant. Technol. Intell. Syst. **1**, 145–150 (2017)
19. Dorodnykh, N.O., Yurin, A.Yu.: A domain-specific language for transformation models. In: CEUR Workshop Proceedings. Information Technologies: Algorithms, Models, Systems (ITAMS 2018), vol. 2221, pp. 70–75 (2018)
20. Mauro, N., Esposito, F., Ferilli, S.: Finding critical cells in web tables with SRL: trying to uncover the devil's tease. In: Proceedings of the 12th International Conference on Document Analysis and Recognition, pp. 882–886 (2013). https://doi.org/10.1109/ICDAR.2013.180
21. Adelfio, M., Samet, H.: Schema extraction for tabular data on the web. Proc. VLDB Endow. **6**(6), 421–432 (2013). https://doi.org/10.14778/2536336.2536343
22. Chen, Z., Cafarella, M.: Integrating spreadsheet data via accurate and low-effort extraction. In: Proceedings of the 20th ACM SIGKDD International Conference Knowledge Discovery and Data Mining, pp. 1126–1135 (2014). https://doi.org/10.1145/2623330.2623617
23. Embley, D.W., Krishnamoorthy, M.S., Nagy, G., Seth, S.: Converting heterogeneous statistical tables on the web to searchable databases. Int. J. Doc. Anal. Recogn. **19**(2), 119–138 (2016). https://doi.org/10.1007/s10032-016-0259-1
24. Kandel, S., Paepcke, A., Hellerstein, J., Heer, J.: Wrangler: interactive visual specification of data transformation scripts. In: Proceedings of the SIGCHI Conference Human Factors in Computing Systems, pp. 3363–3372 (2011). https://doi.org/10.1145/1978942.1979444
25. Hung, V., Benatallah, B., Saint-Paul, R.: Spreadsheet-based complex data transformation. In: Proceedings of the 20th ACM International Conference on Information and Knowledge Management, pp. 1749–1754 (2011). https://doi.org/10.1145/2063576.2063829
26. Harris, W., Gulwani, S.: Spreadsheet table transformations from examples. ACM SIGPLAN Not. **46**(6), 317–328 (2011). https://doi.org/10.1145/1993316.1993536
27. O'Connor, M.J., Halaschek-Wiener, C., Musen, M.A.: Mapping master: a flexible approach for mapping spreadsheets to OWL. In: The Semantic Web – ISWC 2010. LNCS, pp. 194–208 (2010). https://doi.org/10.1007/978-3-642-17749-1_13
28. Mulwad, V., Finin, T., Joshi, A.: A domain independent framework for extracting linked semantic data from tables. In: Search Computing, pp. 16–33 (2012). https://doi.org/10.1007/978-3-642-34213-4_2
29. Da Silva, A.R.: Model-driven engineering: a survey supported by the unified conceptual model. Comput. Lang. Syst. Struct. **43**, 139–155 (2015). https://doi.org/10.1016/j.cl.2015.06.001
30. Berman, A.F., Nikolaichuk, O.A., Yurin, A.Y., Kuznetsov, K.A.: Support of decision-making based on a production approach in the performance of an industrial safety review. Chem. Pet. Eng. **50**(1–2), 730–738 (2015). https://doi.org/10.1007/s10556-015-9970-x
31. Yurin, A.H., Dorodnykh, N.O., Nikolaychuk, O.A., Berman, A.F., Pavlov, A.I.: ISI models, mendeley data, v1 (2019). http://dx.doi.org/10.17632/f9h2t766tk.1
32. https://github.com/tabbydoc/tabbyxl/wiki/Industrial-Safety-Inspection
33. Dorodnykh, N.O., Yurin, A.Yu., Stolbov A.B.: Ontology driven development of rule-based expert systems. In: The 3rd Russian-Pacific Conference on Computer Technology and Applications (RPC 2018), pp. 1–6 (2018). https://doi.org/10.1109/RPC.2018.8482174
34. Grishenko, M.A., Dorodnykh, N.O., Nikolaychuk, O.A., Yurin, A.Yu.: Designing rule-based expert systems with the aid of the model-driven development approach. Expert Syst. **35**(5), 1–23 (2018). https://doi.org/10.1111/exsy.12291

An Approach for Distributed Reasoning on Security Incidents in Critical Information Infrastructure with Intelligent Awareness Systems

Maria A. Butakova, Andrey V. Chernov(✉), and Petr S. Shevchuk

Rostov State Transport University, Rostov-on-Don 344038, Russia
{butakova, avcher}@rgups.ru

Abstract. A class of intelligent situational awareness systems is considered. Such systems are designed to make effective decisions about the occurrence and processing of incidents in critical information infrastructure. Features of critical information infrastructure on an example of automated and information systems in the JSC "Russian Railways" are considered. Possible types of incidents occurring in critical information infrastructure are given. The problem of new types of incidents detection in the critical information infrastructure is formulated. An approach for decision-making on situational awareness in a critical information infrastructure based on distributed reasoning models in uncertain information conditions is proposed.

Keywords: Security incidents · Distributed reasoning · Critical infrastructure · Intelligent awareness systems

1 Introduction

Information management infrastructure of the railway transport corporation JSC "Russian Railways" is the large corporate computer network system in Russia that provides technological processes for transporting goods and passengers by rail. Multifaceted problems the safety ensuring of technological processes and information security in railway infrastructure were considered earlier in papers [1–3]. Criteria for the importance of critical information infrastructure include social, economic, political, environmental significance, as well as significance for ensuring the country's defense and law and order, and state security. These categories and the criteria themselves fully apply to the information infrastructure of railway transport [4].

It should be noted that the main content of security criteria is about the vulnerability of information infrastructure to information leaks. Usually, the sphere of information security includes activities that are aimed at counteracting targeted threats. However, for information management systems in railway transport, the aspects of technological safety considered in the context of the occurrence or actions of the unintended events related to the correct functioning of systems are in particular importance. The information management system operated on the railway transport should not allow the

R. Silhavy et al. (Eds.): CoMeSySo 2019, AISC 1046, pp. 248–255, 2019.
https://doi.org/10.1007/978-3-030-30329-7_23

emergence and development of incidents related to the threat to life and health of people, damage to property and the environment, the economy and the defense capability of the country. Thus, the class of critical information and control systems in rail transport is comprehensive from all points of view.

The use of subsystems for monitoring events and incidents is a common practice for critical information infrastructures. The information infrastructure and the capabilities of the distributed computing environment for the transport processes management in JSC "Russian Railways" currently provide continuous monitoring and recording various kinds of incidents. For operational management tasks at different hierarchy levels, it is not required so much information about incidents, but awareness in the operational situation and the situation as a whole. These circumstances make it possible to identify the needs of critical information infrastructure in situational management. A special role in situational management is played not only by identification, and various events registration associated with the emergence of various situations that lead or may lead to the abnormal modes of systems operation, accidents, failures, equipment breakdowns, and threats to the safety of people, catastrophic consequences for the environment and the like. Further incident handling and decision making, together with forecasting, are relevant to current and priority tasks for the critical information infrastructure of rail transport.

The tasks related to the incidents handling in the critical information infrastructure of the railway transport are currently being carried out manually using computers. The development of intelligent decision-making technologies, the emergence of new information-control systems with intelligent capabilities significantly expands the critical information infrastructures in rail transport. It is evident that intelligent systems in railway transport should provide support for decision-making that does not contradict the primary objectives and tasks of the entire transport system.

In connection with the needs, this paper considers new technologies and the requirements of situational management, a new approach to intelligent situational awareness systems design. These systems are intended to make effective decisions about the occurrence and processing of incidents in the critical information infrastructure JSC "Russian Railways". An important feature of the proposed approach to intelligent situational awareness systems is the possibility of using methods for distributed reasoning to extract new knowledge about incidents occurring in critical information infrastructure.

The paper is structured as follows. Section 2 provides information on the currently available scientific research in the selected field. Section 3 sets out the task of detecting new types of incidents in critical information infrastructure and suggests methods for making decisions about situational awareness in a critical information infrastructure based on distributed reasoning methods. Section 4 contains the features analysis of the critical information infrastructure by example of complex information systems JSC "Russian Railways". Possible types of incidents occurring in critical information infrastructure are given with their ontology models. Section 5 concludes this paper.

2 Previous Work

The topic of situational awareness and situational management in various intelligent management systems is of particular interest for scientific research, in connection with the growing need to create situational centers for transport also. Recording and processing incidents are the main operations in the intelligent decision support systems too. Main types of incidents are related to computer security incidents. Because of this, we detail typical processing stages by the security incidents example. The definition of computer security incident one can find in [5]. Therefore, we can define security incident as the emergence of one or more unwanted, or unexpected security events, with which there is a significant likelihood of business operations compromising and the creation of an information security threats.

Researchers examine security incidents in critical information infrastructures from several points of view, applying different mathematical approaches helping to design situational awareness in intelligent decision support systems. Usually, the first problem concerns to security incident processing are to collect the required information from different sources. The sequence of that collecting contains registering, storing, and sending notifications about incidents occurrence. There are different ways to such processing performing. One of the adopted approaches proposed in [6] presents a workflow based security incidents management. Authors have proposed a modern way to incidents of data collecting in [7] recently. A new idea of the cited paper is based on using distributed intelligent agents, smart objects so called. Smart objects are equipped with sensors and actuators with software-defined algorithms enabling to interact with each other in the local area network and exchange with locally adopted decisions. Virtual smart objects middleware allows communicating approach that is used in Service-Oriented Architectures. A special feature of the discussed approach consists in the ability to maintain the hardware data processing acceleration.

The next problem is choosing of security incident data model [8]. Organizing data about security incidents into taxonomies can be performed as collective ontology creating from several locally distributed ontologies. Usually, collectively distributed ontology engineering cycle ends with united static knowledge database design, but authors have proposed an approach allowing using the evolving distributed ontologies in [9]. Other techniques of incidents handling can be involved, and recent methods rely on intelligent cognitive features also, for example [10].

The final stage of incident processing is reporting and visualization. Researchers pay attention to the problems of quantity and statistical measuring on impacts and consequences in the main part of their papers. One can find basic architectures of cyber-security incident reporting in safety-critical systems in [11]. It is worth to note that not all incident data models agree with further reporting architectures. Results presentation and method complicate incident processing system architecture because of the difference in techniques used. Once more, we underline the pros of our method proposed earlier in [9] allowing designing the central incidents data processing model and decision support in the same formal definitions.

In order to use automated approaches to decision-making in intelligent situational awareness systems, the following section proposes methods for distributed reasoning based on extensions of distributed description logic.

3 Proposed Approach

Situational awareness implies the possibility of obtaining complete and reliable information for making decisions in real time, including characteristics and peculiarities of the situation. Situational awareness is a three-step process [12]:

(1) Perception of elements of the environment.
(2) An understanding of the current situation.
(3) Projection of the future state.

The proposed approach consists of three stages by analogy. Figure 1 shows these stages visually.

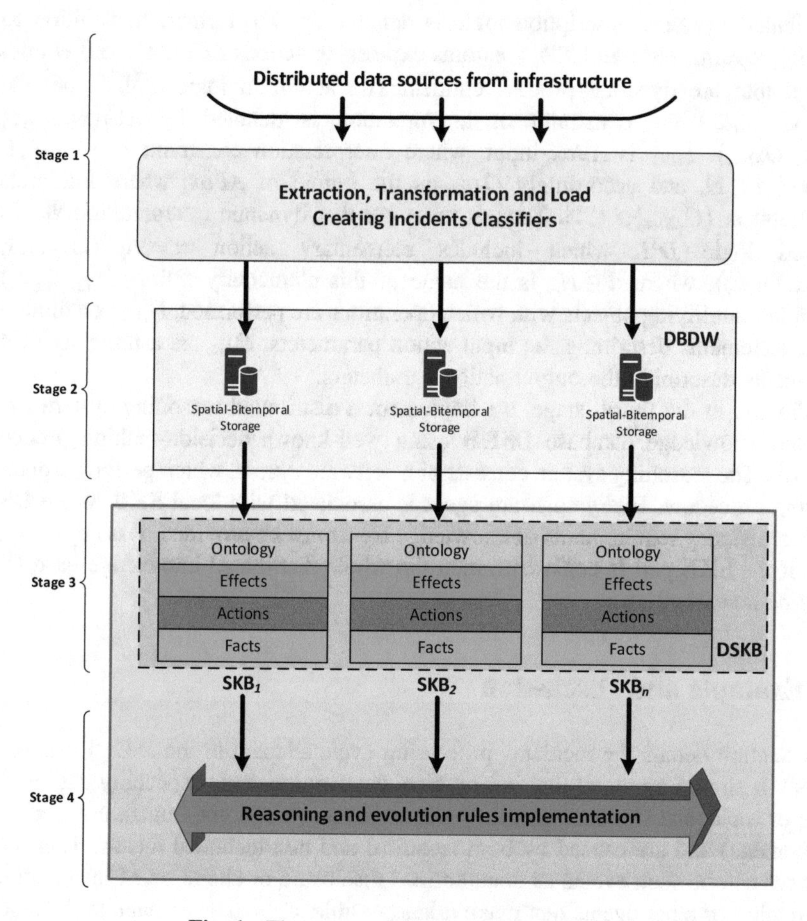

Fig. 1. The stages of the proposed approach

Let's outline these steps briefly. First stage is devoted to acquiring data about incidents from heterogeneous data sources. Because of different nature that data must be converted into classifiers which have the same format. The second stage performs creating bi-temporal data warehouse about incidents. The third stage is the involving of distributed dynamic description logic for ontologies forming. The fourth stage is decision-making support procedures for incident processing by the distributed reasoning system.

The first stage of the proposed approach is the classifiers creating for subject areas. This stage also involves the implementation of standard ETL (Extraction, Transformation Load) procedures.

In the second stage, we create distributed bitemporal data warehouse using models of on-line analytical processing and supporting two-dimensional temporal queries [13].

Next, in the third stage, we describe the formal definition of the distributed situational knowledge base (**DSKB**). In terms of [14], this formalization is performed as follows. **DSKB** includes incidents and consists of situational knowledge bases (**SKB**). Distributed dynamic description logic is denoted as D^3L further. In addition to the existing axioms *ABox* and *TBox*, axioms expressing actions *ACt* (*ACtions*) entities are added, that are dynamic process entities. The action in logic D^3L is denoted as $ACt(x_1,\ldots,x_n)$ the constraint of interpretation is denoted by $ACt(x_1,\ldots,x_n) = \langle I_{ABox}, O_{ABox}\rangle$. I_{ABox} is *ABox* input, where interpretation constraint is $\{I_{ABox}\}, i \in \mathbb{N}$ $\{I_{ABox_i}\}, i \in \mathbb{N}$, and accordingly O_{ABox} is the output of *ABox*, where interpretation constraint is $\{O_{ABox}\}, i \in \mathbb{N}$. Thus, it turns out that dynamic interpretation $\mathfrak{A}$ of considered logic D^3L which includes elementary action $\alpha = ACt(x_1,\ldots,x_n) = \langle I_{ABox}, O_{ABox}\rangle$, where $\alpha \in N_A$ is the name of this elementary action, $(x_1,\ldots,x_n)$ are variables identifying objects with which operations are performed, I_{ABox} is a finite set of *ABox* statements describing the input action parameters, O_{ABox} is a finite set of *ABox* statements describing the output action parameters.

Finally, in the fourth stage, we implement a distributed reasoning system for distributed knowledge database **DSKB** using well-knows decision-making procedures [15, 16]. The reasoning system consists of n software agents, which perform a decision-making procedure. Every software agent is associated with local **SKB** of the **DSKB**. The consistency requirements of knowledge base must be provided. It is apparently the fact, if the **SKB** part is consistent, then the whole distributed knowledge base **DSKB** must be consistent also.

4 Example and Discussion

This Section details the incidents processing cycle adapted to the JSC "Russian Railways". It should be noted that information security incidents especially can be deliberate or accidental (for example, they can be the result of some human error or natural phenomena) and are caused by both technical and non-technical means. Their consequences can be such events as unauthorized disclosure or alteration of information, its destruction or other events that make it inaccessible, as well as damage to the assets of the organization or their theft. Information security incidents that were not reported but identified as incidents cannot be investigated, and no protective measures can be applied to prevent the recurrence of these incidents.

Currently, the processing of the incident in the critical information infrastructure of JSC "Russian Railways" is an interactive, partially automated process with manual data input. It contains the following procedures: "*Incident Registration*", "*Incident Resolving*", "*Incident Closure*", "*Incident Control*", "*Reporting Formation*", "*Process Assessment and Improvement*". The basis for the registration of an infrastructure incident is an event in the serviced critical information infrastructure that led to the hardware errors or software errors, failures in equipment or channels of the data transmission network, engineering systems in the zone of responsibility of the *Main Computing Center* of JSC "Russian Railways", which affected the *Quality of Service Provision*. For requests classified as "*Incident*", the dispatching personnel create a new record and assign them to workgroups according to the routing sheets.

Next, we give several examples of incidents described in the general case by different ontologies; however, having similar semantics. Examples of incidents occurring at users' workstations are the following user requests:

- the computer does not turn on;
- there are computer failures;
- any function of the software does not work;
- error in the work of the automated workplace;
- the transactional self-service terminal, the ticket and cash register equipment, the electronic self-service terminal do not work;
- does not work/does not print the printer;
- there is no network connection of the workstation.

Examples of infrastructure incidents in IT services are the following events:

- inaccessibility of data network equipment in monitoring;
- failure or malfunction of data transmission network equipment;
- software failure or failure of server hardware;
- failure or failure of the application server;
- stopping the process, service or service on the server;
- failure in the working of software and applications;
- the inoperability of a portion of the software functionality;
- massive (more than 3) user requests for one event;
- inaccessibility of a link for a specific IT service.

Note that in the examples given, both for automated workplaces and for infrastructure incidents in the entire IT infrastructure, there are semantically similar incidents. Nevertheless, with manual input, there are significant differences in their wording.

Figure 2 presents a sample of extensive incidents ontology in considered information infrastructure designed by authors with using WebVOWL tool [17].

Next, we must consider the detection of new types of security incidents in critical information infrastructure in more detail. Let the initial ontologies of the subject domains are given, and there is a scheme for generating a new type of incident. It is required, when a new type of incident is detected, to perform ontology integration for a generalized semantic representation of information, to form some general ontology from individual ones, and to provide a thesaurus for the functioning of the situational awareness system.

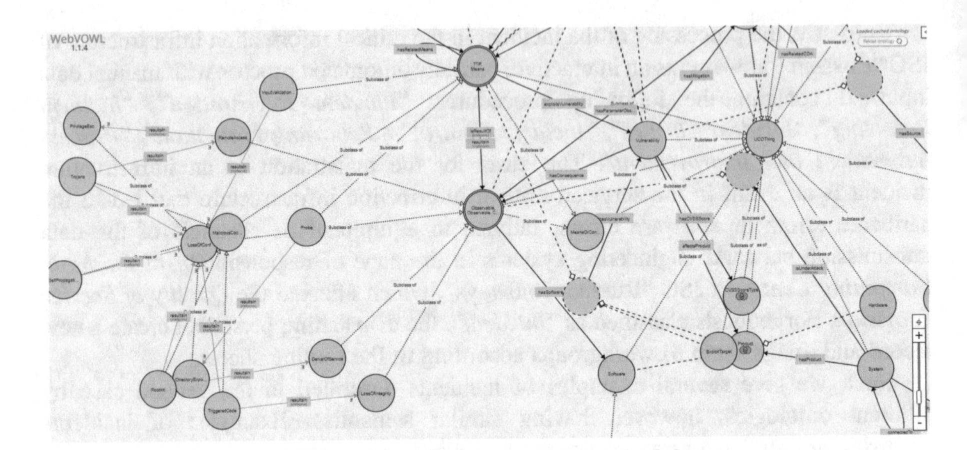

Fig. 2. Sample part of incidents ontology in critical information infrastructure

5 Conclusion

The paper substantiates the need to develop new methods for automating incident detection in critical information infrastructure of railway transport. The most modern and promising way to solve this problem is to create situational awareness systems within the framework of the developed intelligent railway management systems. Based on the experience gained in the development of incident detection systems for information security violation in transport systems, the authors propose the use of distributed reasoning methods for integrated ontologies of subject areas.

Acknowledgment. The reported study was funded by the Russian Foundation for Basic Research, according to the research projects No. 19-07-00329-a, 18-08-00549-a, 17-0700620-a.

References

1. Thaduri, A., Aljumaili, M., Kour, R., et al.: Cybersecurity for eMaintenance in railway infrastructure: risks and consequences. Int. J. Syst. Assur. Eng. Manag. **10**, 149–159 (2019). https://doi.org/10.1007/s13198-019-00778-w
2. Masson, É., Gransart, C.: Cyber security for railways – a huge challenge – Shift2Rail perspective. In: Lecture Notes in Computer Science, vol. 10222, pp. 97–104. Springer, Cham (2017)
3. Braband, J.: Cyber security in railways: quo vadis? In: Lecture Notes in Computer Science, vol. 10598, pp. 3–14. Springer, Cham (2017)
4. Bastow, M.D.: Cyber security of the railway signalling & control system In: 9th IET International Conference on System Safety and Cyber Security (2014). https://doi.org/10.1049/cp.2014.0986
5. Weik, M.H.: Computer Science and Communications Dictionary. Springer, Boston (2000). https://doi.org/10.1007/1-4020-0613-6_3411

6. Belsis, M.A., Simitsis, A., Gritzalis, S.: Workflow based security incident management. In: Lecture Notes in Computer Science, vol. 3746, pp. 684–694. Springer, Berlin (2005)
7. Chernov, A.V., Savvas, I.K., Butakova, M.A.: Detection of point anomalies in railway intelligent control system using fast clustering techniques. In: Advances in Intelligent Systems and Computing, vol. 875, pp. 267–276. Springer, Cham (2019)
8. Belsis, M.A., Godwin, N., Smalov, L.: A security incident data model. In: IFIP Advances in Information and Communication Technology, vol. 86, pp. 481–494. Springer, Boston (2002)
9. Butakova, M.A., Chernov, A.V., Guda, A.N., Vereskun, V.D., Kartashov, O.O.: Knowledge representation method for intelligent situation awareness system design. In: Advances in Intelligent Systems and Computing, vol. 875, pp. 225–235. Springer, Cham (2019)
10. Andrade, R., Torres, J., Cadena, S.: Cognitive security for incident management process. In: Advances in Intelligent Systems and Computing, vol. 918, pp. 612–621. Springer, Cham (2019)
11. Johnson, C.W.: Architectures for cyber-security incident reporting in safety-critical systems. In: Lecture Notes in Social Networks, pp. 127–141. Springer, Cham (2015)
12. Endsley, M.R., Garland, D.G.: Situation Awareness Analysis and Measurement. CRC Press, Boca Raton (2010)
13. Garani, G., Adam, G.K., Ventzas, D.: Temporal data warehouse logical modelling. Int. J. Data Min. Model. Manag. **8**(22), 144–159 (2016)
14. Baader, F., et al. (eds.): The Description Logic Handbook: Theory, Implementation and Applications, 2nd edn. Cambridge University Press, Cambridge (2010)
15. Serafini, L., Tamilin, A.: Local tableaux for reasoning distributed description logics In: Proceedings of DL 2004. CEUR-WS (2004). http://ceur-ws.org/Vol-104/12Serafini-final.pdf. Accessed 25 Apr 2019
16. Serafini, L., Borgida, A., Tamilin, A.: Aspects of distributed and modular ontology reasoning. In: Proceedings of IJCAI 2005 (2005). http://www.ijcai.org/Proceedings/05/Papers/0801.pdf. Accessed 25 Apr 2019
17. Web-based visualization of ontologies. http://vowl.visualdataweb.org/webvowl.html. Accessed 25 Apr 2019

A New Approach Towards Developing a Prescriptive Analytical Logic Model for Software Application Error Analysis

Wong Hoo Meng(✉) and Sagaya Sabestinal Amalathas

Taylor's University, Subang Jaya, Selangor, Malaysia
hoomeng@hotmail.com

Abstract. Software application is heavily being used by today's business operations. However as of the fact, it is no software application in the market can guarantee that it has no software defect escaped during the release. Whenever software application error arises, it can be the cause happened either within the software application layer or any other factor outside the software application layer. The root cause analysis activity becomes more complex in such situation as the analysis activity involves multiple layers. Time consuming is increased to identify the valid error. It leads to the entire analysis duration which can be easily prolonged. This arises the problem statement that, the duration of root cause analysis on software application error carries crucial impact to the service restoration. The objective is to create a logic model to conduct decision making process. This process includes the identification of the root cause in a more accurate manner, and shorten the duration of root cause analysis activity. The design of Analytic Hierarchy Process (AHP) hierarchy will depend on the software application errors and other involved errors found from different layers. These errors become the participants. The participants can be grouped further based on the error categories. Once the hierarchy is constructed, the participants analyze it through a series of pairwise comparisons that derive numerical scales of measurement for the nodes. Then, the priorities are associated with the nodes, and the nodes carry weights of the criteria or alternatives. Based on the priority, the valid error and the preferred resolution can be decided. Many past researches in AHP had been conducted, however there is a knowledge gap of AHP in software application error analysis. Therefore, a Prescriptive Analytical Logic Model (PAL) incorporates with AHP into the proposed algorithm is suggested. This logic model would contribute a new knowledge in the area of log file analysis to shorten the total time spent on root cause analysis activity.

Keywords: Application analysis · Application log · Application debugging · Log file analysis · Log processing · Analytic Hierarchy Process

1 Introduction

In today's world in general, application can be either mobile application or software application which are commonly used by the human. No doubt, these applications are impacting human lives in many ways. Many business company in the industry adopts

R. Silhavy et al. (Eds.): CoMeSySo 2019, AISC 1046, pp. 256–274, 2019.
https://doi.org/10.1007/978-3-030-30329-7_24

Information Technology (IT) as a strategic enabler. These business companies requires applications heavily to sustain their business operations. As for this paper. Of course the focus of this paper falls on software application. The point is, wherever software application becomes crucial for processing the business transactions, it has lower toleration for downtime expected by the company management. This was clearly supported by Labels: Data Center, Downtime, www.evolven.com (2014), and indeed software application downtime can cause the business operation ceased. Business company has option that it can run an IT production support team to provide support service to whichever business-as-usual (BAU) system running in the organization. Another option is that it can engage service provider to provide the same IT support service. Regardless whichever option is chosen, the time spent on conducting root cause analysis on software application error is crucial. Without accurately identifying the valid error, it creates impact to the service restoration to the software application.

As of the fact, identifying the root cause of the software application error is crucial before the resolution is decided and deployed into production environment. Therefore, there are several required actions in the root cause analysis activity. These actions involve collecting related information from different log files and selecting the related log events based on the time event when software application error occurred. However, at most of the time, collecting input information for root cause analysis is time consuming. This statement is supported by Management Logic (2012) stated that "The most time consuming aspect of Root Cause Analysis (RCA). Practitioners must gather the all the evidence to fully understand the incident or failure.". On the other hand, Horvath (2015) had also pointed out that "While the analysis itself can be time-consuming, the chance to mitigate or eliminate the root causes of several recurring problems/problem patterns is definitely worth the effort.". Hence, it is crucial to look for an efficient method to reduce the prolonging time at the root cause analysis activity.

Software application generally has its built-in event logging ability. The purpose of this event logging records the information of what activity is carried out or even what incident is occurred at that exact time. This information includes appropriate debugging information, and later the same information can be analyzed for software application root cause analysis purpose. The concern raised to the required information logging is that how much logging information is accepted as sufficient for software application root cause analysis. In addition, what is the appropriate category for the logging event such as information, error, debug, and fatal should be fetched as the input information to the root cause analysis activity. In the situation that if the extensive event logging level is enabled, this can lead to excessive logging information generated. With that, there are two issues raised. The first issue is that, the performance of software application is reduced by comparing with before and after extensive event logging option is enabled. The second issue is that, the manual analysis activity is becoming much more difficult and even tedious to identify the root cause of the software application error. Therefore, in the software application development process, it is a great concern on how much detail event logging should be logged into the log file. At the same time, the event logging must mitigate the performance impact created to the software application. These mentioned concerns had also been highlighted by Loggly (2017) and Panda (2011).

As per the following Fig. 1, Operating System communicates between software application and assigned resources (such as CPU, Memory, Network, and Hard Disk) on the virtual machine. Software application has to interact with Operating System to obtain allocated server resources to handle software application processing. This is because software application has high dependency on server resources to carry out its execution. Without the server resources, software application cannot execute itself at the software application layer. Further more, software application requires to communicate to its database server through the Local Area Network (LAN) for retrieving and updating software application data.

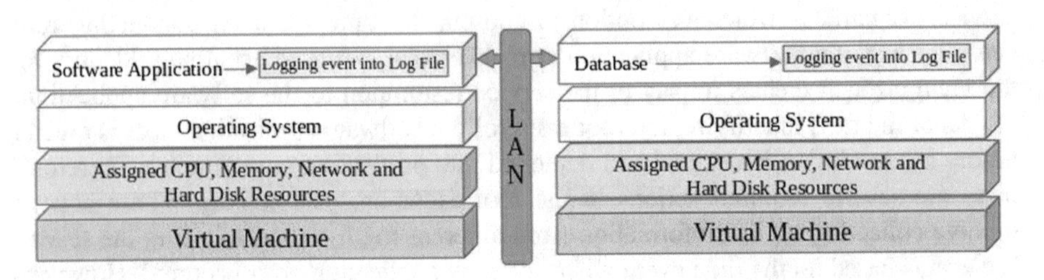

Fig. 1. Software application is required sufficient server resources to execute all its functionality.

To identify the root cause of software application error, it can be prolong under the following challenges.

(a) Software application log is hard to be understand.
(b) Error is occurred beyond software application layer, such as Operating System, Network, and Database layers. Software application log alone is insufficient to identify the actual error.
(c) IT support personnel has insufficient knowledge and experience in performing analysis.
(d) Root cause analysis is conducted manually, crucial information is overlooked.
(e) Historical error events are taking even longer time to be located.

With all the mentioned concerns and scenarios, by depending software application log file alone is not sufficient to conduct software application root cause analysis whenever error is occurred beyond the software application layer. Hence a proposed research is required for establishing a prescriptive analytical logic model. This proposed logic model incorporated the proposed algorithm to conduct the root cause analysis activity. It must target to increase the accuracy for error identification, and to reduce the prolonging time spent on the duration of root cause analysis. Therefore, this is a good potential to contribute new knowledge to the software application analysis.

2 Significant of the Study

The proposed research is not focusing on the time spent of developing the resolution. This is because if the valid error is not identified accurately, then it is very high chances that the resolution may not resolve the real issue. Therefore, the ultimate focus must fall on the root cause analysis and how fast the valid can be identified during the analysis activity is crucial. By reducing the prolonging time spent on root cause analysis, and improving the accuracy of identifying valid software application error. It can resume the software application service to the users in a shorter total time taken. This is because with today's rapid business competitive world, time consuming on analysis and trouble-shooting activities is unacceptable. Furthermore, it is a continuous battle for the IT support to face day-to-day software application error challenge in order to provide reliable up-time for the software application utilized in the business organization. On the other hand, business companies must still continue to utilize their existing software applications (without incur any additional operation budget). At the same time, it must continue to allow the companies to save the investment budget on spending the capital amount to replace all or partial of the software applications. Without introduces any new software application, companies can avoid to re-train their users on using the new software applications. This propose logic model brings the above benefits to business companies which are using software application for their daily operations crucially. The objective is to get the valid error fixed and to mitigate any same error re-occurrence in the near future. Over the years there were various researches had been done at this area such as consolidate the logs or integrate the logs for analysis. In fact, there had been very few attempts to propose an algorithm on root cause analysis beyond the software application layer. To support this, Jan (2001) had showed more detailed log file analysis by comparing the past published techniques whereas other studies were only indicated intentions or suggestions. Moreover, there is no attempt to incorporate Analytic Hierarchy Process (AHP) for decision making on valid software application error. This can be supported by Vaidya and Kumar (2004) that they had reviewed Analytic Hierarchy Process (AHP) with 150 application papers in total. There is no knowledge in applying AHP on identifying valid error. Hence, this is a great potential in this research which brings contribution to business intelligent studies.

3 Research Objective

To mitigate prolonging on conducting root cause analysis activity.

4 Literature Review

In the past research, Stewart (2012) is focusing on debugging real-time software application error using logic analyzer debug macros, whereby Eick et al. (1994) they are focusing on presenting the error logs in a readable manner. Moreover, Peng and Dolores (1993) suggested to focus on error detection in software application at the time of software development and maintenance. However, Salfner and Tschirpke (2015) are focusing on analyzing error logs by applying the proposed algorithms in order to predict future failure. Their software application error analysis approaches focus on software application error log obtained from software application database. Some literature suggested that the software application error analysis would be better if it is built in during the software development process. This approach is still within the same application development boundary without factoring in any other area of concerns. It can cause the software application failure. In another way of explanation, whenever hardware CPU and memory utilization is running high, or even storage disk space is running low, software application logging may not be accurate anymore. Hence, the root cause analysis would not be accurate to identify the real issue to understand the main reason to cause the software application failure. Murínová (2015) had attempted to integrate multiple log files from various software monitoring tool and network devices for better root cause analysis on Web application error. However, there is no proposed model stated in the research. Even until Landauer et al. (2018), they introduced an unsupervised cluster evolution approach that is a self-learning algorithm that detects anomalies in terms of conducting log file analysis. However, this approach is under machine learning rather than AHP. From a different point of view, this approach is good because it can be adopted into the proposed model to detect the software application error. Hence, this is a great potential to propose a logic model to research a new approach towards developing an algorithm for software application error analysis. At the same time to contribute new knowledge in the area of software application root cause analysis using AHP. By comparing the above secondary data with the proposed logic model, it can be noticed that the focus boundary on software application analysis. The technique to identify the root cause of software application is different. They focus on software application boundary whereby the proposed model focuses horizontally on all possible boundaries. In addition, due to the focus boundary is different, it leads to the technique to identify the root cause of software application also different.

5 Research Proposed

With a server (regardless it is physical or virtual) box, you have multiple layers. Most common logs that are required is:-

(a) System monitoring log for server resources such as CPU, Memory and Hard Disk usage.
(b) Network monitoring log for the server such as network communication within the Local Area Network (LAN).
(c) Operating System event log for server.

The illustration of multiple layers for a server can be referred to Fig. 1.

On the other hand, the software application especially for those that are rated at enterprise level category, it involves multiple tiers such as Web Client tier, Web Container tier, Application Container tier, and Database tier. Beside the Wen Client tier, each tier has its own log file. Certain tiers are even having multiple log files based on the software application design. Hence, the proposed research is to develop an algorithm towards to a prescriptive analytical logic model for analyzing software application error. The analysis is based on the logs retrieved from various software applications by referring to Fig. 2.

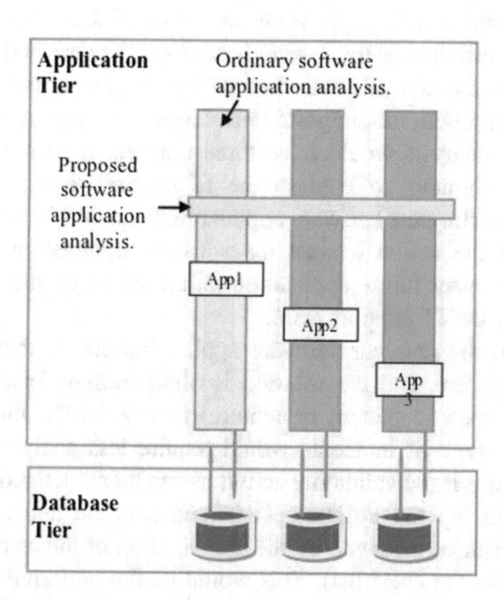

Fig. 2. The proposed software application analysis algorithm will analyze across multiple databases.

5.1 Proposed Research Scope

The proposed scope of this research is to define the algorithm. The algorithm consists of simple and complex analysis inside the prescriptive analytical logic model for software application error analysis. Therefore, by having the proposed logic model in the production environment, the logic model is required to react to software application error when the error is detected in the software application log file. With this logic model, it is also required to retrieve other related log files through various software applications shown as Fig. 2. The proposed algorithm mainly consists of two analysis areas, which are simple and complex analysis to form a prescriptive analytical logic.

5.2 Proposed Simple Analysis

For simple analysis area, it is required to build the predefined logic to handle the common software application error, whereby this model is to guide the system builder on answering a set of predefined questions on common software application errors and carry out the predefined activities only react to these common software application error, for example, restarting the software application process if it is stopped.

5.3 Proposed Complex Analysis

For complex analysis area, it is required to build a logic which collects necessary log events as data from the involved software applications. The collected data will serve as input information at the initial stage. With the collected data, this model will base on the past incidents determined as the system behavior. By combined with the predefined templates, the automated analysis activities will be triggered and finally generate the analysis outcome along with the suggested resolution steps and action to the IT support team. This complex analysis would have three different modes which are "manual", "semi-auto" and "fully-auto" offered to the IT support team. As for the complex analysis area, by predicting the software application behavior, it performs the suggested steps and carry out the action against the software application error based on the analysis. This will prevent future application failure based on the permission given to the offered mode by the IT support team.

By focusing into the re-occur software application errors, these errors occur in a specific pattern or feature, and the solution is often straight forward (can be applied after validating the specific pattern or feature) to resolve the incidents. The human involvement on this type of incidents would require less analysis but more on validating activities. Hence if the validating activities can be predefined into a checklist, the logic model can pick up the ultimate predefined solution and react to the incident automatically. This can be achieved by the combination of the answers (yield from the validating activities in the checklist). This would be the preferred method in the logic model that handles the common software application incidents. We call this logic as simple analysis. The same simple analysis logic can be applied to manage Server (a physical or virtual box running a vendor Operating System) or even Networking devices (such as switch or router) if they have incidents occur in the specific pattern or feature.

The software application errors which have no uniform pattern or feature, for this type of software application errors. The percentage of human involvement is high. This is because the person who handles the incident requires to obtain the software application log files and to search any similar error logged in the past. We call these files and records as input information. With the input information obtained, the person conducts the analysis activities before the person can identify the software application error root cause. Only the preferred resolution steps is agreed then it is applied to resolve the software application error.

For the first time occurring software application error. If both yielding input information activities and analysis activities can be automated. Base on the outcome of the analysis activities, human expects to see a list down of each possible root cause along with the proposed resolution steps in a complete list. Then, the decision is on the person to choose which is the preferred option. If the person chooses to proceed with the suggested resolution steps, then the person will receive the final question. The question is expecting the response from the person, whether agrees to let the automated activities execute the same suggested resolution steps automatically in the future if the same incident occurs again. Of course, this logic has the ability to handle unpredictable software application errors by performing simulated analysis activities comparing with human, we call this logic as complex analysis.

Whenever the complex analysis is triggered. It will pull the related logs based on specific time frame (duration) before and during the software application failure from various application logs. These logs are:-

(a) Software Application logs is the beginning to trigger the root cause analysis,
(b) Configuration management logs is for understanding any recent applied software application patches or Operating System patches,
(c) Performing and capacity monitoring logs is for identifying any hardware resources running insufficient, and
(d) Production support ticketing tool logs is for cross-checking any related issue recently occurred under the predefined database scheme.

These above logs as input information will be utilized crucially for root cause analysis to resolve the software application issue.

Indeed, the simple analysis can be existed independently at the initial stage. However, when the specific number of reoccur incidents hits. The complex analysis will be activated to perform the required analysis activities automatically. It will produce the complete analysis report and suggestion(s). Base on this suggested design, the complex analysis would have a loosely but it is fairly important relationship with the simple analysis. This is because the complex analysis needs to understand how many times the simple analysis has handled the same incidents in the past. This information is crucial to make a decision on suggesting the reasonable resolution steps to the human after the complex analysis produces the analysis report.

5.4 Proposed Algorithm, Architecture Design, and Solution Modeling

The proposed algorithm under the Prescriptive Analytical Logic Model includes the crucial activities in the following Table 1.

The proposed algorithm is derived from the overall architect design of the Prescriptive Analytical Logic Model shown as follows (Fig. 3):-

This architect design mainly is to retrieve the specific time of event logged from the related log files as the input data to the Prescriptive Analytical Logic Model. Then the following root cause analysis activities are carried out:-

1. You can tap on any market monitoring product for detecting Software Application error.
2. Based on the new found Software Application error (Cross check with the first Configuration file) for valid common software application error.
3. Evaluate and identify the suitable weighting to the new found Software Application error.
4. Update the first Configuration file (Append the error into the first Configuration file).
5. Construct AHP metric table with metric variables (each with weight) for decision making – for the Valid Error.
6. Analyze and decide whether the new found error is Alert, Warning or Valid Error.
7. Validate the possible resolution option for the new found error (Cross check with the second Configuration file).

Table 1. The proposed process activity under proposed algorithm.

No.	Algorithm
1	With the granted permission, retrieve and/or integrate various log files obtained from all involved software application log files/databases
2	Identify whether the newly reported software application error is first time occurrence or re-occurrence by cross-checking the database which is associated to the prescriptive analytical logic
3	Identify possible log data and select the necessary log data for analysis under the defined software application error classification
4	**Allocate weight to each possible software application error based on *Analytic hierarchy process* (AHP)**
5	**Shortlist the software application error under the highest weight**
6	Analyze the selected log data for shortlisted software application errors and define possible resolution option
7	**Allocate weight to each possible resolution option based on AHP**
8	**Shortlist the preferred resolution option under the highest weight**
9	Deploy the preferred resolution option to fix the software application error under the predefined condition
10	Store the analysis result and resolution action into a database which is associated to the prescriptive analytical logic for future reference and knowledge base activities

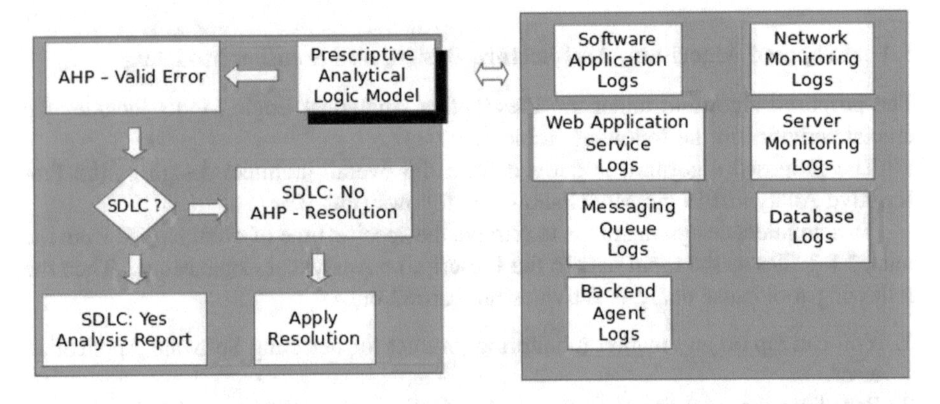

Fig. 3. The architecture design of Prescriptive Analytical Logic Model.

8. Construct AHP metric table with metric variables (each with weight) for decision making – for the Preferred Resolution.
9. The action of Preferred Resolution (whether deploys directly or produces analysis report) is based on the second Configuration file.

Therefore, the solution modeling of the Prescriptive Analytical Logic Model for the process flow design is derived.

The solution modeling of the Prescriptive Analytical Logic Model for the process flow design (Fig. 4):-

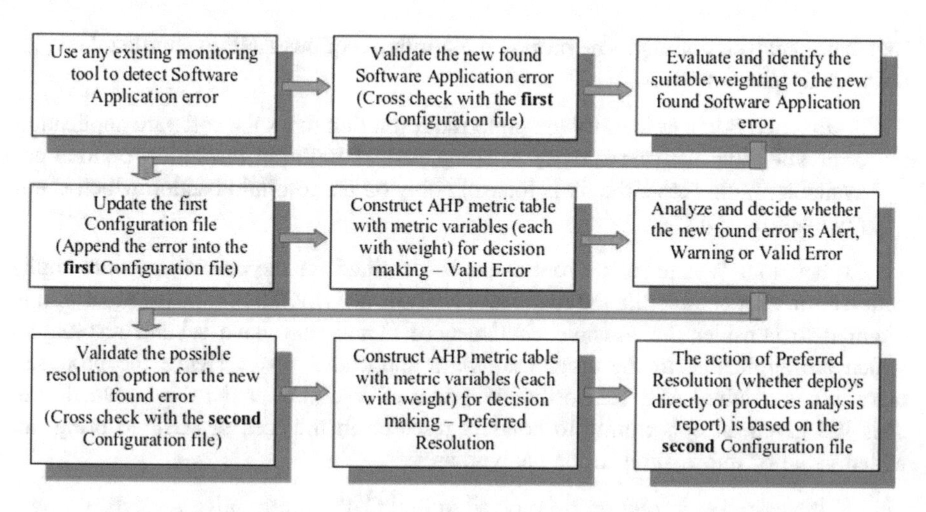

Fig. 4. The solution modeling for the process flow design.

6 Analytic Hierarchy Process (AHP)

AHP was developed by Saaty stated in Wikipedia (2015). The proposed algorithm of the prescriptive analytical logic can adopt the decision making process of AHP to decide the valid software application. Then, followed by using the same decision making process to shortlist the best resolution. This is because by referring to Saaty (1987), the three principles which are the decomposition principle, the comparative judgments, and the synthesizing priorities. In addition, Vaidya and Kumar (2004) had provided the discussion on how to apply AHP under the analytic hierarchy process. This helps to understand how the proposed algorithm applies in the scenario after the valid software application error, and preferred resolution are identified during the root cause analysis activity.

By comparing with those past researches or similar area of researches in the area of software application log file analysis. The AHP was not introduced or applied for handling the activities such as shortlisting the valid software and shortlisting the preferred resolution. In order to have further understanding on why AHP is introduced in this proposed logic model, and how it plays the important role. The proposed algorithm should consider if there is a scenario that multiple software application errors, and warnings occur almost at the same time at a software application execution. Remember that there will be two required activities:-

i. First activity is to identify the valid software application error among multiple software application errors and warnings.

Explanation: When the proposed logic model extracts the software application errors from various software applications for analysis, there must be a technique or process required to filter out software application errors which are determined as low priority or invalid errors, leaving the crucial software application error for conducting root cause analysis activity. The proposed algorithm requires AHP to shortlist the valid software application error.

ii. Second activity is to identify the preferred resolution to fix the software application error after the analysis activity is completed, if multiple resolution options are available. In this case, the fix is focusing only on the resolution action which is not involved in SDLC.

Explanation: Whenever the root cause is identified for the crucial software application error after conducting the root cause analysis activity. The proposed algorithm is required to consider the possible circumstance that it may have several resolutions which are applicable to the crucial software application error. Hence the proposed algorithm is required AHP to shortlist the preferred resolution under this circumstance. This has given an opportunity to conduct research in the area of AHP to bring the added value of information in the analytic area.

iii. In the case that resolution is involved in SDLC, the Prescriptive Analytical Logic Model will provide a detail analysis result as a report to show what root cause is identified and what is the proposed resolution.

Explanation: This is because if the fix (resolution) involved the activity of SDLC, it is out of scope on this proposed research. However, if the fix is straight forward such as restarting the software application service or network connection as the examples, it can be decided by the AHP when the defined impact level stated in the configuration file.

The approach of applying AHP in this proposed research is for decision making activity although AHP has been used in other areas. By comparing with the published articles, the differentiation in this proposed logic model is that AHP is utilized on log file analysis activity for decision making on valid software application error as well as identifying the preferred resolution to the error if possible. On the other hand, the log file analysis area of knowledge currently has no AHP knowledge involved yet. As the result, AHP is utilized in the proposed logic model is a new knowledge contribution and the crucial activities can be illustrated in the following Fig. 5.

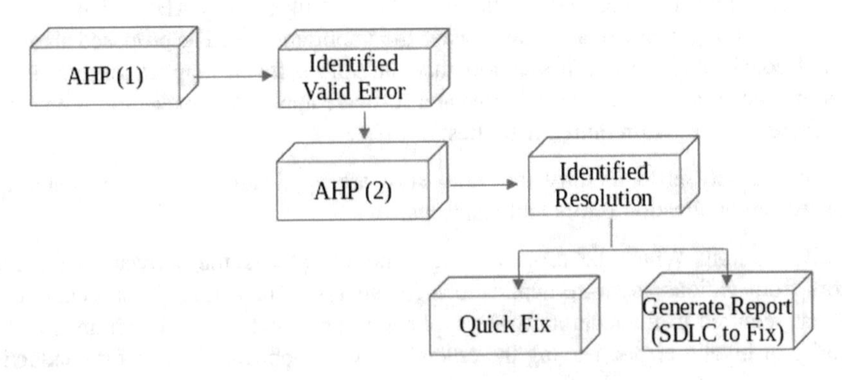

Fig. 5. The AHP activities in Prescriptive Analytical Logic Model.

The following Table 2 contains the important AHP activities under the proposed algorithm.

Table 2. The AHP activities in the algorithm.

Activity	Algorithm
AHP (1) is referring to the following activities	
4	Allocate weight to each possible software application error based on *Analytic Hierarchy Process* (AHP)
5	Shortlist the software application error under the highest weight
AHP (2) is referring to the following activities	
7	Allocate weight to each possible resolution option based on AHP
8	Shortlist the preferred resolution option under the highest weight

Hence, with the Prescriptive Analytical Logic Model, it carries high potential on knowledge contribution as the new logic model has the algorithm to be able to identify the root cause of the software application error more accurately under the AHP processing approach, and deliver the outcome to shorten the analysis duration of conducting root cause analysis activity during the software application downtime.

7 Proposed Prerequisites of PAL

7.1 Multiple Tiers Environment

In this proposed research, it is required to run on a multiple-tier environment which involves Client tier, Web Container tier, Application Container tier, and Database tier. The proposed algorithm will be implemented into a software plug-in component sitting at the logic tier to integrate all the required log event as input data from various software applications and store in a separate location for information retrieval later. A new standard database schema may be required on this research, and it will be applied across all the required databases of the involved software applications. This will help on retrieving log events as data from various databases when the root cause analysis activity is triggered at the logic tier whenever software application error is detected in the main software application log file. Otherwise, by granting necessary permission to retrieve log events as data from the involved software applications should be sufficient.

7.2 Network Time Protocol (NTP) Server

Whenever a software application involves in a multiple-tier environment, it is good to standardize the server time across all the involved servers. The supporting reason is that the proposed Prescriptive Analytical Logic Model is required to collect log data from different software applications based on the specific given time as the primary key, and therefore the server time on all the involved servers are required to have the network time synchronization. As per Masterclock (2016), "Accurate time stamping is key to root-cause analysis, determining when problems occurred and finding correlations. If network devices are out of sync by a few milliseconds or, in extreme cases a few seconds, it can be very difficult for network administrators to determine the sequence of events.".

7.3 Technologies

The proposed prototype will be designed as a software application plug-in component. It can be coded by using either Java Programming or Python. The Java Programming is the preferred programming language as it is a platform independent programming language. Therefore, at the Operating System platform, it required to install Java Runtime Environment (JRE). Then, proceed to configure the Java home path and memory allocation for the Java Virtual Machine (JVM) will do.

8 Proposed Logic Model Design of PAL

Referring to Table 1, the proposed algorithm contains a set of required activities, or we should say the ten activities. However, applying predefined database schema may not be a good idea as it alters the existing database schema of the software applications which is being used by the users in the production mode. Updating the database schema can lead to the impact of change on database structure, and cause the software application queries behaving differently. Indeed, regardless any strong reason, the proposed model design should look into the opportunity that no footprint should be leaving to the involved software applications as the primary concern in this research. Therefore we must look into the safe way or even good suggestion to avoid any disturbance to the involved software applications.

The proposed logic model would serve as a guidance to develop the prototype at the end of the research. This PAL analyzes error without disturbing the existing software applications running in production mode. Which means the PAL is developed as a plug-in tool. This plug-in tool is designed and developed to be an independent software application component, and it is sitting at the logic tier. This plug-in component can integrate into an existing software application to fetch the error from the log file of existing software application. It carries out the analysis activities without interrupt the existing software application. Under such proposed design, it helps to avoid any interference of existing software application execution while the plug-in component can still pull the related log information from software application. Whenever more data or information is required for further analysis which is depending on circumstance, the plug-in component can also pull the related information from various databases such as Configuration Database, Production Support Ticketing System Database, and Application and System Monitoring Database during the analysis process.

Ideally, for the involved software application databases, they are required to inject with the predefined schema. This schema would allow the PAL to initiate Structured Query Language (SQL) queries to retrieve the necessary logged event or related information. Then, based on the yield data or information, it will be served as the input information to the PAL for further analysis activities. However, we must factor in a possible consideration that those databases with certain unforeseen reasons. They are not allowed to inject with the predefined schema. In such situation, the only linkage to relate all the necessary log data across multiple software application databases, which is the "time". With the time of event logged in the software application, we can search the log data from other software application databases based on the given time.

By referring to the proposed algorithm mentioned in Table 1, the PAL is playing the role to conduct reactive action whenever software application error is detected in its log file. Based on the "time" recorded the software application log event, PAL will retrieve all the necessary logs from the involved software applications. With the proposed PAL, it is good to have a simple analysis activity to handle first occurrence software application error. As for the occurrence software application error, a complex analysis is designed to handle deep dive investigation.

For "Simple Analysis", the proposed functionality is for quick fix such as:-

- To restart software application service.
- To kill dead process in the software application process queue, or
- Any straight forward and quick action for resolution.

For "Complex Analysis", the proposed functionality is:-

- To categorize the repeatedly pattern of the software application error in various categories,
- To analyze the root cause, and
- Proactively to select a suitable resolution with suggested resolution steps automatically.

On every software application error is found, the error must be submitted into "Simple Analysis" section for initial analysis. This section first thing first validates whether it is a re-occurrence error, and

- If the error is identified as the first-time encountered software application error, the next action is to validate whether this first-time encountered software application error has any standard resolution step which has been predefined in the resolution list.
  (a) If the standard resolution steps are found in the predefined resolution list, then proceed to apply the resolution steps followed by storing the analysis result and applied resolution action into the knowledge base database. Then, the analysis activity is ended.
  (b) If the standard resolution steps are not found in the predefined resolution list, then the error will be submitted into "Complex Analysis" section for further action.
- If the error is identified that it is not the first-time encountered software application error, then the error will be submitted into "Complex Analysis" section for further action.

As for the “Complex Analysis” section, the software application error is categorized under the predefined categories of possible errors depending on the error aspect. The objective is to focus within a specific category of error and narrow down the root cause. There are changes that error is hard to determine and hard to match into any error category. Under this circumstance we will have a “Other” category to categorize the error into it. On the next analysis activity, we need to look for any past analyzed result for the same issue from the knowledge base database. Base on the current information we obtain which is the specific error category and the past analyzed result for the same issue from the knowledge base database. The analysis process will identify the need of additional required information retrieved from various software application databases. These software application databases are Configuration Database, Production Support Ticketing System Database, and Application and System Monitoring Database, and etc.).

With all the obtained information, the analysis process must identify the possible involved areas with this error. Whether it is solely being fixed at the software application layer, or other layers such as:-

- Hardware capacity layer (i.e. CPU, memory, hard disk),
- Operating System layer,
- Network layer,
- Database layer, or
- Any combination which more than one layer.

Again, the analysis process will identify the need of additional required information retrieved from specific or various log files from different layers. There is also circumstance that when a major software application error occurs, it would also rigger a list of related and even non-related errors or warning messages in the software application log, with such the analysis process will need to retrieve software application log information under a specific time duration from the software application log file as for the input information of the analysis process.

The AHP process is applied to allocate weight to the software application error as well as errors found from different layers under the same “timing”. Therefore, AHP process is playing an important role to allocate weight to each possible error based on the error aspect. Whenever the weight is allocated on each error, the AHP process will shortlist the error under the highest weight. Once the shortlisted error is determined, the AHP process will proceed to evaluate all the possible resolutions based on the weight allocation to each possible resolution. Finally, the AHP will shortlist the resolution based on the highest weight. If the possible resolution is more than one, then the PAL will generate a report by providing the possible resolution option.

The PAL is designed to have two configuration files, one is designed to record down the common errors and room to cater for new found error. Then, another one is designed to record down the resolution steps for the common errors as well as room to cater for human to provide new resolution steps based on the new found error.

These preferred resolution actions under the resolution are manual, semi-automatic, and fully automatic. The PAL must obey the configurations in order to carry out the resolution under the preferred actions. Especially for the “Complex Analysis”, it is required to store the analysis result and the resolution action into a knowledge based database which is associated to the PAL for future reference. Regardless the analysis activity is under simple analysis or complex analysis, the knowledge based is catered for upcoming analysis activity and fine tuning the logic. It is mainly for improving the decision making on selecting the final resolution to be applied to the new found error in the near future.

Lastly, from a training purpose perspective, since the knowledge base of the proposed PAL will play an important role for the entire analysis process during the error analysis activity. It can be utilized as training material to train any new joining IT support engineers to pick up the support experience rapidly. At the same time it benefits for those who are experience support engineer to improve the knowledge of software application support. To ease of retrieving the information from knowledge base, the knowledge base database can be either stored locally at the database server or in the cloud. This is because to access the knowledge based in the cloud, the information is accessible at any time. The following Fig. 6 will explain the flow of the entire PAL process.

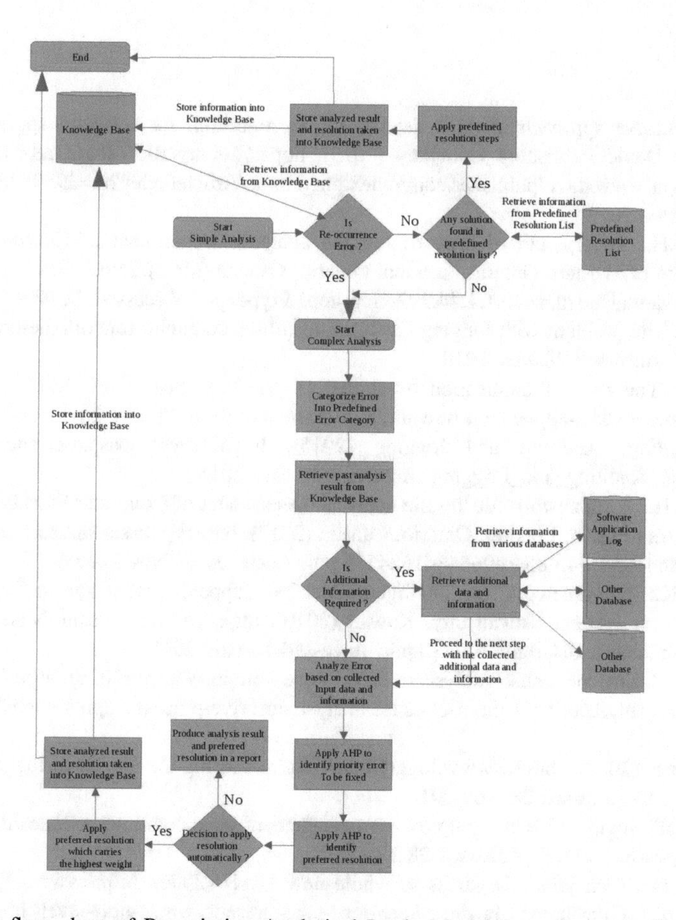

Fig. 6. The flow chart of Prescriptive Analytical Logic Model for software application error analysis.

9 Conclusion

Software application encounters situation that malfunctioning, it can be due to issue with the software application layer or other external factors outside the software application layer. The error can be the factor within the software application defect, or other external factors such as server resources, Operating System, Network communication, Database connectivity and etc. All these external causes affect the accuracy of the software application event logging to be correctly logged into the error log file. Hence, if the root cause analysis is solely depending on software application error log file, it is definitely insufficient to identify the software application error accurately whenever the software application error occurred under the circumstance which it is outside the boundary of software application layer. Therefore, a prescriptive analytical logic model incorporated with AHP targets to reduce the duration of the root cause analysis, and accurately identify the valid error as well as the preferred resolution to the error.

References

Wong, H.M.: An approach towards developing an algorithm for software application error analysis. David Publishing Company (2019). https://doi.org/10.17265/2328-2185/2019.04.006. http://www.davidpublisher.com/index.php/Home/Article/index?id=39805.html. Accessed 10 June 2019

Andrews, J.H.: Theory and practice of log file analysis. Department of Computer Science, University of Western Ontario. London, Ontario, Canada (1998). http://citeseerx.ist.psu.edu/viewdoc/download?doi=10.1.1.44.292&rep=rep1&type=pdf. Accessed 28 Nov 2015

Atwood, J.: The problem with logging (2008). https://blog.codinghorror.com/the-problem-with-logging/. Accessed 28 Nov 2015

Bloom, A.: The cost of application downtime is priceless. StatusCast (2014). http://www.statuscast.com/cost-application-downtime-pricess/. Accessed 28 Nov 2015

Error Handling: Auditing and logging (2015). https://www.owasp.org/index.php/Error_Handling,_Auditing_and_Logging. Accessed 28 Nov 2015

Andrews, J.H.: A framework for log file analysis. Department of Computer Science, University of Western Ontario, London, Ontario, Canada (2017). https://pdfs.semanticscholar.org/9d1a/97988d8b41354cd4bf85ace96648d1684555.pdf. Accessed 28 Nov 2015

Hasan, M.K.: A framework for intelligent decision support system for traffic congestion management system. Kuwait city, Kuwait (2010). http://www.umsl.edu/divisions/business/pdfs/Mohamad%20Hasan%20IDSS.pdf. Accessed 13 Nov 2015

Horvath, K.: Using root cause analysis to drive process improvement (2015). http://intland.com/blog/safety-engineering/using-root-cause-analysis-to-drive-process-improvement/. Accessed 28 Nov 2015

Loggly Inc.: (2017). https://www.loggly.com/blog/measuring-the-impact-of-logging-on-your-application/. Accessed 28 Nov 2015

Management logic (2012). http://www.management-logic.com/toolbox/sales/Root-Cause%20Analysis/Index.html. Accessed 28 Nov 2015

Margulius, D.: Tech jobs take stress to whole new levels (2006). http://www.infoworld.com/article/2655363/techology-business/tech-jobs-take-stress-to-whole-new-levels.html. Accessed 13 Nov 2015

Mercer, E.: How technology affects business operations (2015). http://science.opposingviews.com/technology-affects-business-operations-1659.html. Accessed 28 Nov 2015
Murínová, J.: Application log analysis. (2015) http://is.muni.cz/th/374567/fi_m/thesis_murinova.pdf. Accessed 28 Nov 2015
Ornelas, L.V.: Important of event logging. SANS Institute (2003). https://www.giac.org/paper/gsec/3297/importance-event-logging/105437. Accessed 28 Nov 2015
Panda, A.: High performance and smarter logging (2011). https://dzone.com/articles/high-performance-and-smarter. Accessed 12 Dec 2015
REDDIT. Experienced Dev's: how much of your time do you spend troubleshooting? (2015) https://www.reddit.com/r/webdev/comments/3sldcc/experienced_devs_how_much_of_your_time_do_you/. Accessed 14 Nov 2015
Salfner, F., Tschirpke, S.: Error log processing for accurate failure prediction (2015). www.usenix.org. https://www.usenix.org/legacy/event/wasl08/tech/full_papers/salfner/salfner_html/. Accessed 12 Dec 2015
StackOverflow: How much time do you spend in production troubleshooting? (2015). http://stackoverflow.com/questions/1425069/how-much-time-do-you-spend-in-production-troubleshooting. Accessed 14 Nov 2015
Eick, S.G., Nelson, M.C., Schmidt, J.D.: Graphical analysis of computer log files. Commun. ACM (1994). http://citeseerx.ist.psu.edu/viewdoc/download?doi=10.1.1.43.4832&rep=rep1&type=pdf. Accessed 12 Dec 2015
Rinnan, R.: Benefits of centralized log file correlation. Gjøvik University College (2005). http://citeseerx.ist.psu.edu/viewdoc/download?doi=10.1.1.121.8787&rep=rep1&type=pdf. Accessed 16 Mar 2017
Stewart, D.B.: Troubleshooting real-time software issues using a logic analyzer. InHand Electronics, Inc (2012). http://www.embedded.com/design/debug-and-optimization/4236800/Troubleshooting-real-time-software-issues-using-a-logic-analyzer. Accessed 12 Dec 2015
SuccessFactors: Using technology to increase your business productivity (2015). https://www.successfactors.com/en_us/lp/articles/using-technology-to-increase-your-business-productivity.html. Accessed 28 Nov 2015
Labels: Data Center, Downtime. Downtime, outages and failures - understanding their true costs. Evolven Software (2014). www.evolven.com. http://www.evolven.com/blog/downtime-outages-and-failures-understanding-their-true-costs.html. Accessed 13 Nov 2015
Wikimedia Foundation, Inc.: Application server (2017a). https://en.wikipedia.org/wiki/Application_server. Accessed 28 June 2017
Wikimedia Foundation Inc.: Logfile (2017b). https://en.wikipedia.org/wiki/Logfile. Accessed 28 May 2017
Wikipedia: Analytic hierarchy process. Wikimedia Foundation, Inc (2015). https://en.wikipedia.org/wiki/Analytic_hierarchy_process. Accessed 13 Nov 2015
Peng, W.W., Dolores, R.: Wallace. Software error analysis. NIST Special Publication (1993). http://www.geocities.ws/itopsmat/SoftwareErrorAnalysis.pdf. Accessed 12 Dec 2015
Masterclock: Network time synchronization - why you need an NTP aerver. (2016). https://www.masterclock.com/company/masterclock-inc-blog/network-time-synchronization-why-you-need-an-ntp-server. Accessed 4 Sept 2018
Saaty, R.W.: The analytic hierarchy process – what it is and how it is used. Pergamon Journals Ltd (1987). https://core.ac.uk/download/pdf/82000104.pdf. Accessed 9 Apr 2018
Jan V.: Log file analysis (2001). https://www.kiv.zcu.cz/site/documents/verejne/vyzkum/publikace/technicke-zpravy/2001/tr-2001-04.pdf. Accessed 18 Jan 2017

Vaidya, O.S., Kumar, S.: Analytic hierarchy process: an overview of applications. Department of Mechanical Engineering, Army Institute of Technology, Pune 411 015, India. National Institute of Industrial Engineering (NITIE), Vihar Lake, Mumbai 400 087, India (2004). http://ac.els-cdn.com/S0377221704003054/1-s2.0-S0377221704003054-main.pdf?_tid=39850adc-5d78-11e7-b5bf-00000aab0f26&acdnat=1498815840_e0c9a10c99c46ad30db8da4ef17e817b. Accessed 12 Sept 2018

Saaty, R.W.: The analytic hierarchy process—what it is and how it is used. Math. Model. **9**, 161–176 (1987). https://doi.org/10.1016/0270-0255(87)90473-8. Accessed 12 Sept 2018

Landauer, M., Wurzenberger, M., Skopik, F., Settanni, G., Filzmoser, P.: Dynamic log file analysis: an unsupervised cluster evolution approach for anomaly detection (2018). www.sciencedirect.com. https://www.sciencedirect.com/science/article/pii/S0167404818306333/pdfft?md5=e114106f24719a22492bdc5df17f2742&pid=1-s2.0-S0167404818306333-main.pdf. Accessed 2 Apr 2019

Serbu, R., Marza, B., Borza, S.: A spatial analytic hierarchy process for identification of water pollution with GIS software in an eco-economy environment. MDPI (2016). https://www.google.com/url?sa=t&rct=j&q=&esrc=s&source=web&cd=8&ved=0ahUKEwjTk9bz-cLbAhUFupQKHcXaCmAQFghkMAc&url=http%3A%2F%2Fwww.mdpi.com%2F2071-1050%2F82F11%2F1208%2Fpdf&usg=AOvVaw3k7wSGKCoC_Exqr0vJKHK7. Accessed 12 Sept 2018

Gerdsri, N., Kocaoglu, D.F.: Applying the analytic hierarchy process (AHP) to build a strategic framework for technology roadmapping 2007. http://ac.els-cdn.com/S0895717707001069/1-s2.0-S0895717707001069-main.pdf?_tid=e0165298-5d78-11e7-9000-00000aab0f6b&acdnat=1498816119_34b7b92d988eea0c318b8d23810e293a. Accessed 21 Feb 2019

Chang, D.-Y.: Applications of the extent analysis method on fuzzy AHP (1996). http://ac.els-cdn.com/0377221795003002/1-s2.0-0377221795003002-main.pdf?_tid=e251d09a-7732-11e7-b173-00000aacb35d&acdnat=1501644789_e6ae23503d2e2f883d2413693f63275a. Accessed 12 Sept 2018

Saaty, T.L.: Decision making with the analytic hierarchy process. Int. J. Serv. Sci. **1**(1) (2008). http://citeseerx.ist.psu.edu/viewdoc/download?doi=10.1.1.409.3124&rep=rep1&type=pdf. Accessed 12 Sept 2018

Teehankee, B.L., De La Salle University: The analytic hierarchy process: capturing quantitative and qualitative criteria for balanced decision-making (2009). https://www.researchgate.net/publication/256009323_The_Analytic_Hierarchy_Process_Capturing_Quantitative_and_Qualitative_Criteria_for_Balanced_Decision-Making. Accessed 12 Sept 2018

Triantaphyllou, E., Mann, S.H.: Using the analytic hierarchy process for decision making in engineering applications: some challenges (1995). http://s3.amazonaws.com/academia.edu.documents/31754775/AHPapls1.pdf?AWSAccessKeyId=AKIAIWOWYYGZ2Y53UL3A&Expires=1498819440&Signature=iIMAp3tvmRbBTX8k6rUvRIorb7o%3D&response-content-disposition=inline%3B%20filename%3DUSING_THE_ANALYTIC_HIERARCHY_PROCESS_FOR.pdf. Accessed 12 Sept 2018

Krejˇcí, J., Stoklasa, J.: Aggregation in the analytic hierarchy process: why weighted geometric mean should be used instead of weighted arithmetic mean. Expert Syst. Appl. **114**(2018) 97–106 (2018). https://www-sciencedirect-com.ezproxy.taylors.edu.my/science/article/pii/S0957417418303981. Accessed 12 March 2019

An Analytical Modeling for Boosting Malicious Mobile Agent Participation in Mobile Adhoc Network

B. K. Chethan[1(✉)], M. Siddappa[2], and H. S. Jayanna[3]

[1] Department of Information Science and Engineering, Sri Siddhartha Institute of Technology, Tumkur, Karnataka, India
crsh2019@gmail.com

[2] Department of Computer Science and Engineering, Sri Siddhartha Institute of Technology, Tumkur, Karnataka, India

[3] Department of Information Science and Engineering, Siddaganga Institute of Technology, Tumkur, Karnataka, India

Abstract. The contribution of mobile agents in existing wireless network is basically to cater up the connectivity demands. However, adoption of mobile agents in decentralized system like Mobile Adhoc Network (MANET) is found to be highly vulnerable where there is few evidence of offering security legitimacy of services offered by mobile agents. The security problem becomes highly challenging if adversary dynamically alters the attack tactics. Therefore, the proposed system introduces a novel analytical model which is capable of capturing the malicious behavior of mobile agent and quarantines them from further spreading the intrusion. Utilizing probability modeling, the proposed system introduces a resistance on the basis of intensive evaluation of local and global trust. The study outcome of proposed system shows that it offers better communication performance along with security in contrast to existing communication scheme in MANET.

Keywords: Mobile Adhoc Network · Mobile agent · Security · Attack pattern · Malicious behavior · Probability

1 Introduction

Mobile Adhoc Network (MANET) is one of the integral part of the pervasive computing in recent era as well as future era. It is basically distributed network consisting of mobile nodes without having any access point in its deployment. The participating nodes collaborate itself to the neighbor nodes whose physical distance is lower than the communication ranges of the nodes. In the post PC-era, various independent or a subsystem of Internet of Things (IoT) based applications requires an infrastructure less wireless network, where MANET plays an important role [1]. The MANET working group, many academician and industrial researchers contributed to evolve many proactive and reactive routing protocols to makes the success of the network performance in terms of the data delivery and achieve any source to any destination kind of the communication [2]. Since MANET is a resource constraints network of mobile

R. Silhavy et al. (Eds.): CoMeSySo 2019, AISC 1046, pp. 275–284, 2019.
https://doi.org/10.1007/978-3-030-30329-7_25

nodes with limited power as operated by battery and bandwidth, a non-optimal or unbalance usages of these resources drastically reduces the network performance as well of existing security issues in MANET [3]. The proposed paper discusses about a solution towards resisting the malicious mobile agents without using any conventional encryption approach. The organization of the paper is as follows: Sect. 2 briefs of existing approaches towards usage of mobile agents while Sect. 3 briefs of research problems. Discussion of proposed methodology is carried out in Sect. 4 followed by discussion of system design. Result discussion is carried out in Sect. 5 while summary of proposed paper is carried out in Sect. 6.

2 Related Work

This section briefs of recent approaches towards utilization of mobile agents [4] for catering up various communication demands of MANET. The work carried out by Rohankar [6] throws an interesting usage of agent-based approach over wireless adhoc network using predictive approach over conventional MANET routing scheme. The work of Halim et al. [7] has introduced a selective data forwarding scheme using mathematical model using the role of agent for both routing and monitoring system considering on-demand routing scheme. The study outcome shows better performance in latency and detection. The agent-based mechanism was also used in IoT and cloud ecosystem as MANET finds its ultimate deployment over this. Study considering securing adhoc network using mobile agents was seen in the work of Abosamra et al. [8] where authors have used encryption protocol for securing the communication. Combination of agent usage and trust is another frequently adopted approach for securing communication in MANET. The recent works carried out by García-Magarino et al. [9] have used agent-based approach for facilitating mining massive data. Another recent work of Fortino et al. [10] has discussed how agent-based mechanism can be embedded with smart objects in IoT environment. The most recent study carried out by Wang and Li [11] have presented a secure communication mechanism that selects the agent nodes which constructing routes. These techniques ensure reliable delivery of packet between the agent nodes. The work of Gargees and Scott [12] has presented a technique for facilitating data structurization over IoT media in order to extract pattern of communication. However, such studies are more data oriented and less on security. Adoption of multi-agents was reported to be used in the study carried out by Quan et al. [13, 14] which emphasizes on both fault tolerance and security breach prevention. Agent-based approaches are also used by Harrabi et al. [15] where the authors have used it over vehicular network. Another work carried out by Fortino et al. [16] have discussed that hybridizing the characteristics of agents has multiple benefits of communication system assessment in an IoT. The study carried out by Santos et al. [17] have discussed about the usage of software agents along with incorporation of intelligence. Another recent work of Shehada et al. [18] has adopted trust based scheme along with reputation in order to model a mobile agent focusing on social network. The technique uses an adaptive process in order to evolve up with a decision making system. The work carried out by Hsieh et al. [19] have constructed a design of an agent that using event-driven approach for constructing the communication mechanism in

IoT. However, there is less inclination of the existing schemes to address security problems associated with MANET and are more focus on re-defining communication system on wireless network. The next section briefs of research problems.

3 Problem Description

From the prior section, it can be seen that there has been extensive studies being carried out towards using mobile agents in adhoc environment. However, usage of mobile agents towards secure communication as well as protection for mobile agent itself has not been extensively carried out. This leads a security uncertainty for the communication assisted by the mobile agents. Apart from this, all the security approaches are intensively towards protecting specific forms of attack which is not useful if the attack scenario changes. Therefore, the problem statement of the proposed study can be stated as "*It is quite a challenging task to construct a framework capable of identifying the discrete behavior of mobile agents such that agent-based communication system can be further boosted for MANET communication*".

4 Proposed Methodology

The proposed system introduces mobile agents to assist in bridging the communication demands in MANET but is highly vulnerable to security threats. Therefore, proposed system considers a presence of adversarial model where there is no apriori information about the criticality of attack in MANET by malicious mobile agents. The scheme adopted as a solution is showcased in Fig. 1.

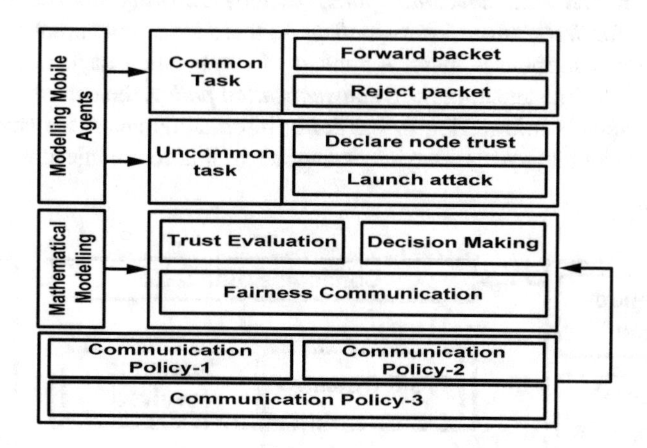

Fig. 1. Schema of methodology for 1st objective

The above Fig. 1 exhibits that the proposed system considers modeling mobile agents with inclusion of two types of task to be executed by them. Owing to presence of common task, it is highly challenging to distinguish regular node from malicious mobile agent. Although, there are distinct charecteristics too, but they are rarely expected to be exhibited by malicious mobile agent in order to resist the chances to get entrapped by the security system. Therefore, the proposed model constructs a simplified mathematical modeling on the basis of probability theory in order to evolve up in a potential and reliable decision making system. The core components of the proposed mathematical modeling are trust evaluation, decision making, and fairness communication. A novel concept of trio communication policy has been constructed which governs all the possibilities of communication behaviour of both the type of nodes. The next section illustrates about the system design and implementation.

5 System Implementation

The proposed system considers a deployment region, number of mobile agents, and proportion of the intruder node. Apart from this parameters, the proposed system also performs initialization of few more parameters e.g. profit assigned to agent for performing, number of resourced involved in performing forwarding of data, profit allocated to agent for performing dropping of node. The next part of the implementation is carried out for estimating number of occurances associated with data forwarding, identified intrusion or data dropping, probability of intrusion, and ambiguous trust factor for all the mobile agents present in deployment region. The proposed system implements three different communication policies as a part of contribution for assisting in better decision making as well as mapping with real-world greedy scenario of the mobile agents. According to *first communication policy*, the current mobile agents will have complete information about the next task to be executed by the mobile agent and vice-versa. In *second communication policy*, a probability is assigned to each undertaking of *first communication policy* thereby allowing the mobile agents to arbitrarily opt for *first communication policy*. As there are infinite number between the probability limits so there is massive number of combination of *first communication policy* for each mobile agent. In *third communication policy*, the current mobile agents don't have complete information or has fuzzy information about the communication policy of target mobile agents. One interesting fact of this communication policy is that

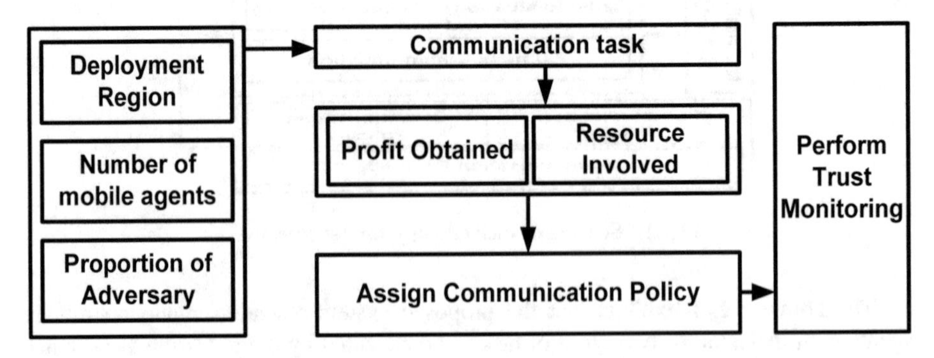

Fig. 2. Initiating communication scheme

tactic adopted by the mobile agent gives indication about the trust factor on the basis of historical information. Therefore, the initial communication is initiated (Fig. 2) followed by consistent trust computation.

The proposed system model primarily takes all the above mentioned information to perform identification of adversary node. The complete system mainly considers third communication policy which lets the source mobile agent performs evaluation of the target mobile agents in order. This evaluation is carried out by computing probability of suspicious mobile agents where the dependent parameters are (i) probability of adversary node and (ii) residual ambiguous trust. The study also uses a threshold*Thr* to explore if the target node bears any malicious intention. The system compares probability of adversary node with first conditional parameter. The computation of the conditional parameter is carried out by using probability where first conditional parameter represents *best chances of data forwarding by target mobile agent* divided by *total chances of both data forwarding and launching attack.*

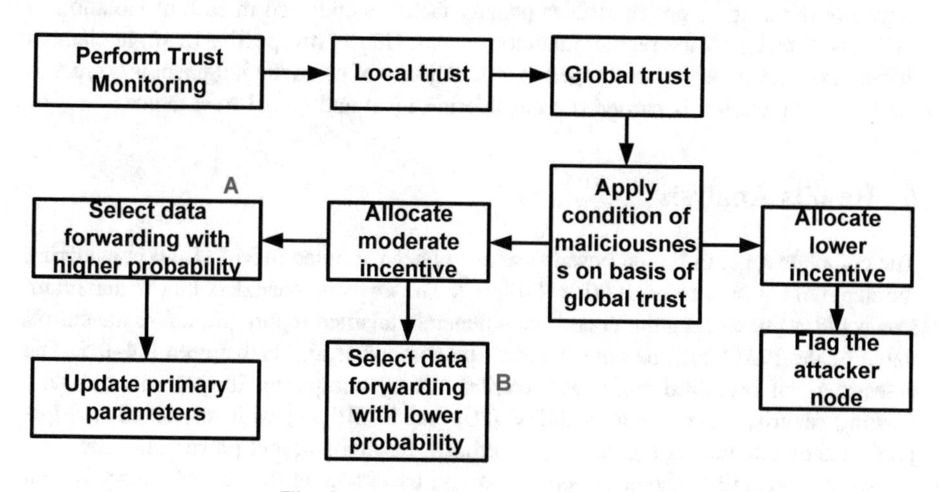

Fig. 3. Identification of malicious node

The formation of the first condition is about framing favorable chances of data forwarding by the mobile agent in present of all the probabilities connected with both data forwarding and launching attack. Therefore, a marking A in Fig. 3 will mean that target node is malicious and has higher probability of launching an attack. Similarly, the marking B in Fig. 3 represents that probability that the target mobile agent is an adversary and that is found less than first conditional parameter. In such positive case, it will mean that the target mobile node could be malicious but it doesn't have any harmful action to be launched at that time. Hence, there are good chances that target mobile node will perform data packet forwarding with certain probability. However, if the condition found to be probability of attack greater than it will mean that there are some good chances to confirm that the target node has malicious intention which can be calculated by different probability. The value of first probability (mark A in Fig. 3) is 1 while the second probability (mark B in Fig. 3) is estimated as favorable chances of attack divided by total attack chances. Finally, the system retains the latest value of occurances and probabilities while transmits to all neighboring mobile agents about the conclusion it draws from evaluation of target mobile agent in terms of flag message to claim the target node is malicious.

Therefore, the proposed algorithm allows the source mobile agents/nodes in MANET to make use of probability on the basis of few operational parameters to estimate the malicious intention of the target mobile agent. The strength of this algorithm is that it offers multiple assessments for the trust factor associated with the target mobile agents where trust score is extracted directly as well as from the neighboring nodes of the target mobile agents. Another interesting part of this algorithm is that even if the target mobile agent is not 100% confirmed to be malicious, but if its current intention is just to forward data packet than it is permitted to do so. Apart from this, there is also possibility about the wrong judgment by the regular agent about the legitimacy of the target mobile agent in communication region. For this purpose, the proposed system should first check the global trust value followed by assessing the degree of threat to be anticipated. A penalty factor is allocated in case of violation of the fairness factor by the regular malicious agent. Hence, irrespective of slight chances that some target nodes by misjudged, but in majority of cases the judgment is correct as the trust computation is carried out considering local and global trust factor.

6 Results Analysis

The complete logic of the proposed system has been scripted in MATLAB considering the deployment area to be $1000 \times 1200\ \text{m}^2$. The analysis considers that is maximum loss could occur to a regular node if they generate falsified report. Therefore the cut-off value of the trust could be considered to be somewhat ideally between 0.4–0.5. The assessment of proposed study was carried out by comparing its performance with existing security scheme e.g. SAODV [20] and SLSP [21] with respect to multiple performance parameters e.g. latency, overhead, throughput, and processing time.

Existing security schemes offers protection by assuming that the adversary is well known and all its actions are very much well defined. This is highly unpractical scenario and MANET when integrated with IoT can invoke adversary with exponentially high dynamicity. A higher controllable environment is developed which ensures that malicious mobile agent doesn't invoke attack and permits them to forward a data packet. This is also a good prevention approach as malicious node will soon drain out of energy which was primarily destined for launching attack. This causes non-significant effect on the throughput resulting in a good balance between the security enhancement and data communication performance. Apart from this the updating of the trust factor of the regular or malicious node is carried out only for the communicating nodes and neighboring nodes. This causes highly reduced overhead effect on the communication in MANET system. This are the practical reasons which causes reduced latency (Fig. 4(a)), reduced overhead (Fig. 4(b)), increased throughput (Fig. 4 (c)), and faster processing time (Fig. 4(d)). Therefore, the proposed system can be claimed of offering simplified solution towards upgrading security features for evaluating trust for mobile agents in MANET.

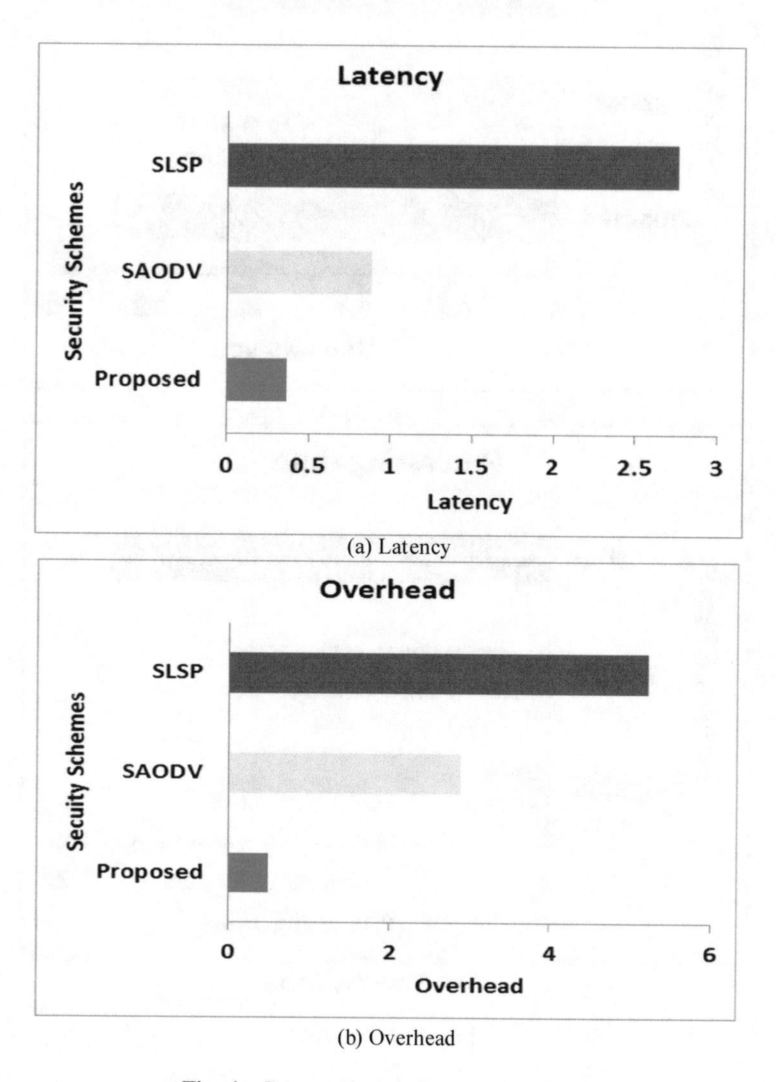

(a) Latency

(b) Overhead

Fig. 4. Comparative performance analysis

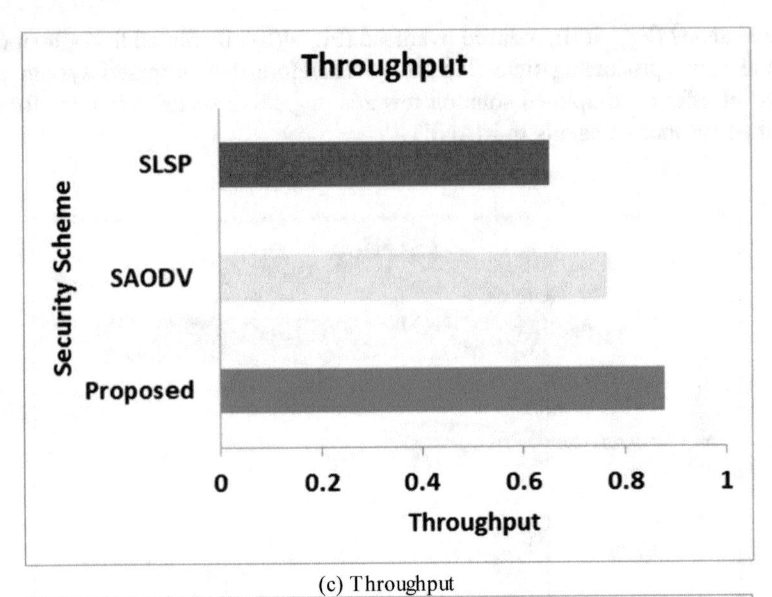

(c) Throughput

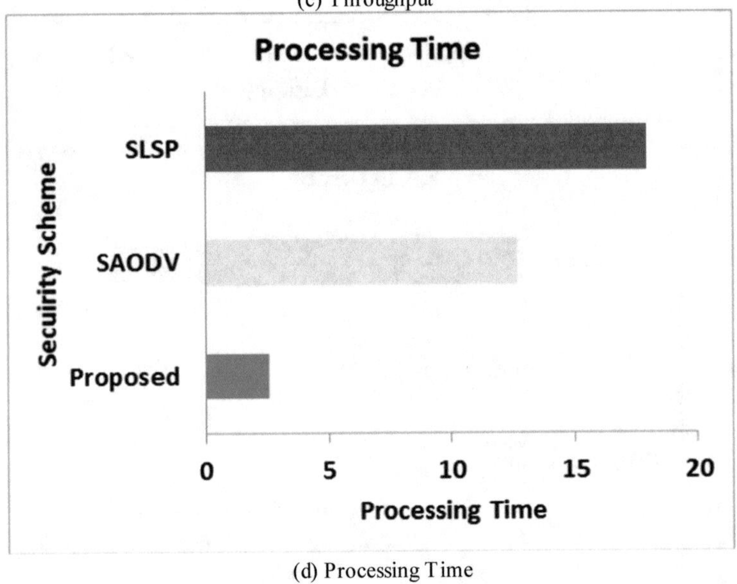

(d) Processing Time

Fig. 4. (*continued*)

7 Conclusion

Existing approaches towards securing communication system in MANET itself is a bigger challenge owing to its decentralization system. Therefore, it becomes a computationally challenging task to offer a proper identification and authentication of mobile agents present in MANET. Therefore, this paper presents a discussion of a sophisticated and simplified model harnessing probability concept in order to offer resistivity against malicious mobile agents. The contribution of the proposed system are (i) a sophisticated adversarial model is developed in order to exhibit that proposed system could identify and resist any intruder and not limited to specific one, (ii) the proposed system has no dependency towards any form of predefined information about the attacker, and (iii) without using conventional encryption mechanism, the proposed identification mechanism is modeled using trust factor only. Hence, the simulation outcome of study is found to offer a good balance between security demands and communication demands in MANET.

References

1. Cooper, C.: Apples cook: 172 million post-PC devices in the last year. Technical report, CNET, New York, NY, USA, March 2012
2. Sarkar, S., Datta, R.: Mobility-aware route selection technique for mobile ad hoc networks. IET Wirel. Sens. Syst. **7**(3), 55–64 (2017)
3. Lauf, A.P., Peters, R.A., Robinson, W.H.: A distributed intrusion detection system for resource-constrained devices in ad-hoc networks. Ad Hoc Netw. **8**(3), 253–266 (2010)
4. Marwaha, S., Tham, C.K., Srinivasan, D.: Mobile agents based routing protocol for mobile ad hoc networks. In: Global Telecommunications Conference, GLOBECOM 2002, vol. 1, pp. 163–167. IEEE, Taipei (2002)
5. Channappagoudar, M.B., Venkataram, P.: Mobile agent based node monitoring protocol for MANETs. In: 2013 National Conference on Communications (NCC), New Delhi, India, pp. 1–5 (2013)
6. Rohankar, R.: Agent based predictive data collection in opportunistic wireless sensor network. Procedia Comput. Sci. **57**, 33–40 (2015)
7. Halim, I.T.A., Hossam, M.A., Fahmy, A.M.B., El-Shafey, M.H.: Agent-based trusted on-demand routing protocol for mobile ad hoc networks. In: 2010 Fourth International Conference on Network and System Security, pp. 255–262. IEEE (2010)
8. Abosamra, A., Hashem, M., Darwish, G.: Securing DSR with mobile agents in wireless ad hoc networks. Egyptian Inf. J. **12**(1), 29–36 (2011)
9. García-Magariño, I., Lacuesta, R., Lloret, J.: Agent-based simulation of smart beds with Internet-of-Things for exploring big data analytics. IEEE Access **6**, 366–379 (2018)
10. Fortino, G., Russo, W., Savaglio, C., Shen, W., Zhou, M.: Agent-oriented cooperative smart objects: from IoT system design to implementation. IEEE Trans. Syst. Man Cybern.: Syst. **99**, 1–18 (2017)
11. Wang, N., Li, J.: Shortest path routing with risk control for compromised wireless sensor networks. IEEE Access **7**, 19303–19311 (2019)
12. Gargees, R.S., Scott, G.J.: Dynamically scalable distributed virtual framework based on agents and pub/sub pattern for IoT media data. IEEE IoT J. **6**(1), 599–613 (2019)

13. Quan, Y., Chen, W., Zhihai, W., Peng, L.: Distributed fault detection for second-order delayed multi-agent systems with adversaries. IEEE Access **5**, 16478–16483 (2017)
14. Wang, D., Wang, W.: Distributed fault detection and isolation for discrete time multi-agent systems. In: Computational Intelligence, Networked Systems and Their Applications, pp. 496–505. Springer, Berlin (2014)
15. Harrabi, S., Jaafar, I.B., Ghedira, K.: Message dissemination in vehicular networks on the basis of agent technology. Wirel. Pers. Commun. **96**(4), 6129–6146 (2017)
16. Fortino, G., Gravina, R., Russo, W., Savaglio, C.: Modeling and simulating Internet-of-Things systems: a hybrid agent-oriented approach. Comput. Sci. Eng. **19**(5), 68–76 (2017)
17. Santos, J., Rodrigues, J.J., Casal, J., Saleem, K., Denisov, V.: Intelligent personal assistants based on internet of things approaches. IEEE Syst. J. **12**(2), 1793–1802 (2018)
18. Shehada, D., Yeun, C.Y., Zemerly, M.J., Al-Qutayri, M., Al-Hammadi, Y., Hu, J.: A new adaptive trust and reputation model for mobile agent systems. J. Netw. Comput. Appl. **124**, 33–43 (2018)
19. Hsieh, H.-C., Chang, K.-D., Wang, L.-F., Chen, J.-L., Chao, H.-C.: ScriptIoT: a script framework for and internet-of-things applications. IEEE IoT J. **3**(4), 628–636 (2015)
20. Lu, S., Li, L., Lam, K., Jia, L.: SAODV: a MANET routing protocol that can withstand black hole attack. In: 2009 International Conference on Computational Intelligence and Security, Beijing, pp. 421–425 (2009)
21. Papadimitratos, P., Haas, Z.J.: Secure link state routing for mobile ad hoc networks. In: Proceedings of 2003 Symposium on Applications and the Internet Workshops, Orlando, FL, USA, pp. 379–383 (2003)

Natural Language Search and Associative-Ontology Matching Algorithms Based on Graph Representation of Texts

Sergey Kuleshov, Alexandra Zaytseva(✉), and Alexey Aksenov

Saint-Petersburg Institute for Informatics and Automation of RAS, Saint-Petersburg, Russia
cher@iias.spb.su

Abstract. The ability to freely publish any information content is causing rapid growth of unstructured, duplicated and unreliable information volumes with irregular dynamics. This significantly complicates timely access to actual reliable information especially in the tasks of the specific scientific topics monitoring or when it is necessary to get quick insight of adjacent scientific fields of interest. The paper contains the description of the technology of text representation as a semantic graph. The algorithmic implementation of proposed technology in the tasks of fuzzy and exploratory information search is developed. The problems of current search technologies are considered. The proposed ontology-associative graph matching approach to post-full-text search system development is capable of solving the problem of document search under conditions of insufficient initial data for correct query formation.

The proposed graph representation of texts allows restricting usable ontology, which in turn gives the benefit of thematic localization of the search region in the field of knowledge.

Keywords: Ontology matching · Associative ontology approach · Natural language processing · Search algorithms · Graph representation

1 Introduction

Enormous data volumes accumulated and being progressively generated by digital society lead to the necessity of intensive development in the field of search technology. The ability to freely publish any information content is causing rapid growth of unstructured, duplicated and unreliable information volumes with irregular dynamics. This significantly complicates timely access to actual reliable information especially in the tasks of the specific scientific topics monitoring or when it is necessary to get quick insight of adjacent scientific fields of interest.

The current search technologies are mainly based on full-text search algorithms. The obvious feature of such approach – to find exact match of search query, becomes a disadvantage when user is not familiar enough with topic terminology to formulate the

R. Silhavy et al. (Eds.): CoMeSySo 2019, AISC 1046, pp. 285–294, 2019.
https://doi.org/10.1007/978-3-030-30329-7_26

correct search queries and thus the results of such search algorithm would be of doubtful relevance [1–3].

In the proposed approach a prepared corpus of texts and search index in given subject area are presented using an associative-ontological approach in the form of oriented loaded graphs representing relations of basic concepts [4–8], including the relationships between key parameters expressed numerically through the use of statistical analysis. This form of presentation is representative for the expert, in contrast to the forms of representation of knowledge obtained by machine learning methods that are "hidden" inside neural networks and are not available for their direct study and modification [9, 10]. The employment of the proposed form of semantic representation allows correction of the automatically generated ontologies using the tool of visualization. Many researches all over the world emphasize the importance of data visualization in different areas of scientific research [11–15].

Recent years the interest in semantic data processing technologies in specific subject areas using ontological models has not weakened, new systems are being created, actively using ontology matching and ontology mapping technologies. For example in the work [16] queries are processed and clustering by extracting content from text description using of NLP-technology. In [17] the mechanism of multi-criteria spatial semantic queries based on elements of ontology and dynamic construction of GeoSPARQL queries are developed. In [18, 19], technological innovations are considered as the way of optimization of the use of vital resources in social biological and economic systems. In order to manage technological innovations, an ontological structure is created and tested, which reflects the current body of knowledge in the subject area and allows identifying links and patterns.

To overcome the described problems, the various search technologies based on semantic search principles are being developed which could be referred as post-full-text search forming a new paradigm of exploratory information search.

This paradigm could be also characterized as "indirect information search" which allows using broadly defined topic as a search query. For example, topic could be defined as a sample document or a collection of documents. But as a new technology it encounters number of problems to be addressed. For the time being there is no unified standards for functionality and interface decisions for a such systems. Another disadvantage is a frequently unobvious or unexpected reaction to users query. This happens mostly in two general cases: the found documents are relevant to topic but contain unfamiliar terminology or documents are irrelevant topic but were found due to homonymy. The first case can give a positive result of new knowledge acquirement but the second case can confuse the user and give a negative result.

This paper proposes a variant of post-full-text search system based on the technology of representing texts as a semantic graph and considering the search task as a selecting of the documents with graph of text matching the graph of search query as well as the associative ontology matching method. The graph of text can be also referred as associative ontology – the ontology reflecting the associative dependencies between the elements of text [4–6].

2 The Algorithm for Text Search Based on Associative Ontology

Let's consider the case when associative ontology was obtained from corpus of texts. To bind corpus of texts to associative ontology the following structure will be used:

$$\left(\bigcup_i \mathrm{G}_i, \mathrm{D}, \mathrm{K}\right), \tag{1}$$

where $\mathrm{D} = \{d_1, d_2, d_3, \ldots, d_n\}$ – the set of documents which are belong to corpus of texts, $\mathrm{K} = \{k_1, k_2, k_3, \ldots, k_n\}$ – the values calculated with use of $\xi_{\mathrm{R}2}$ for every document d_i. All the documents in D are then presented as united graph with set of edges $\mathrm{E_G} = \bigcup_i \mathrm{E}_{2i}$ and vertices $\mathrm{V_G} = \bigcup_i \mathrm{V}_{2i}$ consisting of graphs for every document d_i (Fig. 1).

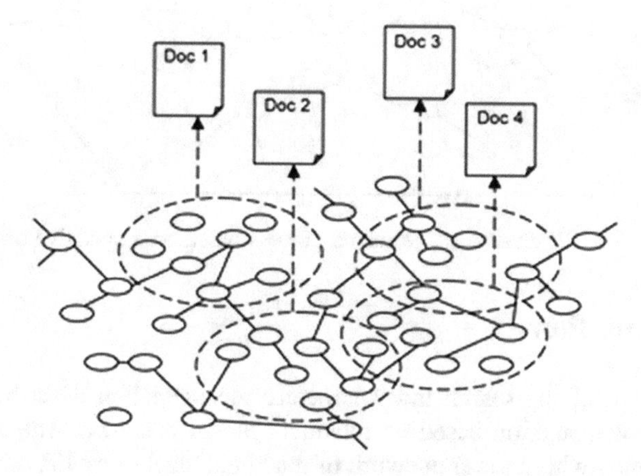

Fig. 1. The illustration of the document relation to semantic ontology region principle

We define the associative search as a process of determination which documents are containing the same semantic relations as search query [4, 7].

Each search query passes through the processes of stop-word removal and lemmatization $w_i \xrightarrow{m} \bar{w}_i$, then the graph with set of edges $\mathrm{E_Q} \subset \mathrm{E_G}$ is being created based on the methods that are represented in previous section. For the words w_i from query Q:

$$\forall \bar{w}_i \in \mathrm{Q} \;\&\; \forall \bar{w}_j \in \mathrm{Q} \Rightarrow e_{\mathrm{Q}}(\bar{w}_i, \bar{w}_j) \in \mathrm{E_Q}. \tag{2}$$

Formally the process of associative search can be defined as the process of getting document subset from corpus of texts: $\forall d, e_{\mathrm{Q}} \in \mathrm{E_Q} \vee e_{\mathrm{Q}} \in \mathrm{E}_d$. Wherein the number of matching elements e_{Q} can be considered as indicator of relevance.

The following cases are possible in search operation $E_R = E_Q \cap E_G$ on query Q:

- $|E_R| = |E_Q|$ – the documents containing all relations from query are exist in corpus;
- $|E_R| < |E_Q|$ – there is no document in corpus of texts containing all relations from query (the created associative ontology incomplete or the search query is incorrect;
- $|E_R| \equiv \varnothing$ – nothing is found on search query (associative ontology contains no relations satisfying search query).

The result of search query processing is a set of documents $\{d_i\}$, $E_{d_i} \cap E_Q \neq \varnothing$, which is being transformed into ordered list (Search Engine Results Page – SERP) by the application of ranking function. The search process is illustrated on Fig. 2.

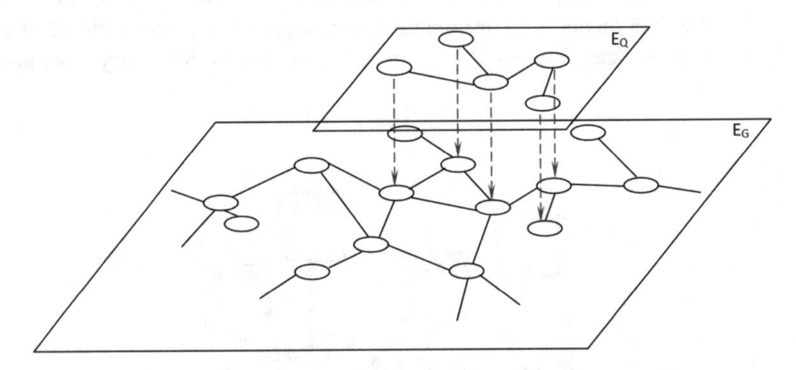

Fig. 2. The illustration of the search process based on associative ontology

3 Implementation

As an illustration of the search index structure representation described in previous section, the implementation based on relational model is shown. The corpus of texts can be represented within the framework of relational algebra by ER-diagram (Fig. 3). The set D corresponds to the document entity, the set corresponds to the link entity [8]. The set corresponds to the word entity.

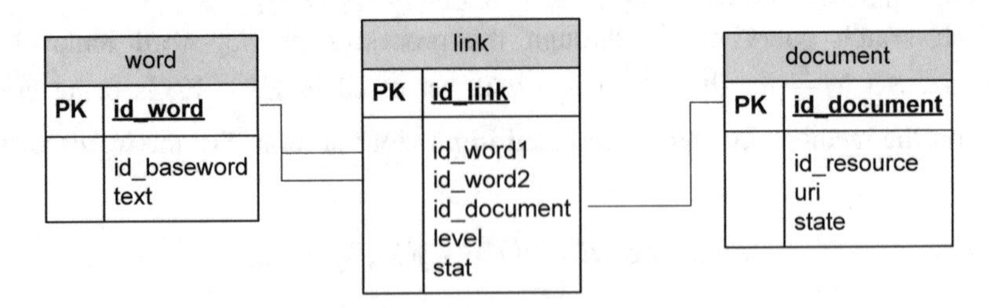

Fig. 3. The illustration of the search process based on associative ontology

The realization of search mechanism using database-management system (DBMS) is implemented as SQL-query. Generalized SQL-query for search phrase consisting of n words is as follows [8]:

```
select id_document from (
select id_document, count(*) as cnt from test where
W_CONDITION group by id_document
) where cnt>N;
where N=(n-1)!,
W_CONDITION="(((id_w1=id_word1)and(id_w2=id_word2))or((id
_w1=id_word2)and(id_w2=id_word1)))or(((id_w1=id_word1)and
(id_w2=id_word3))or((id_w1=id_word3)and(id_w2=id_word1)))
or..."
```

Thus the *n*-word query needs $4(n - 1)!$ comparisons given the W_CONDITION expression.

The speed of such query is limited only by the efficiency of DBMS which is dependent on the speed of hard drives containing DBMS data structures.

To speed up search operations, a number of approaches can be used, including NOSQL, "key-value" base, etc. for storing link table indexes (LINK entity).

The proposed realization gives the ability to execute search operations without the need to store copies of text documents and thus minimizes the demands for data storage.

4 Associative Ontology Matching

Ontology matching is the basis for fuzzy search algorithms based on the methods of entering sub-ontologies into ontology, for example for solving tasks of long-term monitoring in the field of interest.

In this paper we define E_A as the ontology built from analysis of texts for the period of time $t_1 + dt$, and E_B – built from analysis of texts for the period of time $t_2 + dt$, where $t_1 < t_2$. Then we can define, that

$$E_{new} = \frac{E_B}{E_A}$$

is the ontological graph structure, which corresponds to the new trends have appeared during the time period $t_2 - t_1$.

$$E_{back} = \frac{E_B}{E_A}$$

is the ontological graph structure, which corresponds to the concepts and phenomena obsolete during the same time period. Figure 4 illustrates the process of ontology matching.

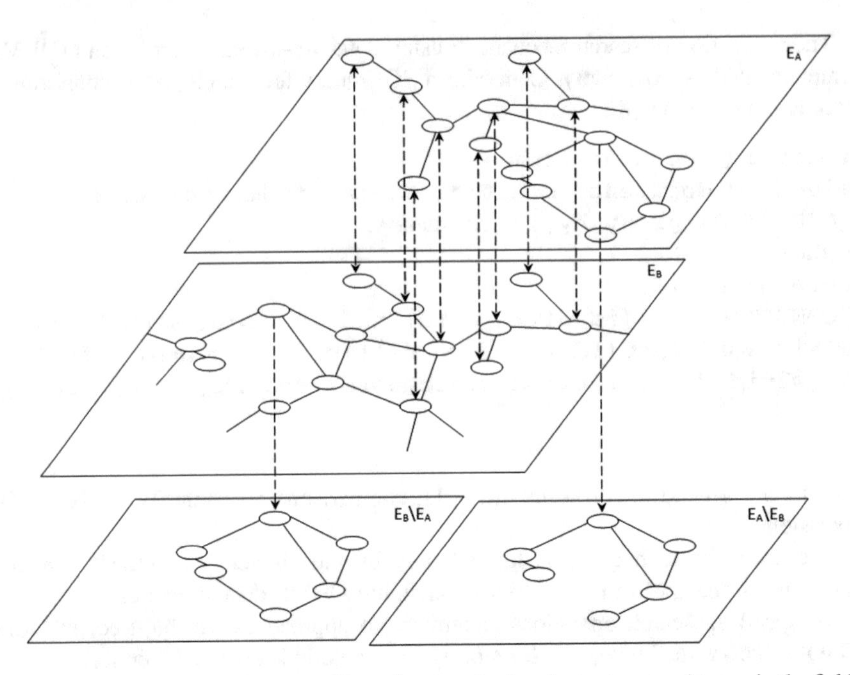

Fig. 4. The process of ontology matching when monitoring the temporary changes in the field of interest

The union of a collection of sets from available sources (anthology) forms the associative ontology E_G

The union of all the associative semantic environments of texts (associative ontologies) in the given field of interest give as the thematic area E_T:

$$\bigcup_i E_i = E_T,$$

where E_i is the associative environment of the text T_i (Fig. 5).

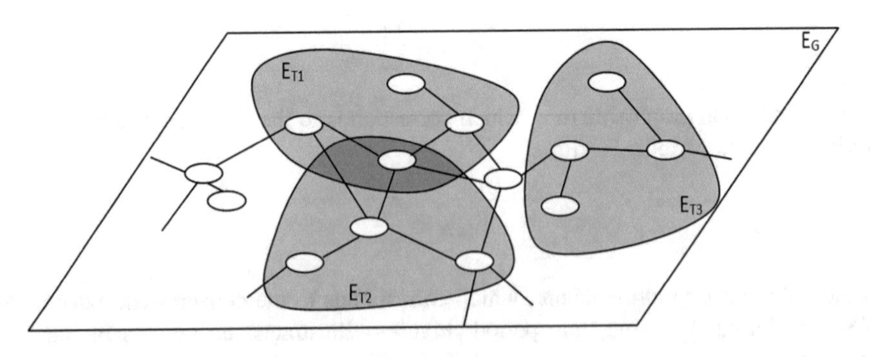

Fig. 5. The thematic areas illustration

We have to note that

$$E_T \subseteq E_G.$$

Figure 6 shows the common structure of method of ontology constructing, comparison and visualization.

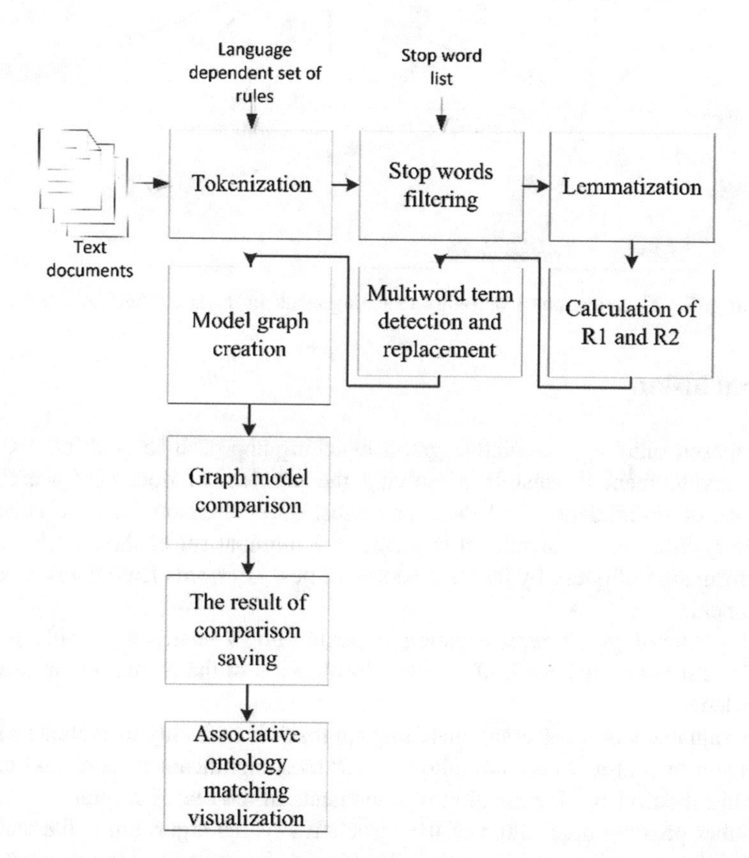

Fig. 6. The common structure of method of ontology constructing, comparison and visualization

The result of the developed algorithm of ontology matching and visualization of the resulting ontology is the structure formed by concatenation of three ontology graphs marked by three colors (red, green, white). The red graph nodes are the out-of-use terms (from E_{back} ontology), the green graph nodes are the new terms, which appeared in the result ontology because of diachrony, white nodes are the terms that present in both ontologies (the stable terms). The example of two ontologies associative matching is shown on Fig. 7.

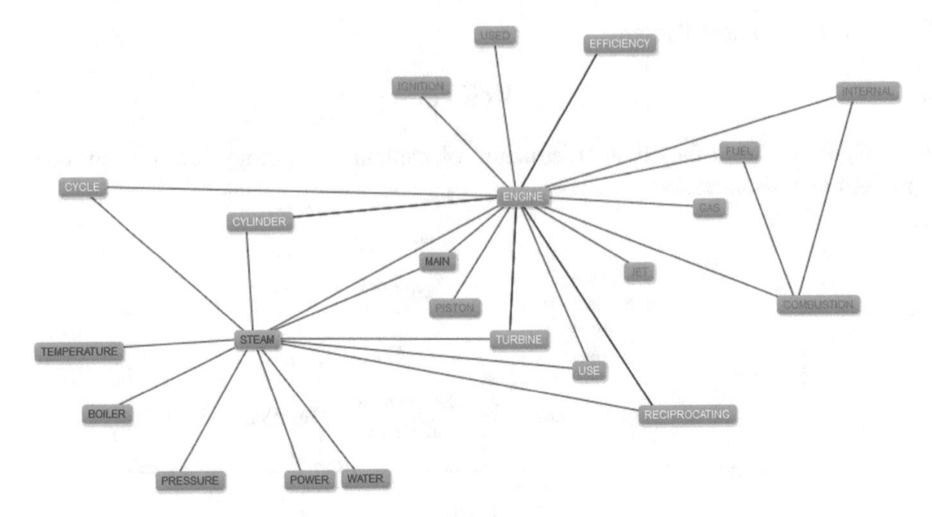

Fig. 7. Visualization of ontology matching result in the some field of interest

5 Conclusion

The proposed ontology-associative graph matching approach to post-full-text search system development is capable of solving the problem of document search under conditions of insufficient initial data for correct query creation. This is paradigm of exploratory information search. In this case, the minimal set of the search keywords allows formation of query by iterative adding of new keywords from query associative environment.

The proposed graph representation of texts allows restricting usable ontology, which in turn gives the benefit of thematic localization of the search region in the field of knowledge.

The valuable feature of graph matching approach is an ability to evaluate similarity of texts and overall text cohesion allowing for most significant parts of text detection and giving the user brief annotation of documents in the search results.

Another possible application of the associative search algorithm is the analysis of ontology changes over a given period. Within the framework of the subject domain under consideration, this makes it possible to create a prediction method based on the results obtained from ontology changes for the next time interval. Developing a technology to visualize the dynamics of changing the significance of concepts over time, studying causal relationships based on the significance of concepts and tools for automating them will further create an empirical model for predicting changes in a particular field of interest.

Acknowledgements. The research is partly supported by the RFBR, project N 16-29-12965\18 and by the budget 0073-2019-0005.

References

1. Kuznecova, Ju.M., Osipov, G.S., Chudova, N.V.: Intellectual analysis of scientific publications and the current state of science. J. Large-Scale Syst. Control **44**, 106–138 (2013). (in Russian)
2. Smirnov, A.V., Pashkin, M., Chilov, N., Levashova, T.: Agent-based support of mass customization for corporate knowledge management. J. Eng. Appl. Artif. Intell. **16**(4), 349–364 (2003)
3. Smirnov, A., Levashova, T., Shilov, N.: Patterns for context-based knowledge fusion in decision support systems. J. Inf. Fusion **21**, 114–129 (2015)
4. Kuleshov, S.V., Zaytseva, A.A., Markov, S.V.: Associative-ontological approach to natural language texts processing. J. Intellect. Technol. Transp. **4**, 40–45 (2015). (In Russian)
5. Zaytseva, A.A., Kuleshov, S.V., Mikhailov, S.N.: The method for the text quality estimation in the task of analytical monitoring of information resources. J. SPIIRAS Proc. **37**(6), 144–155 (2014). https://doi.org/10.15622/sp.37.9. (In Russian)
6. Mikhailov, S.N., Malashenko, O.I., Zaytseva, A.A.: The method for the infology analysis of patients complaints semantic content in order to organize the electronic appointments. J. SPIIRAS Proc. **42**(5), 140–154 (2015). https://doi.org/10.15622/sp.42.7. (In Russian)
7. Kuleshov, S., Zaytseva, A., Aksenov, A.: The tool for the innovation activity ontology creation and visualization. Adv. Intell. Syst. Comput. **763**, 292–301 (2019)
8. Kuleshov, S.V.: The development of automatic semantic analysis system and visual dynamic glossaryies. Ph.D. (Tech) thesises, Saint-Petersburg (2005). (in Russian)
9. Malagrino, L.S., Roman, N.T., Monteiro, A.M.: Forecasting stock market index daily direction: a bayesian network approach. J. Expert Syst. Appl. (2018). https://doi.org/10.1016/j.eswa.2018.03.039
10. Todd, A., Beling, P., Scherer, W., Yang, S.Y.: Agent-based financial markets: a review of the methodology and domain. In: Proceedings of 2016 IEEE Symposium Series on Computational Intelligence (SSCI) (2016). https://doi.org/10.1109/SSCI.2016.7850016
11. Zakharova, A., Vekhter, E., Shklyar, A., Pak, A.: Visual modeling of multidimensional data. J. Dyn. Syst. Mech. Mach. **5**(1), 125–128 (2017). (in Russian)
12. Roshchina, M.K., Il'yashenko, O.Yu.: Data visualization as a management decision-making tool for retailers. In: Materials of SPbPU Science Week Scientific Conference with International Participation, pp. 112–114 (2016). (in Russian)
13. Wang, C., Ma, X., Chen, J.: Ontology-driven data integration and visualization for exploring regional geologic time and paleontological information. J. Comput. Geosci. **115**, 12–19 (2018). https://doi.org/10.1016/j.cageo.2018.03.004
14. Dew, R., Ansari, A.: Bayesian nonparametric customer base analysis with model-based visualizations. J. Mark. Sci. **37**(2), 216–235 (2018). https://doi.org/10.1287/mksc.2017.1050
15. Keim, D., Andrienko, G., Fekete, J., Görg, C., Kohlhammer, J., Melançon, G.: Visual analytics: definition, process, and challenges. In: Lecture Notes in Computer Science (including subseries Lecture Notes in Artificial Intelligence and Lecture Notes in Bioinformatics), vol. 4950, LNCS, pp. 154–175 (2008)
16. Zhang, N., Wang, J., Ma, Y., He, K., Li, Z., Liu, X.F.: Web service discovery based on goal-oriented query expansion. J. Syst. Softw. **142**, 73–91 (2018)

17. Abburu, S.: Ontology driven cross-linked domain data integration and spatial semantic multi criteria query system for geospatial public health. Int. J. Semantic Web Inf. Syst. **14**(3), 1–30 (2018)
18. Cancino, C.A., La Paz, A.I., Ramaprasad, A., Syn, T.: Technological innovation for sustainable growth: an ontological perspective. J. Cleaner Prod. **179**, 31–41 (2018)
19. Kondratyev, A.S., Aksyonov, K.A., Buravova, N.A., Aksyonova, O.P.: Cloud-based microservices to decision support. In: International Conference on Ubiquitous and Future Networks, ICUFN, July 2018, pp. 389–394 (2018). https://doi.org/10.1109/ICUFN.2018.8437015

Measurable Changes of Voice After Voice Disorder Treatment

Milan Jičínský(✉) and Jan Mareš

Faculty of Electrical Engineering and Informatics, University of Pardubice, Studentská 95, 530 02 Pardubice I, Czech Republic
milan.jicinsky@student.upce.cz

Abstract. The main purpose of this paper is to show identification possibilities of voice differences for people whose voice has been influenced by any kind of voice disorder. Introduction of common diseases of vocal cords or larynx is followed by a chapter including ordinary treatment techniques. Even if the surgery ends up well the voice production can be affected in some way. Doctors are mostly able to measure only limited amount of voice characterizing parameters. More precise analysis of subjects within predefined time intervals should lead to more specific results and may prove more efficient. This article presents a different scientific approach which is based on the voice parameterization and analysis. The only thing needed for this kind of research is obtaining of recordings of analyzed subjects (before surgery, soon after that and then for example 2 months later). These recordings can be processed using common audio processing methods and required variables are extracted and saved in form of so-called feature vectors. Some features are expected to change as the result of treatment. Some of used methods are similar to ordinary techniques or they have something in common, but it allows to measure and identify even more variables describing the voice. Diagnostic experience can be supplemented by our software, where many parameters are visualized. But the final decision is still up to the doctor.

Keywords: Vocal cords · Voice disorders · Vocal treatment · Feature vectors · Speech velocity · Log-energy · Zero crossing rate · Spectrum · Cepstrum · Pitch

1 Introduction

Speech processing or voice analysis is a modern branch with many interdisciplinar and intersector applications. It is historically used in security [1], automatic speaker recognition [2] etc. Voice analysis can be important even for medical purposes. More recently, voice captured researchers' attention because of usefulness in order to assess early vocal pathologies, and neurodegenerative and mental disorders [3]. Voice processing as an important part of biomedicine, diagnostics or medical treatment is relatively new field of study. This call aims to summarize and extend the state-of-the-art knowledge of this field bringing together applied mathematics, biomedicine and signal processing. Many research groups deal with a voice as a source of large amount of information about the speaker as sex, age or regional origin [4]. Mehta et al. reports on investigation into the use of a miniature accelerometer on the neck surface below the larynx to acquire and analyze a

R. Silhavy et al. (Eds.): CoMeSySo 2019, AISC 1046, pp. 295–305, 2019.
https://doi.org/10.1007/978-3-030-30329-7_27

large set of ambulatory data from patients with hyperfunctional voice disorders (before and after treatment stages) as compared to matched-control subjects. They have introduced a platform for voice health monitoring that employs a smartphone as the data acquisition platform connected to the accelerometer [5–7]. Speech rhythm abnormalities are commonly present in patients with different neurodegenerative disorders. These alterations are hypothesized to be a consequence of disruption to the basal ganglia circuitry. Therefore, the aim of the study of Rusz et al. was to design a robust signal processing technique that allows the automatic detection of spectrally distinctive nuclei of syllable vocalizations and to determine speech features that represent rhythm instability and rhythm acceleration [8]. When there is anything wrong in airways, larynx or another part of the system responsible for production of speech, the voice character change would by recognizable. Some voice changes can be detected without any physician examination. For example, deepening of the voice, hoarseness or having trouble with pronouncing can be recognized by hearing. If the patient is treated well, his voice production should be improved by the time. If there are any complications during the surgery, the upcoming self-healing effect of human body would be slowed down. More serious impact on the production of voice is expected. And so, the time needed for the whole healing process would be longer. This is the reason why some measurable differences are supposed to be present among the patient recordings.

2 Human Voice

Human voice is one of the earliest ways to express things. Written form of language was invented long time after spoken form. It is necessary to understand the way of voice production in order to analyze it further.

2.1 Voice Production

Creation of speech signal is complicated process. The whole voice producing system can be divided into three subsystems as noted in Table 1. Air pressure system provides air (and its pressure) necessary for vocal cords vibrations. Vibratory system also called voice box converts air pressure to sound waves. Voice box releases air according to the frequency of vocal cords vibrations. It produces so called voiced sound.

Table 1. Voice subsystems [9].

Subsystem	Voice organs
Air pressure system	Diaphragm, chest muscles, ribs, abdominal muscles, lungs
Vibratory system	Voice box (larynx), vocal cords
Resonating system	Vocal tract – throat (pharynx), oral cavity, nasal passages

Finally, resonating system changes voiced sound to the voice form that we all know. This happens due to resonations of speech signal in vocal tract. This is how the voice is produced with its features which differs a little bit for each subject. This fact makes voices of different people distinguishable. The last part is called articulation. Articulation isn't directly related to voice production itself. This part is responsible for creating individual words.

Vocal cords are the key part of the whole voice production system. Column of air moving from lung repeatedly opens and closures a gap between vocal cords. This gap is called glottis. Glottis is open while breathing and closed while swallowing which is shown at Fig. 1. During the speech glottis repeatedly changes phases of opening and closing. The frequency of repetition is related to the speaker and it differs for male, female and children. In the field of audio processing this parameter is called fundamental frequency or just pitch. More detailed information about the whole process of voice production is shown in [9] and [10]. If the speech production process is somehow negatively influenced by any means affecting any part of the voice production subsystem experts have a special term for this state of the system. It is called voice disorders [11].

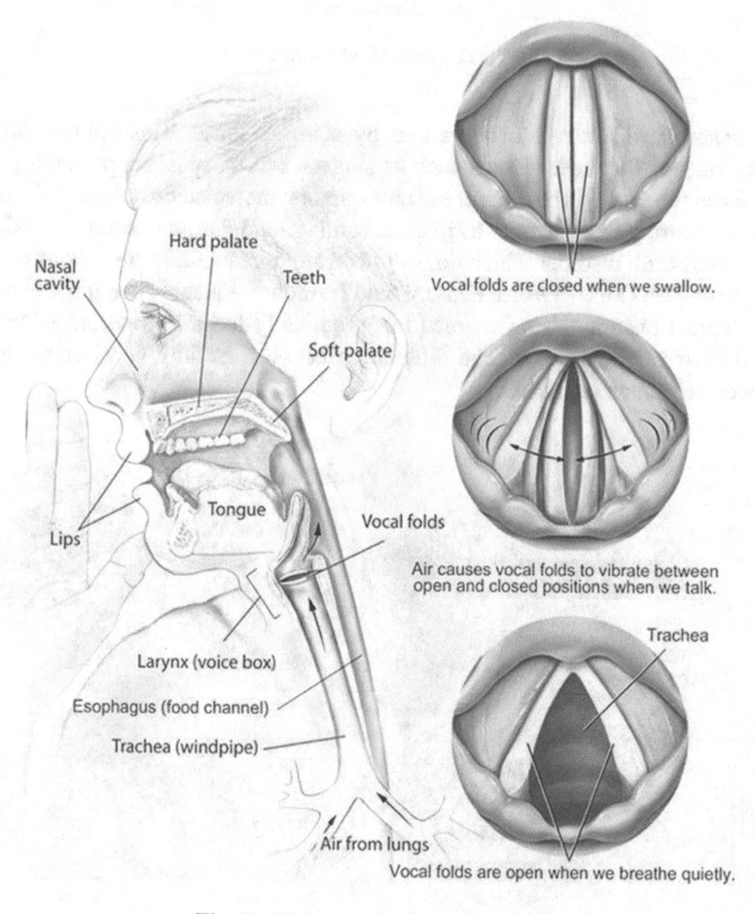

Fig. 1. Voice production system [12].

2.2 Voice Disorders

Any negative changes of human voice are called voice disorders. There are many reasons why the voice disorder of any kind is developed. It can last a few days, or it could be longtime serious problem. It depends on kind of voice disorder. Disorders can be divided into some groups according to the cause of disorder as shown at Fig. 2.

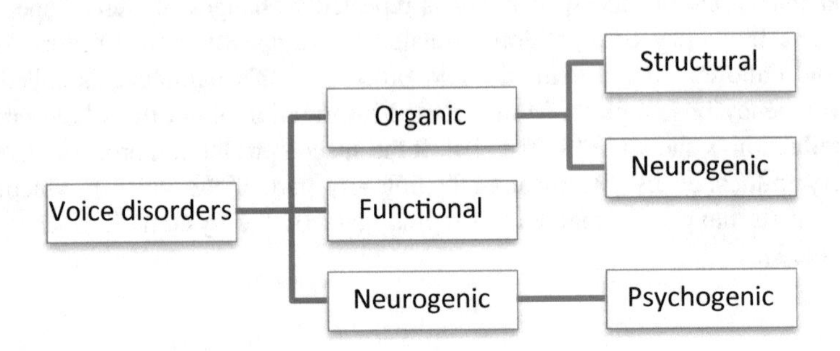

Fig. 2. Division of voice disorders.

For example, disorder can be caused by a voice abuse. This applies either for common folks or for professionals such as singers, actors, speakers presenting a long speech. Even the case of voice tiredness ranks among the voice disorders. This happens to the fans cheering their favorite team or screaming loud during concerts. Luckily, this is only a short-term disorder which doesn't need any treatment. More serious problems could appear if this kind of voice abuse would be more frequent. Then it could result in a physiological change. Using a voice inappropriately leads to swelling of the most impacted parts of vocal cords. The differences between healthy vocal cords and diseased ones are shown at Fig. 3.

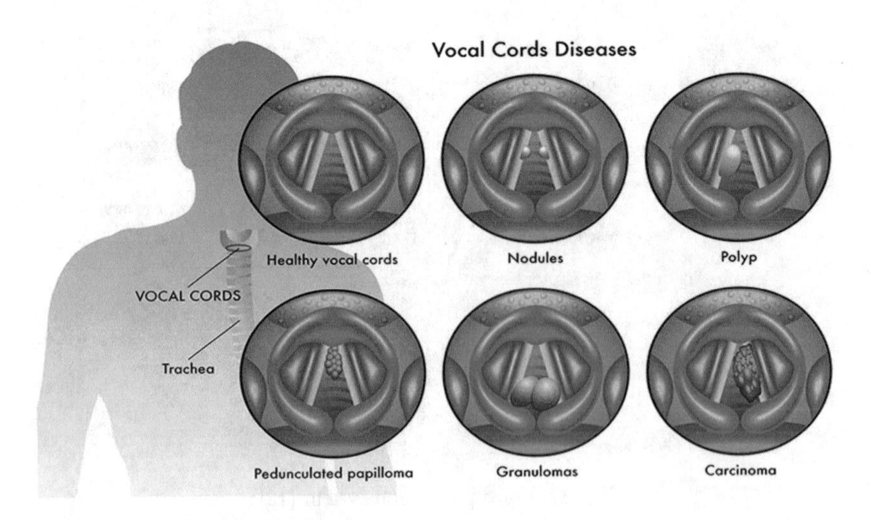

Fig. 3. Diseases of vocal cords [13].

Very common voice disorder which may require a surgery affects vocal cords. Vocal cord nodules and polyps are growths formed on swollen spots of vocal cords. It can be caused by smoking, allergies, loud talking or consuming beverages containing components responsible for drying out the throat and vocal cords. They are in the way of air column passing through the vocal cords and making them vibrate. This changes a character of produced sound. More detailed information about polyps and nodules is available at [13] and [14]. The vibratory system can be also a target of cancer.

3 Diagnosis and Treatment

Nowadays there are ways of examining voice, looking for causes of voice disorders and cure them. Treatment techniques differs depending on what kind and how serious disorder is. The very first step is always finding a way how the patient uses his voice. It is important to know what kind of job he does and what are his hobbies if it's related to vocal usage. This step is usually followed by filling in some specialized questionnaires. These are the tools developed for collecting necessary information about patient's vocal usage. Assessment papers are made based on these two types of questionnaires called Voice Handicap Index (VHI) and Voice-Related Quality of Life (V-RQOL).

Phonation evaluation can be made based on index presented by a Union of European Phoniatricians. This index has a scale from 0 to 6 where the lowest value means normal voice without any problems and the highest value is reserved for absolute loss of voice. Alternatively, there is GRBAS index which is abbreviation for grade, roughness, breathiness, aesthenicity, strain. Each part is evaluated separately. All these information helps a physician to evaluate, how serious can be a patient's voice disorder. The whole process continues by the diagnostics of voice disorder. The mostly used method is called videostroboscopy.

3.1 Videostroboscopic Analysis

As mentioned before, most physicians prefer the videostroboscopy instead of other diagnostic methods. Nowadays it is the-state-of-the-art technique which provides information on the state of vocal box. The procedure requires a flexible fiberoptic telescope which is passed down to the throat. There is a fiberoptic strobe light and camera at the tip of the telescope. The camera magnifies vocal cords and larynx. The patient is asked to say a few phrases or sentences. The physician watches the screen where he can see the image transferring from the throat. This technique monitors vocal folds, larynx and a function of nearby muscles while the patient is speaking. Any changes in the vibratory system, whether they are visual or functional, can be detected this way. The physician looks for any anomalies such as polyps or granulomas and he is watching if the muscles which contributes to the voice production works fine. The whole procedure is digitally captured.

As for the principle of videostroboscopy, human eye wouldn't be able to recognize such fast action as vibrations of vocal cords or opening and closure of glottis. This is the reason why videostroboscopy uses a special technique to distinguish these fast actions. If some periodic movement has the same frequency as the flashing light shining on it then human eye would see stable image. If there is a difference between those frequencies the movement would be visible as slowed down. This allows physicians to see whatever phase they want. Nowadays, there is an automatic synchronization of frequencies implemented thanks to the microphone or electroglottography (EGG).

Next method which also ranks among videostroboscopic analysis is called videokymography (VKG). The only difference between videostroboscopy and VKG is VKG high-speed scanning technique providing almost 8000 images per second. This might prove helpful if any non-periodic or fast action must be captured. The videostroboscopy can be expanded by electromyography (EMG), pneumography and spirometry.

3.2 Phonetogram

Commonly used tool to show the range of analyzed voice is phonetogram. Thanks to the phonetogram the area of voice using can be determined and graphically presented. The patient is asked to read standardize text while his voice is being recorded. Thanks to the calibration it is possible to measure sound pressure level (SPL). As for the explanation of SPL, practically it means that louder sounds are related to the high values of SPL. Using a suitable algorithm mentioned below, fundamental frequency of speaker is also extracted from the speech. The phonetogram is graphical tool that allows to match frequency range with sound pressure level measured in decibels. Color range represents sample density. A tone scale is usually added to horizontal frequency axis. Phoniatricians also agreed on a fact that voice disorders reduce frequency and sound pressure level range. This is the main reason to use phonetogram. The phonetogram can be used for comparison of a long-term changes and for monitoring patient's voice recovery after the surgery. Vocal usage areas may vary a lot.

Disadvantage of using phonetograms is that the microphone reserved for recording of patient's voice needs to be calibrated every single time. The main reason of calibration is that information about SPL needs to be obtained in order to plot phonetogram. Typical parameter values for calibration are SLP 94 decibels and frequency 1 kHz (Fig. 4).

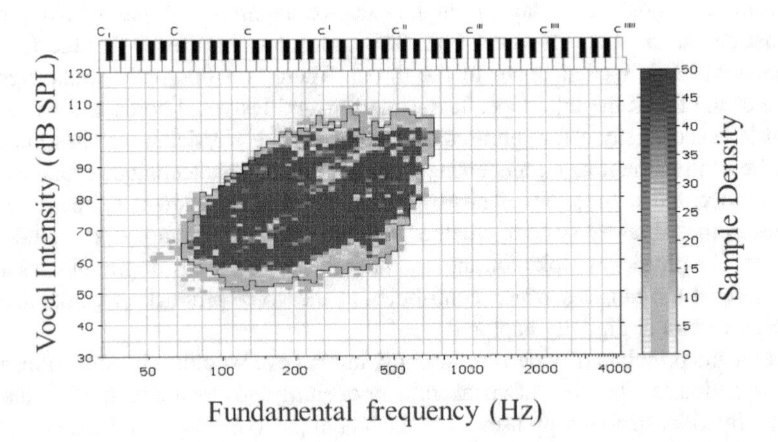

Fig. 4. An example of phonetogram [15].

3.3 Voice Range Profile

Determination of voice range profile or sometimes vocal range profile (VRP) is considered to be the most spread and commonly used technique in the field of otorhinolaryngology. It uses phonetograms of patients to visualize an area of their vocal usage. Physicians can observe several details. Probably the most important part of phonetogram is its shape and contour. For example [16] mentions that the shape of phonetogram can be described as the sum of two overlapping ellipses. Together with density it shows which area corresponds to ordinary speaking. Of course, experienced physician can see the difference between ordinary people and those whose voice has been trained. There are also visible differences between speech and singing. Both elements use different frequency and SPL areas. There is also visible local minimum (considering SPL for high frequencies) corresponding to the voice change from chest to falsetto. This dip is typically located about 390–440 Hz. As for male, this point is usually close to the lower limit and women are supposed to have this minimum about 440 Hz. On the other hand, singers and vocally trained people have this dip only insignificant. Another important parameter derived from contour is enclosed area. It is defined by upper SPL function and lower SPL function. An enclosed area is computed in few steps. The integral of loud phonation is computed followed by a soft phonation integral. Finally, the enclosed area is equal to the subtraction of loud and soft phonation areas. As mentioned before there is an area of phonetogram corresponding to “normal” speech. So, the normal speaking range can be determined. Parameter which takes it into account is called mean speaking fundamental frequency.

3.4 Treatment

As for the treatment techniques, practically there are three ways of curing voice disorder - medical or surgical treatment and voice therapy. Medical treatment supposes using a medicine. Surgical treatment means that any growths affecting the voice will be removed during the surgery. Voice therapy helps patients to understand how to use their voice in right way. Proper vocal training even extends the range of voice. More detailed information about possibilities of voice disorder treatment can be found in [17].

4 Data Obtaining Process

Thanks to the cooperation with university hospital Královské Vinohrady (FNKV) in Prague, experts from the field of otorhinolaryngology are involved. All participants agreed on division of patients into 3 categories with a respect to their diagnosis (surgery of thyroid, half of thyroid and the rest). Rare variants of voice disorders cannot be processed because of lack of data. As for the timeline, patients will be recorded 3–4 times. Stage 0 (before treatment) – patient suffers from voice disorder. Stage 1 (immediately after treatment) – patient undergone a surgery. He/she can be recorded approximately 2 days after. Stage 2 (improvement) – recording during usual check-up approximately one week later. Stage 3 (healthy voice) – at least one month after surgery. The most significant impact on human voice can be recognized between stage

0 and stage 1. Significant improvement appears at stage 2. The last recording stage is voluntary because patients don't suffer from voice disorder any more.

The recording process takes place in specialized silent chambers located in FNKV. These rooms are designed for audiometry procedures. The whole procedure can be interpreted in few steps: microphone calibration, utterance recording and evaluation of examined data. Patients are recorded while they read standardized text. Text is divided into three paragraphs to make reading more comfortable and to prevent mistakes. Usage of a single microphone, having the same acoustic conditions and reading standardized text is beneficial for minimizing negative variability as much as possible. So that only a variability of voice can be analyzed. It means any vocal changes influenced by treatment can be distinguished. As for the necessary facility, omnidirectional DPA microphone with a flat frequency response is a key part. Because of low sound quality of embedded sound cards, external sound card is used for transferring sound to the computer. Figure 5 shows connection scheme. Audacity application records and saves utterances as audio files with wav extension. Finally, audio files are processed by feature extraction algorithm.

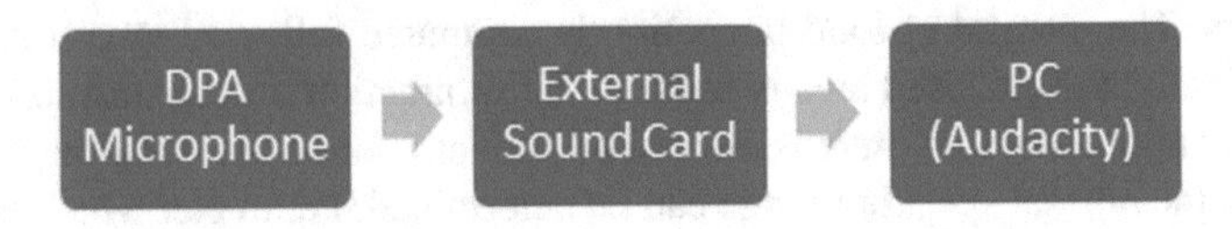

Fig. 5. Simplified connection scheme.

5 Audio Processing Algorithm

The key point of research is to obtain parameters present in speech. For that purpose, an audio processing algorithm was created in Matlab. Although the most of useful information about ordinary speech is contained in the zone from 0 to 4 kHz, all the speech recordings are expected to have sampling rate 16 kHz. Before proceeding to calculation of parameters, the speech signal has to be equally divided into smaller parts called frames. Segmentation of speech signal is absolutely fundamental and nowadays it's practically part of each system processing sound in any way. In this case the frames are 20 ms long (320 samples) and they overlapps for 10 ms.

Parameterization can be made after signal division for each frame. In the field of audio processing the parameters are called features and they are stored in so called feature vectors. Because we can only guess which features would change, a large number of features is used for the experiment. Features can be divided into basic, spectral, cepstral, prosodic and the rest. This paper presents feature vectors containing basic (log-energy, zero crossing rate), spectral (spectral energy coefficients), cepstral (Mel-frequency cepstral coefficients), prosodic (fundamental frequency) and speech velocity defined in [18]. Some of these static coefficients are complemented by their dynamic variant. This applies to log-energy, spectral energy and MFCC. Dynamic coefficients are computed from the static ones. It represents a difference between nearby frames. Hence, they are called delta and delta-delta coefficients (according to the difference order).

All features except MFCC and fundamental frequency are extracted the same way as shown in [18] and [19]. As for the frequency, sometimes also called pitch, the application works on basis of a cross-correlation version of pitch detection algorithm. MFCC extraction is using Fast Fourier Transform, triangular filters and discrete cosine transform to obtain a cepstrum. There is a Matlab2018a function called mfcc which has been used for its complex setting including replacement of the first cepstral coefficient, setting an amount of output coefficients and its optimization for such applications (Fig. 6).

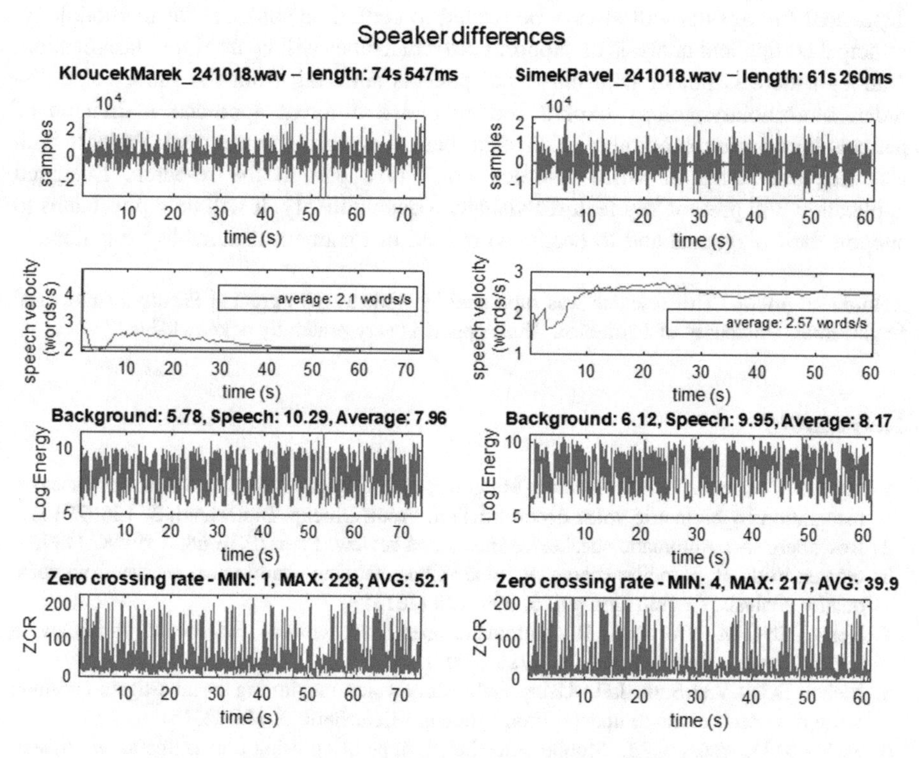

Fig. 6. Interface of the voice processing software.

As for the developed application, it is designed for comparison of two audio tracks. Figure placement corresponds to the division of opened window into two parts. The left side contains information about the first recording while the right side is reserved for the second one. The first line of graphs shows the recordings itself. Values represent sound pressure level SPL (dB). Information including filename and length of the recording is attached above. The second graph shows speech velocity. Third line contains log-energy graph together with average value, background noise level and speech energy level. The last row is reserved for graphs presenting zero crossing rate also with average, minimal and maximal value. Next windows can visualize pitch, spectrogram, MFCC etc. In case of future use results of analysis can be saved and reopened anytime.

6 Further Research

Multidimensional analysis based on feature vectors consisting of 80 features describing speech signal from different point of view and the complexity of used algorithm ensures obtaining significant results. Further research will be aimed on creating a representative dataset and its complete analysis. Only the real data obtained from patients can help us decide which features might be significant and which might not. This kind of approach will be of course text dependent (standardized text) and speaker dependent (the results will always be related to certain person). As for methodology, principal component analysis or support vector machines will be used for classification. Although there is potential of clustering patients suffering from the same voice disorder, otorhinolaryngology experts recommend a different approach – division of patients into 3 groups (surgery of thyroid, half of thyroid and the rest). Patients with less frequent forms of voice disorder won't take part in the research. Designed application will present key features visually a quantitatively. It will help physicians to support their diagnosis and to decide what kind of treatment is suitable for patient.

Acknowledgment. This research was supported by students SGS grant at Faculty of Electrical Engineering, University of Pardubice. This support is very gratefully acknowledged.

References

1. Mazaira-Fernandez, L.M., Álvarez-Marquina, A., Gómez-Vilda, P.: Improving speaker recognition by biometric voice deconstruction. Front. Bioeng. Biotechnol. **3**, 126 (2015)
2. Rosenberg, A.: Automatic speaker verification: a review. Proc. IEEE **64**, 475–487 (1976)
3. Gómez Vilda, P., Rodellar Biarge, V., et al.: Characterizing neurological disease from voice quality analysis. Cognit. Comput. **5**, 399–425 (2013)
4. Benzeghiba, M., De Mori, R.: Automatic speech recognition and speech variability: a review. Speech Commun. **49**, 763–786 (2007)
5. Mehta, D.D., Van Stan, J.H.: Using ambulatory voice monitoring to investigate common voice disorders: research update. Front. Bioeng. Biotechnol. **3**, 155 (2015)
6. Mehta, D.D., Zañartu, M.: Mobile voice health monitoring using a wearable accelerometer sensor and a smartphone platform. IEEE Trans. Biomed. Eng. **59**, 3090–3096 (2012)
7. Mehta, D.D., Zeitels, S.M.: High-speed videoendoscopic analysis of relationships between cepstral-based acoustic measures and voice production mechanisms in patients undergoing phonomicrosurgery. Ann. Otol. Rhinol. Laryngol. **121**, 341–347 (2012)
8. Rusz, J., Hlavnička, J.: Automatic evaluation of speech rhythm instability and acceleration in dysarthrias associated with basal ganglia dysfunction. Front. Bioeng. Biotechnol. **3**, 104 (2015)
9. The voice foundation. https://voicefoundation.org/. Accessed 19 Sept 2018
10. Titze, I.R., Martin, D.W.: Principles of voice production. J. Acoust. Soc. Am. **104**, 1148 (1998)
11. Deliyski, D.D.: Acoustic model and evaluation of pathological voice production. In: Third European Conference on Speech Communication and Technology, Berlin (1993)
12. Voice Changes. http://www.westsidehn.com/what-is-voice/. Accessed 29 Sept 2018
13. Vocal cord polyp. https://uciheadandneck.com/clinical-specialties/vocal-cord-polyp. Accessed 20 Sept 2018

14. Vocal cord nodules and polyps. https://www.asha.org/public/speech/disorders/Vocal-Cord-Nodules-and-Polyps. Accessed 20 Sept 2018
15. Bloothooft, G., Pabon, P.: Vocal registers revisited. In: Sixth European Conference on Speech Communication and Technology, EUROSPEECH, Budapest (1999)
16. Sulter, A.M., Wit, H.P., Schutte, H.K., Miller, D.G.: A structured approach to voice range profile (phonetogram) analysis. J. Speech Hear. Res. **37**, 1076–1085 (1994)
17. Treatment of Voice Disorders. https://www.ent.umn.edu/patients/lions-voice-clinic/voice-problems/treatment-voice-disorders. Accessed 29 Sept 2018
18. Jičínský, M., Marek, J.: New Year's Day speeches of Czech presidents: phonetic analysis and text analysis. In: IFIP International Conference on Computer Information Systems and Industrial Management. LNCS, vol. 10244, pp. 110–121. Springer, Cham (2017)
19. Jičínský, M., Marek, J.: Clustering analysis of phonetic and text feature vectors. In: 2017 IEEE 14th International Scientific Conference on Informatics, pp. 146–151. IEEE, Poprad (2017)

Optimization of the Novelty Detection Model Based on LSTM Autoencoder for ICS Environment

Jan Vavra(✉) and Martin Hromada

Faculty of Applied Informatics, Tomas Bata University in Zlin, Zlin, Czechia
jvavra@fai.utb.cz

Abstract. The recent evolution in cybersecurity shows how vulnerable our technology is. In addition, contemporary society becoming more reliant on "vulnerable technology". This is especially relevant in case of critical information infrastructure, which is vital to retain the functionality of modern society. Furthermore, the cyber-physical systems as Industrial control systems are an essential part of critical information infrastructure; and therefore, need to be protected. This article presents a comprehensive optimization methodology in the field of industrial network anomaly detection. We introduce a recurrent neural network preparation for a one-class classification task. In order to optimize the recurrent neural network, we adopted a genetic algorithm. The main goal is to create a robust predictive model in an unsupervised manner. Therefore, we use hyperparameter optimization according to the validation loss function, which defines how well the machine learning algorithm models the given data. To achieve this goal, we adopted multiple techniques as data preprocessing, feature reduction, genetic algorithm, etc.

Keywords: Neural networks · Industrial control systems · Anomaly detection · Genetic algorithm

1 Introduction

Detection capabilities have become a necessity for a considerable number of individuals and systems. This development has accelerated in recent years due to the rapid evolution of technology. Industry 4.0 and the internet of things (IoT) became real. However, with the complex systems, the data became multi-dimension with millions of records per hour. Hence, nowadays, data scientists face an increasing problem which can be characterized as "big data". Chen et al. [1] discuss the main problems in scientific research and they pointed out a big data opportunities in connection with machine learning. Thus, the big data makes the detection capabilities of every detection system increasingly difficult and computationally demanding task. Furthermore, this development also profoundly influences the anomaly detection task, which is the main objective of the article. Thus, machine learning algorithms need to be adapted to create a reliable and effective anomaly detection system.

The scope of the research paper is dedicated to secure Industrial Control Systems (ICS) against cyber-attacks. ICS is cyber-physical which is commonly used as a part of

R. Silhavy et al. (Eds.): CoMeSySo 2019, AISC 1046, pp. 306–319, 2019.
https://doi.org/10.1007/978-3-030-30329-7_28

the critical information infrastructure. These systems have an important influence on our environment; therefore, every disruption is considered as critical. The acronym ICS is often confused with Supervisory Control and Data Acquisition (SCADA) which is the main subgroup of ICS. However, we decided to use ICS for the purpose of the article. Knapp et al. [2] published the comprehensive publication about network cyber-security of ICS systems, where is also described ICS and SCADA architecture in detail. Moreover, a considerable number of publications considered the emergencing trend in increasing of vulnerabilities of ICS cyber-security. Maglaras et al. [3] concluded emerging of new challenges due to the synergy between the ICS and the IoT.

Anomaly detection is a special task for machine learning. Chandola et al. [4] described trends and applications of anomaly detection systems in a considerable number of fields. A data scientist does not have a balanced dataset in most cases. Even, he mostly does not know all types and sources of anomalies, especially in large and sophisticated systems. Thus, the anomaly detection is essentially connected to the field of novelty detection. It is also classified as a one-class classification problem. Moreover, the main idea is relatively straightforward. The model is created on known data without anomalies via machine learning algorithms. Therefore, each novelty is detected as a deviation from the model and classify as an anomaly. Markou and Singh [5] demonstrated the characteristic of novelty detection and classify main areas of novelty detection.

We decided to use recurrent neural networks in autoencoder architecture to create a predictive model for novelty anomaly detection in the ICS environment. Recurrent neural networks are commonly used for sequence task as text translation, speech recognition, or time series prediction. We decided to use Long Short-term memory (LSTM) networks, which includes internal memory. Thus, we assume their useful application to sequences of ICS network communications for the creation of a cybersecurity detection system. Moreover, the autoencoder architecture provides desirable properties to the anomaly detection model. There are several scientist publications about LSTM autoencoders for anomaly detection [6–9]. However, to the best of our knowledge, there is none about ICS networks cyber-security. In order to address the tuning of hyperparameters of neural networks, we decided to exploit the optimization based on genetic algorithms. The optimization parameter is in this case validation loss function of the recurrent neural network. Moreover, we trying to minimize loss function in this optimization problem.

The rest of the article is organized as follows. In Sect. 2 recurrent neural networks algorithms for anomaly detection are described. Section 3 is dedicated to the description of the genetic algorithm used to solve the optimization problem of presented neural network. The experimental setup is specified in Sect. 4. Section 5 shows the results of the experiments and in Sect. 6 the results of the article are discussed.

2 Anomaly Detection Based on Recurrent Neural Networks

Neural networks are powerful machine learning algorithms for supervised and unsupervised learning. The main idea is based on imitation of the human brain functions. The neural network consists of neurons and synapses, which are the connection between neurons. Each neuron is activated based on activation function, which simulates biological processes in the brain. Moreover, each neural network consists of multiple layers. In each of them is one or multiple neurons based on the problem we want to solve. The neural network architecture may vary; however, there are few rules. The first layer is called an input layer, and the last layer is called an output layer. These layers are known to us. However, there can be multiple hidden layers between the input and output layers. The training of neural networks consists of several steps. Moreover, the whole process is based on matrix multiplication. The first step initializes the input layer of the neural network and propagates forward-propagation through all layers. The following step is dedicated to calculating the error between actual results and predefined results. The error is propagated back via backpropagation, where the optimization task of the loss function via gradient descent is needed to be solved. The last step is the creation of prediction via a trained model. Moreover, the loss function gives information on how well neural network models dataset via the calculation of errors. The errors represent the absolute difference between our dataset and the prediction. Furthermore, validation loss is connected with validation dataset where the loss function is calculated.

A recurrent neural network (RNN) is a particular case of a neural network which is designed for data sequences. They are capable of capturing dynamic time sequences. Moreover, ANN is able to retain sequence information from a predefined time window. Unlike, the traditional ANN where after processing the data point, the information about the sequence is lost. This approach can result in problems for time-related data. Furthermore, RNN can pass information about sequence through multiple layers. Therefore, the internal memory of neuron is commonly used [10].

We decided to exploit the LSTM neural network, which can be classified as a recurrent neural network. The LSTM was introduced by Hochreiter and Schmidhuber in publication LONG SHORT-TERM MEMORY [11]. They highlighted the main advantages of the proposed algorithm. The LSTM is able to bridge very long time lags due to constant error backpropagation within a memory cell and solve vanishing gradient problem. Moreover, LSTM can handle noise, distributed representations, and continuous values and generalize well [11]. The LSTM have become one of the most popular algorithms for a time, depending on the series. Unless of tradition RNN, the LSTM extend memory capabilities which made possible learn distant dependencies over a long period of time. The LSTM unit can be described as memory block than a classic neuron. Furthermore, the architecture of the LSTM unit is shown in Fig. 1.

As can be seen in Fig. 1, the LSTM unit has an input gate, output gate, and forget gate where x(t) is input, h(t) is hidden vector, c(t) is a cell state. The input gate i(t) is dedicated to filtering important inputs from unimportant inputs and controls cell state. Basically, decides which inputs will be used. The output gate o(t) controls the output of the cell; therefore, controls also a hidden vector. Forget gate f(t) control the information which should be used within the cell and define cell state [12].

Autoencoders are often used for the representation learning task. Moreover, autoencoders use so-called "bottleneck", which is represented by one hidden layer in a neural network. The bottleneck is the layer with the least neurons in the network. Furthermore, the bottleneck compresses knowledge and learn the most important attributes of the dataset. The basic architecture of autoencoder is shown in Fig. 2. The

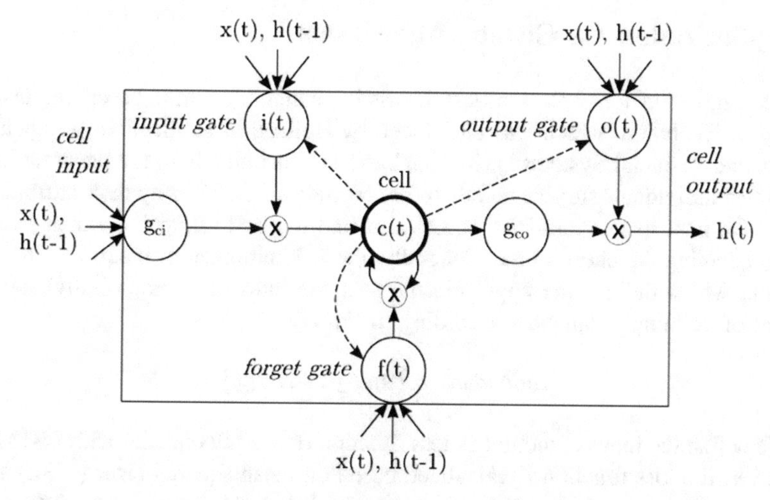

Fig. 1. The basic architecture of the LSTM memory cell [12].

autoencoder can be consist of multiple hidden layers. Moreover, the layers from the input layer to bottleneck are called encoder layers. Encoder transforms input representation into a compressed state. Otherwise, the decoder is represented by layers from bottleneck to output layer. Decoder transforms compressed representation into the original state. In order to train autoencoder, we do not need labels due to the input (number of neurons) should be exactly the same as outputs. Hence, autoencoders are usually used for unsupervised learning tasks. Moreover, the main goal of the autoencoder is to learn data representation. The anomaly detection system based on autoencoder relies on a simple assumption that every deviation from the learned model can be classified as an anomaly.

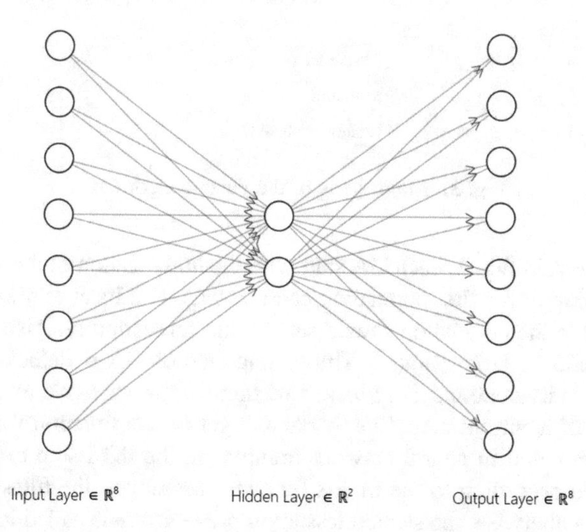

Fig. 2. The basic architecture of the autoencoder.

3 Optimization via Genetic Algorithms

Genetic algorithm (GA) is a robust heuristic search algorithm based on Darwin's Theory of Evolution, which was introduced by Holland in publication "Adaptation in Natural and Artificial Systems" [13]. The basic idea is built about the presumption that only fittest individuals survive and therefore reproduce. That is why their attributes will pass to the next generations. So the GA can find the best fitting individual (solution) which reflecting the chosen task. Generally, the GA minimize or maximize the fitness function, which defines the appropriateness of the individual. Each individual has a number of defining parameters according to Eq. (1).

$$Individual = \{p_1, p_2, \ldots\ldots p_n\} \quad (1)$$

We define the fitness function as loss function of the autoencoder with LSTM cells. Moreover, the loss function is calculated based on mean squared error (MSE) estimator. Every GA is based on the iterative process that is similar to the evolution of the population through the generations. The diagram of one generation is shown in Fig. 3.

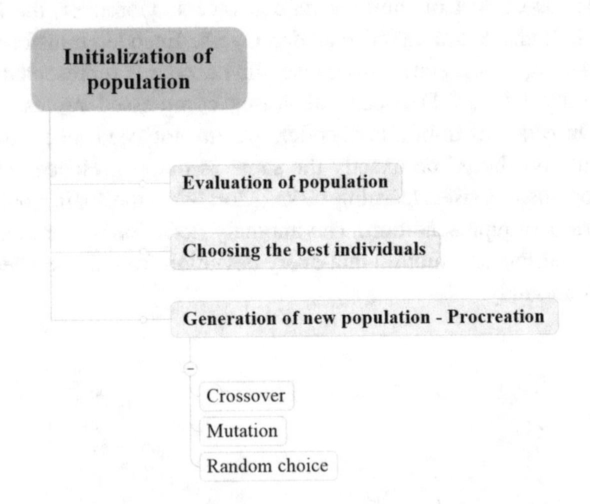

Fig. 3. Major parts of the genetic algorithm.

As can be seen in Fig. 3, each GA start with the initialization of the first generation. The poorly setting of the first population can converge to a local minimum. Therefore, each individual in the population should set parameters randomly. Also, the number of population should be large enough. The second step of GA is dedicated to the evaluation of each individual and get fitness function. In our case, we evaluate a considerable number of recurrent neural networks and get fitness function for each of them, which is loss function in neural network training. In the third step of GA, the population is divided according to the fitness function. Moreover, the fittest group is classified as elites. Otherwise, the second less appropriate group is called losers. However, the first group is used for procreation of offsprings. Each iteration two parent are randomly selected from the predefined group. Moreover, one individual cannot be chosen as parent 1 and 2 at the same time. Then, two offsprings are created according to the random recombination of parents parameters. In order to avoid convergence to local minima or maxima, the mutation is introduced. Mutation randomly selects one of the offsprings parameters and change it. Therefore, a more optimal solution can be found compared to local minima. Furthermore, the second method to avoid local minima is called as a random choice, where the algorithm selects an individual from the losers group. The pseudocode for presented GA is shown in Table 1.

Table 1. Pseudo code of the genetic algorithm.

```
Inputs: parameters, retain, population_size, generation_size, selection_random, mutation
Generation = 0;
Initialization first population (population_size)
    Individual = random choice of parameters
    return (population)

For generation in generation_size:
    For individual in population:
        Evaluate individual (fitness function);
        return (fitness function)
    Divide sorted(fitness function) population based on retain (keep elites);
    Keep random individual which loose (selection_random);

    While elites<population:
        For individual in population_losers
            Take two different parents from population_elites;
            Create two offsprings by the crossover of parents parameters;
            Mutate on of the offspring parameter (mutation);
            return (offsprings)
        update (population)
```

4 Research Methods

Every machine learning algorithm is considerably dependent on proper dataset. The data are the vital cornerstone of each machine learning model. That is why we decide to use labeled presented in the publication: Providing {SCADA} Network Data Sets for Intrusion Detection Research [14]. The authors provide pcap files, datasets, and csv files with labels. Therefore, the dataset provides separated network intrusion detection dataset that includes cyber-attacks and regular operation of ICS system. Moreover, the dataset was generated on ICS system with the specific architecture that is shown in Fig. 4.

We exploited Run1_6RTU dataset of one hour of regular Modbus traffic to create and evaluate machine learning model. We exploited Run1_6RTU dataset of one hour of regular Modbus traffic to create and evaluate machine learning model. The data from pcap file were extracted into csv files. The dataset has 88 dimensions which correspond

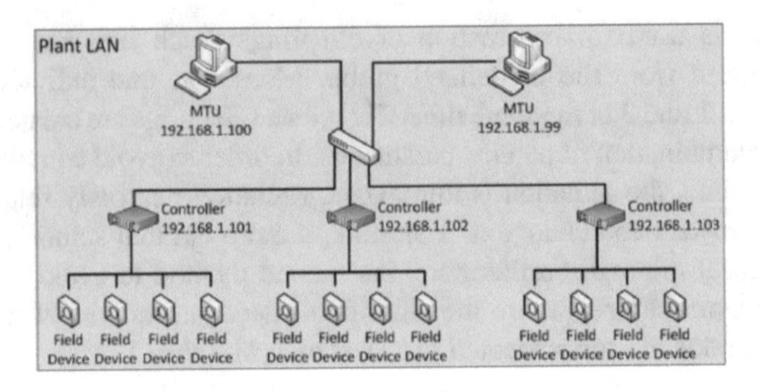

Fig. 4. The architecture of ICS system used for dataset generation [14].

to the number of features and 134 690 records. We divided the dataset into train dataset, which represents 70% of the original dataset and validation dataset, which represents 30% of the original dataset. The validation dataset is dedicated to evaluate the model and calculate the validation loss function. Moreover, the original dataset was preprocessed in order to fit the format which neural network needs. Thus, each example is transformed into a unique numerical form. Additionally, the missing values are replaced by the mean of a specific feature (variable). Lastly, all records were normalized in rage from 0 to 1. The representation of all 87 features in the dataset is shown in Table 2.

Table 2. Extracted features from pcap file Run1_6RTU.

ws.col.Protocol	modbus.reference_num	tcp.options.wscale.multiplier	tcp.analysis.ack_rtt	tcp.connection.fin
ip.len	modbus.request_frame	tcp.payload	tcp.ack	tcp.connection.syn
tcp.flags	tcp.analysis.flags	tcp.reassembled.length	tcp.flags.fin	tcp.flags.ns
tcp.analysis.bytes_in_flight	tcp.checksum.status	tcp.segment.count	tcp.analysis.initial_rtt	tcp.flags.urg
tcp.nxtseq	tcp.connection.sack	tcp.time_delta	mbtcp.unit_id	tcp.options.mss
tcp.flags.ack	tcp.flags.ecn	tcp.window_size_scalefactor	modbus.data	tcp.options.timestamp.tsecr
tcp.hdr_len	tcp.flags.str	_ws.col.Time	modbus.func_code	tcp.options.wscale
frame.time_delta	tcp.option_len.1	_ws.col.Length	modbus.regnum16	tcp.options.wscale.shift
modbus.bitnum	tcp.options.sack_perm	tcp.srcport	modbus.word_cnt.1	tcp.pdu.size
modbus.byte_cnt	tcp.options.timestamp.tsval	tcp.window_size	tcp.analysis.window_update	tcp.reassembled_in
tcp.segment_data	tcp.len	tcp.connection.rst	tcp.urgent_pointer	modbus.bitval
tcp.time_relative	udp.dstport	tcp.flags.cwr	""""ip.dst""""	tcp.flags.reset

(*continued*)

Table 2. (*continued*)

ip.src	mbtcp.len	tcp.flags.res	frame.time	tcp.options
frame.len	modbus.bit_cnt.1	tcp.option_kind	tcp.window_size_value	tcp.reassembled.data
tcp.dstport	modbus.padding	tcp.options.mss_val	tcp.flags.push	tcp.stream
tcp.analysis.push_bytes_sent	modbus.regval_uint16	tcp.port	tcp.analysis.acks_frame	
tcp.seq	tcp.analysis	tcp.segment	udp.srcport	
tcp.flags.syn	tcp.checksum	tcp.segments	mbtcp.prot_id	

Train multiple neural networks on high dimensional dataset would be an excessive time-consuming task. Therefore, the feature reduction algorithm should be used. We decide to exploit the Principal Component Analysis (PCA) algorithm in order to significantly reduce the dimension of the dataset. The new reduced dataset also has variability and information of the original dataset. This multivariate analysis is based on a covariance matrix of features in the dataset. The covariance matrix expresses the relationship between features and standard deviations. Moreover, every principal component can be interpreted as a new feature which was created based on the covariance matrix. PCA start to convert original features into principal components; however, the algorithm also tries to compress as much as possible information into the first principal component. This process continues iteratively, where each subsequently principal component contains less information. Therefore, there is a threshold for principal components after which principal components contains a negligible amount of information. Hence, we decided to use cumulative explained variance in order to identify the threshold. We construct a cumulative explained variance graph which is shown in Fig. 5. As can be seen, approximately 30 principal components represent one hundred percent of the information.

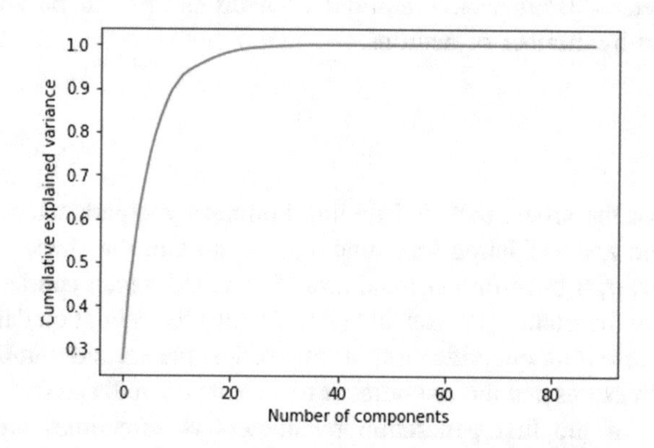

Fig. 5. Cumulative explained variance for principal components.

The settings of GA is a vital part of the optimization task. We decided to create a population that consisted of 40 individuals. The number of generation were set on 35.

Moreover, the mutation rate is fifteen percent, and a random selection is set to five percent due to avoid convergence into local minima. Lastly, we keep forty percent of elites group and remove the rest (losers group) in each generation. The space created by eliminating individuals is filled by offsprings created by elites group.

The GA is searching for the best representation of hyperparameters for our recurrent neural network. We choose the following hyperparameters for modeling space where GA will search for optimal result. The hyperparameters can be seen in Table 3.

Table 3. Used hyperparameters for recurrent neural network.

Number of neurons	{10, 12, 15, 17, 20, 22, 25, 27, 30, 32, 35, 37, 40}
Number of layers	{1, 2, 3, 4, 5, 6, 7, 8, 9, 10, 11, 12}
Batch size	{32, 64, 128, 256}
Number of epochs	{100, 200, 300, 400}
Recurrent dropout	{0.1, 0.2, 0.3, 0.4, 0.5, 0.6, 0.7, 0.8, 0.9}
Dropout	{0.1, 0.2, 0.3, 0.4, 0.5, 0.6, 0.7, 0.8, 0.9}
Activation function	{relu, elu, tanh, sigmoid}
Optimizer	{rmsprop, adam, adagrad, adadelta, adamax, nadam}
Bottleneck - number of neurons	<Number of neurons>
Bottleneck position	<Number of layers>

Almost all hyperparameters are set at the beginning of GA. However, "Bottleneck position" and "Bottleneck - number of neurons" hyperparameters are randomly generated according to hyperparameter selection (Number of layers, Number of neurons) in each iteration. The basic rule is that the "Bottleneck position" has to be inside defined space by number of layers hyperparameter. Moreover, the number determined by hyperparameter: "Bottleneck - number of neurons" has to be smaller than the number assigned by number of neurons.

5 Results

The main goal of the article included finding optimal hyperparameters according to a genetic algorithm and validation loss function. To do that the shape of data must be changed. Moreover, a three-dimensional array is created which can be represented as (1884, 50, 30) for the training dataset and (808, 50, 30) for validation dataset, where the first number representing quantification of time series, the second number representing the number of timesteps and the last number represents the reduced dimension by PCA. The adjustment of the first generation parameters is sometimes crucial. The low diversity leads to convergence into local minima. This is usually caused by mishandled with the generation of parameters or a small number of individuals in the population. Furthermore, the hyperparameters setting of the first population is shown in Fig. 6.

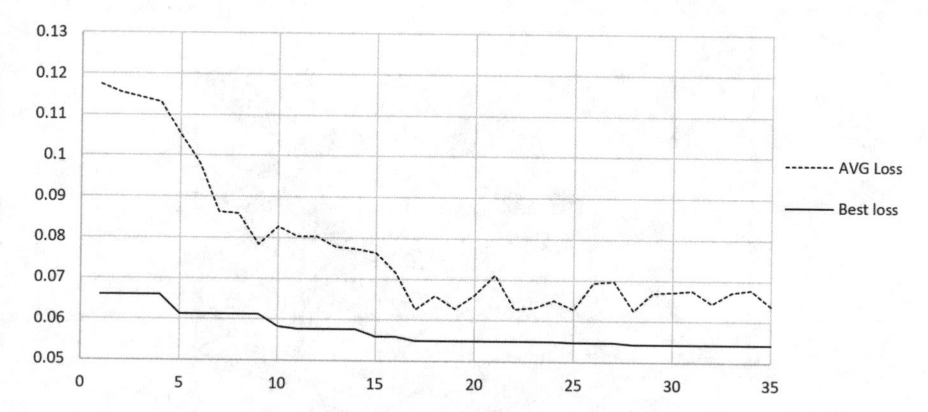

Fig. 6. The fitness function of all generations.

The Fig. 6 shows the random distribution of parameters across all individuals in the first population. Furthermore, the red bold lines represent the individual in the elites group (top 40%) designed for the reproduction. Otherwise, the rest of individuals (worst 60%) represent losers group which is shown as grey thin lines. The most of parameters from Table 3 are represented in the first population. However, "bottleneck position" and "bottleneck - the number of neurons" are underrepresented parameters, especial for higher numbers. Moreover, the first generation will be more discussed in the conclusion section.

The GA lasted 35 generations with a population of forty individuals (recurrent neural networks). The average fitness and for each generation is shown in Fig. 7. The graph progression shows us the development GA (average loss) over 35 generations. The highest decline of the fitness function is registered for the first generations (in the first quarter). However, there are slightly increased between the 10 generation and 15 generation. Lastly, the generation converged into a minimum, and since then it has been oscillating around the final value, which is minimum average loss values closed to the best loss value.

The population converged into one best solution (generation 17) according to the fitness function minimalization, where validation loss reach 0.0626 and the loss reach 0.0549. The hyperparameters of the best generation are nearly identical across all individuals. Moreover, the hyperparameters of the best individual can be seen in Table 4.

The hyperparameters from Table 4 were used to recreate the recurrent neural network due to the utilization of best the best result. Moreover, the process of recurrent neural network training is captured in Fig. 8 via loss function. The loss function of the neural network starts converged approximately after 222 epochs with the loss function value 0.0601.

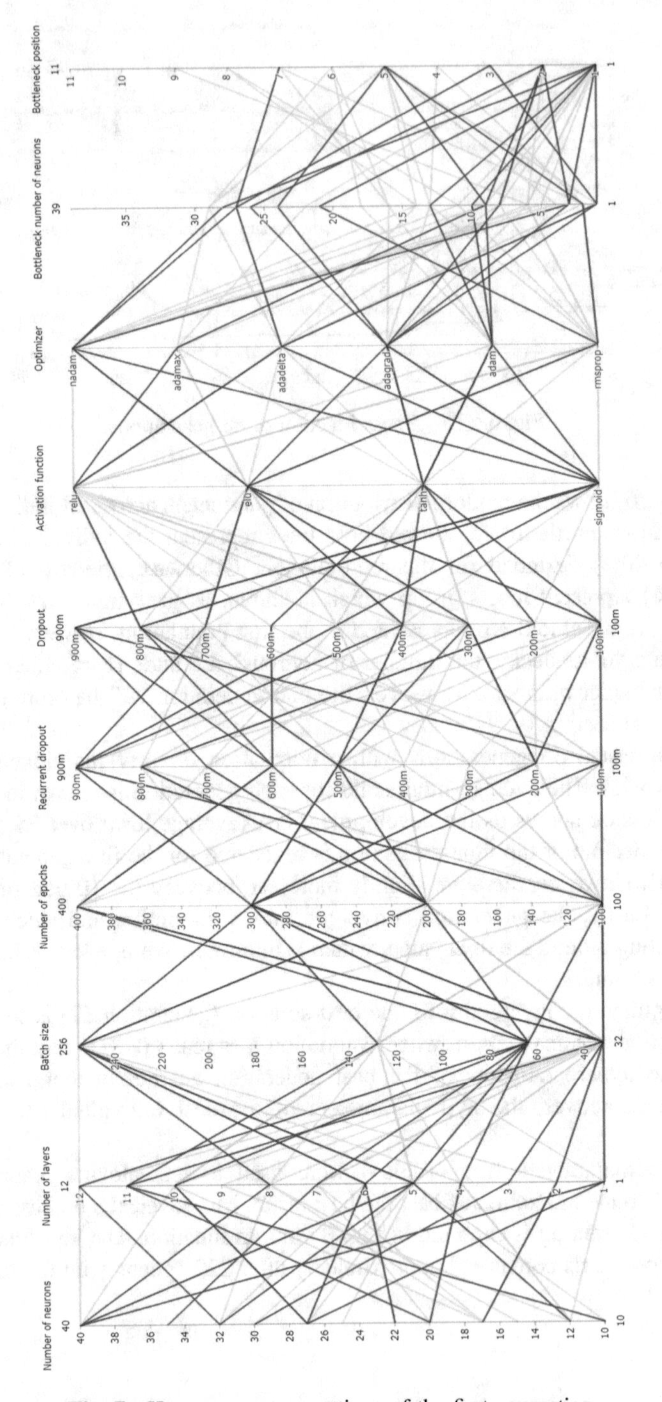

Fig. 7. Hyperparameters settings of the first generation.

Table 4. Best hyperparameters for LSTM autoencoder.

Number of neurons	27
Number of layers	1
Batch size	32
Number of epochs	300
Recurrent dropout	0.1
Dropout	0.1
Activation function	elu
Optimizer	nadam
Bottleneck - number of neurons	21
Bottleneck position	1

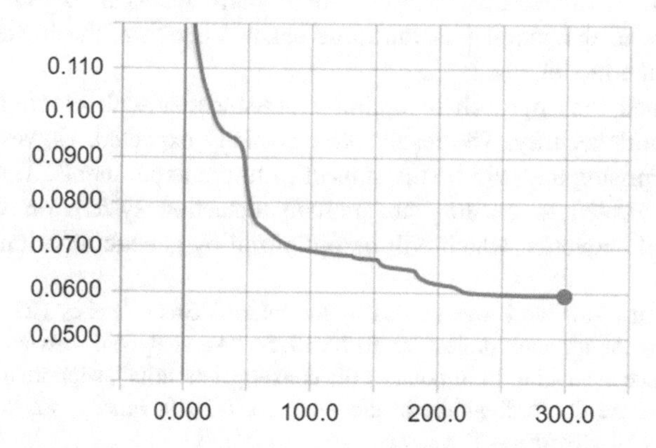

Fig. 8. The loss function of the best hyperparameters setting.

6 Conclusion

The article was constructed with a simple purpose to evaluate the possibility of hyperparameter optimization for anomaly detection system in the ICS environment. Thus, the recurrent neural network LSTM is exploited for a time series of industrial network communication. Therefore, we try to calculate the best fitting model via optimization. GA is one of the possible ways how to achieve this task.

Each recurrent neural network was trained on dataset of shape (1884, 50, 30) and validated on dataset of shape (808, 50, 30). Moreover, the 30 features were extracted from 87 industrial network features which represent the main properties and therefore the behavior of the monitored network. Thus, we use GA to optimize the hyperparameters of LSTM autoencoder in ten-dimensional space. The layout of the first generation is randomly selected from predefined hyperparameters. However, "bottleneck position" and "bottleneck - the number of neurons" are underrepresented parameters, especial for higher numbers of neurons or layers. This anomaly is caused by the dependence of the hyperparameters on number of layers and number of neurons.

Additionally, these recursive parameters are based on double randomization. That is why they are partially unrepresented hyperparameters.

The GA converged into minima after 17 generations. Moreover, the hyperparameters in Table 4 represent the best setting according to the loss function. The architecture is defined with only one hidden LSTM layer which is relatively interesting and should be subjected to further investigation. Moreover, one hidden layer contains 21 neurons. Additionally, the dropout and recurrent dropout, which are responsible for overfitting prevention were both sets on 0.1. This corresponds to the anticipated values. Otherwise, the number of epochs were set on 300 epochs, which indicated relatively swift convergence of the model. The elu activation function is unexpected choice since elu function is designed for deep neural networks. However, elu better generalization abilities have manifested in the experiment LSTM neural networks. Moreover, nadam optimizer is expected solution due to its strong capabilities in loss reduction. Lastly, the lowes batch size of value 32 was chosen. Additionally, this is an expected result due to the interference of the samples in the same batch. Therefore, the lower batch size is better to train the model.

We presented the approach to optimize autoencoder with recurrent neurons for industrial network security. The results were partially expected. However, part of the results was surprising and will be the subject to future experiments. The aim of future research is dedicated to creating an anomaly detection system on the top of the recurrent neural networks, which will be optimized by genetic algorithms.

Acknowledgment. This work was funded by the Internal Grant Agency (IGA/FAI/2019/002) and supported by the research project VI20172019054 "An analytical software module for the real-time resilience evaluation from point of the converged security", supported by the Ministry of the Interior of the Czech Republic in the years 2017–2019. Finally, we thank Lemay and Fernandez [14] who provides ICS datasets.

References

1. Chen, C.L.P., Zhang, C.-Y.: Data-intensive applications, challenges, techniques and technologies: a survey on Big Data. Inf. Sci. **275**, 314–347 (2014)
2. Knapp, E.D., Langill, J.T.: Industrial Network Security: Securing Critical Infrastructure Networks for Smart Grid, SCADA, and Other Industrial Control Systems. Syngress, Waltham (2014)
3. Maglaras, L.A., et al.: Cyber security of critical infrastructures. ICT Express **4**(1), 42–45 (2018)
4. Chandola, V., Banerjee, A., Kumar, V.: Anomaly detection: a survey. ACM Comput. Surv. (CSUR) **41**(3), 15 (2009)
5. Markou, M., Singh, S.: Novelty detection: a review—part 1: statistical approaches. Sig. Process. **83**(12), 2481–2497 (2003)
6. Marchi, E., et al.: A novel approach for automatic acoustic novelty detection using a denoising autoencoder with bidirectional LSTM neural networks. In: 2015 IEEE International Conference on Acoustics, Speech and Signal Processing (ICASSP), pp. 1996–2000. IEEE (2015)

7. Park, D., Hoshi, Y., Kemp, C.C.: A multimodal anomaly detector for robot-assisted feeding using an LSTM-based variational autoencoder. IEEE Robot. Autom. Lett. **3**(3), 1544–1551 (2018)
8. Malhotra, P., et al.: LSTM-based encoder-decoder for multi-sensor anomaly detection. arXiv preprint arXiv:1607.00148 (2016)
9. Kiran, B., Thomas, D., Parakkal, R.: An overview of deep learning based methods for unsupervised and semi-supervised anomaly detection in videos. J. Imaging **4**(2), 36 (2018)
10. Lipton, Z.C., Berkowitz, J., Elkan, C.: A critical review of recurrent neural networks for sequence learning. arXiv preprint arXiv:1506.00019 (2015)
11. Hochreiter, S., Schmidhuber, J.: Long short-term memory. Neural Comput. **9**(8), 1735–1780 (1997)
12. D'errico, F., et al. (eds.): Conflict and Multimodal Communication: Social Research and Machine Intelligence. Springer, Heidelberg (2015)
13. Holland, J.H.: Adaptation in Natural and Artificial Systems: An Introductory Analysis. Holland, JH (1975)
14. Lemay, A., Fernandez, J.M.: Providing {SCADA} network data sets for intrusion detection research. In: 9th Workshop on Cyber Security Experimentation and Test, CSET 2016 (2016)

Self-organising Clusters in Edge Computing

Luís Nogueira[1] and Jorge Coelho[1,2](✉)

[1] School of Engineering (ISEP), Polytechnic Institute of Porto (IPP), Porto, Portugal
jmn@isep.ipp.pt

[2] LIACC Research Centre, University of Porto, Porto, Portugal

Abstract. Computation-intensive applications generally require a high computing capacity for data processing and storage that cannot be easily offered by a single Internet of Things (IoT) device. Such limitations can be successfully addressed by offloading processing and storage from resource-constrained devices to more powerful ones. In this context, edge computing is emerging as a valuable approach, since it allows data to be stored and processed closer to where it is created instead of sending it across long routes to data centres or clouds.

We are interested in supporting spontaneous and opportunistic behaviour in this new dynamic environment, where computational power and storage capacity can be offered from the edge, with low latency and high bandwidth, by enabling cooperation between a subset of available edge nodes. We argue that cluster formation is necessary when a single node cannot execute a specific service fulfilling the imposed non-functional requirements, and it may also be beneficial when groups perform more efficiently when compared to a single's node performance.

1 Introduction

By 2020, it is expected that 20.8 billion connected devices will exist [5], demanding automated mechanisms to overcome various aspects of their management. A significant percentage of Internet of Things (IoT) devices have limited resources, particularly related to their battery life, computation power, and storage capacity, facts that yield an extra complexity in their management. To overcome these obstacles, it is necessary to develop novel and efficient mechanisms.

Computation offloading is recognised as a promising solution to improve the satisfaction of Quality of Service (QoS) requirements in resource-constrained devices by migrating a part or an entire application to a remote server in order to be executed there. Various models and frameworks have been proposed to offload resource intensive components of applications to the cloud for a more efficient execution [14,16,24]. However, offloading to the cloud is not always a solution, because of the high WAN latencies, mainly for applications with real-time restrictions.

R. Silhavy et al. (Eds.): CoMeSySo 2019, AISC 1046, pp. 320–332, 2019.
https://doi.org/10.1007/978-3-030-30329-7_29

The proliferation of IoT and the success of rich cloud services have pushed the horizon of a new computing paradigm, edge computing, which calls for processing the data at the edge of the network [13]. Edge computing has the potential to address the concerns of response time requirement, battery life constraint, bandwidth cost saving, as well as data safety and privacy. There are hundreds of use cases where reaction time is the key value of the IoT system, and consistently sending the data back to a centralised cloud prevents that value from happening [27]. Therefore, edge computing is attracting an increasing attention from academic researchers [1,21,26].

In particular, one of the key design issues in edge computing is resource allocation. While conceptually simple, it is challenging to make optimal decisions on what and where to offload since many factors are coupled and the solution space is very large. Various approaches have been proposed to tackle this resource allocation problem in many kinds of scenarios [1,20,21,26], focusing on either single-user or multi-user cases, but usually the mobile device is assumed to be associated with only a single edge server. With the expectation of small-cell/femtocell base stations been massively deployed in future networks, a resource-constrained device should be able to choose to offload its tasks to multiple nearby edge servers [8].

In this paper we address distributed processing in scenarios where data is heavily processed in a set of dynamically chosen edge nodes, maximising the satisfaction of non-functional requirements on several dimensions. We seek a generic model that enables a self-organising distributed service allocation, *i.e.*, without a central authority distributing the services among nodes. Given a set of services to be executed, a distributed environment must seek the maximisation of the associated QoS constraints. The nodes shall reach efficient service allocation by themselves, seeking a maximal outcome. This will be achieved via the formation of a temporary group of individual nodes, which, due to its higher flexibility and agility, is capable of effectively respond to new, challenging, requirements. We call these groups *self-organising clusters*.

Searching for an optimal resource allocation with respect to a particular goal has always been one of the fundamental problems in QoS management [25]. However, as the complexity of open distributed systems increases, it is also increasingly difficult to achieve an optimal resource allocation that deals with both users' and nodes' constraints within an useful and bounded time [6]. Note that if the system adapts too late to the new resource requirements, it may not be useful and may even be disadvantageous [19].

Our proposal is to quickly establish an initial, sub-optimal, solution according to the set of QoS constraints that have to be satisfied. Then, if time permits, the initial solution is gradually refined until it finally reaches its optimal value or the available deliberation time expires. At each iteration, a new set of Service Level Agreements (SLAs) is found with an increasing utility to the user's request under negotiation but these successive adjustments get smaller as the QoS optimisation process progresses. This approach is based on a previously one proposed in [22] for open real-time environments.

Scheduling algorithms such as the Capacity Sharing and Stealing (CSS) scheduler [23] were developed to efficiently handle the execution of tasks with a variable or unknown execution behaviour, handling overloads with additional capacity that is available from two sources: (i) by reclaiming unused allocated capacity when jobs complete in less than their budgeted execution time; and (ii) by stealing allocated capacities to non-isolated servers used to schedule sporadic best-effort jobs. This way, nodes with less workload can reclaim more time to find a solution and, as such, potentially present a better service solution.

2 System Model

Consider a distributed system with several heterogeneous nodes at the edge, each with its specific set of resources R_i. Service provisioning is based on Service Level Agreements (SLA), which is a set of non-functional properties specified and negotiated between the user and the service provider.

We are primarily interested in dynamic scenarios where new services can appear while others are being executed, the processing of those services has associated QoS constraints, and service execution can be performed by a cluster of nodes. Due to these characteristics, resource availability is highly dynamic as services enter and leave the system at anytime.

It is assumed that a service S can be executed at varying levels of QoS to achieve an efficient resource usage that constantly adapts to the devices' specific constraints, nature of executing tasks and dynamically changing system conditions. There will be a set of independent tasks to be executed, resulting from partitioning the resource intensive service S. Correct decisions on service partitioning must be made at run time when sufficient information about workload and communication requirements become available [29], since they may change with different execution instances and users' QoS preferences.

When users express their expectations from services, they identity functional and non-functional characteristics of required services. In addition, they have to identify which of the QoS criteria are more important than the others. A number of approaches with a variety of architectures and algorithms have been proposed to solve the problem [2,7,18,25]. Ontologies provide a structured framework for modelling the concepts and relationships of a domain of expertise [9,15,28]. Having such a QoS characterisation of a particular application domain, users and service providers are now able to define service requirements and proposals in order to reach an agreement on service provisioning.

Users provide a single specification of their own range of QoS preferences Q for a complete service S, through the relative decreasing importance ($K = 1 \ldots n$) of a set of n QoS dimensions, ranging from a desired QoS level $L_{desired}$ and the maximum tolerable service degradation, specified by a minimum acceptable QoS level $L_{minimum}$, without having to understand the individual tasks that make up the service. For each QoS dimension, a relative decreasing importance order of attributes is also specified ($i = 1 \ldots attr_k$), where k is the number of attributes of dimension K. Each Q_{kj} is a finite set of quality choices for the

j^{th} attribute of dimension k. This can be either a discrete or continuous set. Please note that k and i are not the identifiers of dimensions and attributes in a domain's QoS description, but their relative position in user's service request.

Formulating a service request based on a qualitative, not quantitative, measure through a relative decreasing order on quality dimensions, their attributes, and accepted values, allows the user to encode the relative importance of the new service's performance at the different QoS levels without the need to quantify every quality tradeoff with absolute values. At the same time, from the service provider perspective, the reward of proposing a particular SLA will depend on the number, and relative importance, of the QoS dimensions being served closer to the user's desired QoS level.

3 Cluster Formation

A self-organising cluster formation process should enable the selection of individual edge nodes that, based on their own resources and availability, will constitute the best group to satisfy the user's QoS requirements associated with a resource intensive service. By best group, we mean the group formed by those nodes which offer service closer to the user's desired QoS level.

Whenever the user's node N_i cannot provide service within the user's acceptable QoS levels Q_i, it broadcasts a cooperation request to execute service S_i to edge nodes. The set of tasks that can be remotely executed is determined by a task partition/allocation scheme that dynamically considers the tradeoff between local execution requirements and communication costs [29]. The cooperation request includes a description of each remote task T_i, the user's QoS constraints, and a timeout Δ_t for the reception of service proposals.

Every edge node N_j which is able to satisfy the request, formulates a service proposal according to a local QoS optimisation algorithm (see Sect. 4 for details), and replies to node N_i with both its service proposal P_{ji} and its local reward R_j, resulting from its proposal acceptance. For now, it suffices to say that the local reward is an indicator of the node's local QoS optimisation level, according to the set of services being locally executed and their QoS constraints. How each node measures its local reward will be detailed in Sect. 4.

It is clear that different groups of nodes will have different degrees of efficiency in the service's cooperative execution performance due to different capabilities of their members and their current state. As such, the cluster's members selection must be determined by the proximity of their service proposals with respect to the expressed user's multi-dimensional QoS constraints. Each admissible proposal[1] P_i is then evaluated by determining, for each QoS dimension, a weighted sum of the differences between the user's preferred values and the values proposed in P_{ji}, using Eq. 1. Recall that the user's QoS constraints are presented in a decreasing preference order.

[1] A proposal is admissible if it can satisfy all QoS dimensions within the user's acceptable QoS levels.

$$closeness(P_i) = \sum_{k=1}^{n} w_k * distance(Q_k) \quad (1)$$

where n is the number of QoS dimensions and $0 \leq w_k \leq 1$ is the relative importance of QoS dimension k, Q_k, to the user, and can defined as

$$w_k = \frac{n-k+1}{n} \quad (2)$$

The degree of acceptability of each proposed attribute's value when compared to the request one is given by Eq. 3, considering continuous and discrete domains.

$$distance(Q_k) = \sum_{i=1}^{attr_k} w_i * |acceptability(Prop_{ki}, Pref_{ki})| \quad (3)$$

where $attr_k$ is the number of attributes in dimension k.

In Eq. 3, the function $acceptability(Prop_{ki}, Pref_{ki})$ quantifies, for an attribute i, the degree of acceptability of the proposed value $Prop_{ki}$, when compared to the user's preferred value $Pref_{ki}$ and is defined as

$$acceptability = \begin{cases} \dfrac{Prop_{ki} - Pref_{ki}}{max(Q_k) - min(Q_k)} & \text{if continuous } Q_{ki} \\ \dfrac{pos(Prop_{ki}) - pos(Pref_{ki})}{length(Q_k) - 1} & \text{if discrete } Q_{ki} \end{cases} \quad (4)$$

If attribute i has a continuous domain, this quantification is a normalised difference between the proposed value and the preferred one. For discrete domains, Eq. 4 considers the preferences attached to $Prop_{ki}$ and $Pref_{ki}$ by using their relative position in the application's QoS requirements specification.

In [17] the authors use the notion of a quality index, defining a bijective function that maps the elements of a discrete domain into integer values. We use a similar approach, by mapping the position (index) of that attribute in the domain's specification into $Prop_{ki}$'s and $Pref_{ki}$'s scoring values. When the domain's QoS description defines the possible values for some attribute of a QoS dimension Q_k by a set of intervals, Q_k in Eq. 4 must relate to the particular interval where $Prop_{ki}$ is found. In a similar fashion, if the user expresses his set of acceptable values for an attribute of dimension Q_k considering a set of intervals, $Pref_{ki}$ should be the first value on the $Prop_{ki}$'s interval and the relative decreasing order of importance of that interval to the user must be considered. The best proposal for each of the service's tasks is thus the one that presents the lowest distance to the user's quality preferences in all QoS dimensions.

In order to be useful in practice, cluster formation must try to quickly find a sufficiently good initial solution and gradually maximise its improvement at each iteration. As such, a particular attention is devoted to the selection of the next service proposal to be evaluated from the set of received proposals, rather than depending on the order of proposals' reception. The proposed algorithm, described in detail in Algorithm 1, uses each node's local reward as a heuristic to guide the cluster's formation process. Clearly, nodes with a higher local reward have a higher probability to be offering service closer to this particular user's request under negotiation since the utility achieved by all services being locally executed is higher. Then, for each remote task $T_i \in S$, the next candidate proposal P_{ki} to be selected from the set of received proposals P_i is the one sent by the node N_k with the greatest local reward R_k.

Algorithm 1 allows the global QoS optimisation process to return many possible approximate answers for a given input of service proposals to be evaluated. It can be interrupted at any time, providing a solution and a measure of its quality, which is expected to improve as the run time of the algorithm increases.

Algorithm 1. Cluster formation

Let $E_{P_{ki}}$ be the evaluation value of proposal P_{ki}, sent by node n_k for task τ_i
Let $Best_{\tau_i}$ be the best proposal for task τ_i
Let R_k be the local reward of node n_k
Let R_{best} be the local reward of the node which has proposed $Best_{\tau_i}$ for task τ_i
Let C be the formed cluster

1: Start with an empty cluster C $\quad \triangleright\ C = \emptyset$
2: **for** each task $\tau_i \in S_i$ **do**
3:     **for** each received k^{th} proposal $P_{ki} | P_{ki} \in P_i, max(R_k)$ **do**
4:         $E_{P_{ki}} = closeness(P_{ki})$
5:         **if** $(Best_{\tau_i} - E_{P_{ki}} > \alpha)$ or $(0 < Best_{\tau_i} - E_{P_{ki}} < \alpha$ and $R_k > R_{best})$ **then**
6:             $Best_{\tau_i} = E_{P_{ki}}$
7:             Update cluster with node n_k for task τ_i $\quad \triangleright\ C = C \setminus (n_j, \tau_i) \cup (n_k, \tau_i)$
8:         **end if**
9:     **end for**
10: **end for**
11: **return** cluster C

The quality of each generated cluster is measured by using the evaluation values of the best proposals for each service's task. At each iteration, Eq. 5 returns the quality of the achieved solution. For an empty set of proposals the quality of the cluster is zero. Note that the quality of the cluster is also zero, if there are not any proposals for one or more remote tasks $T_i \in S$.

$$Q_{cluster} = \left\lfloor \frac{|cluster|}{|S|} \right\rfloor * \sum_{i=1}^{|cluster|} \frac{1 - Best_{P_i}}{|cluster|} \tag{5}$$

After determining an initial cluster, the algorithm continues, if time permits, to evaluate the remaining proposals as it tries to improve the quality of the current solution. It is possible that some other node, while achieving a lower local reward, can still propose a better proposal for the specific user's request under negotiation at the expense of a greater downgrade of previously accepted services.

Each node's local reward is also used to improve a global load balancing. Consider two proposals whose evaluations differ at most by a significance value α (this value is set by default to 0.05 but it can be modified by the user or by the framework). For a particular user, the perceived utility will be equally acceptable if any of those nodes is selected for participating in the new cluster. Selecting the node with a higher local reward from two similar service proposals, not only maximises service for a particular user, but also maximises the global system's utility.

The algorithm terminates when all the received proposals are evaluated or if it finds that the quality of a cluster cannot be further improved because the local reward of each node that belongs to the current cluster is maximum.

4 Service Proposal Formulation

All entities that participate in a clustered service execution negotiation must provide sufficient resources to propose a SLA within the user's acceptable QoS levels. It is therefore the responsibility of each individual node to map the user's QoS constraints to local resource requirements, and then reserve resources accordingly [4,11,12,25].

Requests for task execution arrive dynamically at any node and are formulated as a set of acceptable multi-dimensional QoS levels in decreasing preference order. To guarantee the request locally, the node executes a local QoS optimisation algorithm that tries to maximise the satisfaction of the new service's QoS constraints as well as to minimise the impact on the current QoS of previously accepted services. To achieve such goal, we propose a QoS negotiation mechanism that, in cases of overload, or violation of pre-run-time assumptions guarantees a graceful degradation, dynamically determining promised QoS levels within the user's accepted QoS values for each QoS dimension and according to local resources' availability. Offering QoS degradation as an alternative to task rejection has been proved to achieve higher perceived utility [2].

The high complexity of determining the best set of SLAs, taking into account both the users' QoS preferences and node's resource availability, makes it beneficial to propose an algorithm that can trade the achieved solution's quality by its computational cost in order to ensure a timely answer to events. The proposed approach clear splits the formulation of a new set of SLAs in two different scenarios. The first one, detailed in Algorithm 2, involves serving the new task without changing the QoS level of previously guaranteed tasks. The second one, detailed in Algorithm 3, due to the lack of resources, demands degrading the currently provided level of service of the previously accepted tasks in order to accommodate the new requesting task.

When a new service request arrives, Algorithm 2 starts by maintaining the QoS levels of previously guaranteed tasks and by selecting the worst requested QoS level, for all dimensions, for the new arrived task. The goal is to quickly find a feasible initial solution that can later be improved if time permits. Note that this is the SLA that has a higher probability of being feasible, without requiring any modification on the current QoS of previously accepted services.

Algorithm 2. Maximise the QoS level of the newly arrived task τ^a

Let $\mathcal{T}$ be the set of all previously accepted tasks.
Let τ^a be the newly arrived task.
Task τ^a has associated a set of user's defined QoS constraints Q_i.
Each Q_{kj} is a finite set of n quality choices for the jth attribute, expressed in decreasing preference order, for all k QoS dimensions.
Let σ be the determined set of SLAs, updated at each step of the algorithm

1: Define SLA_{τ_a} for task τ^a by selecting the lowest requested QoS level $Q_{kj}[n]$, for all the j attributes of all k QoS dimensions
2: Keep the current QoS level for each task $\tau_k \in \mathcal{T}$
3: Update the current set of SLAs σ $\triangleright$ $\sigma = \sigma \cup SLA_{\tau_a}$
4: **while** $feasibility(\sigma) =$ **TRUE do**
5: **for** each jth attribute of any k QoS dimension in τ_a with $Q_{kj}[m] > Q_{kj}[0]$ **do**
6: Determine the utility increase by upgrading attribute j to $Q_{kj}[m-1]$
7: **end for**
8: Find attribute x with maximum utility increase
9: Define SLA'_{taua} for task τ_a by upgrading attribute x to the $Q_{kj}[m-1]$'s level
10: Update the current set of promised SLAs σ $\triangleright$ $\sigma = \sigma \setminus SLA_{\tau_{min}} \cup SLA'_{\tau_{min}}$
11: **end while**

The algorithm continues, in Algorithm 3, to improve the quality of that initial solution, conducting the search for a better feasible solution in a way that maximises the expected improvement in the solution's quality. With spare resources, the algorithm incrementally selects the configuration that maximises the increase in obtained reward for the new task. When QoS degradation is needed to accommodate the new task, the algorithm incrementally selects the configuration that minimises the decrease in obtained reward for the new set of tasks, which includes the new arrived one.

Note that the algorithm may produce an unfeasible set of SLAs due to local resource availability. As such, since a service proposal can only be considered useful within a feasible set of configurations, the algorithm, if interrupted, always returns the last found feasible solution. Nevertheless, each intermediate configuration, even if not feasible, is used to calculate the next possible solution, minimising the search effort.

The algorithm terminates when the time for the reception of proposals has expired (this time is sent in the cooperation request), when it finds a set of QoS levels that keeps all tasks feasible and the quality of the solution can not be further improved, or when it finds that, even at the lowest QoS level for each task, the new set is not feasible. In this case the new service request is rejected. When it is not possible to find a valid solution for service execution within available time, then no proposal will be sent to the requesting node, and the node continues to serve existing tasks at their current QoS levels.

The algorithm iteratively works on the problem of finding a feasible set of SLAs [3] while maximising the users' satisfaction and produces results that improve in quality over time. Instead of a binary notion of the solution's correct-

Algorithm 3. Find local minimal service degradation to accommodate τ^a

Let $\tau^e \in T$ be the set of tasks whose current QoS level can be changed.
Let τ^p be the set of tasks whose current QoS level cannot be changed.
Each task $\tau_i \in \tau^e$ has associated a set of user's defined QoS constraints Q_i.

12: **while** $feasibility(\sigma) \neq$ **TRUE do**
13: **for** each task $\tau_i \in \{\tau^e \cup \tau^a\}$ **do**
14: **for** each jth attribute of any k QoS dimension in τ_i with $Q_{kj}[m] > Q_{kj}[n]$ **do**
15: Determine the reward decrease by downgrading attribute j to $Q_{kj}[m+1]$
16: **end for**
17: **end for**
18: Find task τ_{min} whose reward decrease is minimum by downgrading attribute x of the QoS dimension y to $Q_{yx}[m+1]$
19: Define $SLA'_{\tau_{min}}$ for task τ_{min} with the new value $Q_{yx}[m+1]$
20: Update the current set of promised SLAs σ ▷ $\sigma = \sigma \setminus SLA_{\tau_{min}} \cup SLA'_{\tau_{min}}$
21: **end while**
22: **return** the new local set of promised SLAs σ

ness, the algorithm returns a proposal and a measure of its quality. The quality of each generated configuration Q_{conf}, given by Eq. 6, considers the reward achieved by the service proposal configuration for the new arriving task r_{T_a}, the impact on the provided QoS of previous existing tasks and the value of the previous generated feasible configuration Q'_{conf}. Initially, Q'_{conf} is set to zero.

$$Q_{conf} = \begin{cases} \left(r_{T_a} * \frac{\sum_{i=0}^{n} r_{T_i}}{n} \right)^{(1-Q'_{conf})} & \text{if feasible} \\ Q'_{conf} & \text{if not feasible} \end{cases} \tag{6}$$

The reward of executing a task at the dynamically determined QoS level depends on the number, and relative importance, of the QoS dimensions being served closer to the user's desired QoS level. Equation 7 computes the reward r_{T_i} achieved by a specific SLA for task T_i by measuring the distance between the user's desired and the node's proposed values.

$$r_{T_i} = \begin{cases} 1 & \text{if task is being best served in all QoS dimensions} \\ 1 - \sum_{j=0}^{n} w_j * penalty_j & \text{if } Q_{jk} < Q_{best_j} \end{cases} \tag{7}$$

In Eq. 7 *penalty* is a parameter that decreases the reward value. This parameter can be fine tuned by the user or the framework's manager according to several criteria and its value should increase with the distance to the user's preferred values.

Using the utility achieved by each proposed SLA it is possible to determine a measure of the node's global satisfaction resulting from the acceptance of the new service request. For a node N_j, the local reward R_j achieved by the set of proposed SLAs is given by

$$R_j = \frac{\sum_{i=1}^{n} r_{T_i}}{n} \quad (8)$$

Note that unless all tasks are executed at their highest requested QoS level there is a difference between the determined set of SLAs and the maximum theoretical local reward that would be achieved if all local tasks were executed at their highest QoS level. This difference can be caused by either resource limitations, which is unavoidable, or poor load balancing, which can be improved by sending actual local rewards in service proposals, and selecting, for proposals with similar evaluation values, those nodes that achieve higher local rewards, as discussed in the previous section.

5 Conclusions

Resource-constrained devices may need to offload computation to other nodes at the edge of the cloud in order to fulfil certain services. Given a set of tasks to be executed with associated QoS constraints, we consider situations where a service is assigned to a self-organising cluster of edge nodes. Service allocation to several nodes is necessary when the processing cannot be performed by a single node or when a single node performs it inefficiently.

Various groups of nodes may have different degrees of efficiency in service execution performance due to different capabilities of their members. As such, service allocation should be done with respect to those differences. Our cluster formation algorithm selects the nodes that offer the solution that includes values closer to user's multi-dimensional QoS requirements. Those requirements are described through a semantically rich QoS specification interface for multidimensional QoS provisioning, allowing the user and applications to define fine-grained service requests.

The increased complexity of dynamic open scenarios may prevent the possibility of computing optimal local and global resource allocations within a useful and bounded time, as the optimal level of deliberation varies from situation to situation. It is therefore beneficial to build systems that can tradeoff the computational cost for the quality of the achieved solution. We propose that nodes start by negotiating partial, acceptable service proposals that are latter refined if time permits, in contrast to a traditional QoS optimisation approach that either runs to completion or is not able to provide a useful solution. At each iteration, the proposed QoS optimisation tries to find a new feasible set of QoS levels with an increasing utility. At any given time, a complete solution for service execution exists, and the quality of that solution is expected to improve overtime.

We are currently evaluating the effectiveness of the proposed approach through extensive simulations using a broad collection of profiles, chosen to cover the spectrum into which real-world applications and both resource-constrained and their more powerful neighbour edge devices would fall or likely exhibit. The main goal of the conducted evaluations is to determine the performance profile of the proposed algorithms [10]. Preliminary results clearly demonstrate that the proposed algorithm for cluster formation achieves a solution's quality of 83% $\pm$ 6% of its optimal solution at only 20% of its completion time. Similar results can also be found for the proposal formulation algorithm. For nodes with spare resources, at only 20% of the computation time, the solution's quality for the new arrived task is near 70% $\pm$ 5% of the achieved quality at completion time. When QoS degradation is needed to accommodate the new task, its service proposal achieves 87% $\pm$ 4% of its final quality at 20% of computation time.

Acknowledgments. This work was partially supported by LIACC through Programa de Financiamento Plurianual of FCT (Portuguese Foundation for Science and Technology).

References

1. Abbas, N., Zhang, Y., Taherkordi, A., Skeie, T.: Mobile edge computing: a survey. IEEE Internet of Things J. **5**(1), 450–465 (2018)
2. Abdelzaher, T.F., Atkins, E.M., Shin, K.G.: QoS negotiation in real-time systems and its application to automated flight control. IEEE Trans. Comput. **49**(11), 1170–1183 (2000). Best of RTAS 1997 Special Issue
3. Albers, K., Slomka, F.: Efficient feasibility analysis for real-time systems with EDF scheduling. In: Design, Automation and Test in Europe, pp. 492–497, March 2005
4. Berthold, H., Schmidt, S., Lehner, W., Hamann, C.J.: Integrated resource management for data stream systems. In: Proceedings of the 2005 ACM Symposium on Applied Computing, pp. 555–562. ACM Press (2005)
5. Cheng, C., Lu, R., Petzoldt, A., Takagi, T.: Securing the internet of things in a quantum world. IEEE Commun. Mag. **55**(2), 116–120 (2017)
6. Chishiro, H., Yamasaki, N.: Practical imprecise computation model: theory and practice. In: IEEE 17th International Symposium on Object/Component/Service-Oriented Real-Time Distributed Computing, pp. 198–205, June 2014
7. Dastjerdi, A., Buyya, R.: A taxonomy of QoS management and service selection methodologies for cloud computing. In: Cloud Computing: Methodology, Systems, and Applications, pp. 109–131. CRC Press, October 2011
8. Dinh, T.Q., Tang, J., La, Q.D., Quek, T.Q.S.: Adaptive computation scaling and task offloading in mobile edge computing. In: 2017 IEEE Wireless Communications and Networking Conference, pp. 1–6, March 2017
9. Dobson, G., Lock, R., Sommerville, I.: QoSOnt: a QoS ontology for service-centric systems. In: 31st EUROMICRO Conference on Software Engineering and Advanced Applications, pp. 80–87, August 2005

10. Dolan, E.D., Moré, J.J.: Benchmarking optimization software with performance profiles. Math. Program. **91**, 201–213 (2002)
11. Fukuda, K., Wakamiya, N., Murata, M., Miyahara, H.: QoS mapping between user's preference and bandwidth control for video transport. In: Proceedings of the 5th International Workshop on Quality of Service, New York, USA, pp. 291–302 (1997)
12. Goebel, V., Plagemann, T.: Mapping user-level QoS to system-level QoS and resources in a distributed lecture-on-demand system. In: IEEE Computer Society (eds.) Proceedings of the 7th IEEE Workshop on Future Trends of Distributed Computing Systems, p. 197 (1999)
13. Hu, Y.C., Patel, M., Sabella, D., Sprecher, N., Young, V.: Mobile edge computing - a key technology towards 5G. Technical report, European Telecommunications Standards Institute (2015). https://www.etsi.org/images/files/ETSIWhitePapers/etsi_wp11_mec_a_key_technology_towards_5g.pdf
14. Khan, M.A.: A survey of computation offloading strategies for performance improvement of applications running on mobile devices. J. Netw. Comput. Appl. **56**, 28–40 (2015)
15. Kim, H.M., Sengupta, A., Evermann, J.: MOQ: web services ontologies for qos and general quality evaluations. Int. J. Metadata Semant. Ontol. **2**(3), 195–200 (2007)
16. Kumar, S., Tyagi, M., Khanna, A., Fore, V.: A survey of mobile computation offloading: applications, approaches and challenges. In: 2018 International Conference on Advances in Computing and Communication Engineering (ICACCE), pp. 51–58, June 2018
17. Lee, C., Lehoczky, J., Siewiorek, D., Rajkumar, R., Hansen, J.: A scalable solution to the multi-resource QoS problem. In: Proceedings of the 20th IEEE Real-Time Systems Symposium, pp. 315–326 (1999)
18. Li, C., Li, L.: Utility-based QoS optimisation strategy for multi-criteria scheduling on the grid. J. Parallel Distrib. Comput. **67**(2), 142–153 (2007)
19. Liu, J.W.S., Bettati, R.: Imprecise computations. Proc. IEEE **82**(1), 83–94 (1994)
20. Mach, P., Becvar, Z.: Mobile edge computing: a survey on architecture and computation offloading. IEEE Commun. Surv. Tutor. **19**(3), 1628–1656 (2017)
21. Mao, Y., You, C., Zhang, J., Huang, K., Letaief, K.B.: A survey on mobile edge computing: the communication perspective. IEEE Commun. Surv. Tutor. **19**(4), 2322–2358 (2017)
22. Nogueira, L., Pinho, L.M.: Iterative refinement approach for QoS-aware service configuration. In: From Model-Driven Design to Resource Management for Distributed Embedded Systems. IFIP, vol. 225, pp. 155–164. Springer (2006)
23. Nogueira, L., Pinho, L.M.: A capacity sharing and stealing strategy for open real-time systems. J. Syst. Archit. - Embed. Syst. Des. **56**(4–6), 163–179 (2010)
24. Noor, T.H., Zeadally, S., Alfazi, A., Sheng, Q.Z.: Mobile cloud computing: challenges and future research directions. J. Netw. Comput. Appl. **115**, 70–85 (2018)
25. Rajkumar, R., Lee, C., Lehoczky, J., Siewiorek, D.: A resource allocation model for QoS management. In: Proceedings of the 18th IEEE Real-Time Systems Symposium, p. 298. IEEE Computer Society (1997)
26. Roman, R., Lopez, J., Mambo, M.: Mobile edge computing, Fog et al.: a survey and analysis of security threats and challenges. Future Gener. Comput. Syst. **698**, 78–680 (2018)

27. Shi, W., Cao, J., Zhang, Q., Li, Y., Xu, L.: Edge computing: vision and challenges. IEEE Internet of Things J. **3**, 1–1 (2016)
28. Tran, V.X.: WS-QoSOnto: a QoS ontology for web services. In: 2008 IEEE International Symposium on Service-Oriented System Engineering, pp. 233–238, December 2008
29. Wang, C., Li, Z.: Parametric analysis for adaptive computation offloading. In: Proceedings of the ACM SIGPLAN 2004 Conference on Programming Language Design and Implementation, pp. 119–130. ACM Press (2004)

IT Technologies and Automation as Factors in Increasing Labor Productivity at Industrial Enterprises

Dmitry Malyshev(✉)

Ural State University of Economics, Ekaterinburg, Russia
malyshevds@usue.ru

Abstract. The fact that increased business efficiency is currently important and intrinsically means that companies are forced to search out reserves related to labor productivity growth and are primarily focused on industrial automation, cost reduction, introduction of information systems intended to monitor changes in parameters and carry out their detailed analysis. These issues are particularly important for industrial enterprises that come very close to the depletion of traditional reserves related to labor productivity growth. As business volumes grow, the effect of such changes also tends to increase. Using the example of several large Russian companies, this article describes the potential of labor productivity growth by introducing IT technologies that, for example, enable enterprises to manage labor resources more effectively. This area of activity can be implemented within HR analytics. There is a hypothesis that the increased application of IT technologies to analyze data within the company and make management decisions will lead to including all necessary data not only about human resources, but also about people directly or indirectly associated with the company, in these processes. The fact that the recommendations in question are rather simple makes it possible to use their potential for other organizations as well, irrespective of their sectorial affiliation and business volume.

Keywords: Labor productivity · Information technologies · Automation · Business process optimization

1 Introduction

Today, many Russian companies are concerned about the issue of labor productivity growth. They have to face it not only because of the directions given by the Russian President and specified in his recent decrees, but also due to such factors as competition, sanctions and negative economic conditions. Despite their opportunity to invest resources in business development, all companies can always deal with process optimization, resource efficiency growth and cost reduction.

One of the most important areas of activity is to automate production and introduce IT systems for corporate management. And if automation affects industrial processes, then digitalization primarily covers activities performed by workers and gives an opportunity to take more effective and carefully weighed management decisions.

R. Silhavy et al. (Eds.): CoMeSySo 2019, AISC 1046, pp. 333–340, 2019.
https://doi.org/10.1007/978-3-030-30329-7_30

In light of this, we set a goal of assessing the potential of labor productivity growth within the company by automating production and introducing IT technologies, developing a system of measures in this respect and analyzing an economic effect of their implementation using the example of a large Russian industrial company.

2 Theoretical Aspects of Labor Productivity Growth and Its Potential

As mentioned above, Russian medium-sized and large companies are concerned about the issue of labor productivity growth since such directions were presented in the Presidential Decree "On the national goals and strategic objectives of the development of the Russian Federation for the period up to 2024" dated May 7, 2018. According to this document, an annual increase of not less than 5% should be ensured in labor productivity within middle-sized and large enterprises of basic non-resource-based sectors of economy by 2024.

According to the findings of McKinsey & Company, Russian companies are characterized by low labor productivity parameters when compared to other developed countries. This is in great part due to the following reasons:

- inefficient labor organization;
- non-transparent and excessive management;
- out-of-date capacities and production methods;
- rare application of a comprehensive approach to the planning of territorial development;
- lack of professional skills;
- underdeveloped financial system.

Thus, these factors, among other things, prevent companies from achieving the goal in view. What solutions can companies use to overcome these factors? In our opinion, it is necessary to search out solutions from among well-tried areas, without shifting responsibility on other entities. What is meant here is the use of scientific labor organization developments that are more and more associated with such tools as "Lean production", 6 Sigma, BPM brought into the Russian practice from abroad, and, however, are not limited only to them, especially if the Soviet scientific management developments are taken into account.

Our review shows that this area of focus is actively highlighted by scientific literature. In a point of fact, information on the number of scientific articles in journals included in the Russian Science Citation Index (RSCI) and devoted to scientific labor organization is presented in Table 1.

Table 1. Information on the number of publications in scientific journals included in the RSCI and devoted to the subject of scientific labor organization

Year	2014	2015	2016	2017	2018
Number of publications	281	459	536	572	383

*publications were selected according to the keyword "Scientific labor organization", "SLO" in their titles, keywords in abstracts from e-library.ru

As is seen from the table, there has been a significant increase in the scientific community's interest in this subject since 2014, with an at least 2 times growth in the last three years. Fewer publications for 2018 are due to a time lag in the RSCI indexation.

Every year at least several works considering various problems of this area are published in Russian top-rated journals. Scientists are interested in macro, meso and micro-assessments of this indicator in different branches of economy. Among them there are the articles by Rofe [12, pp. 3–13], Golovanov [4, pp. 89–96], Dolzhenko that use the example of the banking industry [5, p. 22], etc. Almost all of them have conclusions about low labor productivity levels, but this thesis is often left without comments.

One of the modern landmark works dedicated to this subject is the study by Gimpelson and Voskoboynikov [3]; they clearly show that informal employment has a considerable influence on labor productivity growth rates that are still low in the country. Their colleague from National Research University Higher School of Economics, Kapelyushnikov, analyzes labor productivity and shows that its growth at the beginning of the 2000s is in many respects due to the cheapening of workforce [8].

The adequate assessment of labor productivity rates within Russian enterprises remains by far the most acute problem, which is aptly noted by Kiselkina in her statement "the theory of labor productivity has many problems associated with measuring this indicator" [9, p. 43]. This problem is particularly acute at the micro-level, since various approaches to labor productivity assessment do not allow the comparison of different enterprises, even those occupied with similar activities.

This subject is more and more actively touched upon in foreign scientific journals included in Scopus and Web of Science. Since 2015, there has been an increase in such publications from 15 to 32 in 2017. However, our analysis of full-text versions of these works shows that the Russian practice acts in them as an object of study, but not a subject, in other words, there is an increase in publications where Russian productivity parameters are analyzed, but those which present new theoretical and methodological approaches to their assessment from the perspective of labor economics on behalf of the Russian science are still absent.

There are several works that compare labor productivity using the example of a number of industrial companies from Russia and China [1, 16], where our country's economy is compared to that of Eastern European countries [14]. There are other works that highlight problems associated with labor productivity [15].

Let us analyze the factors that somehow affect labor productivity at Russian industrial enterprises. Some problems that manifest themselves at the macro-level of economic relations are specifically attributable to Russia. The market is not fully based on the services that can be transferred to outsourcing companies. For example, regions have no specialized organizations with qualified human resources (competent to deal with equipment and machinery), which determines the necessity of repairing, maintaining and adjusting equipment by efforts of manufacturing plants.

One more difference concerns oxygen production with the use of their own capacities in the Russian industrial practice. Business owners do not take risks because increased security will be required due to the fact that oxygen is a dangerous substance as its production is not completely free from possible accidents and breakdowns. A key reason for high production maintenance costs consists in bad quality and low outsourcing volumes. Russian enterprises are notable for their own vehicles and related infrastructure (transport workshops and their workers).

The warehouse system specific to our enterprises is usually hypertrophied because of many factors, including but not limited to: specific aspects of accounting, managerial accounting and supervision in the Russian Federation because of which enterprises are forced to use a multilevel network of warehouses (central warehouses, workshop warehouses, etc.).

A key problem related to the Russian production practice is as follows: (a) low process automation level; (b) considerable investments required for automation projects.

As for negative cultural aspects and their consequences, it is possible to note negligence and bad faith demonstrated by Russian suppliers. To minimize related risks, companies have to carry out incoming inspection. In some cases equipment and spare parts suppliers are very far from core production processes, but their work is slow and services are expensive.

Where do things stand in terms of labor organization within enterprises? Due to the fact that robots, mechanisms and related maintenance services are rather high-priced, human labor is definitely cheaper and therefore employers are not interested in its replacement. Requirements established by laws and supervisory authorities in relation to labor protection and industrial safety leave a trace on the regulation of industrial processes and lead to the presence of corresponding jobs across the enterprise.

3 Research Methodology for Factors Affecting Labor Productivity and Its Results

The main research method is comparative data analysis as in the case of three industrial enterprises, as well as factor analysis of the influence of various significant factors on general labor productivity parameters within the enterprise. Managers working for the enterprises in question were individually interviewed, as well as business conditions were compared within the case-study method.

To analyze the factors that affect labor productivity in the Russian conditions, we will consider the three Ural enterprises located in the various regions of the Russian Federation. Due to the confidentiality of the data under study, we will restrict ourselves to the quantitative and qualitative characteristics of these organizations, will not identify them by names or owners, but will confirm that the case in hand is the companies located in the Ural region. Investments intended for their development are currently being brought from large Russian businesses.

The labor productivity parameters demonstrated by the enterprises are given in Table 2.

Table 2. Labor productivity and efficiency parameters demonstrated by the copper-producing companies [7]

Parameters	Enterprise 1		Enterprise 2		Enterprise 3	
	2016	2017	2016	2017	2016	2017
Receipts, mln. RUB	17,410	21,158	10,020	11,560	5,252	5,799
Copper, thous. t	135.5	136.2	77.1	77.6	37.5	40.3
Average staff number, person	2,936	2,930	3,315	3,372	1,631	1,611
Average pay, RUB	44,952	46,617	44,398	45,993	41,342	45,085
Payroll, mln. RUB	1,583.7	1,639.1	1,766.2	1,861.1	809.1	871.6
Labor productivity, thous. RUB/person	5,930	7,221	3,023	3,428	3,220	3,600
Labor productivity, t/person	46.2	46.5	23.3	23.0	23.0	25.0
Own-product real wage, RUB/t	11,688	12,034	22,907	23,983	21,577	21,627

Source: prepared by the author.

The data analysis allows us to conclude that the key condition making a difference in labor productivity levels consists in different approaches to industrial organization and maintenance. This is exactly why companies have different productivity parameters.

As is seen from the table, labor productivity and efficiency parameters taken from Russian companies manufacturing identical products can significantly differ from each other. This is due to different equipment units and modernization levels. For example, it is noticeable that Enterprise 1 is more productive and progressive. Besides, the three companies in question have absolutely different organizational structures. To exclude any price factor, it is better to calculate labor productivity in natural terms.

The table has no labor productivity data reflecting the pre-crisis period (2007), but let us note that in this very year it was significantly lower in natural terms (by 20–200%), therefore, the crisis actually forced these three companies to optimize their industrial processes.

In addition, the data show an important fact: labor productivity depends not only and not just on the number of employees. The introduction of extra modern capacities gives a dramatically better result.

The factor analysis of labor productivity growth as in the case of Enterprise 1 is as follows:

Total labor productivity growth: 46.5/21.9 = 2.1269

Production volume growth due to the introduction of new equipment units and modernization: 136.2/92.8 = 1.468, which corresponds to 69% of labor productivity growth.

Decrease in the number: 4,246 − 2,930 = 1,316 or 31%, which in direct proportion affects labor productivity growth.

At the same time, a decrease in the number amounts to 631 persons or 15% as a result of modernization and restructuring (there is no staff contribution to labor productivity).

The staff optimization that directly affects labor productivity amounts to 685 persons or 16%.

Thus, the new equipment, modernization and restructuring make up 69% + 15% = 84% of the total labor productivity growth.

This leads to the following important conclusion: direct staff optimization reserves without modernization and high-performance equipment are limited, they constitute no more than 20% according to our estimates.

4 Automation and Digital Technologies for Labor Productivity and Staff Performance Growth

Summing up our analysis, we can draw some conclusions.

What are the key opportunities for labor economy? Companies confront problems related to industrial automation and IT systems for corporate management. The parameters presented by Russian scientists show that there are still nearly 20% of workers performing manual operations. Our analysis confirms that there are such reserves when it comes to the industrial organizations under study (see Table 2).

We can affirm with certainty that this is the case in many Russian companies. It is the easiest thing to refer to special climatic conditions, high business overregulation and unrealized labor automation potential, but, as a matter of fact, the industrial management level of Russian companies is low and requires more auxiliary and administrative human resources.

In other words, the potential of industrial automation and IT systems has not yet been realized in the context of labor productivity parameters. The crisis of 2007–2008 and its consequences in the following years prevented Russian companies from implementing and realizing such solutions.

Let us note that activities related to core processes appear to be more effective in case of modernization and industrial automation. However, there are also some problems here. According to Russian managers, workers involved in operating new equipment often handle it in a negligent manner. Equipment failures are numerous, but further analysis is required to assess the influence of various factors on their occurrence: quality control for equipment, climatic conditions, workplace culture. Previous studies show that such processes as scientific labor organization and the effective use of informatization and automation facilities are poorly developed within enterprises. It should, however, be noted that this situation is also connected with a lack of developments in this area.

Based on the issues under discussion, Russian managers are recommended to succeed in the following endeavors:

(1) Reviewing business processes for their optimization. Some part of functions in plant management is performed according to certain traditions and rules because there is a great number of departments with employees having close family ties with managers. Management companies are recommended to invite experts and conduct an independent audit of processes within the company. Particular emphasis should be placed on the industrial automation of all supporting processes, optimization of processes that create value for the enterprise, including the compulsory assessment of such processes and their effect from an economic point of view.
(2) Giving new impetus to the implementation of outsourcing processes for all supporting subdivisions, first of all accounting departments, overhaul workshops, etc. Increasing the number of IT services that can provide cloud solutions for such spheres as accountancy, human resource management, legal services.
(3) Introducing the LEAN system (as well as its separate elements, KANBAN, 5S, etc.) at workplaces [2] and [13]. Developing the culture of labor-saving activity, stimulating careful treatment towards equipment and machinery. It is necessary to give new impetus to the development of innovation activity and review commendation rules and regulations for labor-saving solutions.
(4) Centralizing administrative functions at the level of management companies that carry out only separate supervisory functions at the moment.
(5) Developing labor protection systems within enterprises, for example, by means of red labels (registration of labor protection problems within certain sites and appointment of specialists responsible for their resolution). This area of activity could be attributed to labor unions, but with the use of relevant human resource management technologies, for example, crowdsourcing [6].
(6) Creating scientific management laboratories for enterprises that could act as a platform for testing progressive administrative and manufacturing methods to increase labor productivity within enterprises.
(7) Including the key indicator "Labor productivity gain within the area of responsibility" decomposed between the subordinate levels of management in the list of key performance indicators for guidance to workshops and structural subdivisions.

5 Conclusion

The results of our study allowed us to conclude that labor productivity as an indicator used to assess industrial efficiency can also be used to determine development trends for Russian businesses. Our analysis shows that some part of people working for Russian enterprises does not manufacture products, even at the high level of labor intensity. The result of their work is to ensure specific working conditions for other workers who are engaged in real production, for example, fuel and required materials production, heat supply, transport services, etc. Yes, it is possible to optimize and reduce the number of unproductive employees, but reserves of such labor productivity growth constitute no more than 20%, thus there is a need to search out reserves in some other areas.

References

1. Bhaumik, S.K., Estrin, S.: How transition paths differ: enterprise performance in Russia and China. J. Dev. Econ. **82**(2), 374–392 (2007)
2. Bukhalkov, M.I., Kuzmin, M.A.: Organizational and economic bases of lean production. Organ. Prod. **43**(4), 63–68 (2009)
3. Voskoboinikov, I.B., Gimpelson, V.Ye.: Labor productivity growth, structural changes and informal employment in the Russian economy, 47 p. Higher School of Economics, Moscow (2015)
4. Golovanov, A.I.: Labor productivity as a foundation for the growth of the Russian economy. Tomsk State Univ. Bull. Econ. **4**(16), 89–96 (2011)
5. Dolzhenko, R.A.: Methodical approaches to the assessment of staff productivity. Ration. Comp. Ind. **10**, 21–25 (2012)
6. Dolzhenko, R.A., Bikmetov, R.I.: Possibilities of using crowdsourcing by trade unions for the development of labor protection. News Univ. Ser. Econ. Finance Prod. Manag. **2**(36), 27–36 (2018)
7. Dolzhenko, R.A., Emelyanov, A.A., Dushin, A.V., Voronov, D.S.: Analysis of the expected resources of copper scrap. Stud. Russ. Econ. Dev. **29**(2), 174–181 (2018)
8. Kapelyushnikov, R.I.: Productivity and remuneration: a little simple arithmetic, 40 p. Higher School of Economics, Moscow (2014)
9. Kiselkina, O.V.: Problems of measuring labor productivity, VEPS, 4 (2015). https://cyberleninka.ru/article/n/problemy-izmereniya-proizvoditelnosti-truda. Accessed 23 May 2019
10. Kuchina, E.V., Tashchev, A.K.: Methodological approaches to the assessment of labor productivity at the micro level. Vestnik SUSU. Ser.: Econ. Manag. **2**, 42–47 (2017)
11. Kokovikhin, A.Yu., Dolzhenko, S.B., Dolzhenko, R.A., et al.: Management and staff economics: studies. In: Kokovikhin, A.Yu. (ed.) Ministry of Education and Science of the Russian Federation, the Urals. state econ un-t, Ekaterinburg, 340 p (2017)
12. Rofe, A.I.: Influence of forms of labor organization on its performance. Labor Soc. Relat. **2**, 3–13 (2011)
13. Yamilov, R.: Lean manufacturing: LEAN-processing in the Russian context. News High. Educ. Inst. Ser. Econ. Finance Prod. Manag **1**, 120–126 (2015)
14. Kharcheva, I., Kontsevaya, S., Tinyakova, V.: Analysis of the salary and labour productivity at the enterprises of the dairy industry of Russia and Eastern Europe. In: 25th International Scientific Conference on Agrarian Perspectives - Global and European Challenges for Food Production, Agribusiness and the Rural Economy: Czech University of Life Sciences, Faculty of Economics & Management, Prague (2016)
15. Lobova, S.V., Popkova, E.G., Bogoviz, A.V.: Labor productivity in Russia: reality and alert. In: 3rd International Conference on Advances in Education and Social Science (ADVED), Istanbul, Turkey, pp. 962–967 (2017)
16. Zhao, J., Tang, J.: Industrial structure change and economic growth: a China-Russia comparison. China Econ. Rev. **47**, 219–233 (2018)

The Similarity of European Cities Based on Image Analysis

Zdena Dobesova(✉)

Department of Geoinformatics, Faculty of Science, Palacky University, 17. listopadu 50, 779 00 Olomouc, Czech Republic
zdena.dobesova@upol.cz

Abstract. This article presents the finding of similar cities in Europe from data set Urban Atlas. Basic categories of landuse describe each city. One hundred cites were selected as a basic data set according to size. For finding the similarity, the trained neural network was used. A neural network is part of embedded add-ins Image Analytics in Orange software. One embedder in Orange was selected for the presented purpose. Finally, the hierarchical clustering was used for image descriptors received form neural networks. As a result, the couples of most similar cities is presented in the article. The cities are similar according to the patterns of urban fabrics or green areas patterns or shapes of some areas.

Keywords: Data mining · Neural network · Similarity · City structure · Geoinformatics · Urban Atlas · Orange

1 Introduction

The finding of similarity belongs to the data mining task. The history and evolution of cites are different, and they are under the influence of the political, industry and importance drivers. Moreover, the terrain arrangement, soil and surrounding influence the shape of the city. The existence of the river, mountains are determining on extension of the city during the years.

Some research tries to compare cities according to size and the structure determined for living, industry, commercial zones and leisure. Also, current urban comparisons often focus primarily on urban street networks. Work called A typology of street patterns [1] published in 2014 at the Institute of Theoretical Physics in France Compares 131 European and American cities based on street network metrics on which they apply a hierarchical clustering method. The resulting groups of cities subsequently characterizes and compares. They were faced with the problem of too high a similarity of street networks in most cities. To solve this problem, they focused on the city block geometry which street networks create. As a result of this work, cities were divided into four groups according to the shape and size of the city blocks.

Another similar work published in 2018 works with a street network for 27,000 US cities available through Open Street Map [2], relying on graph theory. Network metrics such as average edge length, crossover density, average node connectivity is calculated for multi-scale urban street networks [3]. The author compared individual territories by

R. Silhavy et al. (Eds.): CoMeSySo 2019, AISC 1046, pp. 341–348, 2019.
https://doi.org/10.1007/978-3-030-30329-7_31

these calculated metrics. None of the presented works did assume the landuse and function of urban parts in the cities or surroundings. Our research tried to consider city landuse.

The data set is described in Sect. 2. Section 3 is a description of used software Orange and steps of processing. The findings of similarities are described in Sect. 4.

2 European Urban Atlas as a Source Data

Copernicus Land Monitoring Service offers the dataset European Urban Atlas. Urban Atlas provides pan-European comparable landuse and land cover data for Functional Urban Areas (FUA). The European Urban Atlas was designed to compare landuse patterns amongst major European cities, and hence to benchmarking cities in Europe. It uses images from satellites to create reliable and comparable high-resolution vector maps of urban land in a cost-efficient manner [4]. There older data set is from 2006, and the newer dataset is from 2012. Some data are newer, e.g. from the 2016 year. Urban Atlas offers over 800 Functional Urban Area and their surroundings (more than 50.000 inhabitants) for the 2012 reference year. The data are freely downloadable in vector format together with metadata, map and color legend for categories of landuse (Fig. 1).

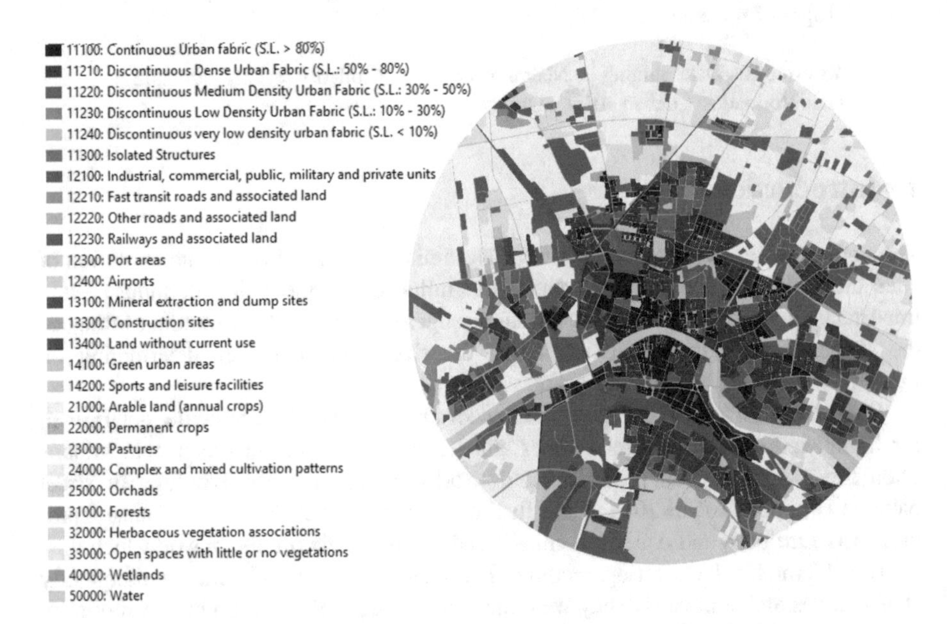

Fig. 1. Categories of landuse in Urban Atlas and city Pisa in Italy.

The land is divided to nearly 30 hierarchical classes: *Urban fabric with different density, Industrial and commercial areas, Transit and railway areas, Water, Green and sports and leisure facilities, and other green areas like forests* [5]. Each category has its color on the map. The tones of red colors are for urban fabric; blue tones are for water and wetlands; yellow tones are for arable land (pastures) and green tones for green urban areas (forests, vegetation, sports areas); grey tones are for transit roads and railways. Moreover, specific areas like construction sites, ports and airports are present (Fig. 1).

We selected 100 cities from the Urban Atlas. The selection was based on the number of the inhabitant in the interval between 50 to 200 thousand that has the largest representation in Urban Atlas. The number of inhabitants was taken from Eurostat statistics [6]. Our selection covers all countries in Europe equally. Because the source data contain not only cities but also the surroundings with other suburbs, cities and villages (whole FUA) we selected the central city manually by a circle that covers only the main city in FUA. The central point of the city was set primarily to the historical center or the buildings with a concentration of government, commercial activities or traffics [7]. The process of data preparation – the setting of a central point, definition of a circle for clipping of the source vector data were processed in GIS software Arc Map for Desktop v. 10.6 [8]. Finally, the export from GIS to the color images was made for all 100 cities. To automate the preprocessing of data, the Python script was utilized. All images were a source data set for similarity findings.

3 Orange and Neural Networks

The *Orange* is an open source machine learning software with good data visualization for novice and expert. Orange offers interactive data analysis workflows with a large toolbox [9]. The add-ons *Image Analytics* simplifies the loading of images and through deep network-based embedding enables their analysis. Embedding represents images with a feature vector, allowing the use of Orange's standard widgets for clustering, classification or any other kind of feature-based analysis [10].

The *Image Embedding* widget reads images and uploads them to a remote server. This widget offers seven embedders (neural networks), each trained for a specific task. For the map processing, the most suitable is *Painters embedder*, which is trained to predict painters from artwork images. The second suitable is VGG-19 embedder: 19-layer image recognition model trained on ImageNet. Deep learning models calculate a feature vector - image descriptors in the form of an enhanced data table with additional columns. We experimented with all embedders, and finally, the Painters embedder was chosen.

The workflow in Orange depicts the whole processing of data about cities (Fig. 2). After the calculation of image descriptors, the node *Distances* calculates distances between image descriptors by Cosine metric. The last node *Hierarchical Clustering* produces a dendrogram. For the linkage in hierarchical clustering, the Ward method was set. The levels in dendrogram reveal the group of most similar cities. The workflow also contains the *Data Table* with image descriptors and *Image Grid* to display all 100 cities in the grid to express proximity. The dendrogram was used as the result for the evaluation of similar cities. The lowest level displays the most similar cities in join branches (Appendix 1).

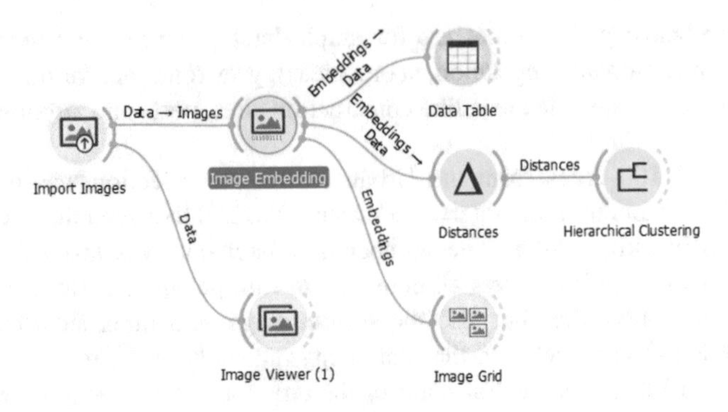

Fig. 2. The workflow in Orange.

4 Results

The hierarchical clustering identifies the four basic groups of similar cities. The first blue group (Appendix 1) are cities that only with a partial cover of circular extent. They are namely ports like Livorno in Italy and Cadiz and Almeria in Spain. All those cities are specific, and the similarity is not high.

Interesting couple of two similar is cities Bern in Switzerland and Maribor in Slovenia at the lowest level in the second red group in the dendrogram. The selected couple is in Fig. 3. These two cities are similar, namely by the curve shape of surrounding categories *arable lands*, *pastures* and *green areas*.

The second couple is from the third green group. They are Cambridge and Warwick from the United Kingdom. The rectangular shape of the urban fabric and landuse like a mosaic of small areas is typical for both cities. Also, the discontinuous urban fabric is dominant than continuous urban fabric in centers for cities in this group. Category *pastures* are dominant in the surroundings of both cities. The third couple in Fig. 3 are from the same green group. They are Le Mans from France and Enschede from the Netherlands. The typical is the equal spreading of small areas of category *Industrial, commercial, public, military and private units* (violet color) for these two cities. Also, the category C*ontinuous urban fabric* (with high density >80%) is dominant than D*iscontinuous medium density urban fabric*.

The last orange group is the biggest in the dendrogram (Appendix 1). For comparison, the couple of two similar Czech towns was selected. The first is České Budějovice and the second is Hradec Králové. For these two towns are typical the dispersion of *Continuous urban fabric with high density* into numerous part of the city that is touched by *Industrial, commercial, public, military and private units*. The category of the *Continuous urban fabric* is typical for the old historical part that is dominant in city centres (like in Bern). In these towns, the continuous fabric is distributed to small kernels over the whole area of the city.

The presented comparison of couples of towns shows that some towns are similar according to the structure of urban fabric or according to the types of surroundings. Also, the shape of areas is influential to the findings of similarity. Some structure of cities is typical for some countries (Cambridge and Warwick from the United

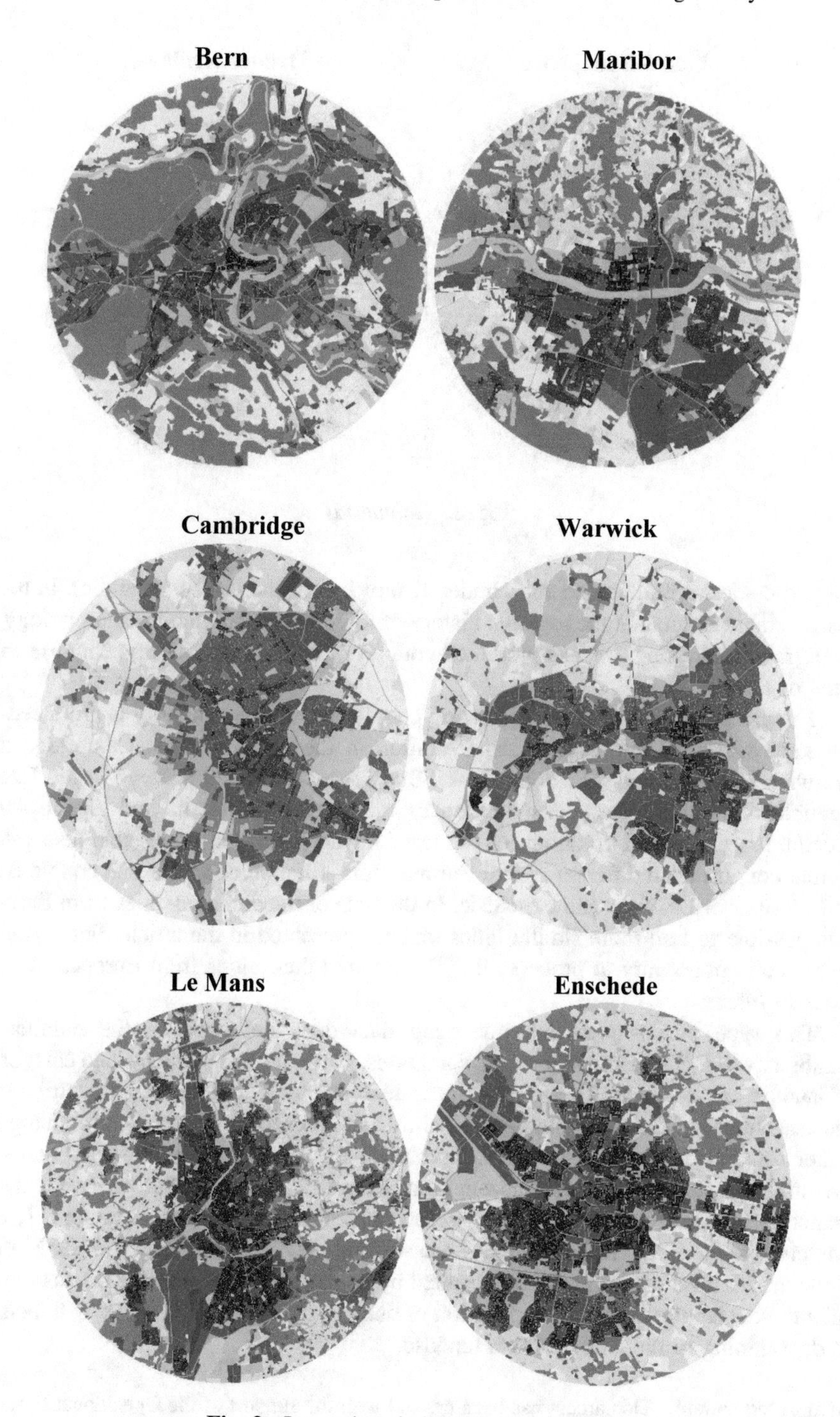

Fig. 3. Interesting similar couples of cities.

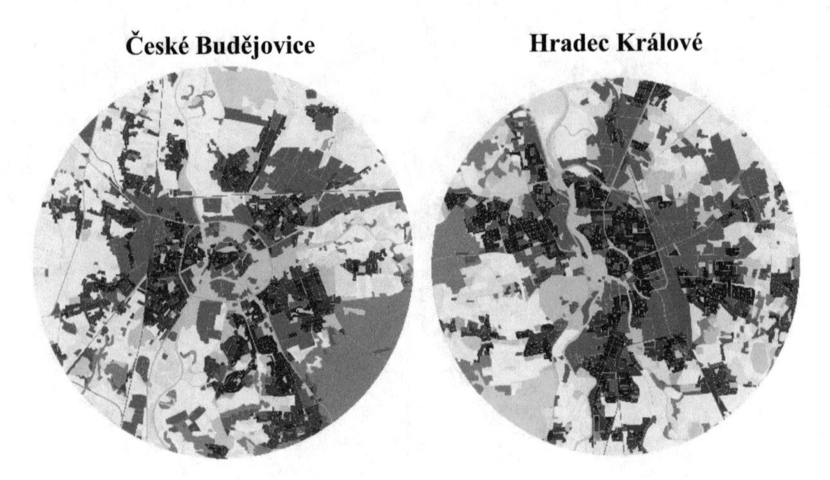

Fig. 3. (*continued*)

Kingdom, České Budějovice and Hradec Králové from the Czech Republic). In those cases is more influential the regional history of town evolution than the morphology of the terrain. Interesting are also presented couples of cities from different countries and parts of Europe.

The presented results show interesting couples where the similarity is produced by the same historical and economical evolution in the same country: Cambridge and Warwick in the United Kingdom, České Budějovice a Hradec Králové in the Czech Republic. In those cases, the evolution of urban grew is not limited like in some specific morphological condition of the terrain. Surprisingly it is possible also found similar couples from different part of Europe, from different countries like couple Bern and Maribor or Le Mans and Enschede. In the case of one hundred cities from Europe, it is possible to find more similar cities that are presented in the article. Furthermore, there is an opportunity to process all 800 FUA and their cities from European Urban Atlas in future.

This type of unsupervised processing data does not consider the number of inhabitants or the area size of landuse categories. Only the arrangements and categories of landuse are assumed. The using of deep learning by neural networks contribute to the assessment of the urban structure of cities. The presented results are promising for father research in the area of urban structures. Beside them, the presented steps of investigation can be used as a lecture for students at Palacky University for study branch Geoinformatics like other practical examples presented in the literature [11, 12]. Students are familiar with Urban Atlas dataset, and they attend the course Data Mining at the master level of study which is granted by the author of the article. This case study suits to be a lecture about the application of deep learning neural network as a method of data mining in the area of urban landuse.

Acknowledgement. This article has been created with the support of the Operational Program Education for Competitiveness – European Social Fund, project CZ.1.07/2.3.00/20.0170 Ministry of Education, Youth, and Sports of the Czech Republic.

Appendix 1

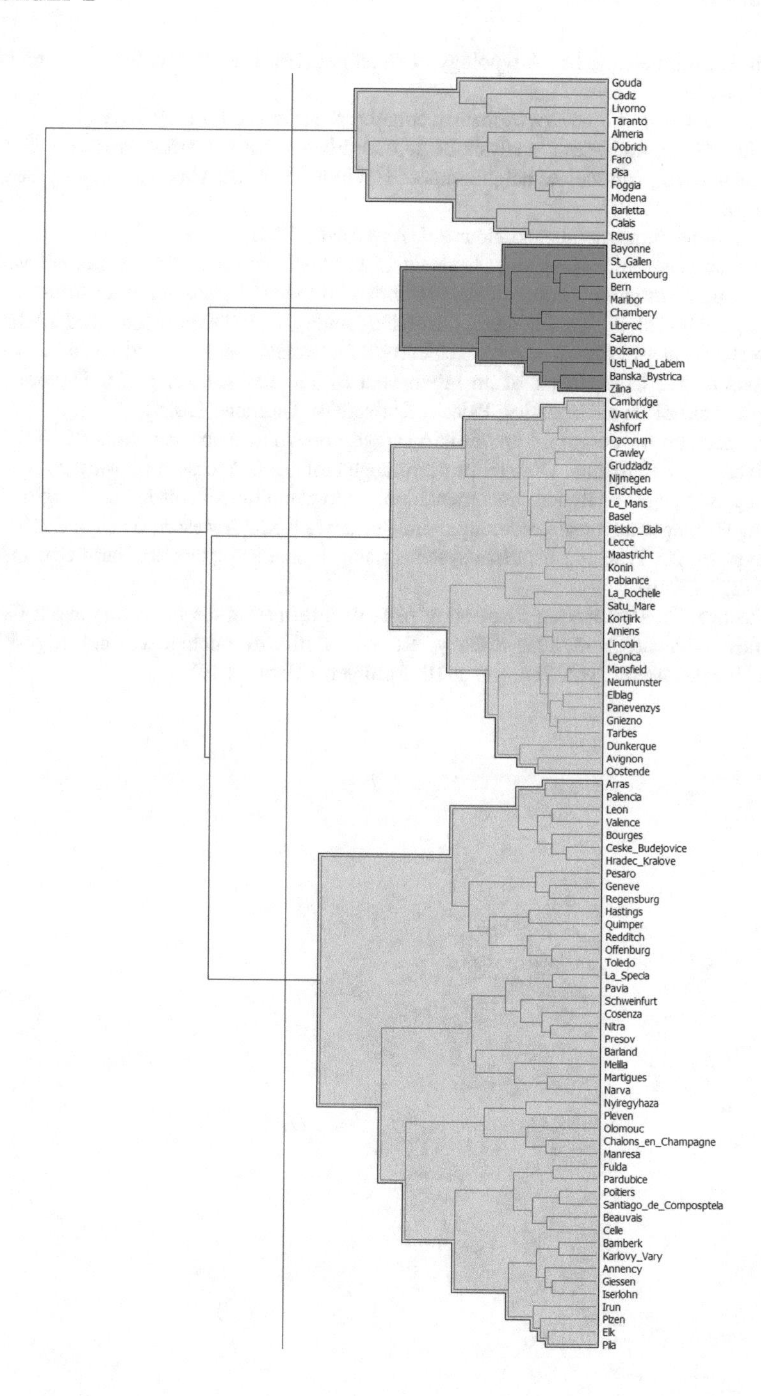

Gouda
Cadiz
Livorno
Taranto
Almeria
Dobrich
Faro
Pisa
Foggia
Modena
Barletta
Calais
Reus
Bayonne
St_Gallen
Luxembourg
Bern
Maribor
Chambery
Liberec
Salerno
Bolzano
Usti_Nad_Labem
Banska_Bystrica
Zilina
Cambridge
Warwick
Ashforf
Dacorum
Crawley
Grudziadz
Nijmegen
Enschede
Le_Mans
Basel
Bielsko_Biala
Lecce
Maastricht
Konin
Pabianice
La_Rochelle
Satu_Mare
Kortjirk
Amiens
Lincoln
Legnica
Mansfield
Neumunster
Elblag
Panevenzys
Gniezno
Tarbes
Dunkerque
Avignon
Oostende
Arras
Palencia
Leon
Valence
Bourges
Ceske_Budejovice
Hradec_Kralove
Pesaro
Geneve
Regensburg
Hastings
Quimper
Redditch
Offenburg
Toledo
La_Specia
Pavia
Schweinfurt
Cosenza
Nitra
Presov
Barland
Melilla
Martigues
Narva
Nyiregyhaza
Pleven
Olomouc
Chalons_en_Champagne
Manresa
Fulda
Pardubice
Poitiers
Santiago_de_Composptela
Beauvais
Celle
Bamberk
Karlovy_Vary
Annency
Giessen
Iserlohn
Irun
Plzen
Elk
Pila

References

1. Louf, R., Barthelemy, M.: A typology of street patterns. J. Roy. Soc. Interface. arXiv:1410.2094v1 (2014)
2. Open street map. https://www.openstreetmap.org. Accessed 03 May 2019
3. Boeing, G.: A multi-scale analysis of 27,000 urban street networks: every US city, town, urbanized area, and Zillow neighborhood. Environ. Plann. B: Urban Anal. City Sci. arXiv:1705.02198 (2018)
4. Urban Atlas. https://land.copernicus.eu/. Accessed 05 May 2019
5. European Environment Agency: Updated CLC illustrated nomenclature guidelines. https://land.copernicus.eu/user-corner/technical-library/corine-land-cover-nomenclature-guidelines/docs/pdf/CLC2018_Nomenclature_illustrated_guide_20170930.pdf. Accessed 13 Jan 2019
6. Eurostat database. https://ec.europa.eu/eurostat/data/database. Accessed 08 Jan 2019
7. Janousek, M.: Comparison of an urban area by circular sectors, p. 75. Diploma theses, Department of Geoinformatics, Palacky University, Olomouc (2019)
8. Esri: ArcMap for desktop. http://desktop.arcgis.com/en/arcmap/. Accessed 05 May 2019
9. University of Ljubljana: Orange. https://orange.biolab.si/. Accessed 05 May 2019
10. Orange3 image analytics documentation. https://buildmedia.readthedocs.org/media/pdf/orange3-imageanalytics/latest/orange3-imageanalytics.pdf. Accessed 05 May 2019
11. Dobesova, Z.: Teaching database systems using a practical example. Earth Sci. Inform. **9**, 215–224 (2016)
12. Dobesova, Z.: Discovering association rules of information dissemination about Geoinformatics university study. In: Silhavy, R. (ed.) Artificial Intelligence and Algorithms in Intelligent Systems, vol. 764, pp. 1–10. Springer, Cham (2019)

The Compare of Solo Programming Strategies in a Scrum Team: A Multi-agent Simulation Tool for Scrum Team Dynamics

Zhe Wang(✉)

Lincoln University, Lincoln, New Zealand
Zhe.wang@lincolnuni.ac.nz

Abstract. Scrum is an agile framework within which people can address complex problems, while productively and creatively delivering products of the highest possible value. A strategy is a way that people going to solve problems based on the existing situation. Team strategy research is different from team composition and how it affects the performance. Team composition can affect the performance is an existing knowledge, without doing simulation, we still can know how personality and capability can affect its performance. But strategy and task allocations methods are a further research go beyond that, particular in an environment, such as scrum, that has an aim for why the team needs to be composed. With the same team, the same task but different strategy can cause significant various outcome for each sprint. This is the way that agent-based modelling is more useful than just say that team composition can affect its work. And based on the current information, to do investigation on how team composition can affect performance will needs various teams, but how strategies can affect team performance will only need the same team to be compared. Scrum is the major motivation that why team needs a strategy to work, the purpose of this strategy is not to investigate how team composition can affect the work but more about how to use the existing affect from the team composition to get the Scrum success rate enhanced, as in real world, we could not change the team as the company only has the team to do the job, but we can change the way that they are doing the job.

Keywords: Scrum team dynamics · Multi-agent based simulation · Pair programming · Solo programming · Team strategy · Task allocations

1 Introduction

An agent is a proactive, reactive and intelligent piece of software working on task and pursues its goals based on its current beliefs, desires and intentions [9]. Agents can work together as a multi-agent system to enable them to collaborate and cooperate with each other. These features are very useful for modelling human activities as each agent can be endowed with different properties and behaviors. A Scrum development team is a social group that aims to deliver valuable software using the Scrum approach. A Scrum team is empowered to organize and manage its own work and its goal is to work collaboratively to deliver value at the end of a sprint. Each team member has

R. Silhavy et al. (Eds.): CoMeSySo 2019, AISC 1046, pp. 349–375, 2019.
https://doi.org/10.1007/978-3-030-30329-7_32

personal attributes, such as capability, character and experience that will affect the team performance. Multi-agent modelling can be used to capture the team dynamics in a Scrum environment. This model can be used to analyse how team dynamics affect the delivery of software. Agile is a concept of creating software through iterative and incremental process. Beck et al. [10] describes agile as "Individuals and interactions over processes and tools; Working software over comprehensive documentation; Customer collaboration over contract negotiation; Responding to change over following a plan [10]".

[11] indicated that analyzed the best advantage of agent based modelling is to simulation the objects in a micro-view rather than an macro-view. The micro-view means the agent can focus on individual behavior that effect on the overall system behavior. Bottom-up based agent design is the only way to simulation agent in micro-view that can observe individual agent behavior and its affection on the whole team. The current scrum team dynamics simulation is designed in this way, which is bottom-up based simulation approach that applied. However, no one has done this before, the top-down based approach in agent-based modelling is the main stream, for its convenient, more effective but only can provide macro-view on the system behavior, as only one agent is necessary. The bottom-up based agent system will cost more design and implementation effort as we are modelling the system in a more detailed and distributed way that include multi-agents, which is further complex than top-down based design.

The bottom-up based design will be the true multi-agent system simulation. And why it has affections and relation with the team strategy design for Scrum? Because it can considerate individual behavior on the team performance, as Scrum team is cross-function team, each member will focus on its own task and make contributions. It is not the team that working on each particular task, it is each team member will take tasks and working. Each agent will need to collaborate with other agents and tasks. Those features can only be captured through real multi-agent design. otherwise there will be actually no team dynamics been simulated, in top-down based approach, system agent working for everything, there will be actually only one team member in all.

Mostly importantly, why micro based simulation is important is because, the agent interactions with each other and tasks will cause various affections on the working strategy that randomly happened, it can provide more space for agent's own decision making and influence from other agents and tasks. As in Scrum, the whole team is working together to achieve the same goal. That goal can make affection on each agent and will be part of the agent perception to make decisions. This can only be simulated through real multi-agent system. As in Scrum, the scrum goal is actually the most important thing for the Scrum team, there is nothing more important than this. Paired team can only be research through multi-agent design, to see the affections of two agents.

Secondly, the agent design in micro view will considerate each agents' own knowledge structure and strategy structure. As the strategy design is based on the view of each agent, so that every agent will have own reasoning structure. why I should take this task? Why I should pair or assistant this agent? does it fit with my goal? Does my goal fit with the Scrum team goal? Those questions will push the agent to do decision based on its own strategy. In this way ontology is also designed in our simulation system to make each agent can have knowledge reasoning.

Here team strategy is actually task allocations methods. In top-down based research, you can only apply task allocations methods to the system agent. However, in bottom-up based approach, the task allocation can be applied to each agent, and there is no necessary that each agent should or must apply the same strategy, this is not real against the real world. Different type of agent should have their own working strategy that applied, this is real world. Some agent would like to try hard task, even its capability is little lower. Some agent would be taking as easy task as possible, even it can do further hard task. Various agent to take various task allocations methods can only be designed based on bottom-up based real multi-agent system. The real multi-agent system can simulate Scrum in real world, this contribution cannot be ignored. It will directly affect the team strategy and task allocations that designed. And it's even will be the most critical component that take the team strategy and task allocations evaluation into reality. This has not been done by any related work in such detailed level that capture almost every aspect of Scrum and can support very complex team strategy and task allocation evaluation compare with the real world.

The remainder of the paper is structured as follows. Section 2 discusses related work. The multi-agent design and the roles of each agents and the processes are described in Sect. 3. The experimental setup, the experiments and the analysis are presented in Sect. 4 and finally Sect. 5 concludes.

2 Related Work

A system dynamic model was implemented to investigate the long-term impact of pair programming on the software evolution aspects [2]. It advocated that pair programming will increase of the cost of the project in the earlier stage but improves the quality of the delivered software compared to solo programming on a long-term basis. This kind of improvement is based on the skills improvement of the pair during the development process which can enhance the quality of the software delivered, which in turn reduces the cost of testing and number of corrections to the product. Noori and Kazemifard developed an agent-based model on pair programming in agile projects [1]. They argued that the different personalities and characters will affect the work of pairs. They modelled the developer through agents and found that some good combinations of specific personalities were well suited for pair programming. The results showed that personality plays an important role in the formation and utility of a pair. For example, when the expertise of both individuals is high, the best pairing is introvert-extrovert. When both individuals are extrovert, the best pairing is low-high or medium-high expertise. This work focused on people selection in scrum teams. Several studies were also conducted analyses the impact of pair programming such as [3, 4] and [5]. These studies found that expert-expert pair and novice-novice pair are not better than expert-novice pair in pair programming as knowledge transfer takes place between various levels. The benefit of pair programming can enhance the quality of product in the perspective of paired team as it can provide higher quality products than solo programming. However, [3] found that there is greater difference between novice-novice pair and solo novice and less difference between expert-expert pair and solo expert. This paper discusses the implementation of paired programming in a Scrum team and describes how paired programming can impact the performance of the Scrum team. Pair programming is used to increase morale, increase teamwork, increase knowledge transfer, enhance learning and knowledge creation [7]. The knowledge transfer in pair programming team should happen between developers with different levels of knowledge [8]. However, Swamidurai and Umphress [6], argued that the static pair programming is not as good as dynamic pair programming for education because static pairings will restrict knowledge sharing to a relatively small scale, while the dynamic pair enlarges the knowledge share scale, which is positive for course education in the university classroom. This paper discusses various pairing strategies and describes how these pairings affect the completion of the software development process.

3 The Multi Agent System Design for Scrum Team Dynamics Simulation

An overview of the Scrum process is shown in Fig. 1. Scrum use a product backlog and sprint backlog to manage the requirements of a project. By reviewing the backlog in each sprint with the product owner, it can be revised to meet changes. A Scrum is initiated by the creation of the product backlog by the product owner. The user stories are prioritised in the product backlog based on their business value. The Scrum team will then have planning meeting to estimate how many sprints are needed to complete these user stories. During the sprint process, a daily meeting will be held to discuss problems, progress, achievements and difficulties of the current sprint. This daily stand-up meeting helps team members to achieve the goals of the sprint. Tasks will be verified once done and user stories will be verified by the product when completed. At the end of each sprint, a sprint review meeting is held to discuss the performance of the team, and what has been delivered.

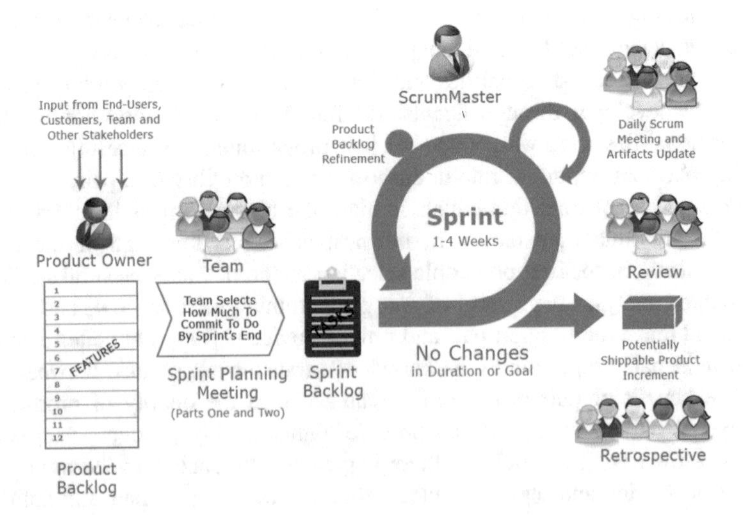

Fig. 1. The scrum process [12]

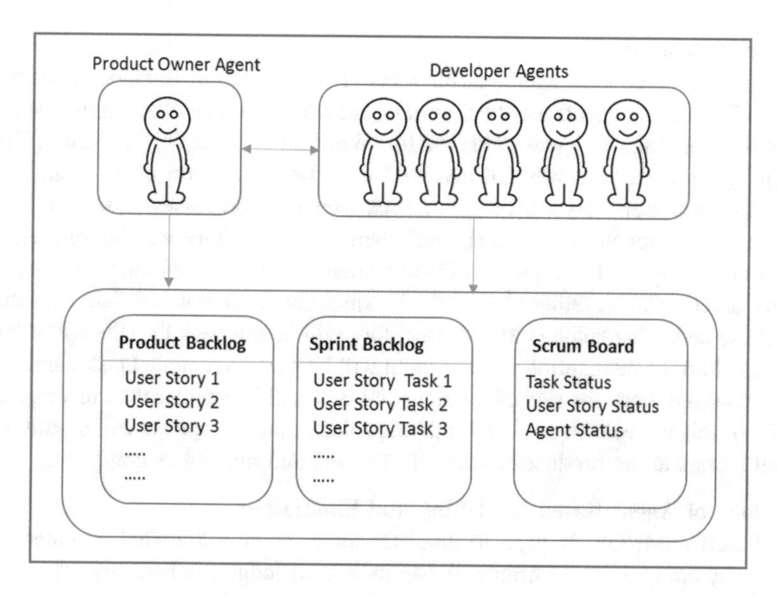

Fig. 2. Scrum simulation MAS architecture

From the current work of [13–15] the simulation system can be described as those components.

The Developer Agent

The developer agent represents a software developer in the Scrum team. Each developer agent has capability value between 1 and 10 which represents its technical capability level. A value of 1 indicates a beginner and 10 indicates an expert. This capability value is used to determine how much time is required by the individual agent to complete a given task. Each agent has a status to indicate their current activities (working, verifying, idle). Working indicates that the agent is currently working on task, verification indicates that the agent is verifying a task that has been completed by another agent and idle indicates that the agent is free and available to take on work.

User Stories and Tasks

Each user story is usually composed of 1–5 tasks. Each task has a size which indicates how much effort is needed to complete the task (the more complex the task is, the larger its size). In this model, it is assumed that a single task is worked on by a single developer agent. A task has a state with 4 possible values, to do (task has not been worked yet), in progress (task currently being worked), to be verified (task completed but has not been verified) and done (task is completed and verified). A task that has been completed must not be verified by the same agent who worked on it.

The Scrum Board

The Scrum board shows the progress of the sprint for each time tick by showing the status of the agents, the tasks and the user stories. In this simulation, it is assumed that a sprint runs for 10 time-ticks to represent a week with five working days. Thus, one day is represented using 2-time ticks.

The Sprint Process

When a sprint starts, each agent selects a task from the Sprint Backlog to work on as shown in Fig. 2. The agents use the Scrum Board to update each other on their status and the status of the task. An agent status can be "Working", "Verifying" or "Idle". Task are verified by an agent who was not involved with the completion of the task. The task verification and user story verification are two separate processes. When all the tasks belonging to a User Story are completed, then that User Story will be verified by the Product Owner agent. If the Product Owner agent is satisfied with the user story, it will be mark as completed. Otherwise, if the Product Owner is not satisfied with the user story, those tasks that belongs to that user story will be put back into the sprint backlog and worked on the next sprint, each of them will be given a remedial task name such as Review task(number), the size of the remedial task will be smaller than its original task size. To simulate the software development project, multiple sprints will be run until all the user stories in the product backlog are verified and marked as complete.

Advantage of Agent Based Modelling and Simulation

Agent based modelling is used to analysis some existing knowledge under various situation, by apply what-if formula. A common knowledge can be analyzed by various what-if to see its valid range and affections. Any strategy is used to tackle a type of problem; however, it is still very good to see the range of the effectiveness of that strategy in a particular range as the evaluation and verification contribution for that strategy in the scientific computing way which is agent-based modelling. This kind of contribution is specifically coming from agent-based modelling methodology.

As we considerate task rang and agent range. Scrum is a complex process, how complex it is will be based on how complex the real world is, that means the requirement from the user story. China has the largest market for software engineering, and needs significant requirement on software, as the population in China is significant big, so that the user stories for software system will be very complex, the Scrum team strategy that applied in different ways will affect the progress of the team.

In our simulation system, the team working strategy is designed based on individual agent's decision making. Based on agent can do or cannot do. Those individual agents finally form the whole team strategy. This is actually a bottom-up based strategy, as each agent make decision and then affect the whole team work. A bottom-up based strategy design will maximum the benefit of agent-based modelling, compare with top-down formed strategy, which system agent (center agent) decide most of the things for the team. This is very different not only in system design, but also in strategy design itself.

The bottom-up based agent strategy design can capture many instantly happened reactions, such as agent is ill, agent is leaving, agent find a new job, etc. Scrum is a complex process, anything can happen during Scrum, if the strategy design is not bottom-up based then it cannot capture those randomly instantly happened events, which will be not real compare with the real world. And also, can reduce the strategy reliability as less factors are considerate by top-down based strategy. In top-down based strategy, there will be no interactions between agents.

However, the center agent still can play critical role in Scrum agent-based modelling, because we still need center agent when conflict between team agents happens.

The system agent can also play as the scrum master and product owner to judge the workload of the team and quality during sprint review.

The sprint retrospective can only be simulated based on bottom-up based agent design, as each agent needs to do evaluation on itself and pair programming can also be better performed based on bottom-up based agent design, it needs help from system agent as well, for example both agents' wants to pair with a very good agent. Then system agent needs to make decision or make it conflict.

The bottom-up based agent design contribution many design features that Scrum modelling and simulation particular needed and this has not been achieved by any agent based modelling simulation approach in such detailed level.

The System Agent Design
(See Fig. 3).

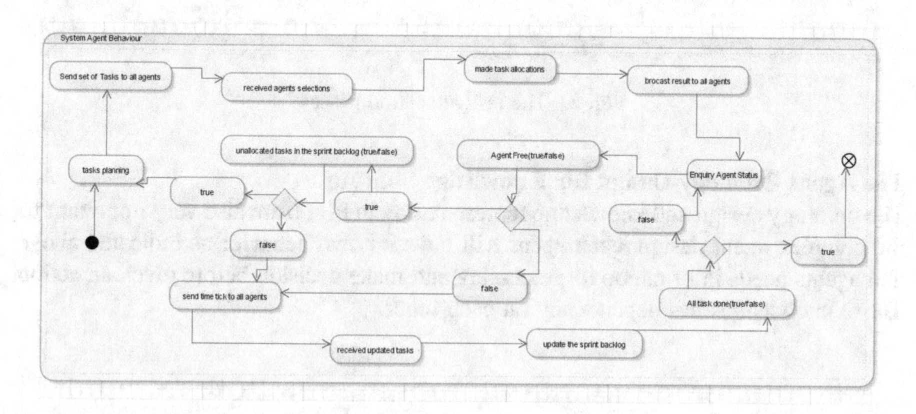

Fig. 3. The system agent design

The Developer Agent Design
(See Fig. 4).

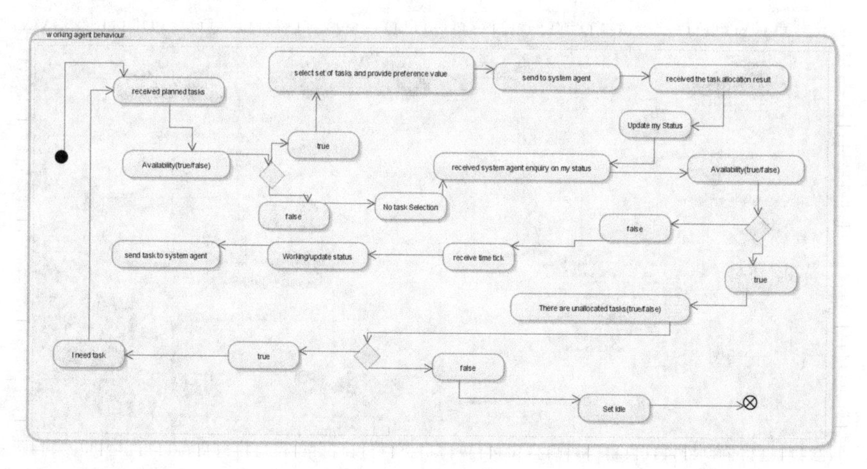

Fig. 4. The developer agent design

The Task Allocation Process

(See Fig. 5).

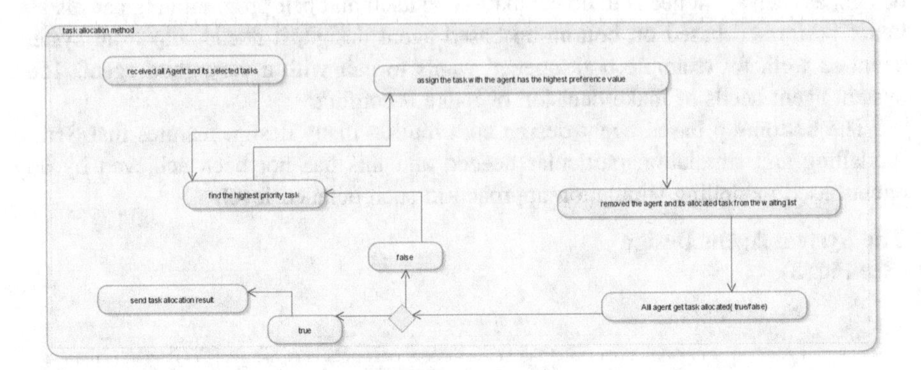

Fig. 5. The task allocation process

The Agent Ontology Design for Knowledge Sharing

The ontology design and knowledge representation in Fig. 6 are also very important for the complex agent design, each agent will have its own behavior as indicated above, and agents needs information to processing and make decision before given an action. Those knowledges has dependency on each other.

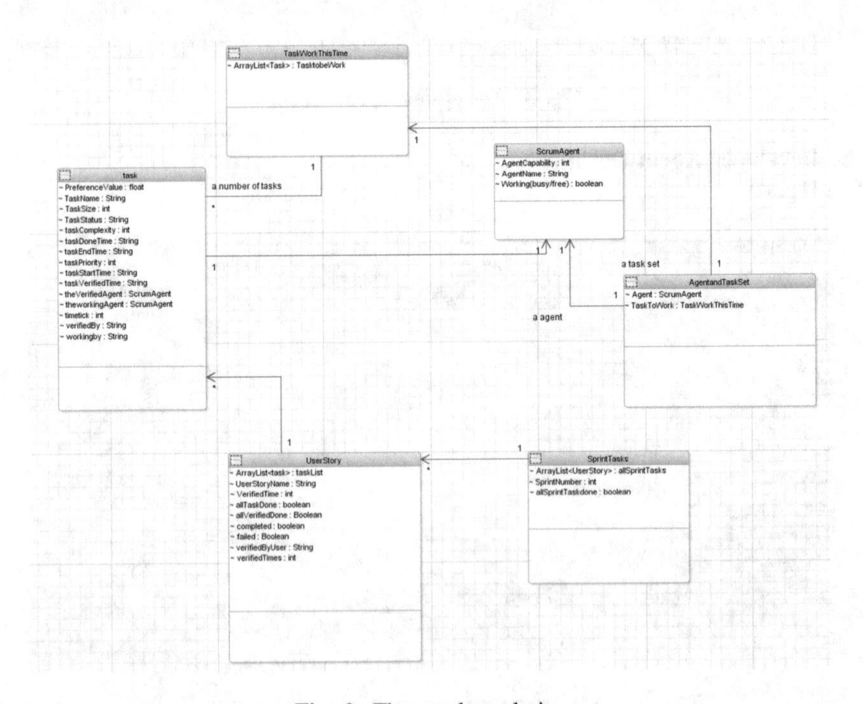

Fig. 6. The ontology design

The Designed Developer Agent Interface

(See Fig. 7).

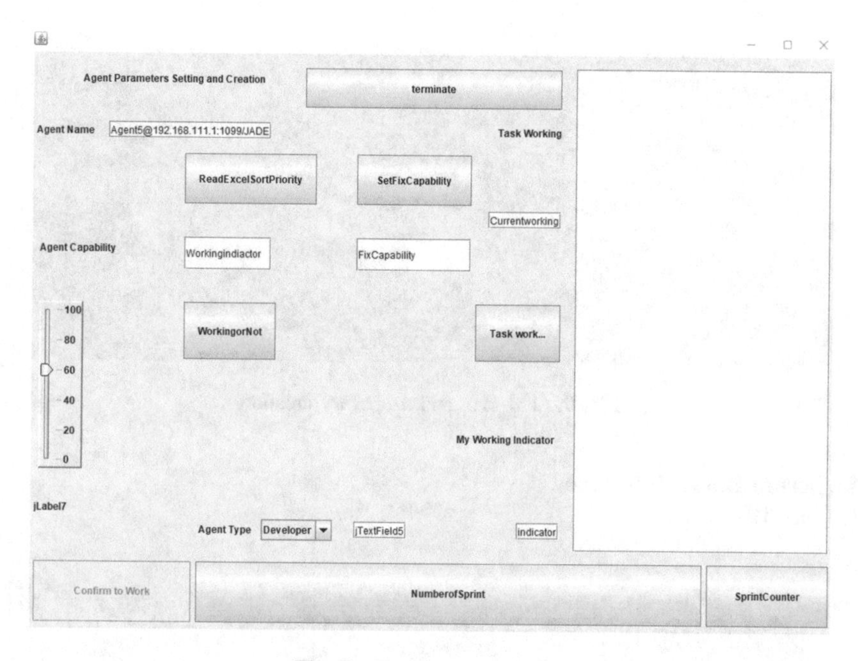

Fig. 7. The developer agent

The Simulation System Agent Interface

(See Fig. 8).

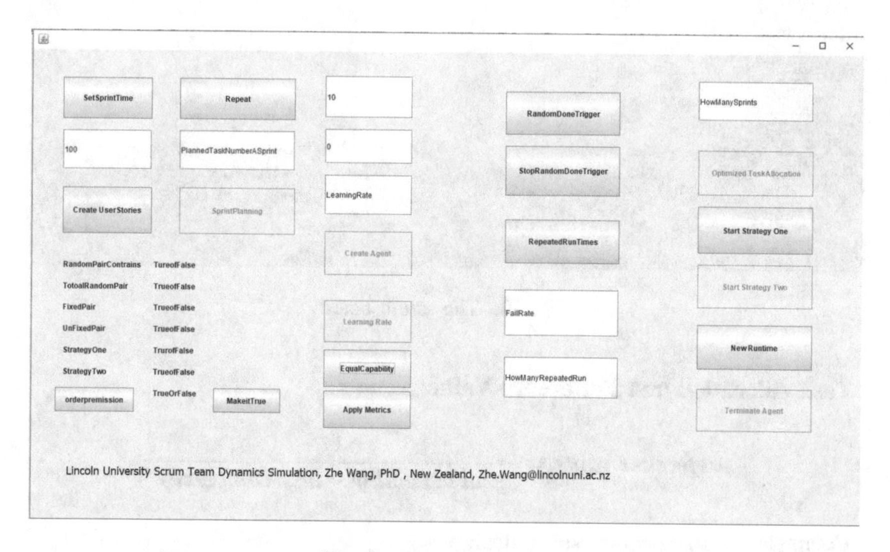

Fig. 8. The system agent

The User Story and Task Creation Interface

(See Fig. 9).

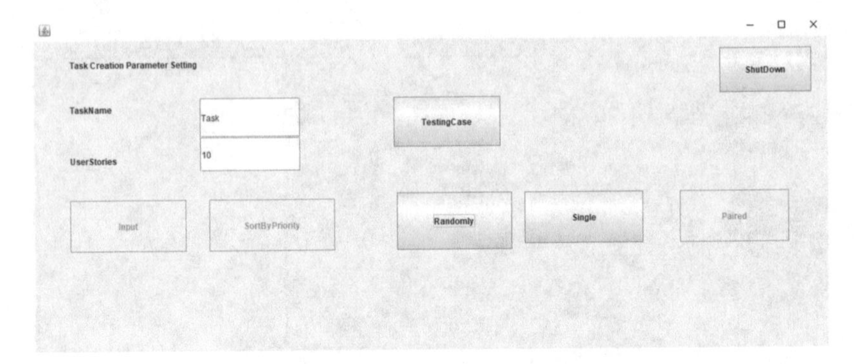

Fig. 9. The task and user story creation

The Scrum Board Interface

(See Fig. 10).

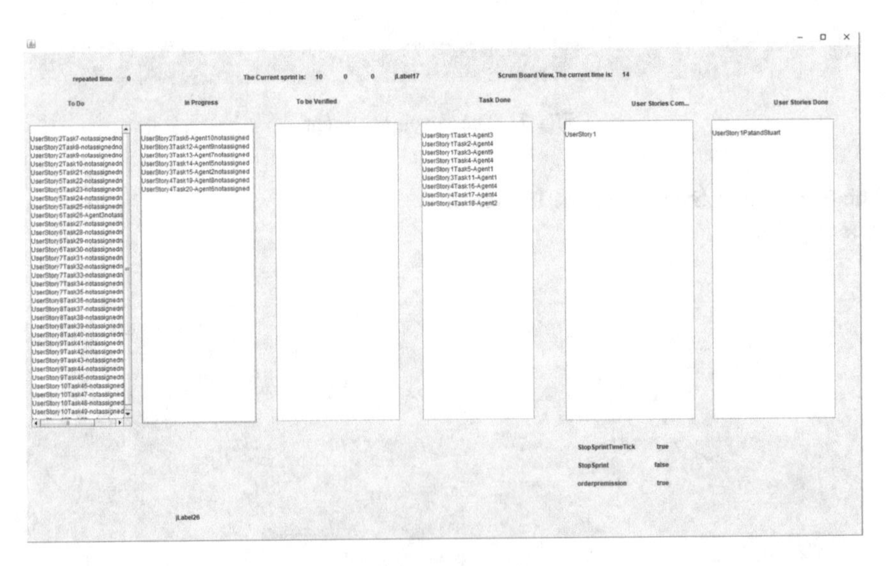

Fig. 10. The scrum board

Task Allocation and Preference Value

$$preference\ value = \frac{1}{agent\ capability - task\ complexity}$$

Formula 1 - preference value design

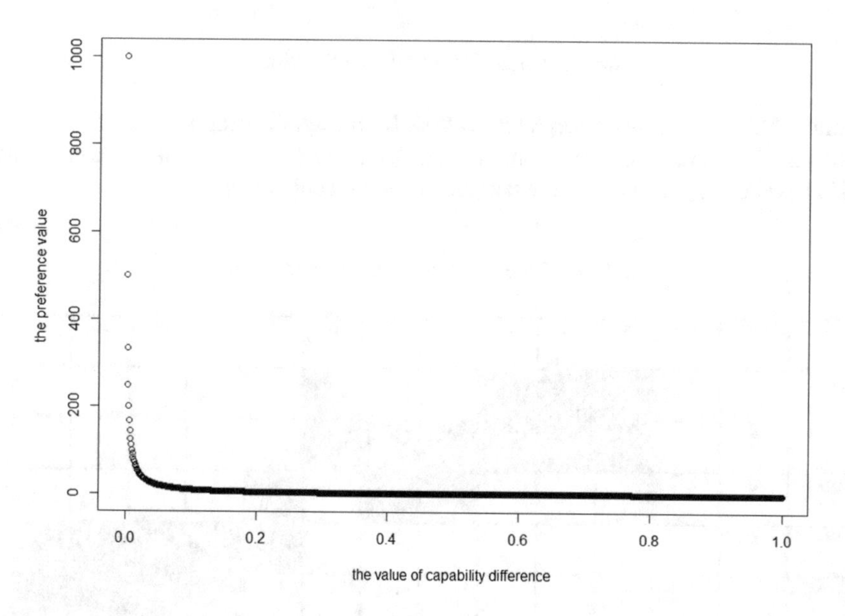

Agent6	Agent7	Agent8	Agent9	Agent10	task	Agent1	Agent2	Agent3	Agent4	Agent5
-5	-4	-3	-2	-1	0	1	2	3	4	5

Fig. 11. The preference value design

The above diagram Fig. 11 shows the task is always around by agent who can do it, which located in the right side, and agent who can not do it, which located in the left side. However, the level of can do or can not do is very different for each agent. We can measure the ranking of very strong such as agent5 and very low capable of working, such as agent6 in the above diagram. In the very benefit of not waster team resource, it is the best to allocate the task to agent 1 to work on the task based on the above diagram, this is the theoretical model of why introduce the preference value calculation for task allocation. The more closed to the task with positive value, the more chance to get the task allocated than other agent who also want the task.

The task size will indicate how many time ticks needed to complete the task based on agent who is in the right side of the task, if the agent is in the left side of the task, it will cost the agent, more time to complete and the algorithm is:

When $\textbf{\textit{agent capality}} < \textbf{\textit{task complexity}}$

$$\textbf{\textit{time needed for a task}} = \frac{\textbf{\textit{task complexity}}}{\textbf{\textit{agent capability}}} \times \textbf{\textit{task size}}$$

When $\textbf{\textit{agent capality}} \geq \textbf{\textit{task complexity}}$

$$\textit{time needed for a task} = \textit{task size}$$

Formula 2 time need for completing a task by an agent design
Based on the above task allocation, we will have the following task allocation and working process by agents as an example shows in Tables 1 and 2.

Table 1. Time tick table and task allocation

	SIZE	time	1	2	3	4	5	6	7	8	9	10	11
Task1	3		A2	A2	A2			A1					
Task2	5		A1	A1	A1	A1	A1	A2					
Task3	2					A2	A2		A1				
Task4	4								A2	A2	A2	A2	A1
Task5	3									A1	A1	A1	A2

Table 2. Task and agent information

	size	Preference Value	Task complexity	Agent Capability	Agent Name
TASK1	3	0.5	4	6	Agent2
TASK2	5	0.3333	4	7	Agent1
TASK3	2	0.25	2	6	Agent2
TASK4	4	0.5	4	6	Agent2
TASK5	3	0.2	2	7	Agent1

There are three types of team composed and task distributed against the team level as shown in Table 3. Which the first type indicate that all agents are able to work on all the task. The second type indicate that some agents are able to work on the tasks and some are not. And the third type indicate that all the agents are low than the task required skills level. Those types are definitely required various task allocation methods to carry on the sprint delivery to achieve the sprint goal for the scrum team, they would need to figure out how to optimize the team task allocation so as to maximums the task delivery in quality on time and budget.

Table 3. The three types of task and team relations

Complex value/ capability value	1	2	3	4	5	6	7	8	9
Type1	Task1	Task2	Task3	Task4	Task5	Task6	Agent1	Agent2	Agent3
Type2	Task1	Agent1	Task2	Task3	Agent2	Task4	Task5	Task6	Agent3
Type3	Agent1	Agent2	Agent3	Task1	Task2	Task3	Task4	Task5	Task6

- **all agent is able to working on all the tasks**

In this case all agent are able to working on all the tasks as shows in Table 5, and then the preference value is the critical information on task allocation based on the task priority, those lower capability agent will have higher preference value which will get task allocated first and then the higher capability agent to take the rest of tasks based on task priority as well. The working process shows in Table 4.

Table 4. Time tick table and task allocation for the first 10 tasks

TASK	size	Time1	Time2	Time3	Time4	Time5	Time6
Task1	1	A9	A8				
Task2	2	A10	A10	A8			
Task3	4	A4	A4	A4	A4	A7	
Task4	1	A8	A9				
Task5	5	A1	A1	A1	A1	A1	A6
Task6	3	A6	A6	A6	A2		
Task7	5	A5	A5	A5	A5	A5	A7
Task8	3	A3	A3	A3	A6		
Task9	3	A2	A2	A2	A8		
Task10	4	A7	A7	A7	A7	A6	

Task and agent information:

$$\textit{preference value} = \frac{1}{\textit{agent capability} - \textit{task complexity}}$$

Table 5. Agent capability and task complexity comparison

Task complexity:

complexity	1	2	3	4	5	6	7	8	9	10	11	12	13	14
	T1	T8	T5	T9	T2									
	T3	T10												
	T4													
	T6													
	T7													

Capability	1	2	3	4	5	6	7	8	9	10	11	12	13	14
					A9	A10	A8			A1	A5	A3	A2	A7
						A4				A6				

	Preference Value	Agent Name
Task1	0.25	Agent9
Task2	1	Agent10
Task3	0.200000003	Agent4
Task4	0.166666672	Agent8
Task5	0.142857149	Agent1
Task6	0.111111112	Agent6
Task7	0.100000001	Agent5
Task8	0.100000001	Agent3
Task9	0.111111112	Agent2
Task10	0.083333336	Agent7

- **some low capability agent cannot work on some tasks**

In this case some low agent cannot work on some tasks as shows in Table 7 and working process shows in Table 6.

Table 6. Time tick table and task allocation for the first 10 tasks

TASK	Size	T 1	T 2	T 3	T 4	T 5	T 6	T 7	T 8	T 9	T 10	T 11	T 12	T 13
Task 1	2	A 5	A 5	A 2										
Task 2	1	A 2	A 1											
Task 3	1	A 1	A 2											
Task 4	5	A 3	A 3	A 3	A 3	A 3	A 2							
Task 5	2	A 4	A 4	A 1										
Task 6	2			A 5	A 5		A 3							
Task 7	4			A 4	A 4	A 4	A 4	A 3						
Task 8	4				A 1	A 1	A 1	A 1	A 4					
Task 9	2				A 2	A 2		A 4						
Task 10	5							A 2	A 2	A 2	A 2	A 2		A 1

Table 7. Agent capability and task complexity comparison

Task complexity:

complexity	1	2	3	4	5	6	7	8	9	10	11	12	13	14
	T 1	T 6	T 8	T 3	T 5									
	T 2		T 9	T 4										
	T 7		T 10											
	T 6													
	T 7													

Agent Capability:

Capability	1	2	3	4	5	6	7	8	9	10	11	12	13	14
		A 5			A 1									
					A 2									
					A 3									
					A 4									

4 Method and Analysis

- Solo Programming

(1) Strategy one: agent can only work on the task it can do and without doing pair programming

Table 8. Strategy one example

1.	time	1	2	3	4	5	6	7	8	9	10	11	12	13	14	15	16	17	18	19	20
task	size																				
1	4	A2	A2	A2	A2		Av3														
2	2	A1	A1			Av2															
3	9	A5	A5	A5	A5	A5	A5	A5	A5	A5	Av2										
4	10	A4	A4	A4	A4	A4	A4	A4	A4	A4	A4	Av2									
5	5			A1	A1	A1	A1	A1	Av3												
6	3							A2	A2	A2	Av3										
7	4		A3	A3	A3	A3	Av2														
8	10								A1	A1	A1	A1	A1	A1	A1	A1	A1	A1	Av3		
9	9										A5	A5	A5	A5	A5	A5	A5	A5	A5	Av3	
10	5											A4	A4	A4	A4	A4	Av3				

level	1	2	3	4	5	6	7	8	9	10
	Task7		Task4	Task1	Task2				Task6	Task3
			Task5	Task8	Task10					
			Task9							
		Agent3			Agent4			Agent1	Agent2	Agent5

In the above case in Table 8, it shows that agent3 can only work on task7, so that in the first round task allocation, which the system agent send top 5 task to the 5 agents to select, agent3 can select nothing, so that it will be idle in the first round task section and begin to work at time tick 2 which the second round task selection begin, then it start to work. Otherwise, during the first-round task section, agent3 will task task5 and that will take longer time for it to complete. Task5 is completed by Agent1 who is more capable and verified by agent3.

(2) Strategy Two: agents can work on the task which needs higher capability but takes longer time and without doing pair programming

This is an example in Table 9 shows that how low capable agent can work on high complex task, but task longer time, if the strong agent at this time is not available. This really maximum the working efficient of the team, but really delay the whole process of a sprint can do. I set all low agent working on high complex task will get a 0 preference value and it will take the agent taskcomplexity/agentcapability times time to complete the task. For example, if the task complexity is 10 and the agent capability is 1, then it will take 10 times time to finish the task. During draw of this time table, for the reason some tasks really takes significant time to complete, so that the middle part which repeat the same process is ignored, this is a more concise chart.

Table 9. Strategy two example

	time	1	3	4	8	9	10	11	12	13	14	15	16	20	21	29	30	36	37
task	size																		
1	3	A3	A3			Av5													
2	8	A5	A5	A5	A5			Av3											
3	10	A4	A4	A4	A4	A4	A4	A4	A4	A4	A4	A4	A4	A4	A4	A4	A4	A4	Av3
4	5	A1	A1	A1	A1	A1	A1	A1	A1	A1	A1	A1	A1	A1	A1	A1	A1	A1	Av5
5	8	A2	A2	A2	A2	A2	A2	A2	A2	A2	Av5								
6	7			A3	A3	A3	A3				Av2								
7	4						A5	A5	A5	A5		Av2							
8	7								A3	A3	A3	A3	A3	A3	A3	A3	Av5		
9	1											A5	Av2						
10	5												A5	A5	Av2				

level	1	2	3	4	5	6	7	8	9	10
			Task9	Task1		Task7	Task3	Task2		Task8
				Task6			Task4	Task5		
				Task10						
	Agent1	Agent4		Agent3	Agent2					Agent5

- **keep the task and agent settings to be exactly the same and testing the various strategies**

  1. java seeds are used to guarantee that any random number generated is fixed on different time and different machine based on its defined seeds.
  2. I generate 50 tasks which is 10 user stories for the project, task in each user story remain identical in both size and complexity as randomly setup, as shown in Fig. 12 for task size and Fig. 13 for task complexity.
  3. The agent team is composed my 10 agents randomly setup and remain identical in each experiment

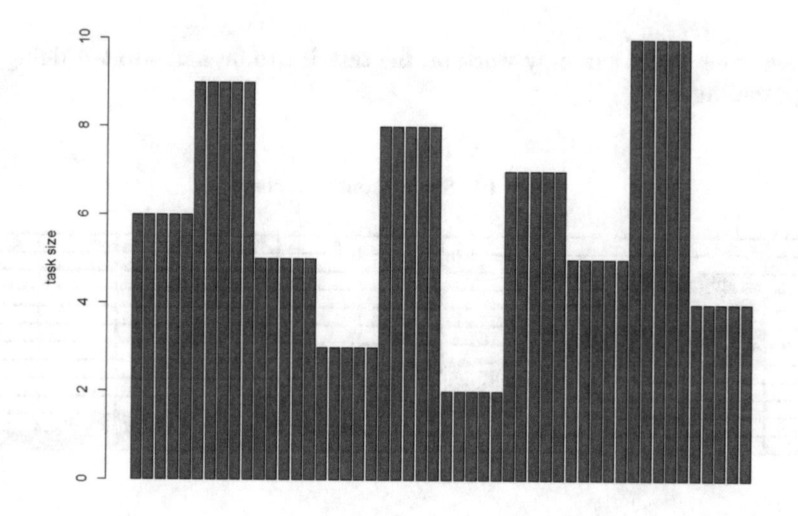

Fig. 12. Task size

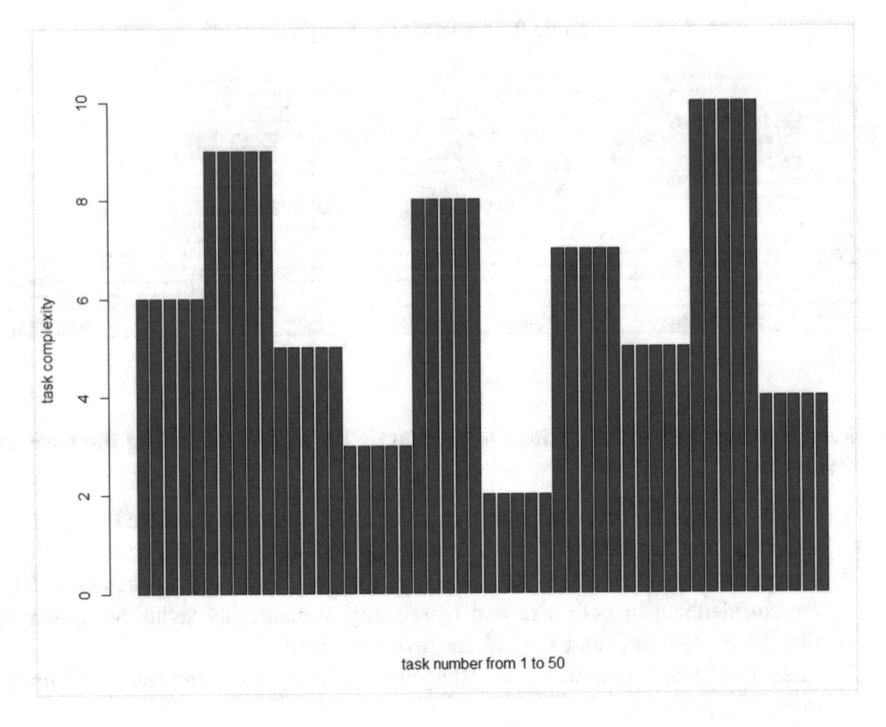

Fig. 13. Task complexity

Strategy one: agent can only work on the task it can do and without doing pair programming

Table 10. Strategy one case study

time	1	6	7	8	15	16	17	24	25	26
task										
1	Agent7	Agent7	Agent3							
2	Agent5	Agent5	Agent1							
3	Agent2	Agent2	Agent5							
4	Agent10	Agent10	Agent9							
5	Agent9	Agent9	Agent7							
6			Agent2	Agent2	Agent2	Agent6				
7			Agent10	Agent10	Agent10	Agent2				
8				Agent9	Agent9	Agent9	Agent6			
9						Agent10	Agent10	Agent10	Agent6	
10							Agent2	Agent2	Agent2	Agent10

As the agent take very longer time to complete a task shows in Table 10, so that I ignore most of the time ticks, but just show the major step of each task start and end.

Table 11. Strategy one case study agent and task table

level	1	2	3	4	5	6	7	8	9	10
	Agent1	Agent6	Agent4		Agent3		Agent7	Agent5	Agent2	Agent10
					Agent8					Agent9
						Task1			Task6	
						Task2			Task7	
						Task3			Task8	
						Task4			Task9	
						Task5			Task10	

The task above Table 11, its size is the same with its complexity, for testing purpose. Obviously, the task can have various size, and various complexity to be measured by the java random seeds.

Strategy Two: agents can work on the task which needs higher capability but takes longer time and without doing pair programming

Table 12. Strategy two case study

time	1	6	7	17	18	27	28	41	43	82	83
task											
1	Agent7	Agent7	Agent5v								
2	Agent5	Agent5	Agent9v								
3	Agent2	Agent2	Agent10v								
4	Agent9	Agent9	Agent7v								
5	Agent10	Agent10	Agent2v								
6	Agent8	Agent8	Agent8	Agent8	Agent2v						
7	Agent6	Agent6	Agent6	Agent6	Agent6	Agent6	Agent6	Agent6	Agent2v		
8	Agent4	Agent4	Agent4	Agent4	Agent4	Agent4	Agent9v				
9	Agent1	Agent1	Agent1	Agent1	Agent1	Agent1	Agent1	Agent1	Agent1	Agent1	Agent4v
10	Agent3	Agent3	Agent3	Agent3	Agent5v						

As the agent take very longer time to complete a task shows in Table 12, so that I ingore most of the time ticks, but just show the major step of each task start and end.

Table 13. Strategy two case study agent and task table

level	1	2	3	4	5	6	7	8	9	10
	Agent1	Agent6	Agent4		Agent3		Agent7	Agent5	Agent2	Agent10
					Agent8					Agent9
						Task1			Task6	
						Task2			Task7	
						Task3			Task8	
						Task4			Task9	
						Task5			Task10	

The task in above Table 13, its size is the same with its complexity, for testing purpose. Obviously, the task can have various size, and various complexity to be measured by the java random seeds.

Comparisons

Strategy one:

Agent can only work on the task it can do and without doing pair programming, the time needed for each task shows in Fig. 14.

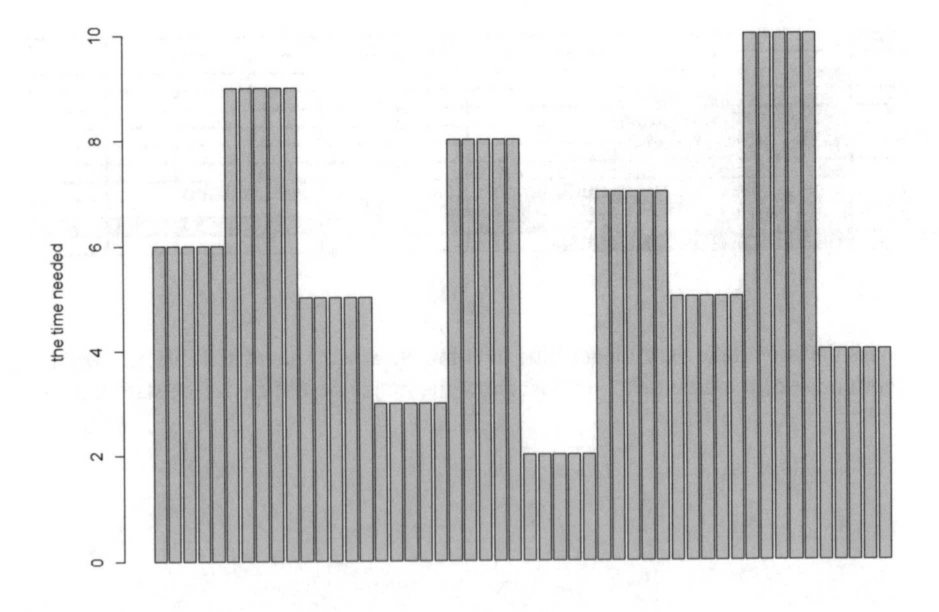

Fig. 14. Strategy one performance on task required time

Agent can only work on the task it can do and without doing pair programming, task end time for each task shows in Fig. 15.

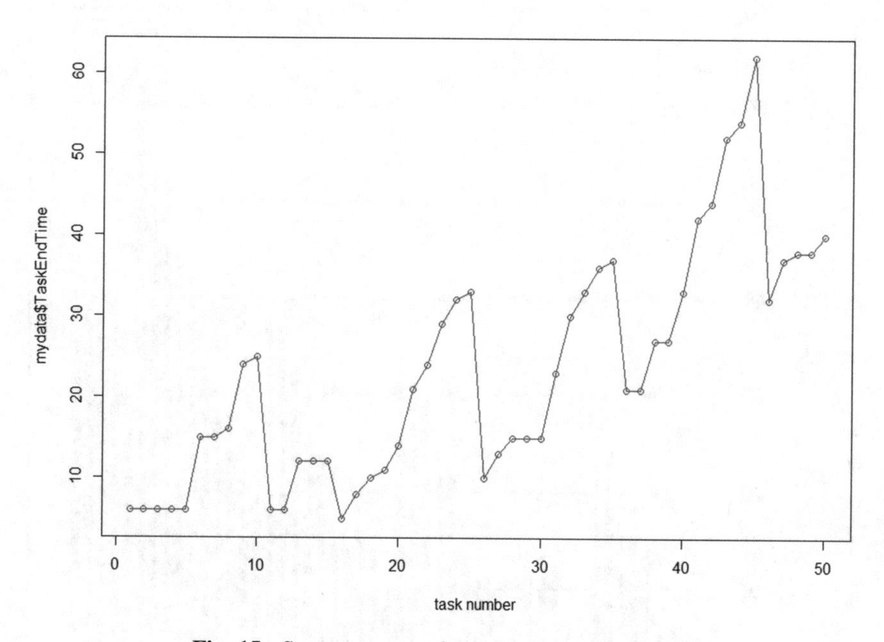

Fig. 15. Strategy one performance on task end time

Figure 16. shows under strategy one, the team performance on each task for every details, including task start time, end time and verified time.

Strategy Two:

Agents can work on the task which needs higher capability but takes longer time and without doing pair programming, the time needed for each task shows in the figure below in Fig. 17.

Agents can work on the task which needs higher capability but takes longer time and without doing pair programming, task end time for each task shows in Fig. 18.

Figure 19 shows under strategy two, the team performance on each task for every details, including task start time, end time and verified time.

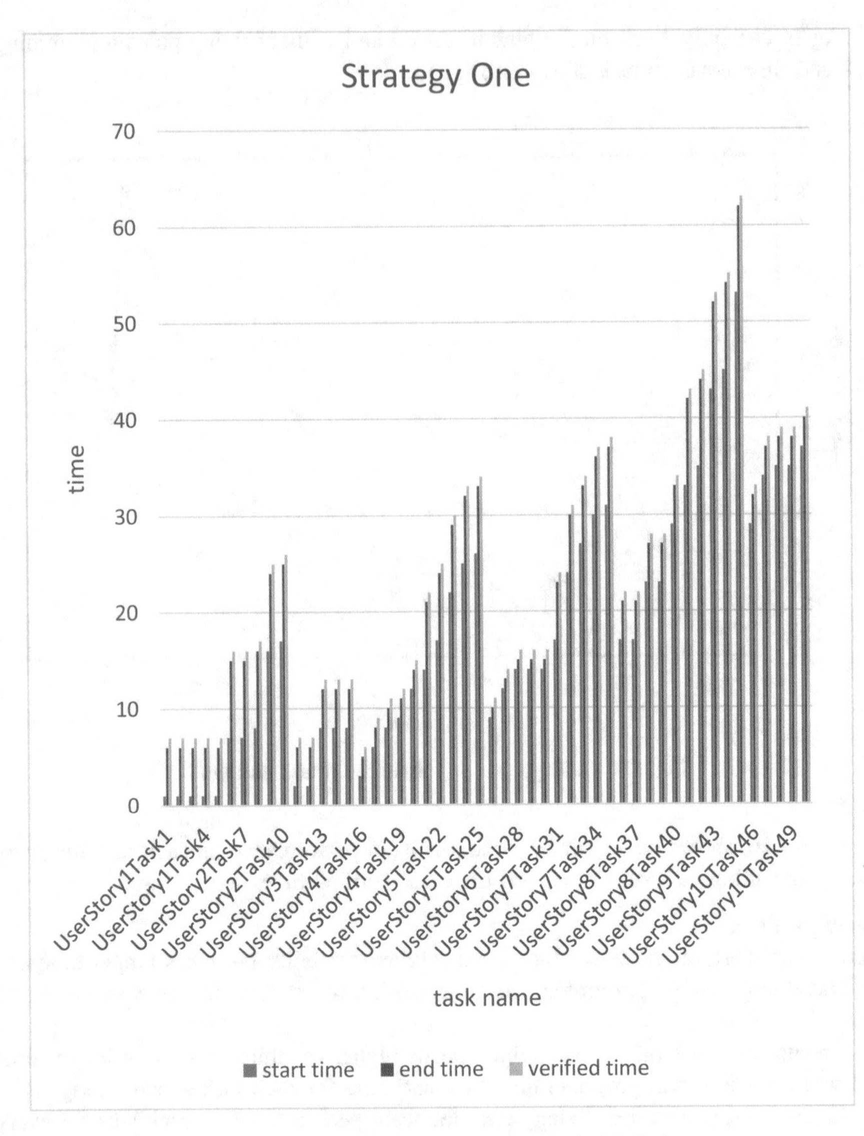

Fig. 16. Strategy one performance on each task detailed progress

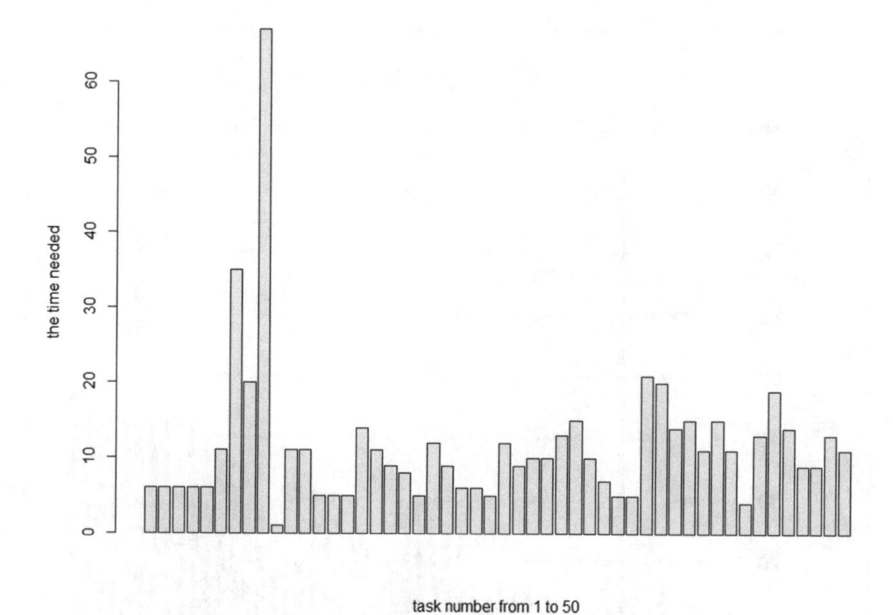

Fig. 17. Strategy two performance on task required time

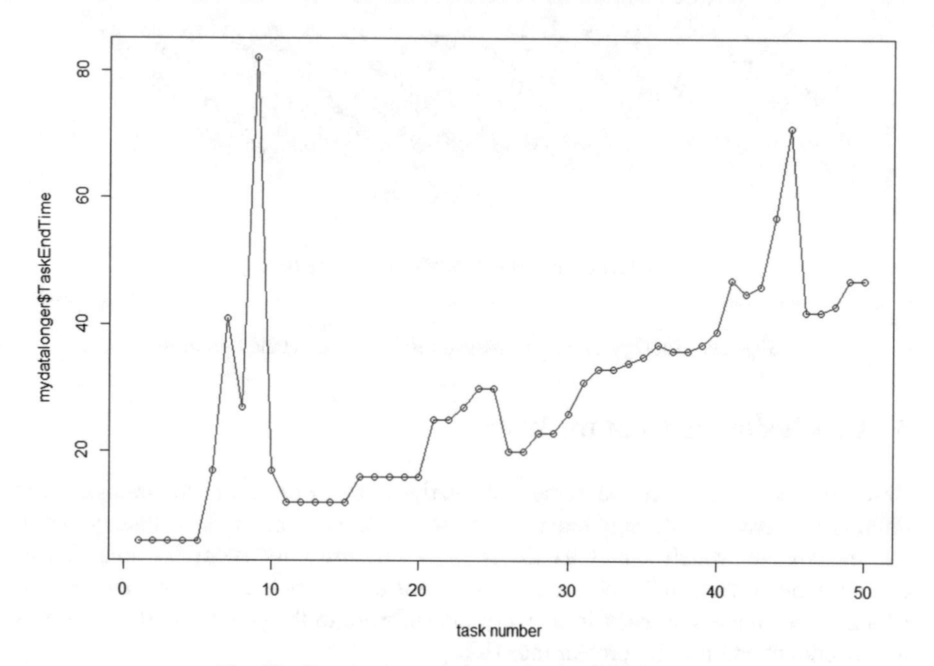

Fig. 18. Strategy two performance on task end time

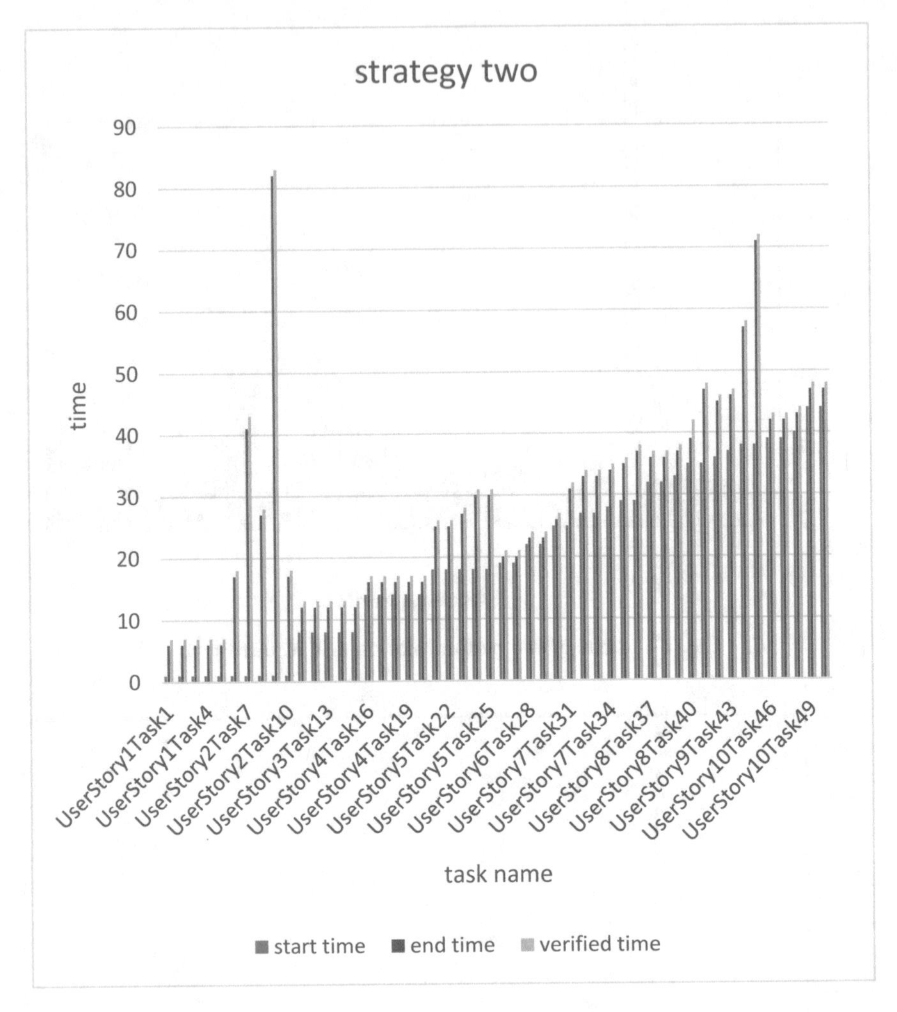

Fig. 19. Strategy two performance on each task detailed progress

5 Conclusion and Future Work

Based on the same team and same task, various strategies can result in significant different outcome, the strategy one performance much more better than strategy two in this experiment, which shows us the strategy one use only about 60 time tick to complete the project, while the strategy two needs 80 time tick to complete the project, which cause significant delay in delivery the software to the enterprise, that will cost higher budget and run the project into risk.

The current simulation software tool is advanced compare with any type of software process simulation system, the agent-based modelling is designed and implemented based on real agent, which each agent has its own thread and decision control to collaboration and interactions within the environment. The simulation system also captures the feature of Scrum and support scrum team dynamics in any level of measurement and scale modelling as it follows the general rules of Scrum modelling. Which makes it can be used for any particular simulation validation and modelling verification in the real world.

The current published papers show its innovations contributions in team strategies, task allocations and published through IEEE and ACM [13–15]. The software tool is also a very important contribution to the scrum-based modelling and simulation. More team strategies and task allocations can be developed based on the evolved complex agent-based simulation software. All this will provide me many and more chance to engage with the software industry in China as well.

The scrum is a project that needs fast delivery of the software, which means the time box for sprint design is very important, otherwise why it needs sprint. And most importantly, the sprint review and retrospective are the most important aspect of scrum, that's guarantee the quality of scrum. Its needs limited time to produce the most qualified software which is the core ideal of Scrum, any Scrum team will aim at this.

Scrum is a very special software engineering process, as Scrum particular dealing with extremely difficulty software projects, it is not used to do easy software which all the user stories are clear, and the team can instantly work on all the user requirements and complete all the task without any further understanding required. Scrum is used to design for complex software projects which both the Scrum team and the product owner are not sure how to design and coding, it advocate sprint review and sprint retrospective is for this reason, it provides more chance and opportunities for the Scrum team and product owner to have frequently meeting on how to proceed the projects in high quality and really delivery the software that needed, as this kinds of software is critical products. The design of the sprints, the frequency of the sprint review and planning are critical for the team to carry on Scrum, which is the core ideal of why Scrum is existing, and why it can be used to solve the problem of software engineering.

The Scrum team dynamics and task allocations are however depends on the Scrum design, that design is the Scrum process design, various scrum process will result in different Scrum result based on the same team and the same task, for example how the team deal with user story understanding and updating, how the team deal with high workload in a sprint, how the team deal with sprint review and retrospective, those will affect the team performance significantly, rather than just the team itself. The team however will be affected by itself, such as personality which many researchers has confirmed on this. But the team will definitely be affected by the software project it is working on as well, that why we need Scrum which is a specially software process to deal with special software projects.

The team dynamics are not just comes from the team itself, it also comes from the Scrum environment which the team is located, agent based modelling are perfectly can deal with agent interactions with environment and agent interaction with agents, the agents needs collaboration and team work to solve the challenge from the scrum environment, in this process, the agent's behaviour and decision making will be very important in pairing and task allocations. The agent is interacting with environment through the task its working on and the team its involved in. Obviously, the whole team has the same goal, which is the Scrum success rate in software delivery. Scrum is a context, agent working inside it, agent can affect the Scrum through its behaviour, in the meantime Scrum can decide the agent behaviour. Scrum and agent are integrated together, rather than to be thinking separately. Further strategies will be developed and tested.

Acknowledgement. Sincerely, Thanks for Dr Patricia Anthony and Dr. Stuart Charter at Lincoln University, New Zealand to support this PhD research, also thanks for Pro. Guojian Cheng at Xi'an Petroleum University, China. Dr Gang Li at Deakin University, Australia and Professor Longbing Cao at University of Technology Sydney provides funding in the related data analysis and machine learning research which I was doing my invited research at UTS, Australia. I also thanks to Edinburgh Napier University, United Kingdom where I get my Msc in Advanced Software Engineering. They are all my best Supervisors support me to growth and become more and more professional.

References

1. Noori, F., Kazemifard, M.: Simulation of pair programming using multi-agent and MBTI personality model. In: 2015 Sixth International Conference of Cognitive Science (ICCS), Tehran, pp. 29–36 (2015). https://doi.org/10.1109/cogsci.2015.7426665
2. Wernick, P., Hall, T.: The impact of using pair programming on system evolution a simulation-based study. In: Proceeding of the 20th IEEE International Conference on Software Maintenance (ICSM 2004). IEEE (2004)
3. Lui, K.M., Chan, K.C.: Pair programming productivity: novice-novice vs. expert-expert. Int. J. Hum.-Comput. Stud. **64**(9), 915–925 (2006)
4. Vanhanen, J., Korpi, H.: Experiences of using pair programming in an agile project. In: 40th Annual Hawaii International Conference on System Sciences, HICSS 2007, Waikoloa, HI, pp. 274b–274b (2007). https://doi.org/10.1109/hicss.2007.218
5. Han, K.W., Lee, E., Lee, Y.: The impact of a peer-learning agent based on pair programming in a programming course. IEEE Trans. Educ. **53**(2), 318–327 (2010). https://doi.org/10.1109/te.2009.2019121
6. Swamidurai, R., Umphress, D.: The impact of static and dynamic pairs on pair programming. In: 2014 Eighth International Conference on Software Security and Reliability-Companion (2014)
7. Williams, L., Shukla, A., Anton, A.I.: An initial exploration of the relationship between pair programming and Brooks' law (2004)
8. Plonka, L., Sharp, H., Linden, J., Dittrich, Y.: Knowledge transfer in pair programming: An in-depth analysis. Int. J. Hum.-Comput. Stud. **73**, 66–78 (2015)
9. Wooldridge, M.: An Introduction to Multi-agent Systems. Wiley, Hoboken (2002)
10. Beck, K., Beedle, M., Van Bennekum, A., Cockburn, A., Cunningham, W., Fowler, M., Grenning, J., Highsmith, J., Hunt, A., Jeffries, R., Kern, J.: Manifesto for agile software development (2001). http://agilemanifesto.org/
11. Gilbert, N., Troitzsch, K.: Simulation for the Social Scientist, 2nd edn. Open University Press, Milton Keynes, (2005). ISBN 0335216005
12. Lin, J., Yu, H., Shen, Z., Miao, C.: Studying task allocation decisions of novice agile teams with data from agile project management tools. In: ASE 2014, Vasteras, Sweden, 15–19 September 2014. ACM (2014). 978-1-4503-3013-9/14/09

13. Wang, Z.: The impact of expertise on pair programming productivity in a scrum team: a multi-agent simulation. In: 2018 IEEE 9th International Conference on Software Engineering and Service Science (ICSESS), Beijing, China, pp. 399–402 (2018). https://doi.org/10.1109/icsess.2018.8663874
14. Wang, Z.: Estimating productivity in a scrum team: a multi-agent simulation. In: Proceedings of the 11th International Conference on Computer Modeling and Simulation, ICCMS 2019, pp. 239–245. ACM, New York (2019). https://doi.org/10.1145/3307363.3310985
15. Wang, Z.: Teamworking strategies of scrum team: a multi-agent based simulation. In: Proceedings of the 2018 2nd International Conference on Computer Science and Artificial Intelligence (CSAI 2018), pp. 404–408. ACM, New York (2018). https://doi.org/10.1145/3297156.3297179

Characteristics of Updating Route Information of Networks with Variable Topology

Vyacheslav Borodin, Vyacheslav Shevtsov, Alexander Petrakov, and Timofey Shevgunov(✉)

Moscow Aviation Institute (National Research University), Moscow, Russia
doc_borl@mail.ru, {vs,nio4}@mai.ru, shevgunov@gmail.com

Abstract. The task of modeling the process of updating route information using broadcasting is considered for data networks with a changing topology and dynamic routing. The effectiveness of the routing information transfer processes is assessed by two indicators: the refresh time and the completeness of routing information update. To determine the performance indicators, the used model takes into account access protocols, the number of network nodes, and the network adjacency factor is used. The article discusses two access protocols, namely, the sequential one and the random one with listening to the carrier. To improve the efficiency of updating, two methods have been proposed – the forced balancing of route information and the selective distribution of route information. The forced balancing means automatic update of the data on the connection of the reverse channel upon receipt of data on the connection of the direct channel. In some cases, this procedure allows to reduce the refresh time by more than two times and, in addition, increase the completeness of updating the route information. The essence of the method of selective distribution consists in transmitting data by the nodes only about routes whose length does not exceed a certain threshold. For a wide class of networks, this means that each node can transmit data only about its adjacent nodes. This method is effective at relatively high values of the adjacency coefficient and lead to a decrease in the refresh time by more than ten times.

Keywords: Variable topology networks · Mash networks · Sensor networks · Routing algorithms · Manet networks

1 Introduction

The change in the network topology and data transmission routes allows one to use such networks in the absence of the line of sight between the network nodes, in case of environmental characteristics change (e.g. radio propagation conditions), or alteration in the mutual arrangement of nodes or their states. These properties can be fully realized by means of adaptive dynamic routing based on the use of up-to-date information about the current state of communication channels and network nodes. The interest in such networks has stimulated the development of the appropriate technologies, in particular, the emergence of adaptive networks, networks with a mesh

R. Silhavy et al. (Eds.): CoMeSySo 2019, AISC 1046, pp. 376–384, 2019.
https://doi.org/10.1007/978-3-030-30329-7_33

topology (mesh radio networks), wireless decentralized self-organizing networks (Manet), and others [1–7].

A large number of works is devoted to the task of network routing [8–15]. In particular, a large number of routing protocols has been proposed. It is shown that the routing problem in the general case can be divided into two tasks, such as collecting route information and determining the best route. The choice of the optimal route is determined on the basis of the route metrics – the numerical characteristics of the communication channels or nodes used in the route.

The analysis shows that the routing information in general can be represented in the form of two components: structural information (or adjacency matrix) and metric information (metrics) about the characteristics of network nodes. In the simplest case, structural information contains only information about the adjacency of the nodes, more generally, information about the characteristics of communication channels between nodes.

The necessary route information can be obtained by exchanging data of route information between nodes. Since the routing information is transmitted at a finite rate, the nodes use delayed, not quite relevant, data for solving the problem of routing in the general case.

The proposed work is devoted to analyzing the mechanisms for transmitting routing information and determining the characteristics of transmission, in particular, the completeness of the update and the time for complete update of the routing information.

2 Data Update Process Model

To solve this problem, a simulation model of the network is used [16, 17]. Below, there are results for the basic network, in which the network contains N identical nodes, which can function as sensors, forming packets for transmission, routers, and actuators.

The network is defined using its adjacency matrix (AM), in which the parameters of the communication channel are determined for each two nodes. Hereinafter, we will assume that the adjacency matrix is symmetric and the network is fully connected, so there is a route between any pair of its nodes. The relative degree of the node $R_d(A)$, which is equal to the number of nodes N_{ad} adjacent to the node A with respect to the total number of nodes, is a characteristic of the network connectivity. In the base model, the coefficient R_d of all nodes will be considered the same for all nodes and equal to $R_d = N_{ad}/(N-1)$. Each node contains its own local adjacency matrix, which is to be used to determine the route.

In the general case, if the nodes are not adjacent, the transmission of packet from the outgoing node A, the connection initiator, to the incoming node B, the data recipient, is possible only through intermediate nodes (relay nodes or transit nodes): $C_1, C_2, \ldots, C_n$. A sequence of nodes $M\{\mathrm{A}, \mathrm{C}_1, \mathrm{C}_2, \ldots, \mathrm{B}\}$, through which the packet is transmitted, will be called a route. The number of channels connecting the source node A to the recipient node B is called a route length. The algorithm for sending structural information is designed to periodically complete update of the routing information in each local AM and consists of the following step.

At the initial time, all local matrices are set to zero. The transfer of route information is carried out in packets. The format of the transmitted packet contains a sequence of data blocks of the same size. The first address block and the remaining information blocks are functionally allocated. In turn, the address block contains the required address information, and the information blocks contain the corresponding metric information.

It is assumed that one information block may contain from 8 to 12 metrics, depending on the type of the metric.

According to the access rule to the radio channel chosen, one of the nodes broadcasts a packet containing the identifier of the transmitting node. This packet comes to all its adjacent nodes. Further, in accordance with the access algorithm, each node transmits data blocks containing the identifier of the transmitting node and all data blocks received from other nodes. To reduce the volume, only blocks containing new, not previously sent information are transmitted. Thus, each node gradually accumulates data on adjacent nodes and the structure of the entire network. The exchange process will be completed if the nodes do not contain new information about adjacency matrices.

To transmit service routing information, we will consider the following discipline protocols of the node maintenance.

Time Division Multiple Access (TDMA). Each node is cyclically allocated a temporary channel where it can transmit one information packet. If the message is long, it is broken into packets that are transmitted sequentially in their own temporary channel. For its effective implementation, the sequential access requires the knowledge of the total number of the nodes in the network and the ability to assign temporary synchronous channel between each of them. When using sequential access, it is assumed that the size of each time window for the data transition from one node is fixed and generally contains several information blocks.

Carrier Sense Multiple Access with Collision Detection (CSMA/CD). Each node listens to the channel and occupies the channel through a random time interval after its release. Each node transmits the entire amount of the available routing information in one packet. The packet format contains a sequence of blocks, however, all information blocks except the first one do not contain the address block. The packet size is variable and is determined by the amount of the routing information prepared for transmission. It is also assumed that the signal propagation time is negligible, or at least substantially less than the time of the data block transmission.

- The following indicators has been chosen to characterize the quality of the distribution:
- The refresh time T_r, which is the time expressed in the number of gaps, from the moment the distribution starts to the moment it is completed. The distribution completion is an event where each node has updated its routing information or the status of all nodes, in the case of multi-step metrics, or adjacent node status information for a one-step metric. The duration of the gap is equal to the time spent on the transmission of one address block over the channel.

- The current distribution time t is the time from the beginning of the distribution to the current point in time.
- The refresh coefficient K_r which is defined as the number of fully updated AM in relation to the total number of nodes.

3 Results and Discussion

The comparative results for the distribution time and the updated adjacency matrices coefficient obtained via modeling carried out for two access protocols are shown in Figs. 1 and 2 correspondingly.

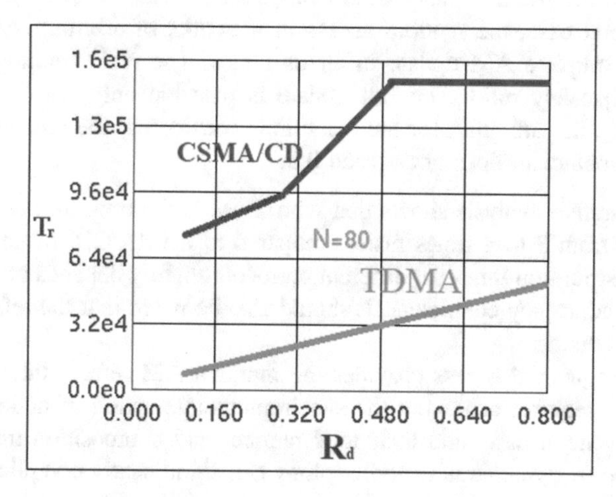

Fig. 1. The dependence of the distribution time of structural information for sequential (TDMA) and random (CSMA/CD) access

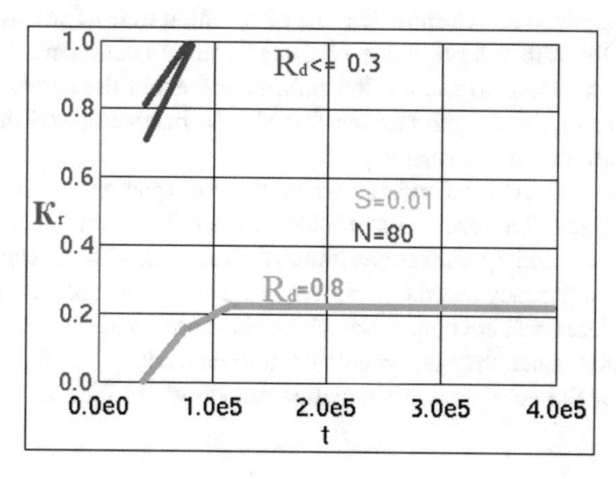

Fig. 2. Dependence of the updated adjacency matrices coefficient K_r on the distribution time.

The simulation results allow one to draw the following conclusions regarding the process of routing information distribution.

- The amount of information transmitted varies widely in the process of distribution. The maximum amount of information transmitted falls appears in the initial part of the distribution.
- At the time of distribution, the access control parameters have significant values. Thus, for TDMA, this control parameter is the size of the information packet. If it is small, then the cost of transmitting address information increases. If the size of the information packet is large, then a part of the time interval will be idle, since the amount of the transmitted routing information decreases. For CSMA/CD, the control parameters are the initial traffic, transmission frequency of information packets, and the packet transmission delay after the channel is released.
- In the general case, the random access in networks of arbitrary structure cannot provide a complete AM update in all its nodes. The main limiting factor is the network adjacency ratio. The full update is possible only with relatively small values of R_d, in particular, for the conditions considered, the critical value for the adjacency coefficient does not exceed 0.3.

The comparative analysis shows that with TDMA, the time for sending structural information is from 3 to 4 times faster compared to CSMA/CD. Moreover, the ratio between the distribution times for different protocols slightly depends on the number of nodes and the adjacency coefficient. It should also be noted that the refresh time goes up with R_d increasing.

Thus, the sequential access provides the minimal AM refresh time. However, its implementation requires achieving the synchronous interaction of nodes weakly connected in the general case, and their total number and composition may vary in the operating mode. A dynamic network topology can significantly complicate the overall organization of the sequential access.

The random access with listening to the carrier is free of the above-mentioned technical limitations. However, as it was highlighted, the AM refresh time is long compared to CA and, moreover, an incomplete update of route information is possible.

The results obtained above show that the distribution time of structural information is large, especially with a large value of the adjacency coefficient. One approach to shorten the refresh time is to use a priori information about the network structure. We will consider two of these methods: the forced AM balancing and the selective distribution of the structural information.

If we use forced AM balancing, when the information about the connection between nodes A and B is received at a node C, the node C automatically updates the local AM adjacency, adding both connection (A, B) and (B, A) to it. Thereby, the forced balancing of the adjacency matrix is performed. Since each node performs the operations, only the data on direct connection channels is new, and the data on reverse links are not transmitted, since they are generated automatically.

The results of the comparative simulation are shown in Fig. 3.

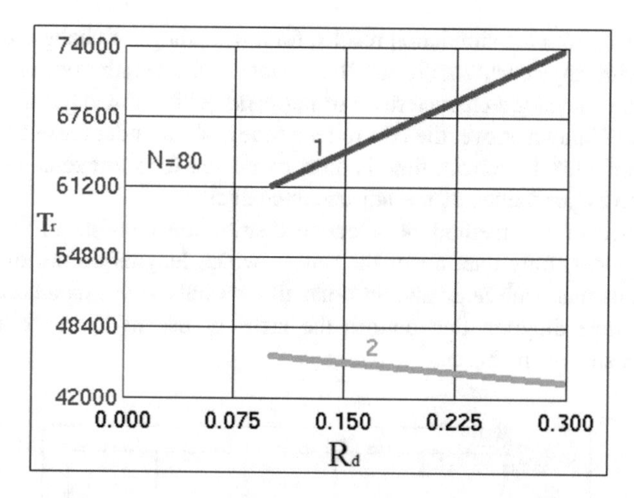

Fig. 3. Distribution time depending on R_d without the balancing (1) and with the balancing used (2).

Therefore, the considered symmetry of the adjacency matrix allows one to reduce the distribution time more than twice. It should be also noted that the optimal values of the access parameters for the forced balancing do not differ from the case of regular distribution. In addition, the balancing with MALC protocol generally leads to an increase in the number of updated adjacency matrices as compared to the regular distribution and, consequently, to an increase in the value of the coefficient K_r.

The second method of the refresh time reduction is the selective distribution method, which is based on the fact that the number of retransmissions in packet transmission is limited during the routing. As an example, Fig. 4 shows a dependence of the average route length L_r on the adjacency coefficient for a different number of network nodes.

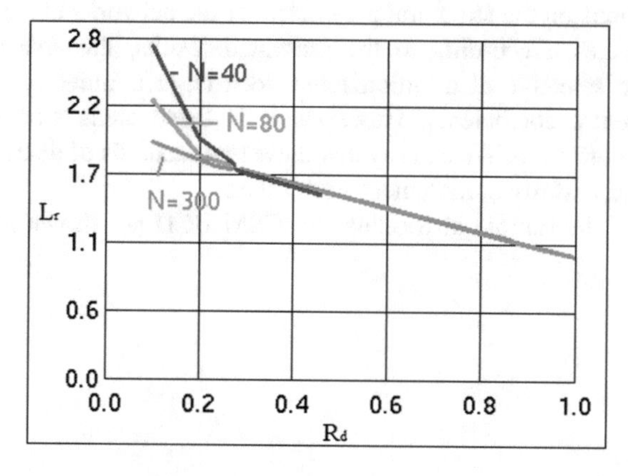

Fig. 4. The route length L_r against the adjacency coefficient R_d for different numbers of nodes N.

As it follows from the simulation results, for a wide range of changes in the number of nodes and the adjacency coefficient, the average route length does not exceed two. Relatively long routes are characteristic of networks with a low adjacency coefficient. However, it was shown above, the AM refresh time is short under these conditions. The effect of minimizing the refresh time is most expressed for short route lengths, which corresponds to larger values of the adjacency coefficient.

The essence of the method of selective distribution consists in the transfering, between the nodes, only data about the routes, which lengths do not exceed two. In fact, this means that each node should transmit data only about its adjacent nodes.

The exchange diagram that defines the areas of use of the bulk and selective distribution is shown in Fig. 5.

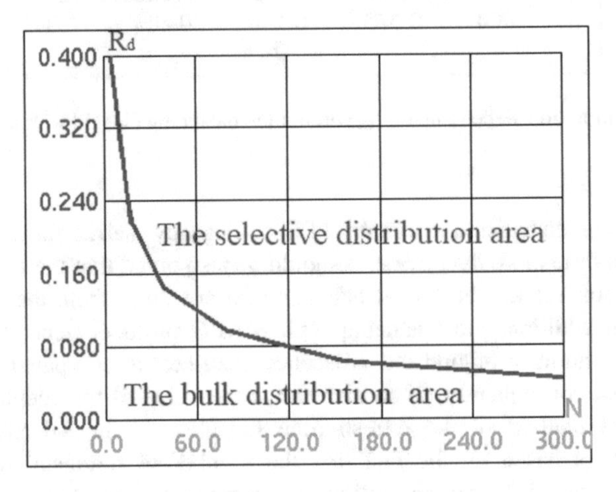

Fig. 5. Exchange diagram for determining the use of bulk and selective distribution

As it follows from the presented results, the selective distribution does not impose any rigid restrictions on the structural parameters of the networks except for the limited length of the routes. According to the presented results, the area of the effective utilization of the selective distribution refers to a large number of nodes and large values of adjacency coefficients. However, it is these areas where mass mailing methods are least effective. We can say that these two methods of distribution, bulk and selective, can successfully complement each other.

The results of the sample distribution for CSMA/CD are shown in Fig. 6.

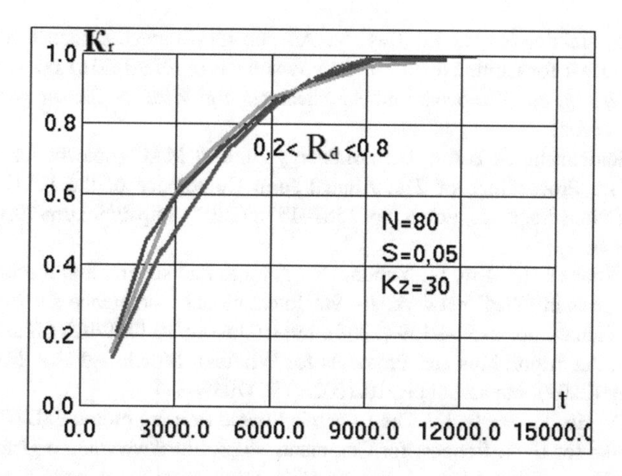

Fig. 6. The dependence of the share of created AM on the time of selective distribution for different values of the adjacency coefficient

The simulation results show that selective mailing leads to a significant decrease in the time of filling adjacency matrices. In particular, for $N = 80$, the formation time of the adjacency matrices is reduced by almost a hundred times.

4 Conclusion

The paper discusses the results of simulation modeling of the local AM update processes at the nodes belonging to an adaptive network where the sequential and random access algorithms are to be used. It is shown that the sequential discipline has a shorter refresh time, however, its implementation requires the synchronous operation of all nodes, which technically can be difficult to implement. The alternatives reducing the refresh time, which takes into account the features of the network topology, have been offered. It is shown, that under certain conditions, the effect caused by the application of these methods can significantly improve the characteristics of routing protocols.

Acknowledgement. The work was supported by state assignment of the Ministry of Education and Science of the Russian Federation (project 8.8502.2017/BP).

References

1. Prehofer, C., Bettstetter, C.: Self-organization in communication networks: principles and design paradigms. IEEE Commun. Mag. **43**(7), 78–85 (2005). https://doi.org/10.1109/MCOM.2005.1470824
2. Bai, F., Helmy, A.: A survey of mobility models in wireless ad hoc networks. In: Wireless Ad Hoc and Sensor Networks, Chap. 1, pp. 1–29. Kluwer Academic Publishers (2004)
3. Dressler, F.: Self-Organization in Sensor and Actor Networks. Wiley, Chichester (2007)

4. Perkins, C., Belding-Royer, E., Das, S.: Ad hoc On-Demand Distance Vector (AODV) routing. Request for Comments 3561. http://www.ietf.org/rfc/rfc3561.txt
5. Karl, H., Willig, A.: Protocols and Architectures for Wireless Sensor Networks. Wiley, Chichester (2005)
6. Ye, W., Heidemann, J., Estrin, D.: An energy-efficient MAC protocol for wireless sensor networks. In: Proceedings of 21st Annual Joint Conference of the IEEE Computer and Communications Societies, vol. 3, pp. 1567–1576 (2002). https://doi.org/10.1109/INFCOM.2002.1019408
7. Fang, W., Shanzhi, C., Xin, L., Yuhong, L.: A route flap suppression mechanism based on dynamic timers in OSPF network. In: 9th International Conference for Young Computer Scientists, Hunan, pp. 2154–2159 (2008). https://doi.org/10.1109/ICYCS.2008.17
8. Boukerche, A.: Algorithms and Protocols for Wireless, Mobile Ad Hoc Networks. Wiley, New Jersey (2009). https://doi.org/10.1002/9780470396384
9. Johnson, D., Hu, Y., Maltz, D.: The Dynamic Source Routing protocol (DSR) for mobile ad hoc networks for IPv4. Request for Comments 4728. http://www.ietf.org/rfc/rfc4728.txt
10. Pillai, A., Kumar, A., Ajina: A survey of network security in mobile ad-hoc network. IOSR J. Comput. Eng. **17**(5), 29–36 (2016)
11. Boukerche, A.: Algorithms and Protocols for Wireless, Mobile Ad Hoc Networks. Wiley, New Jersey (2009)
12. Koucheryavy, A.: State of the art and research challenges for USN traffic flow models. In: 16th International Conference on Advanced Communication Technology, Pyeongchang, pp. 336–340 (2014). https://doi.org/10.1109/ICACT.2014.6778977
13. Jacquet, P., Clausen, T.: Optimized Link State routing protocol (OLSR). Request for Comments: 3626 (2003). http://www.ietf.org/rfc/rfc3626.txt
14. Ogier, R., Templin, F., Lewis, M.: Topology Dissemination Based on Reverse-Path Forwarding (TBRPF). Request for Comments: 3684 (2004). http://tools.ietf.org/html/rfc3684
15. Perkins, C., Belding-Royer, E., Das, S.: Ad hoc On-Demand Distance Vector (AODV) routing. Request for Comments: 3561 (2003). http://tools.ietf.org/html/rfc3561
16. Borodin, V.V., Petrakov, A.M., Shevtsov, V.A.: Simulation model for the study of adaptive sensor networks. Trudy MAI **100**, 21 (2018). http://trudymai.ru/eng/published.php?ID=93398
17. Borodin, V.V., Petrakov, A.M., Shevtsov, V.A.: The analysis of algorithms of routing in a communication network groups of unmanned aerial vehicles. Trudy MAI **87**, 16 (2016). https://www.trudymai.ru/eng/published.php?ID=69735

Prediction-Based Fast Simulation with a Lightweight Solver for EV Batteries

Donggu Kyung and Inwhee Joe(✉)

Hanyang University, 222 Wangsimni-ro,
Seongdong-gu, Seoul 04763, South Korea
{kdg8800, iwjoe}@hanyang.ac.kr

Abstract. In this paper, we propose a fast simulation method using a lightweight solver for EV batteries. In CPS, the simulation time should be reduced for real-time simulation by minimizing the overhead. In order to reduce the simulation time, the number of simulation steps needs to be decreased by a variable step size. To control the step size, a lightweight solver is introduced to predict the event as soon as possible before actual simulation. Through the prediction, a large step size can be used if there is no event, while a small step size can be used if there is an event. The simulation results show that our prediction-based method reduces the simulation time significantly, compared to the conventional non-prediction-based method.

Keywords: EV battery · CPS · FMI · Lightweight solver · Fast simulation

1 Introduction

CPSs (Cyber-Physical System) are composed of physical parts and cyber parts, distributed and connected on the network. The physical machine is designed as a cyber model to control the system or predict the results. EVs (Electric vehicles) can be designed as a cyber model through CPS. An EV designed with a Cyber model can simulate automatic operation or battery condition in a CPS environment to produce optimal results. CPS provides various services such as real time, robustness, stability, safety and security to provide reliable results. In particular, it is important to reduce overhead in order to support real-time in a network environment [1].

FMI (Functional Mock-up Interface) is a standard for simulation, which supports co-simulation in a distributed environment. Each component is designed as an FMU (Functional Mock-up Unit) and can be simulated through a given interface. FMU helps the simulation designer to co-simulate multiple components through a given interface, even if they do not understand each component model. Simulation is the process of solving the ODE (Ordinary Differential Equation) defined in the FMU. To solve the ODE, the FMU computes the result of the simulation through the Solver [2].

This work was supported by the National Research Foundation of Korea (NRF) grant funded by the Korea government (MSIT) (NO. 2019R1A2C1009894).

R. Silhavy et al. (Eds.): CoMeSySo 2019, AISC 1046, pp. 385–392, 2019.
https://doi.org/10.1007/978-3-030-30329-7_34

The step size is the interval at which the simulation proceeds. The smaller the step size, the more frequently you can see the results of the simulation. As you progress through the simulation, events can occur. To find the event point, you have to simulate with a small step size [3]. However, when the simulation is performed with a small step size, the calculation time becomes long. In this paper, we use a lightweight solver such as the Euler Method to predict the event to reduce the simulation time. The interval where the event does not occur is simulated with a large step size to reduce the simulation time. At the same time, the interval where the event occurs is simulated with a small step size to find the exact event point.

2 FMI (Functional Mock-up Interface)

The FMI is a simulation standard for co-simulation, and MA (Master Algorithm) performs simulation through FMU's API. The FMU is designed as a black box, the MA only accesses the inputs and outputs. Traditional simulations require the simulation designer to understand all the components, but the FMI does not require expert knowledge of the components because the simulation designer only accesses the inputs and outputs of the FMU. MA calls doStep of the FMU to proceed with the simulation. The current time t and step size h are needed to invoke doStep. The FMU that called doStep proceeds from t to t + h and outputs the simulation result [2].

2.1 Event Detection Using Rollback

There are two types of events: Time Events and State Events. Time Events defines the time at which an event occurs and processes the event at that time. The Time Events knows the time when the event occurs, so it can be predicted and the event point can be found easily. The State Events defines the conditions under which the event occurs and handles the event if the simulation result satisfies the condition. Since the State Events only knows the conditions under which the event occurs, it can not be predicted because the timing of the event can not be known accurately [2].

For example, ZCD (Zero-Crossing Detector) defines a state event when the event indicator variable z crosses zero. When the simulation proceeds from t to $t+h$, the ZCD determines the event through the $z(t) \times z(t+h) \leq 0$ condition. If the event condition is satisfied, the ZCD detects that an event has occurred between time t and time $t+h$. But we only know that t' satisfying $z(t') = 0$ is located $t < t' < t+h$. To know the exact time of the event, we need to find t' that satisfies $z(t') = 0$ by running the simulation with a small step size h' satisfying $h' < h$. Bisection search reduces the step size to $h' = h/2$ and repeats the simulation with a small step size. This can be repeated to find a time satisfying $z = 0$.

Rollback is required to repeat simulations at small step sizes. To simulate with a new step size h', we need to rollback the simulation time from $t+h$ to t. To roll back the simulation time, we need to store and load the state of the FMU. For this, FMI 2.0 supports rollback through fmiGetFMUstate and fmiSetFMUstate. For rollback, the MA backs up the FMU state by calling fmiGetFMUstate before calling doStep. Then call doStep to run the simulation and see if rollback is required. If rollback is required, the MA calls fmiSetFMUstate to return the state of the FMU to the previous state [4].

2.2 Numerical Integration Method for Ordinary Differential Equations

Simulation is the process of solving ODE (Ordinary Differential Equations), and simulation results can be obtained through numerical integration.

$$x(t) = x_0 + \int_{t_0}^{t} \dot{x}(\tau)d\tau \tag{1}$$

The derivative $\dot{x}$ is defined as an integral function, x_0 is the initial state, and t_0 is the initial time. At time t, $\dot{x}(t)$ is calculated using current state $x(t)$ and current input $u(t)$.

$$\dot{x}(t) = f(x(t), u(t), t) \tag{2}$$

The ODE solver is responsible for solving the ODE using integration algorithms. The ODE solver takes the current time t and the step size h and computes $x(t+h)$ and proceeds the simulation time from t to $t+h$.

$$t_{n+1} = t_n + h \tag{3}$$

There are Euler Method and Runge-Kutta Method in the integral algorithm that calculates ODE.

$$x(t_{n+1}) = x(t_n) + h \cdot \dot{x}(t_n) \tag{4}$$

The Euler method can be calculated using Eq. (4). The Euler method calculates the integral using the slope of the tangent at a given initial value. The Euler method is a first-order method and has a simple calculation formula, which is easy to implement and has a high computation time. However, there is a problem that the accuracy is not high as the cost is low.

$$x(t_{n+1}) = x(t_n) + \frac{1}{6}(k_1 + 2k_2 + 2k_3 + k_4) \tag{5}$$

$$k_1 = h \cdot f(t_n, y_n)$$

$$k_2 = h \cdot f\left(t_n + \frac{h}{2}, y_n + \frac{k_1}{2}\right)$$

$$k_3 = h \cdot f\left(t_n + \frac{h}{2}, y_n + \frac{k_2}{2}\right)$$

$$k_4 = h \cdot f(t_n + h, y_n + k_3)$$

The fourth order Runge-Kutta method can be calculated by the formula (5). The Runge-Kutta method can calculate the next step value $x(t_{n+1})$ as the sum of the weighted average of the incremental values of the size h of the four intervals at the current value $x(t_n)$. The Runge-Kutta method has high accuracy but cost is expensive and computation time is slow [5].

3 Master Algorithm for Event Prediction with a Lightweight Solver

When the simulation is performed with a small step size, the accuracy of the simulation increases, but the calculation time becomes long. However, if you are running a simulation with too large a step size, you may lose accuracy and miss an event point. In order to solve this problem, the simulation is performed in a small step size in the section where the event occurs, but in the section in which no event occurs, the simulation can be performed in a large step size.

Conventional non-prediction-based simulation performs the simulation with the default step size. When the result of the simulation is obtained, it is confirmed whether or not the event occurs. If an event occurs, the simulation proceeds with a small step size to find the event point after the rollback to the previous time. The step size can be set to the minimum value, or it can be calculated using the bisection method or the regula falsi method [6]. In this paper, the simulation was performed with a small step size using bisection method. Bisection method reduces the step size in half and finds the event point (Fig. 1).

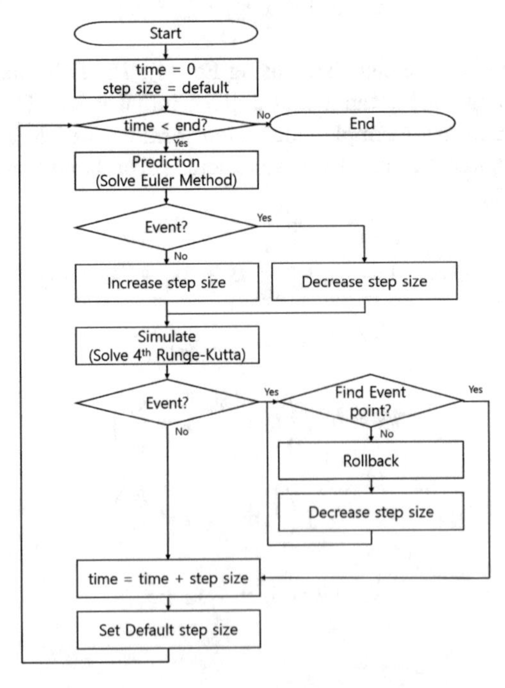

Fig. 1. Event prediction using lightweight solver

The non-prediction-based simulation performs the simulation with the default step size in the interval where no event occurs. Running the simulation every time an event does not occur leads to overhead. In order to prevent this problem, if the event does not occur, the simulation is performed with a large step size. On the contrary, if the event occurs, the simulation should be performed with a small step size. You need to know if an event occurs to determine a variable step size. In this paper, we use the lightweight solver to predict the simulation results and determine the step size according to the presence of events.

The proposed prediction-based simulation predicts the simulation result using the light solver and adjusts the step size according to the presence or absence of the event. First, predict the result before proceeding to the simulation with the default step size. Prediction is calculated quickly using the Euler method, a lightweight solver. Then, it confirms whether the event occurs within the relevant interval through the calculation result.

If the event does not occur, set a step size larger than the default step size. When increasing the step size, use the lightweight solver to check if the event occurs. After increasing the step size, proceed with the simulation. The simulation uses the heavy solver Runge-Kutta method for high accuracy. If there is no event as simulation result, repeat the above steps after increasing the current time by step size.

If an event occurs, the simulation proceeds with a step size smaller than the default step size. Simulation can be done with a small step size so that it does not cross the event point. However, unlike the prediction result, if the event point passes, the simulation proceeds with a small step size after the rollback. In this case, the simulation is performed using a heavy solver to check the correct result. The step size was reduced to half using the bisection method to find the event point.

The major difference between the non-prediction-based simulation and the prediction-based simulation is the addition of a prediction step using a lightweight solver. The non-prediction-based simulation processes the event after the simulation with the default step size. However, the prediction-based simulation checks the event through the prediction step before proceeding with the simulation. The step size is determined according to the presence or absence of the event, and the simulation proceeds with a variable step size rather than the default step size. In the prediction-based simulation, a variable step size is used, so that a simulation can be performed with a large step size for an interval without an event and a small step size for an interval with an event.

4 Simulation Results

In the CPS, the EV can simulate and predict the state of the battery. We modeled a lithium-ion battery to simulate EV's battery. Simulation and modeling were performed using QTronic's FMUSDK. We used version 2.0 of FMI to use Rollback [7].

The initial SOC (State Of Charge) of the battery is 100% and the maximum capacity is set to 18 Ah. The current was adjusted to repeat the discharge and standby of the battery. During discharge, the current was set to 18 A for 720 s and rested at 0 A for 600 s. Repeating the above cycle, SOC was monitored to manage remaining battery capacity [8].

For scenarios that manage the battery, an event is defined as a point where the SOC decreases by 100% to 10%. The simulation end time is set to 5000 s and the default step size is set to 1 s.

The non-prediction-based simulation simulates a step size of 1 s in all time zones. The event check was performed after the simulation, and if the event occurred, the simulation was repeated with a small step size after the rollback. The step size was reduced by half using the bisection method. After finding the event point, we recovered the step size to 1 s and proceeded to the simulation.

Prediction-based simulation proceeded with the prediction step before proceeding with the simulation. In the prediction step, we used the lightweight solver to check whether the event occurred. If the event does not occur, the simulation is performed by increasing the step size up to 10 times. In the interval where the event occurred, the bisection method was used to reduce the step size by half. In order to find the exact event point, exact simulation result is needed. Therefore, simulation is performed using heavy solver, and the simulation is repeated with small step size by repeating rollback.

At the event point where the SOC is 90%, we confirmed that both algorithms repeat the rollback and proceed to a small step size. In the interval where no event occurs, the non-prediction-based simulation proceeds step size by default of 1 s. However, the prediction-based simulation has been simulated for up to 10 s of step size.

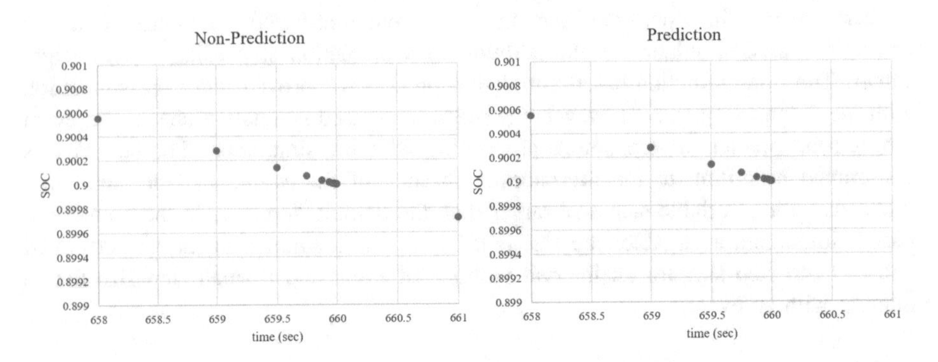

Fig. 2. Detection of a state event through rollback and bisection methods

As shown in Fig. 2, at the event point where the SOC is 90%, it is confirmed that the two MA repeat the rollback and proceed to a small step size. Two MA were simulated by reducing the step size in half using the bisection method. After repeating the small step size and finding the event, we recovered the step size.

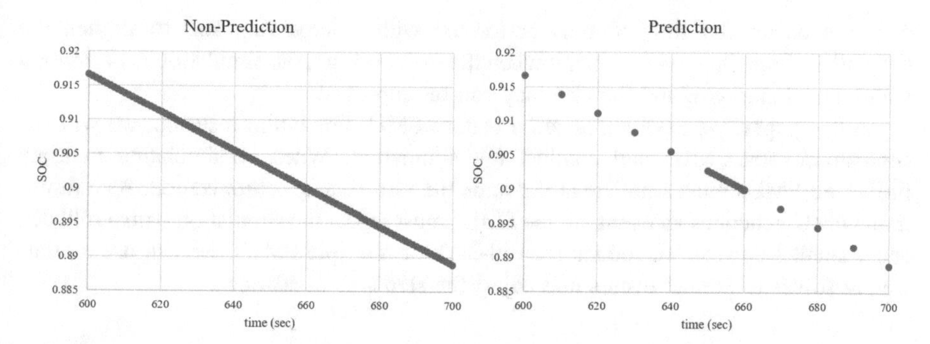

Fig. 3. Comparison of the step size between non-prediction and prediction methods

If there is no event as shown in Fig. 3, the non-prediction-based simulation proceeds step size by default of 1 s. However, the prediction-based simulation was simulated for a maximum step size of 10 s.

As shown in Fig. 4, the non-prediction-based simulation has been executed more than 5,000 steps since the simulation should be performed at 1 s step size in all the sections. However, the prediction-based simulation increased the step size in the absence of the event, resulting in about 700 steps. In addition, as the number of steps decreases, the simulation calculation is also reduced. As a result, we confirmed that the simulation time of the prediction-based simulation is reduced compared to the non-prediction-based simulation.

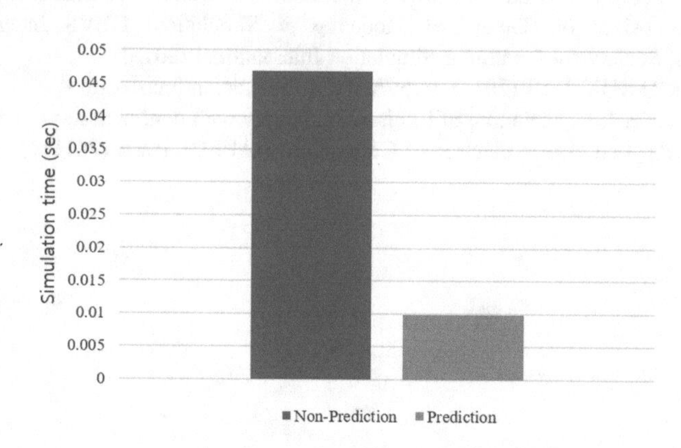

Fig. 4. Performance comparison for non-prediction and prediction methods

5 Conclusions

We modeled and simulated lithium-ion batteries with an FMU that conforms to the FMI standard for EV simulation in CPS. To satisfy the real-time requirements, it is important to reduce the overhead. The prediction-based simulation uses light solvers to predict the outcome and identify events in order to reduce the overhead. If the event does not occur, the simulation is performed with a large step size to shorten the simulation time. On the other hand, when the event occurs, the simulation is performed with a small step size and the accuracy can be improved.

In this paper, we have simulated a battery FMU. For future research, we will co-simulate several FMUs and conduct EV simulation. When co-simulating multiple FMUs, all FMUs must simulate at the same time for time synchronization. As a result, if one FMU generates an event, all the FMUs must repeat the simulation with a rollback and a small step size. To reduce the roll-back of multiple FMUs, we can use a light-weight solver to predict events and adjust the step size in advance.

References

1. Khaitan, S.K., McCalley, J.D.: Design techniques and applications of cyberphysical systems: a survey. IEEE Syst. J. **9**(2), 350–365 (2014)
2. FMI-standard: Functional Mock-up Interface for Model Exchange and Co-Simulation Version 2.0 (2014)
3. Cremona, F., et al.: Step revision in hybrid co-simulation with FMI. In: 2016 ACM/IEEE International Conference on Formal Methods and Models for System Design (MEMOCODE). IEEE (2016)
4. Broman, D., et al.: Determinate composition of FMUs for co-simulation. In: Proceedings of the Eleventh ACM International Conference on Embedded Software. IEEE Press (2013)
5. Butcher, J.C., Goodwin, N.: Numerical Methods for Ordinary Differential Equations, vol. 2. Wiley, New York (2008)
6. Galtier, V., et al.: FMI-based distributed multi-simulation with DACCOSIM. In: Proceedings of the Symposium on Theory of Modeling & Simulation: DEVS Integrative M&S Symposium. Society for Computer Simulation International (2015)
7. QTronic: FMU SDK 2.0.3 (2014). http://www.qtronic.de/en/fmusdk.html
8. Yao, L.W., Aziz, J.A.: Modeling of Lithium Ion battery with nonlinear transfer resistance. In: 2011 IEEE Applied Power Electronics Colloquium (IAPEC). IEEE (2011)

Thermal Process Control Using Neural Model and Genetic Algorithm

Daniel Honc(✉), Petr Doležel, and Jan Merta

University of Pardubice, 532 10 Pardubice, Czech Republic
daniel.honc@upce.cz

Abstract. Predictive Controller of a laboratory thermal process is presented in the paper. Process model is approximated by a neural network. On-line optimization is done by a genetic algorithm. Control algorithm is tested on the laboratory thermal process and compared to the standard control methods like predictive controller with the transfer and state-space linear model and the quadratic programming optimization method or a PI controller.

Keywords: Predictive controller · Neural network · Genetic algorithm · PI controller · Thermal process

1 Introduction

Model Predictive Control (MPC) is one of the Advanced Process Control (APC) methods used in the industry. In simple words, it is an optimization algorithm placed above the standard Digital Control Systems (DSC) ensuring optimal control with respects to the technological but also to the economic conditions and restrictions. DSC layer (mostly PID controller) is present because of the reliability, accuracy and safety of the control. The main task for APC methods is to find the optimal working point (nonlinear static optimization) and to drive the system to this point (dynamical optimization) especially for Multi-Input Multi-Output (MIMO) systems. Usually the APC generates the set-points to DSC.

MPC strategy consists of finding the optimal control actions for the whole prediction horizon based on the optimization and the dynamical model of the controlled system [1–3]. The model is crucial for the successful application. It could be an open loop control strategy. From practical point of view a receding horizon concept is introduced – the feedback is used according to the new measurements (actualization of the state of the controlled system and measurements of the actual disturbances if available). Every sample time the whole procedure is repeated and only the first (actual) control action from the vector of calculated optimal control actions is used. The criterion defining the control goals is tailored to the given control task. Typical requirement in industry is to maximize profit, production volume, to minimize input costs, amount of by-products etc. Nonlinear optimization method must be used if the model is nonlinear [4–6]. In most of the real situations the linear approximation is usually sufficient for the industrial applications. Quality of the model is good around the given working point and worst in the remote areas. Even if the process is more nonlinear it

R. Silhavy et al. (Eds.): CoMeSySo 2019, AISC 1046, pp. 393–403, 2019.
https://doi.org/10.1007/978-3-030-30329-7_35

depends how tight control is required. For slower control (relatively to the controlled process behavior), the approximate gain and the dominant time constant can be enough. Linear models used in MPC are usually the final impulse and step response models, the transfer functions and state-space models. The result model being linear is that the originally nonlinear optimization problem can be solved as a linear or quadratic programming task with very effective algorithms [1–3].

Key parts of the MPC are the cost function formulation, the dynamical process model and the numerical optimization method. Increasing of the on-line optimization speed caused by the computational power increase and new numerical methods allows to use MPC for bigger and faster processes. There is also potential for using the nonlinear models, complex cost functions and nonlinear optimization methods. Theoretically, there is no need of DCS layer – the MPC controller can calculate directly the control actions instead of the set-point for PID controllers.

Authors have experience with the predictive control of a nonlinear plant using piecewise-linear neural model and quadratic programming [7]. A little bit of curiosity and for educational reasons as well, authors want to try to use a nonlinear neural model of the controlled system and a genetic algorithm to minimize the quadratic cost function. Applications with the genetic algorithm and predictive controller can be found e.g. in [8–12]. We will control a real laboratory thermal process (heated metal bar by Peltier element), discuss results and compare them with the standard MPC methods or PI controller.

The paper is structured as follows. Introduction is in Sect. 1. Control strategy is formulated in Sect. 2. In Sect. 3 controlled system is described. Identification experiments are presented in Sect. 4 and control experiments are in Sect. 5. Conclusions are given in Sect. 6.

2 Control Strategy

The paper describes MPC design and laboratory application with process model in the form of neural network and on-line optimization done by the genetic algorithm.

2.1 Cost Function

Standard cost function is considered with the penalization of future control errors and control increments

$$J = \sum_{j=1}^{N_2} r_j(\hat{y}(k+j) - w(k+j))^2 + \sum_{j=1}^{N_u} q_j(\Delta u(k+j-1))^2 \tag{1}$$

where

N_2	is prediction horizon,
N_u	is control horizon,
r_j, q_j	are penalization parameters,

$\hat{y}(k+j)$ is prediction of controlled variable,
$w(k+j)$ is desired value (set-point) and
$\Delta u(k+j-1)$ is control increment (change, move).

The control horizon can be shorter than the horizon for the set-point following. We can reduce the computational complexity by supposing some control increments at the end of the control horizon to be zero – the control variable will freeze at the last calculated control action. Penalization of control increments is considered to avoid the steady-state control error – this would be the case if the absolute control actions were penalized.

2.2 Process Model and Its Identification

In the context of this contribution, identification is considered as a statistical approach that provides a model of a dynamical process from the measured data. Many different identification techniques are available and the suitable one must be chosen according to the specific requirements.

Usually, linear models in the form of ARX (Auto-Regressive model with eXogenous input) or ARMAX (Auto-Regressive-Moving Average model with eXogenous input) equation are used for MPC [13]. However, different approach is provided in this paper.

Dynamical neural models are generally able to approximate even highly complex and nonlinear processes very precisely since feedforward neural networks (FFNNs) are proven to be universal approximators [14]. On the other hand, a black-box-like structure of the neural model restricts a bit the conventional use for the controller design. Since the optimization is performed by a variant of genetic algorithm, no strict conditions are required to the model form. Therefore, a dynamical neural model can be preferably implemented for this case.

There are several possibilities available to design a neural model of the process. One of them assumes the process description by the following nonlinear discrete-time difference equation

$$\begin{aligned} y_s(k) = \psi[&y_s(k-1), \ldots, y_s(k-n), \\ &u(k-1), \ldots, u(k-m)], m \leq n \end{aligned} \tag{2}$$

where

$\psi(.)$ is a general nonlinear function,
y_s is calculated output from the process,
u is input to the process and
n is the order of the difference equation.

The aim of the identification is to design FFNN, which approximates the function $\psi(.)$. In other words, FFNN must be designed and trained to provide outputs as close to process outputs as possible. A general diagram of a dynamical neural model is illustrated in Fig. 1.

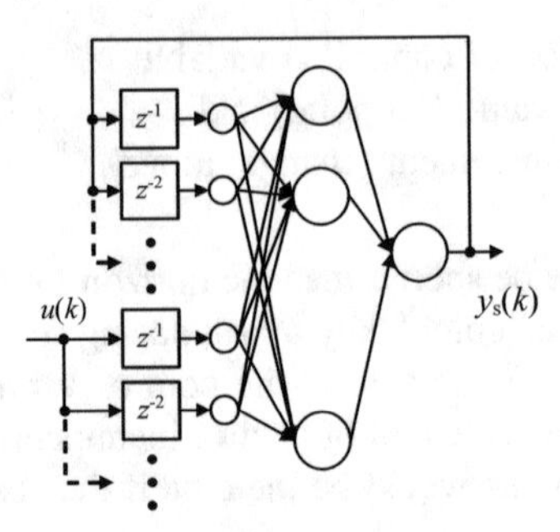

Fig. 1. Dynamical neural model.

A dynamical neural model design generally consists in training and testing set acquisition, neural network training and pruning, and neural model validating. These steps have already been defined comprehensively many times e.g. in [15, 16].

2.3 Optimization Method

Genetic algorithms are suitable for constrained, nonlinear optimization and non-convex optimization problems [8]. Genetic algorithms belong to evolutionary algorithms, they use population of candidate solutions. They are inspired by the Darwinian evolution using the crossover and mutation genetic operators. A more detailed description of this technique can be found e.g. in the book [17].

Papers where genetic algorithms have been used for MPC control are e.g. [8–10]. Authors suggested a variety of methods to deal with the time constraint of the predictive controller, such as using shifted individuals from previous time steps. Authors applied genetic algorithms to find a feasible descent solution rather than the optimal solution at each sampling time in [9]. Genetic algorithm for thermal process control is published in [11]. Genetic algorithms were used also in predictive control to select the control and prediction horizons [12].

Authors used a population-based genetic algorithm with real alphabet to optimize the cost function at each time step. The genes of individuals directly represent the sequence of control actions. The fitness function equals to the cost function.

When generating the initial population, we were inspired by the method of shifted individuals suggested in [8] and [9]. We do not generate individual control actions from the full range of possible control actions, but we derived individuals of the initial population from the shifted best individual from the previous time step. From the previous time step we remove the first control action to move the entire sequence of the control actions. Missing last control action is duplicated from the previous control action. The remaining individuals of the initial population are generated as random deviations of the individual genes of this modified individual within a given range.

As a selection mechanism we used tournament selection. We used the one-point crossover and the mutation that adds a random deviation to the selected gene within a given range. For termination of the run of the genetic algorithm we use maximum number of generations and maximum elapsed time because of time constraint of the controller.

3 Controlled System

Proposed algorithm is tested at a temperature control system GUNT RT040 [18] – see Fig. 2.

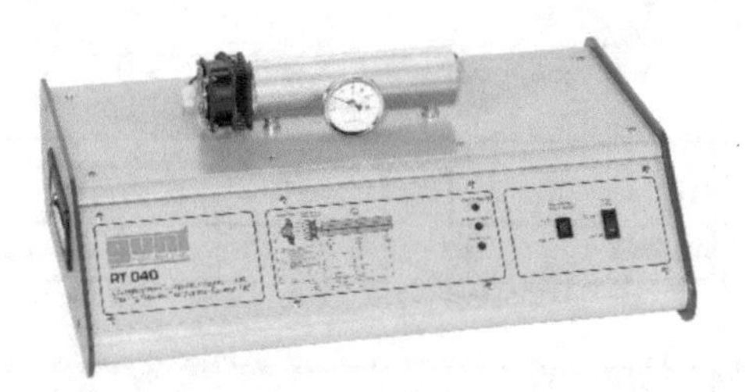

Fig. 2. RT 040 training system [6].

The metal rod is heated or cooled using the Peltier element (signal Y). The bar temperature is measured in three positions by the temperature sensors (X1, X2 and X3 signals). The fan (signal Z) on the Peltier element cooler can be switched on or off – see Fig. 3.

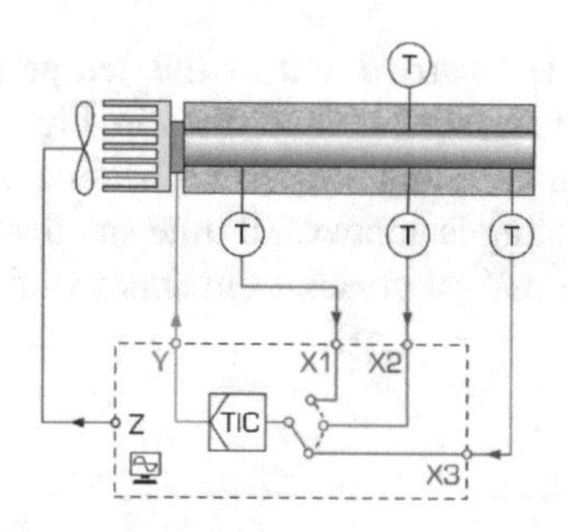

Fig. 3. RT 040 training system diagram [6].

Voltage (control signal of the Peltier element) in the rage from −5 to 5 V is used as a control action u (manipulated variable). Farthest sensor X3 (from Peltier element) is used as a temperature measurement point for the controlled variable y. RT 040 uses USB multifunction LabJack U12 data acquisition card [19] with drivers support to different environments. Measurement and control software is written as a MATLAB script and run on a PC – see Fig. 4. The same script is used for the simulation and for the control experiments. The simulation can be done in real time or as fast as possible. The real-time simulation allows to test whether the sample time is enough long for the certain algorithm and used hardware – load means what percentage of the sampling period is consumed with the measurement, control algorithm, actuation and plotting. The predicted controlled variable and the calculated future optimal control action are displayed – red line starting from the current time instant. The user can check whether the algorithm with given parameter works fine or must be stopped and retuned.

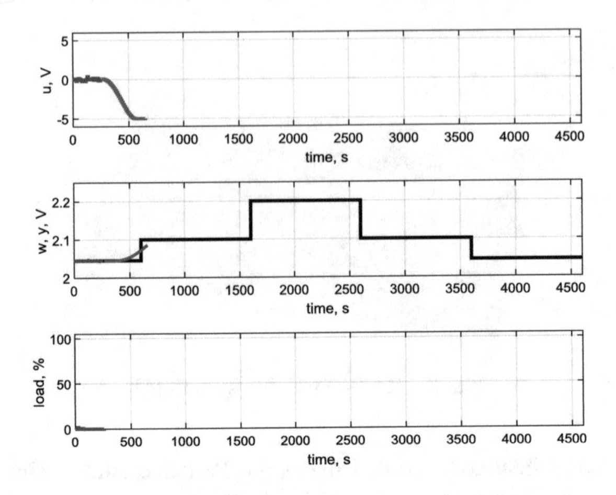

Fig. 4. Control software window in MATLAB.

4 Identification Experiments

Identification experiments are carried out – the temperature response to 14 step changes, each 1 h long is measured – blue lines in Fig. 5. Fan is on during all the experiments. Positive *u* means cooling, negative *u* heating. From the response, it can be seen that the gain by the cooling is approx. 10 time smaller than by the heating. This is usual nonlinearity of the thermal processes – efficiency of the cooling is smaller than of the heating.

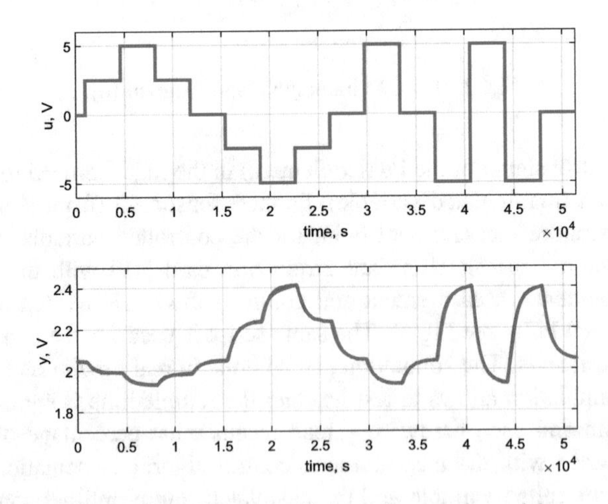

Fig. 5. Identification with neural network model.

The process is approximated with the dynamical neural model – red line in Fig. 5. The procedure of the neural model design is not described here, since it is a standard procedure. At the end of this procedure, the neural network inside the neural model consists of four inputs ($m = 2$, $n = 2$), five neurons with the hyperbolic tangent activation functions in the hidden layer, and one output neuron with the linear (identical) activation function. The sampling time is set to 10 s.

The model uses absolute u and y. This model is used by all simulations like a simulation model and by the neural network and genetic algorithm predictive controller for predictions – they are calculated in a cycle and used for the cost function evaluation.

Second order continuous-time transfer function parameters are identified for the predictive controllers with transfer-function and state-space model – Eq. (3) and red line in Fig. 6.

$$F(s) = \frac{-0.048}{(758s + 1)(91s + 1)} \tag{3}$$

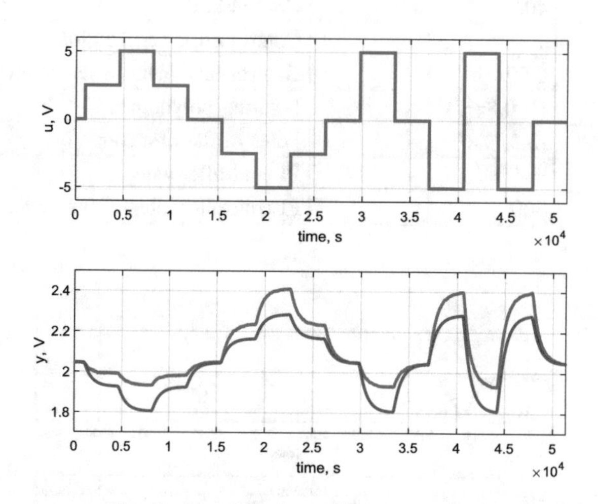

Fig. 6. Identification with linear model.

Working point $u = 0$ V, $y = 2.045$ V is used by the identification. The system has negative gain and practically one dominant time constant. Continuous-time transfer function is transformed into the discrete-time transfer function and state-space model for the predictive controllers.

5 Control Experiments

Four control experiments were carried out. The most important experiment is the experiment with the neural network and genetic algorithm – see Fig. 7. For comparison with the standard control methods another three experiments are measured – predictive controller with the transfer function model and quadratic programming – Fig. 8, predictive controller with the state-space model and quadratic programming – Fig. 9 and PI controller – Fig. 10. Algorithms for both predictive controllers can be found in [20]. The parameters of the controllers are in Table 1.

Table 1. Parameters of the controllers.

Symbol	Value	Dimension	Meaning
T_S	10	s	Sample time
N_2	40	s	Prediction horizon
N_u	40	s	Control horizon
r	1	–	Control error penalization
q	0.001	–	Control increment penalization
C	$(1-0.8z^{-1})^2$	–	Filtering polynomial
P	[0.8, 0.8]	–	Poles of the observer
r_0	−100	–	PI controller gain
T_i	300	s	PI controller integral time constant

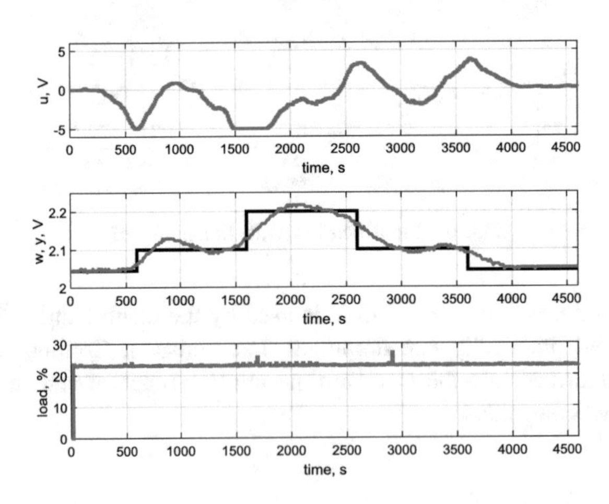

Fig. 7. Control with neural network and genetic algorithm.

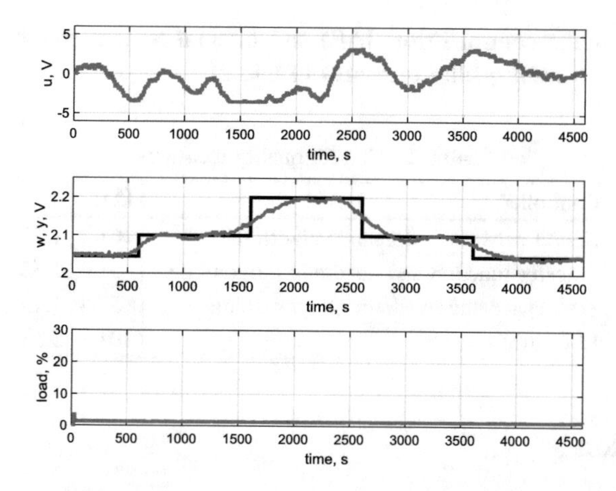

Fig. 8. Control with transfer function and quadratic programming.

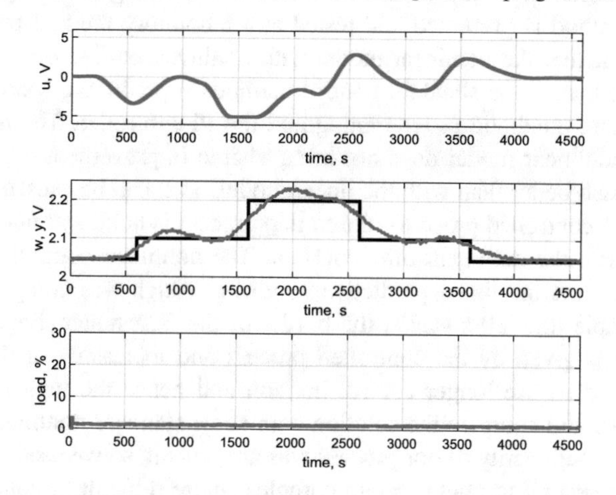

Fig. 9. Control with state-space and quadratic programming.

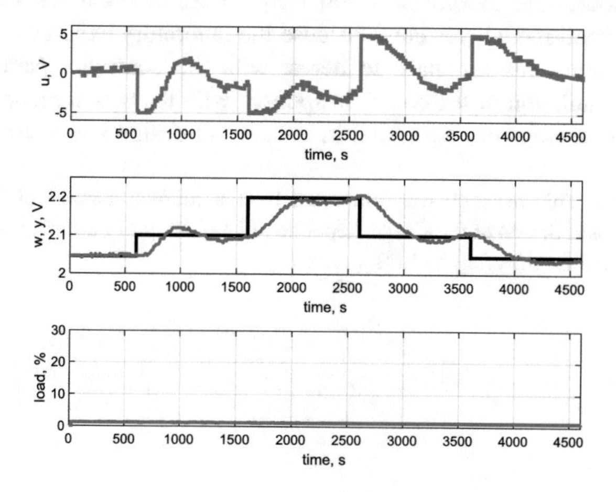

Fig. 10. Control with PI controller.

Integral Absolute control Error (IAE) and Integral Square control Error (ISE) is calculated as the control quality measures (Table 2).

Table 2. Control quality measures.

Controller	IAE	ISE
Neural network and genetic algorithm	61.0	1.75
Transfer function and quadratic programming	51.8	1.32
State-space and quadratic programming	58.4	1.43
PI controller	101.1	5.53

6 Conclusion

Predictive controller with the neural network model and genetic algorithms as the optimization method is presented and tested at a laboratory thermal process. Authors were curious whether the nonlinear model with nonlinear on-line optimization will be better or worse than some standard control methods – predictive controllers with the linear models and quadratic programming and the PI controller. The result is, that in our case, the nonlinear model does not bring visible improvement even if the identification was much better than with the linear model. This can be caused by the relative simplicity of the controlled process – controlled process is nonlinear but practically first order system with the dominant time constant. The nonlinear controller burdened the computer more than the linear predictive controller which was comparable to the PI controller. Sample time 10 s was at the border of the fastest samplings we can use – horizon length is given by the controlled process and increasing of the sample time leads to the need to use longer control horizon and hence the higher computational demands. We should compare the solution with some standard nonlinear MPC control algorithm. The nonlinearity of our process was not crucial, so we used the linear MPC. It would be interesting to control more complex, more difficult to control, more nonlinear process because there should be better to see the advantage of the NMPC over the other methods. Our system was too easy to control. On the other hand, the advantage of a good model was clear because the laboratory experiments took quite a long time and there was no space to iterate with the controller parameters tuning. Therefore, the simulation of the control responses with the neural model and expected measurement noise was very beneficial in the control design and testing phase.

Acknowledgment. This research was supported by Institutional support of The Ministry of Education, Youth and Sports of the Czech Republic at University of Pardubice and SGS grant at Faculty of Electrical Engineering and Informatics.

References

1. Rossiter, J.A.: Model-based Predictive Control – A Practical Approach. CRC Press, Boca Raton (2003)
2. Camacho, E.F., Bordons, C.: Model Predictive Control. Springer, Heidelberg (2007)
3. Maciejowski, J.: Predictive Control with Constraints. Prentice Hall, Upper Saddle River (2002)
4. Allgöwer, F., Findeisen, R., Nagy, Z.K.: Nonlinear model predictive control: from theory to application. J. Chin. Inst. Chem. Engrs. **35**(3), 299–315 (2004)
5. Magni, L., Raimondo, D.M., Allgöwer, F.: Nonlinear Model Predictive Control - Towards New Challenging Applications. Lecture Notes in Control and Information Sciences, vol. 384. Springer, Heidelberg (2009)
6. Qin, S.J., Badgwell, T.A.: An overview of nonlinear model predictive control applications, In: Nonlinear Predictive Control, pp. 369–392. Springer, Heidelberg (2000)
7. Honc, D., Doležel, P., Gago, L.: Predictive control of nonlinear plant using piecewise-linear neural model, In.: Proceedings of the 21st International Conference on Process Control, pp. 161–166 (2017)
8. Onnen, C., Babuška, R., Kaymak, U., Sousa, J.M., Verbruggen, H.B., Isermann, R.: Genetic algorithms for optimization in predictive control. Control Eng. Pract. **5**(10), 1363–1372 (1997)
9. Chen, W., Li, X., Chen, M.: Suboptimal nonlinear model predictive control based on genetic algorithm, In.: Third International Symposium on Intelligent Information Technology Application Workshops, IITAW 2009, pp. 119–124. IEEE (2009)
10. Blasco, X., Martinez, M., Senent, J., Sanchis, J.: Generalized predictive control using genetic algorithms (GAGPC). An application to control of a non-linear process with model uncertainty. In: International Conference on Industrial, Engineering and Other Applications of Applied Intelligent Systems, pp. 428–437, Springer, Heidelberg (1998)
11. Stojanovski, G., Stankovski, M.: Model predictive controller employing genetic algorithm optimization of thermal processes with non-convex constraints. In: Genetic Algorithms in Applications. InTech (2012)
12. Mohammadi, A., Asadi, H., Mohamed, S., Nelson, K., Nahavandi, S.: Optimizing model predictive control horizons using genetic algorithm for motion cueing algorithm. Expert Syst. Appl. **92**, 73–81 (2018)
13. Bobál, V.: Digital self-tuning Controllers: Algorithms, Implementation and Applications. Springer, London (2005)
14. Hornik, K., Stinchcombe, M., White, H.: Multilayer feedforward networks are universal approximators. Neural Netw. **2**(5), 359–366 (1989)
15. Haykin, S.: Neural Networks: A Comprehensive Foundation. Prentice Hall, Upper Saddle River (1994)
16. Korbicz, J., Janczak, A.: A neural network approach to identification of structural systems. In: Proceedings of IEEE International Symposium on Industrial Electronics, Poland, pp. 98–103, vol. 1 (1996)
17. Goldberg, D.E., Holland, J.H.: Genetic algorithms and machine learning. Machine Learn. **3** (2), 95–99 (1988)
18. RT 040 Training system: temperature control, HIS. https://www.gunt.de/en/products/mechatronics/automation-and-process-control-engineering/simple-process-engineering-control-systems/training-system-temperature-control-hsi/080.04000/rt040/glct-1:pa-148:ca-83:pr-1045. Accessed 12 Jan 2019
19. U12 Series. https://labjack.com/products/u12.Accessed 12 Jan 2019
20. Honc, D., Sharma, K.R., Abraham, A., Dušek, F., Pappa, N.: Teaching and practicing model predictive control. IFAC-PapersOnLine **49**(6), 34–39 (2016)

Adaptation of Open Up in the Scrum Framework to Improve Compliance in Scope, Risk Management and Delivery Times in Software Development Projects

Omar L. Loaiza(✉) and José M. De León

Universidad Peruana Unión, Carretera Central Km. 19.5, Lurigancho-Chosica, Lima, Peru
{omarlj,magdielaguilar}@upeu.edu.pe

Abstract. This study consisted of adapting artifacts and practices from the Open Up development method in Scrum to improve compliance in scope, risk management and delivery times in software development projects. The practices (administrative, technical and deployment) and artefacts of Open Up were adapted in planning events, Sprint and Scrum retrospectives. Improvements were measured by indicators with the following results; in scope: delivery deadlines met (66% to 95%) and missed milestones (80% to 0%); in risk management: number of risks (from 4 to 3) and risk identification (from 4 to 9) which means a reduction in incidence from 100% to 33%; and, in time compliance: project delay (53% to 13%) and in delayed tasks (42.50% to 31.75%). The effects of the adaptation were: the development of plans was improved, the control of the evolution of the architecture of the product was improved as well as the deliveries were better planned. This study is relevant because it allows orienting software development projects where there is a need to follow plans based on prescriptive and normative contexts of an organization and at the same time have an agile approach. The results conclude that the adaptation improved the project with a surpassed learning curve; it is useful in prescriptive and agile development environments. It is suggested to replicate the study under similar conditions and/or other indicators.

Keywords: Scrum adaptation · Hybrid development methods · Open Up development method · Scope compliance · Time compliance · Risk management

1 Introduction

1.1 Research Context

Organizations increasingly handle a greater demand to satisfy with products and services to their clients, creating a sensation of constant organizational adaptation, which involves carrying out changes in their internal processes and in those that are related with their businesses [1]. It is clear that this has an impact on software development projects, making requirements management and development a very tedious task [2]. It

R. Silhavy et al. (Eds.): CoMeSySo 2019, AISC 1046, pp. 404–418, 2019.
https://doi.org/10.1007/978-3-030-30329-7_36

becomes more difficult when the scope [3], one of the most important artifacts, is provided and/or managed by the client, as it conditions and varies the estimation of time, causing risks for the project, which in the long run has effects on the quality of the product.

In software development, projects need to be managed by development methods or frameworks that allow for this purpose. Those managed by agile approaches, among them Scrum, are sensitive to frequent changes in requirements, caused not only by business conditions or the influence of organizational culture but also by the learning curve assumed by trial and error approaches in development teams as Schwaber maintains [4], by causing management in scope and time to be significantly affected [5], which shows that the management of agile projects, as in Scrum, has aspects to overcome as expressed by Sommerville [2] and Batra et al. [6], as well as the results reported by Bista, Karki and Dongol [7], and Ghezzi [8]; so that carrying out projects managed by a plan and at the same time swiftly developed is, nowadays, progressively a good option [2].

It is good to point out that Boehn and Turner [9] maintain that the success of a project requires a dose of agility but at the same time a quota of planning so as not to be placed at the extremes; where the center of gravity for greater success lies in the balance managed by risks [9]. There can be some danger in misinterpreting the agile manifesto [10], on the one hand, an extreme can be found in appreciating the agile elements that postulate themselves more than traditional elements; on the other hand, to act strictly under very structured forms would be very restrictive and harmful for the success of the projects.

Cooper suggests that there are situations in which hybrid methods are used [11], with agile aspects due to their flexibility in the face of existing changes in projects with a certain dose of traditional methods because they provide stability [12]. The advantage of providing prescriptive answers to an agile working method/framework is that the latter does not lose its adaptability but gains answers; on the other hand, following a merely prescriptive method is more cumbersome and less ductile for medium and small projects, because most of its assumptions (artefacts and practices) have to be fulfilled in order to be effective.

The purpose of this research study was to apply the adaptation of the Open Up development method within the Scrum framework in a software development project, which, due to its characteristics, required an agile handling but at the same time with prescriptive responses (adaptable to the norms and restrictions of the environment) in the planning to improve the fulfillment of the management of scope, time and risks in a State organization of the health sector in Peru. This application experience is relevant, as it can serve as a reference for similar realities, expanded and/or exptrapolated with other indicators.

1.2 Scope, Time and Risk Management in the Open Up Method

Open Up is an agile philosophy [13] software development method that applies the iterative and incremental development strategy, although more prescriptive than Scrum. This method is based on the life cycle of software development in four stages:

inception, development, construction and transition. In addition, Open Up is composed of roles, practices and artifacts.

The development of this method is based on the division of the planning of a project into iterations, and these into minor tasks (micro increments) to produce software deliveries to the customer. The practices are divided into three categories: administrative, technical and deployment [13]; those of an administrative nature make it possible to delimit the scope of a plan and its subsequent monitoring; those of a technical nature are activities that have a direct impact on the evolution of the architecture in the life cycle, which foster or produce artefacts until a software is delivered; while deployment practices have to do with the release and implementation of the software in the client's environment. It should be noted that planning reduces risks.

The artefacts that show the work of the Open Up process are: software and documentation [13]. These artifacts have to do with: architecture development, deployment, product development, development environment and customer management, testing and requirements management, and project management.

The artifacts of requirements management and project management serve to delimit the scope – this translates into a schedule that indicates the time horizon of the project – in the marking of milestones, that indicate if deadlines are met, identify the delays and produce deliverables in reasonable times, and additionly the team can incorporate the conscious management of risks.

1.3 Scrum Framework and Its Challenges in Project Management

Scrum is a generic framework, plausible to be used for a myriad of scenarios [3, 14, 15]. Its basic structure can be seen in Fig, 2. This framework starts from a Product backlog, which is a set of requirements that must be developed; this product stack is divided into smaller sub-sets called Sprint Backlog; these two artifacts are carried out in agile and empirical planning events, where it is estimated how long it will take to implement the requirements and the cost of effort. Each Sprint Backlog is executed in an event called a Sprint, which ends when a partial version of the final product is delivered. In Scrum, the team coordinates before the Sprints, during day meetings, and after the Sprints to elaborate lessons learned (Sprint Retrospecrive) and a technical revision of the product (Sprint Review). Scrum's focus is on the product and not so much on the process, hence its empirical character [16].

On the other hand, Scrum is a framework adopted in the world of software development [17], hence its empirical character and the reason why adaptive studies are done; it is also unprescriptive and highly flexible [18]. In spite of its popularity and flexibility, it may be too specific, just like a custom-made suit [19], and may even seem incompatible with some realities [15, 20] that by their nature and normativity require it.

In this regard, Amir et al. [19] mentions that Scrum is intolerant of the transition from prescriptive environments to agile environments and viscerversa because it is generic; on the other hand, its emphasis is on achieving the product and dealing little with architecture [16], which generates disconnects between software deliveries and requirements, which explains its weak capacity to handle the scope of software projects.

For their part, Ashraf and Aftab [15] report adaptations where there is an emphasis on improving equipment at the level of practices and artefacts and varying the product backlog techniques, but a marked weakness persists for managing scope and time. There are few notions of how to manage the development of an architecture and risk.

Meanwhile, Hajjdiab and Taleb [21] found that the palpitating challenges in agile software development lie in the management of requirements throughout the process, the implementation of a process in the sprint, weakness in decision making derived from poor planning as well as the evaluation of project performance through indicators.

In addition, Wan, Zhu and Zeng [22] argue that success in agile projects is achieved by balancing [9] the product focus with some degree of adaptation in the management of software projects [2, 6] i.e.: the need to explain artifacts, be guided by a basic schedule, follow an order, etc.; this involves taking measures to synchronize the coherence between requirements and deliveries.

In relation to risks, changes in the project environment [9] foster a disturbing change in the expectations that projects need to manage in their development process, so inexperience and self-learning in the management of Scrum's flexibility can be counterproductive.

This highlights the need to continue experimenting in adaptations of practices and artefacts in areas of software development where the empirical evidence is still insufficient to cover all situations by the permanent development of the software industry and the changing reality that involves organizations of all types and sizes. These initiatives should aim at improving project management on various types of indicators by trial and error.

2 Methods

This research study was carried out in 4 stages (see Fig. 1) and took into consideration the normative prescriptive context of the organization where the project was executed and expert opinion for the adaptation.

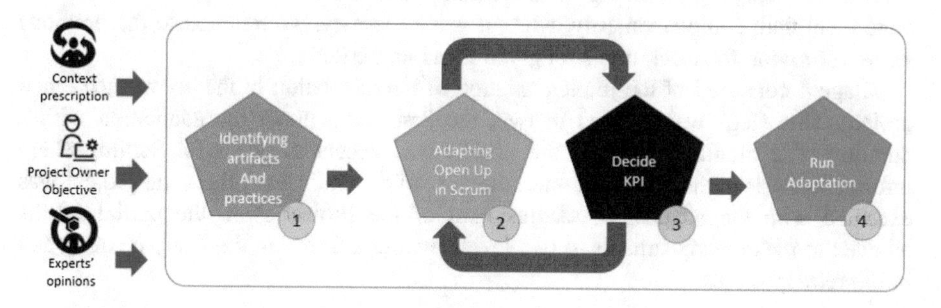

Fig. 1. Metodological sequence of the investigation.

In the first stage, the artefacts and practices (administrative, technical and deployment) of the Open Up development method to be adapted within the Scrum framework were identified through the application of two expert surveys. The inclusion criteria for the experts were: (1) professionals from other development teams or staff who worked previously in the study site, as well as people with similar realities to the project, (2) experience in development projects that have taken into account prescriptive methods such as RUP or something similar and within the Scrum framework, and (3) who have participated in projects of organizations with a tendency to use their organizational norms. The number of experts consulted and reached was 20 due to the inclusion criteria.

As for the structure of the instruments, the Open Up practice is divided into three sections: administrative practices with 6 items, technical practices with 6 items and deployment practices with 2 items; with respect to the instrument of artefacts, there was a relationship of 29 items. The questions for these two instruments were extracted by reviewing the Open Up documentation [13]. Each instrument contains questions, whose options consider the relevance or nonrelevance based on experts' experience, i.e. whether one or more of the 29 artefacts and 15 practices of Open Up should be adapted to Scrum; each question has options on a scale of 1 to 5, where 1 is "not necessary" and 5 is "very necessary"; so that the scope, delivery times and risk management can be met. The result of this stage is an identified set of artifacts and practices to be adapted in the next stage.

In the second stage, the artifacts and practices of Open Up in Scrum were adapted. Thus, Open Up activities and artefacts were adapted to Scrum events, moving through the Scrum framework under the following criteria: (1) to comply with normative prescriptions for planning and monitoring, (2) to promote teamwork, (3) to facilitate the flow of information, (4) to measure the progress of what has been planned, and (5) to allow for the identification of risks.

For the third stage, the indicators for measuring scope, time and risks were determined based on the work of Montero, Omieva and Palacin [23] and can be seen in Tables 4, 5 and 6. Then, after their reviewing the documentation, the documentary sources necessary for recording the information and making the respective calculation were identified; complementarily, when it was necessary, we returned to the previous stage to provide feedback and strengthen ideas and criteria.

Stage 4 consisted of the implementation of the adaptation in the above-mentioned project. This stage was divided in two, the first one without the adaptation with a duration of 2 months, in which the project was executed only with Scrum, where information was gathered for the measurement of the indicators; then, the project was executed with the adaptation, readjustment of the Project Plan, the partial results allowed to make readjustments to two levels, within a Sprint in the Daily meeting and in the retrospectives.

3 Results

3.1 Identification of Artifacts and Open Up Practices to Be Adapted in Scrum

In the results of the survey of experts, of the 29 Open Up artefacts, 7 were considered to have the greatest importance in managing time and scope in projects in the life cycle (Table 1). The development team that applied the adaptation matched the experts' scores when they were interviewed, taking into account state regulatory requirements [6].

Table 1. Open Up artifacts selected to adapt

Artifacts	Mean	Some what necessary	Necessary	Very necessary
Project Plan	4.50	5%	40%	55%
Risk list	4.00	20%	50%	25%
System-wide requirements	4.30	10%	50%	40%
Implementation	4.25	10%	55%	35%
Build	4.35	10%	45%	45%
Back out Plan	3.85	35%	45%	20%
Release	4.55	15%	60%	25%

For the experts, the practices in Table 2 guide the process by means of a Project Plan to comply with deliveries stipulated in short periods that are a normative requirement of the state health entity in Peru [6]. The practice of life cycle risk assessment addresses risks between tasks, ensuring that there are not many biases and that there is better processing in the request for changes, improving adherence to the schedule; the effect is seen in the outputs of iterations (sprints) when continuous integration becomes less complicated and better planned by having a better controlled evolutionary architecture.

Table 2. Selected Open Up practices to be adapted

Practices	Mean	Some what necessary	Necessary	Very necessary
Iterative development	4.30	10%	50%	40%
Risk-value lifecycle	4.00	20%	60%	20%
Release planning	3.80	35%	50%	15%
Whole team	3.95	25%	55%	20%
Tool selection	3.75	30%	50%	10%
Concurrent testing	3.65	45%	40%	15%
Continuous integration	3.85	30%	55%	15%
Evolutionary architecture	3.70	45%	35%	20%
Evolutionary design	3.70	45%	40%	15%
Production release	4.25	15%	45%	40%

Now that the artefacts and practices of Open Up have been selected, it is time to indicate the type of participation they have in the adaptation guided by the characteristics of the project and the objectives of this study (see Table 3). Some elements do this to collaborate with scope (S) through a plan, others with scope and time (S, T) to monitor how the plan is being met in project development; and finally, some assist in identifying and addressing risk management (R). It should be noted that the practices and artefacts add value to what Scrum does, do not alter the events, i.e., help the development team to focus the project between that which is planned by the prescriptive context of the project and the agility needed for deliveries. In the case of artifacts, these are constituted as inputs or outputs (see the direction of the arrows in Fig. 2) for Scrum events; while practices are constituted as empirical adaptations to Scrum events, hence the arrow of an artefact points towards an event, as can be seen in Fig. 2.

Table 3. Participation of Open Up practices and artifacts in Scrum

Open Up elements	Scrum events					
	Sprint backlog	Sprint planning meeting	Sprint	Daily meeting	Sprint review	Sprint retrospective
Artefacts						
Project plan	S	S				
Risk list		R				
System-wide requirements	S					
Implementation			ST			
Build			ST			
Back out plan					ST	
Release		S		ST	ST	
Practices						
Iterative development			ST			
Risk-value lifecycle	R	R	R	R	R	R
Release planning		S				
Whole team				ST		ST
Tool selection			ST			
Concurrent testing			ST			
Continuous integration			ST			
Evolutionary architecture			ST			
Evolutionary design			ST			
Production release					ST	

3.2 Adaptation of Artifacts and Practices of Open Up in the Scrum Framework

Since Scrum is a general purpose framework, Open Up practices and artefacts identified in Tables 1 and 2 have been inserted, and can be seen in Fig. 2. The artefacts constitute the input or output of Scrum activities (product backlog, sprint backlog, sprint, etc.); while the practices are activities that must be taken into account when executing events (sprint, daily meeting, sprint review, etc.).

With regard to the Product backlog, this was fed by a set of functional requirements (FR) raised from the application of an interview guide with key actors in the affiliate coverage process in the areas of health care and funeral services provided by the state entity which is the focus of the study; these FR were taken to a Requirements-Activity matrix to align business activities with the FR and that the team formulate the Use Cases of the System. These Use Cases are the input to construct the Product Backlog; and at the same time it serves as input for the elaboration of the Project Plan next to the Release planning, given that the contracting entity needs it to agree to the calendarization of programmed deliveries with certain conditions and thus to satisfy proceedings and administrative terms. The sprint backlog is fine-tuned and detailed based on information and restrictions of the Project Plan; the Release planning and the Risk List is inserted in the tasks of each sprint backlog to develop delivery tasks and manage risks that affect the schedule, costs and other aspects.

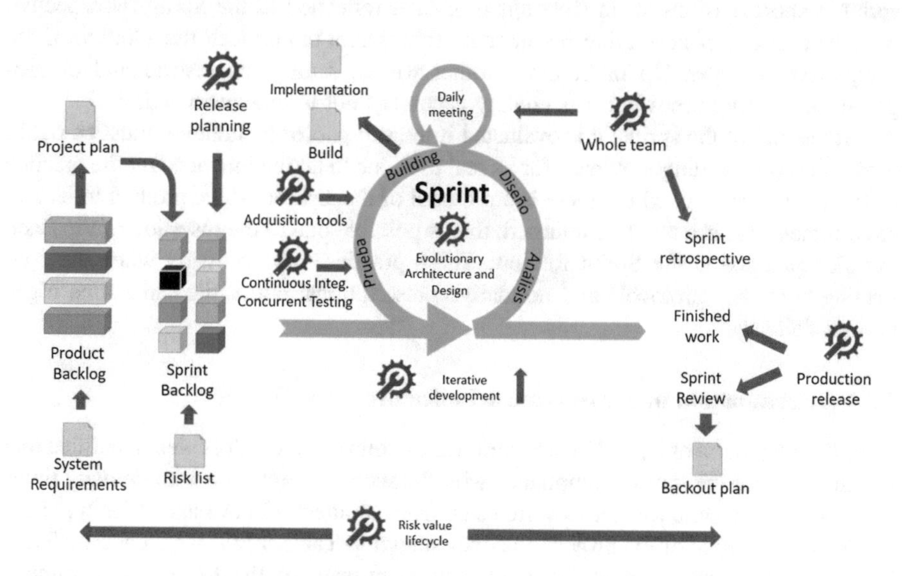

Fig. 2. Adaptation of Open Up practices in Scrum.

In the sprint, practices and artifacts of Open Up have also been added. The sprint considers fundamental activities for the development of software that come from the life cycle of software development [4, 24]: analysis and design (for the design of the architecture), construction (programming of the architecture) and testing.

The iterative development activity ensures that the guidelines of the Project Plan are present in each sprint; therefore, as a way of collaboration, a practice is inserted that monitors how the architecture and the characteristics evolve throughout part of the development life cycle (analysis, design, building and testing) in order to verify their traceability with the FR. The practice of obtaining tools is inserted within the building section, so that it guides how the developers have them depending on the type of license; the building section outputs are a functionality or implemented sub part (main files of the source code, libraries, etc.) and the compiled software, either a funconality or sub system that is considered operational to be tested. Now, when testing is done, it is relevant to run continuous integration actions guided by internal testing in the section that corresponds to a developer or that covers more systemic sections in concurrent and collaborative environments such as Github.

Additionally, the practice of the risk life cycle (identification of all objectively possible risks according to reality and the treatment of them by iterations) crosses the pre and post sprint events as well as during them in a complementary way, without losing sight of the results and situations occurred in each iteration and improvement; the risks are seen in the daily meetings and in the Sprint review and Sprint retrospective, so that decisions can be made to obtain the planned software products.

It is also worth noting the management of the team to organize itself daily in each sprint in short rensions, or in meetings of greater reflection as the Sprint retrospective with the purpose of generating productivity throughout the project; the addition of the perspective of Open Up in Scrum in other words, setting more structured or programmed meeting times but it is equally necessary not to take away agility.

At the end of the sprints it is evaluated if the sub part of an iteration must be put to production (proceeding to release for its real use) due to taking into account the practice of Production release, where there is a protocol of disposition of the product to its real environment; or, if it must be returned; this is possible due to the insertion of the Back out plan practice in the Sprint Review. These practices are necessary when the contracting party has protocols and normative prescriptions, but at the same time urges prompt delivery.

3.3 Determination of Measurement Indicators

Once the adaptation of Open Up in Scrum was established, indicators were identified that would allow for measuring compliance with the scope, compliance with delivery times and risk management. Indicators were taken from Montero, Onieva and Palacin [23].

As for the scope of the project, the ones shown in Table 4 [23] were chosen. These indicators were measured at the frequency mentioned in the table, and assume a behavior over time in the execution of the project.

Table 4. Indicators of the scope of the project

Indicator	Formula	Desired trend	Period of capture	Measuring frequency
Delivery deadlines met	$\frac{\text{Completed deadlines}}{\text{Initial project deadlines}}$	Neutral or positive	Punctual	Monthly
Failed milestones	$\frac{\text{Failed milestones}}{\text{Total milestones}}$	Negative	Month	Monthly

As far as timekeeping is concerned, they are listed in Table 5 [23]:

Table 5. Project time indicators

Indicator	Formula	Desired trend	Period of capture	Measuring frequency
Project delay	$\sum_i Delays$	Negative	Exercise to date	Monthly
Overdue tasks	$\frac{\text{Overdue tasks}}{\text{Current tasks}}$	Negative	Punctual	Week

Finally, the indicators for risk management are in Table 6 [23] and they are:

Table 6. Indicators of project risks

Indicator	Formula	Desired trend	Period of capture	Measuring frequency
Risks	Number of risks	Positive	Punctual	Trimonthly
Possible risks	$\frac{\text{Number of possible risks}}{\text{Number of total risks}}$	Negative	Punctual	Monthly

Once the indicators were established, it was possible to apply the adaptation of Open Up in Scrum in the focus location of this study.

3.4 Results of the Execution of the Adaptation in the Management of the Scope, Risks and Delivery Times

The results correspond to the implementation of stage 4 of the research study:

Results of Project Scoping Indicators

The indicators used to measure compliance with the scope were analyzed and studied according to the stipulations of the Project Plan and Release planning versus the dates recorded in the artifacts throughout the process, in order to determine if there is a variation in compliance with deadlines and milestones. The results are shown in Table 7:

Table 7. Results of indicators in compliance with the scope

Indicator	Month					
	Scrum		Adaptation			
	1	2	3	4	5	6
Deadlines met	1	5	7	10	14	18
Theoretical deadlines	3	4	3	2	4	3
Delivery deadlines met	33%	71%	70%	83%	88%	95%
Failed milestones	2	1	4	2	1	0
Total milestones	3	2	5	7	5	4
Failed milestones	66%	50%	80%	29%	20%	0%

In month 3, the percentage of compliance decreased, and this is due to the fact that when the implementation of the adaptation began, the members of the team did not understand in depth what they had to do. They had inconveniences, when making the modifications to the project plan. The demands of the Open Up artifact allowed them to glimpse elements such as: software delimitation, requirements matrix, among others, for which the project plan had to be reconstructed again. In addition, team members were not clear about how the contingency plans provided by the risk-related artifact worked. In addition, they learned how to make plans to roll back Scrum's retrospectives. Almost from month 4 on, the team got used to properly fulfilling its responsibilities, therefore, the percentage of deadlines met increased.

Regarding the milestones, in the first month of the implementation of the adaptation (month 3), there were 80% of milestones failed due to having to pass through a learning curve, the practice of Open Up called Whole Team was used to improve the productivity factor with the team structure and collaboration within the team at the Daily meeting and Sprint retrospective. This synergy allowed the missed milestones to diminish.

Delivery Time Results

Table 8 shows the behavior of the percentage of delayed tasks during the time in which Scrum was implemented (month 1 and 2), this shows that regularly, there were delayed tasks. It is interesting to note that as the practices adapted to Scrum, initially (month 3) there is an increase in the delay, but as in other cases, when these are mastered and better understood in light of the Project Plan, there is a decreasing trend in project delays.

Table 8. Results of the indicator on overdue tasks

	Month	% Week 1	% Week 2	% Week 3	% Week 4	% Monthly	Average delay
Scrum	Month 1	75%	67%	50%	25%	53%	42.50%
	Month 2	50%	20%	33%	20%	32%	
Adaptation	Month 3	60%	50%	60%	66%	59%	31.75%
	Month 4	25%	33%	33%	25%	29%	
	Month 5	40%	33%	14%	20%	26%	
	Month 6	20%	14%	0%	13%	13%	

It should be noted that the introduction of the practice of Release planning contributed to the reduction of delays, because it balanced the introduction of planning tasks of the deliveries foreseen in the Project Plan in each Sprint with the delivery of pending tasks, causing that the end of the month was not overloaded. This situation was overburdened at the beginning of the adaptation, hence 59% of tasks were delayed, but then improved because there was a practice that made the percentage of compliance visible in daily meetings and encouraged decision making as an agile practice.

During the use of Scrum there was an average 42.5% delay in tasks, while with the implementation of the adaptation this average decreased to 31.75%, representing an improvement of 10.75%. It is also interesting to note (see Table 8), that the first weeks of each month the percentage of delay is greater than the subsequent ones, due to the progressive improvement in the adjustment in the execution of activities due to Release planning, although the tendency of week 1 in each month is downward.

The delay of the tasks had a domino effect on the overall delivery of the project. However, it was also exceeded from month 3 onwards (see Table 9).

Table 9. Project delay indicator results

Scrum		Adaptation			
Month 1	Month 2	Month 3	Month 4	Month 5	Month 6
53%	32%	59%	29%	26%	13%

Results in Risks

The indicators used to measure risk management were analyzed and studied according to their behavior over time as shown in Table 10:

Table 10. Results of risk-related indicators

Indicator	Scrum		Adaptation			
	Month 1	Month 2	Month 3	Month 4	Month 5	Month 6
Number of possible risks	4	5	3	1	3	3
Number of total risks	4	6	6	4	8	9
Incidence of possible risks	100%	83%	50%	25%	38%	33%

During the application of Scrum, the number of possible risks and the number of total risks were very similar; however, when implementing the adaptation, the trend of possible risks has a tendency to remain and the trend of total risks to increase. Indeed, during the development of the project this was what happened, since, thanks to Open Up practices in the life cycle of risk value, among others and to the selected Open Up artefacts, it was possible to first detect more risks that were hidden in months 1 and 2, thus increasing the total number of risks. The possible risks were managed thanks to the introduction of Release planning tasks and risk identification and management activities in the sprints and Daily meeting respectively.

The application of adapted practices helped to detect more risks. The 5 extra risks detected in the second quarter versus the first, represent an improvement of 31.25%, which evidences a positive trend of the indicator.

4 Discussions

Scrum is very popular worldwide, adapts to equipment development needs and enables faster deliveries to the customer [17]. However, when there are contexts that are a little more normative than others or simply that need to be guided by a plan [8], or when the scope of the project suddenly crosses diverse organizational areas, it is insufficient in planning and management given its agile and more operative character [11], so it is beneficial to move to alternatives that combine agility with planning as shown in this article [2].

In this sense, regarding adaptation, 7 artefacts and 10 practices were introduced to insert practices in planning, management and risk management, as well as to manage the above in Scrum with the software development life cycle [9]. This adaptation allows for the execution of development projects in contexts where the software delivery can be carried out in the shortest time [10] in an iterative and incremental way, and at the same time fulfills normative prescriptions [5]; and, to reinforce the original Scrum events. As evidenced by the exposure of the results, there was a learning curve that the team went through between month 2 and the beginning of month 3 to understand adaptation by trial and error [4]. Measuring indicators allowed the development team to monitor project performance.

It is important to point out that this experience is not universal, but rather pertinent and useful communication that supports the tendency to shift to hybrid development methods, at times when agile projects face challenges of improvement and maturation such as: support in the management of activities, inconsistency between what the client asks and what he/she receives, achievement and monitoring of goals, compliance with regulations and standards of the client, etc. [17]. The above challenges trigger a chain of risks, which when managed, aims to improve compliance in scope and delivery times in software development projects.

The results evidenced in the fulfillment of the scope and time suggest that the fulfillment of commitments improves gradually, thanks to the help provided by the project plan [11], because it promotes empowerment in the team to manage tasks [11], the management of risks and the formalization of agile practices sometimes of very free will [8], as well as to evaluate the releases during the Sprints to their endpoints in the daily meetings and the retrospectives, and of being necessary to return to the Sprint internal nonconformities of the team by means of the Back out plan; similar improvements were reported when RUPs with Scrum were applied in Brazilian government entities [20]. The considerations stated at the beginning of this paragraph allowed for a sustained increase in compliance from month 3 onwards and a decrease in missed milestones as shown in Table 7; a reduction in delays from 42.50% to 31.75% when only executed with Scrum and then with adaptation respectively, the effect of which is directly seen in the decrease in overall project delay from 53% to 13%. Finally, the introduction of release planning and its management in sprints helps to visualize risks in time, because it shows what should have been in the outputs. Table 10 summarizes this improvement.

5 Conclusions

The adaptation of the Open Up software development method within the Scrum framework, through selected artifacts and practices, leads to the result that compliance with the scope, time and risk management in a software project was improved. This adaptation is convenient when the context of the project requires quick deliveries but at the same time needs to comply with normative prescriptions that guarantee its development and therefore must be contemplated in the sprints. It must also be said that the artefacts and practices introduced strengthened the usual Scrum events as well as improved synergy in the development team.

Furthermore, as far as indicators are concerned, these may vary according to the needs of each project. The indicators of Montero, Onieva and Palacin [23] were a good starting point and in this research study they served to cover three basic needs in every project (scope, time and risk) [5] which were mentioned in the objective of this work. It would be interesting to replicate this research study with the same indicators and other additional ones to reinforce the results of this study, and to be able to factor efforts into a new hybrid model of software development, or, in any case, into a useful model goal for similar realities, given that the regulatory framework and prescriptive spectrum may vary from project to project.

References

1. Broy, M.: The leading role of software and systems architecture in the age of digitization. In: Gruhn, V., Striemer, R. (eds.) The Essence of Software Engineering. Springer, Cham, pp. 1–23 (2018)
2. Sommerville, I.: Ingeniería De Software. Pearson, Madrid (2011)
3. Puarungroj, W., Boonsirisumpun, N., Phromkhot, S., Puarungroj, N.: Dealing with change in software development: a challenge for requirements engineering. In: 2018 3rd Technology Innovation Management and Engineering Science International Conference (TIMES-iCON), pp. 1–5 (2018)
4. Schwaber, K., Sutherland, J.: The scrum guide (2016)
5. Walczak, W., Kuchta, D.: Risks characteristic of agile project management methodologies and responses to them. Oper. Res. Decis. **23**(4), 75 (2013)
6. Batra, D., Weidong, X., VanderMeer, D., Dutta, K.: Balancing agile and structured development approaches to successfully manage large distributed software projects: a case study from the cruise line industry. Commun. AIS **27**(8), 379–394 (2010)
7. Bista, R., Karki, S., Dongol, D.: A new approach for software risk estimation. In: 11th International Conference on Software, Knowledge, Information Management and Applications (SKIMA), vol. 2017, pp. 1–8, December 2017
8. Ghezzi, C.: Formal methods and agile development: towards a happy marriage. In: The Essence of Software Engineering. Springer, Cham, pp. 25–36 (2018)

9. Boehm, B., Turner, R.: Balancing Agility and Discipline, 7th edn. Addison Wesley, Boston (2009)
10. Manifesto for agile software development (2001). http://agilemanifesto.org/iso/es/manifesto.html
11. Cooper, R.G.: Agile-stage-gate hybrids. Res. Technol. Manag. **59**(1), 21–29 (2016)
12. Jiménez, E., Orantes, S.: Metodología Híbrida para Desarrollo de Software en México. CiCIG **1**, 5 (2012)
13. Eclipse Process Foundation, "Open Up." https://wiki.eclipse.org/EPF_Wiki_User_Guide
14. Kniberg, H.: Scrum Y Xp Desde Las Trincheras, 1st ed. InfoQ (2007)
15. Ashraf, S., Aftab, S.: Latest transformations in scrum: a state of the art review. Int. J. Mod. Educ. Comput. Sci. **9**(7), 12–22 (2017)
16. Ciric, D., Lalic, B., Gracanin, D., Palcic, I., Zivlak, N.: Agile project management in new product development and innovation processes: challenges and benefits beyond software domain. In: 2018 IEEE International Symposium Innovation Entrepreneurship, pp. 1–9 (2018)
17. VersionOne: 11th annual state of agile report. 12th annual state agil (2018)
18. Sharma, S., Hasteer, N.: A comprehensive study on state of Scrum development. In: International Conference on Computing, Communication and Automation (ICCCA), pp. 867–872 (2016)
19. Amir, M., Khan, K., Khan, A., Khan, M.N.A.: An appraisal of agile software development process. Int. J. Adv. Sci. Technol. **58**, 75–86 (2013)
20. De Oliveira, F., Pereira, R.: Adaptação e implantação da metodologia Scrum para projetos ágeis numa Autarquia Federal. Rev. Gestão Tecnol. **16**(2), 260–276 (2016)
21. Hajjdiab, H., Taleb, A.S.: Adopting agile software development: issues and challenges. Int. J. Manag. Value Supply Chain. **2**(3), 1–10 (2011)
22. Wan, J., Zhu, Y., Zeng, M.: Case study on critical success factors of running scrum. J. Softw. Eng. Appl. **06**(02), 59–64 (2013)
23. Montero, G., Onieva, L., Palacin, R.: Selection and implementation of a set of key performance indicators for project management. Int. J. Appl. Eng. Res. **10**(18), 39473–39484 (2015)
24. Kniberg, H., Skarin, M., de Mary Poppendieck, P., Anderson, D.: Kanban y Scrum–obteniendo lo mejor de ambos. InfoQ (2010)

Automated Extraction and Visualization of Spatial Data Obtained by Analyzing Texts About Projects of Arctic Transport Logistics Development

A. V. Vicentiy[1,2](✉), V. V. Dikovitsky[1], and M. G. Shishaev[1,3]

[1] Institute for Informatics and Mathematical Modeling, Subdivision of the Federal Research Centre "Kola Science Centre of the Russian Academy of Science", 24A, Fersman st., Apatity, Murmansk region 184209, Russia
alx_2003@mail.ru, dikovitsky@gmail.com, shishaev@arcticsu.ru

[2] Apatity Branch of Murmansk Arctic State University, Lesnaya st. 29, Apatity, Murmansk region 184209, Russia

[3] Murmansk Arctic State University, Egorova st. 15, Murmansk 183038, Russia

Abstract. This paper considers the problem of extraction and presenting geospatial information from natural language texts in a form that is convenient for human perception and analysis. The information technology for extraction and visualization of geo-data from natural language texts for the automatic construction of cartographic interfaces is proposed. For text analysis, we used syntactic, morphological and semantic methods. For visualization, we used existing geocoding services and geoservices. The paper also describes some projects for the development of transport logistics in the Arctic. Text descriptions of the "Integrated Development of the Murmansk Transport Hub" project, the "Belkomur" project and the "Northern latitudinal way" project were used to create thematic corpus of documents and approbation of technology. The result of the work of information technology is shown on the example of the "Integrated Development of the Murmansk Transport Hub" project. A brief assessment of the possible applications of the developed information technology is given.

Keywords: Analysis of natural language texts · Geovisualization · Extracting facts from texts · Cartographic interface · Arctic transport logistics · Northern Sea Route

1 Introduction

The main purpose of this research is the development of information technology for extraction and visualization of geo-data from natural language texts. We use the methods of syntactic, morphological and semantic analysis of texts in order to identify toponyms and other objects of the real world that are geographically referenced, mentioned in the texts.

R. Silhavy et al. (Eds.): CoMeSySo 2019, AISC 1046, pp. 419–433, 2019.
https://doi.org/10.1007/978-3-030-30329-7_37

The task of identifying toponyms in texts is closely related to such research areas as Named entity recognition (NER) and Automatic content extraction (ACE). In turn, the tasks solved within the NER framework are subtasks for automatic extraction of structured information from unstructured and semi-structured machine-readable texts. The common goal of NER-systems is to find and classify references to named entities in unstructured texts. In this case, as a rule, pre-defined categories are used. For example, the names of organizations, proper names, locations, various codes, etc. [1]. The tasks solved within the framework of ACE are aimed at developing modern methods and technologies for extracting information from text and multimedia sources. In particular, the three main tasks of ACE are: identifying the objects mentioned in the text (organizations, locations, vehicles, etc.), relations between the objects (role, part, closeness, etc.) and events (interaction, movement, creation, etc.) [2].

Recently, text mining methods are being actively developed for automatic extraction of the most important information (text meaning) from large corpuses of texts. One of these methods is thematic modeling. This method allows to extract the thematic structure of documents. In particular, it is used to create means of summarizing texts and improving the general understanding of the contents of the text corpus as a whole. In work [3] the system of visual analytics is presented. This system allows the use of interactive visual analysis (IVA) techniques to understand the meaning of large collections of documents based on thematic modeling. The paper [4] describes the technology of semantic analysis and visualization of semantic models of text documents. These technologies are based on the use of various methods of extracting formal semantics from the thematic corpus of documents. On the basis of the proposed technologies, three methods of visual express content analysis of documents are implemented. These methods allow to quickly understand the meaning of the analyzed documents.

The authors of the paper [5] use large amounts of geo-tagged text data obtained within the urban environment as a data source. For a joint analysis of text data and their location, the authors offer a visual approach to the integration of text and a map of the metro. Metro lines are used to divide the city into several districts and determine the boundaries of each district. Text is formed from keywords extracted from geo-tagged text data. In the process of visualization, the text is attached to the corresponding areas of the city in accordance with geotags. Thus, the user is provided with an interactive visualization, which allows to obtain additional information about geo-referenced city objects. This visualization technology can also be used to support the study of patterns of text distribution, the study of urban spaces and navigation in urban environments.

Most of the information generated today is unstructured data. Most of this data is presented in the form of texts of arbitrary shape and structure (free-form texts). Examples of such texts are news articles, blogs, reports, posts on social networks, etc. Considering the fact that almost 60% of all data are in one way or another related to geospatial data [6], research into the identification of geodata in texts and the processing of this data remains relevant. It should be noted that in recent years significant advances have been made in the field of geographic information retrieval (GIR) in the text. The monograph [7] presents recent advances in GIR, the development of spatial-oriented search engines and meeting the needs of users in geographic information. Despite this, modern geo-information systems (neither research nor commercial) are not capable of automatically detecting, structuring, storing and visualizing geo-information found in free-form texts. In addition, many studies focus mainly on either retrieving or presenting location information in text. At the same time, further research and visualization of the data obtained through the study of their spatial, temporal and semantic characteristics is quite rare. In this regard, information technology for extraction and visualization of geo-data from natural language texts for the automatic construction of cartographic interfaces proposed in our work is relevant and is aimed at meeting the information needs of users when solving specific tasks.

This technology is useful for presenting data on specific projects to users who do not fully understand the spatial location of various objects of the Arctic transport and logistics system due to its complexity, length and a large number of components. A special case of the use of technology may be its use in decision support information systems. In this case, a visual representation of a geo-referenced project may be useful for a decision maker (DM) to quickly understand the "geographical meaning" of the relative position of control objects.

In addition, the visualization of spatial data is well suited to support the solution of tasks of monitoring and evaluating the degree of implementation of Arctic transport logistics projects. Instead of studying hundreds of pages of project documentation, the user can simply compare the cartographic image, built on the basis of text analysis of current reports, with the plans for the project. Project management of the creation of the Arctic transport and logistics infrastructure based on traditional means is also difficult due to their large scale, a large number of executors, the parallel execution of various types of interrelated works, the territorial distribution of these works and other factors. In such circumstances, the automatic construction of cartographic visualization allows to get a holistic view of the project status and identify potential problems of the project. The cartographic image formed on the basis of the analysis of the text description allows for a preliminary visual analysis of the availability of transport and logistics infrastructure of specific locations. And also to make assumptions about the needs of the macroregion in those or other objects of the transport and logistics infrastructure and the degree of connectivity of the Arctic transport system with the transport system of other regions.

2 Materials

2.1 The Main Data Sources and Characteristics of Data Sources

Natural language texts were used as the main source of data for analysis. All used texts can be divided into several groups. In particular, one of the groups is represented by various official documents and regulations. These documents are created, as a rule, by official representatives or authorities and large business structures.

Distinctive characteristics of documents of this group are, as a rule, a large volume, the presence of the internal structure of the document, references to other official documents and regulations, the presence of special terminology, a large length and a complex structure of sentences. In addition, official documents and regulations often change over time. The relevant authorities may make various amendments and changes to such documents. All these characteristics influence the result of the analysis of such documents.

The second group of texts used for analysis are texts from the official sites of specific projects for the development of Arctic transport logistics. For example, the site of the project "Belkomur" - http://www.belkomur.com.

This type of text is distinguished by a simpler language of presentation, since these texts are aimed at a wide audience of the Internet. As a rule, these texts have a small volume, the internal structure of texts is often poorly defined, texts can contain a large number of links to other Internet resources (including multimedia). Sentences usually have an average length and relatively simple structure. The number of specific terms is relatively small, so that most of the site's audience can understand the meaning of the text without having to refer to special explanatory dictionaries.

The third group of texts is texts from social networks. Many modern projects, authorities (regional governments, ministries, governors, etc.), and large business structures, have official accounts in popular social networks. For Russia, the most relevant are such social networks as Facebook, Twitter and the social network widespread in Russia, VKontakte.

The peculiarity of these texts are the following: a small amount of messages (posts), the presence of links to other information resources, the almost complete absence of special terminology (special terms are replaced by commonly used words that are close to each other, or are given descriptively), the small length and simple structure of both separate sentences and posts in general.

However, despite the relatively small volume compared with other types of texts, texts from social networks differ in the speed of publication (up to several times a day) and high "concentration of information" in the text. At the same time, in most cases, one post is devoted to any one fact or event. Therefore, posts in social networks are especially important for monitoring the current status of various projects of the arctic transport logistics.

In the fourth group of texts we included "all other" relevant texts available for analysis. In particular, such texts may also include texts obtained from "unofficial sources", but noteworthy in the context of the text analysis task being solved. Examples of such texts can be various internal or official documents used within organizations implementing arctic transport logistics projects, various technical information (drawings, reports, technical tasks, etc.). In addition, such documents include materials from various professional conferences, meetings, etc. As well as transcripts of official interviews, video conferences, radio broadcasts, etc.

Thus, the analysis uses heterogeneous data sources - natural language texts. A partial list of the main characteristics of these texts can be represented as follows:

- text volume;
- text structure;
- links to other sources of information;
- number of special terminology;
- length and structure of sentences;
- other text characteristics (text variability; publication operativeness; accuracy or level of trust in the source, etc.).

2.2 Major Development Projects of the Arctic Transport Logistics

In recent decades, the topic of Arctic Transport Logistics (ATL) has become more and more popular. The growing interest of official authorities, business structures and scientists to this topic is due to global climate change and the possibility of using the Northern Sea Route (NSR) to transport large volumes of cargo [8].

However, it is impossible to provide a large amount of traffic without an appropriate transport and logistics infrastructure (modern ports, railways and stations, terminals, warehouses, logistics hubs and much more). Therefore, the Russian Federation, as a country through which the Northern Sea Route passes, is investing huge funds in the creation of new and modernization of the existing Arctic transport and logistics infrastructure.

At present, the transport system of the Russian Federation is highly unevenly developed. The transport infrastructure is most developed in the European part of the country, while Siberia, the Far East and the Arctic regions of the country practically do not have a full-fledged system of transport communications. This situation leads to the transport isolation of these regions and significantly complicates their development and the development of any industrial and economic activity. Even in such regions as the Murmansk and Arkhangelsk regions, where the transport infrastructure is relatively well developed, the major problems are considerable deterioration and obsolescence of fixed assets, as well as inconsistency with modern logistics standards and international requirements for cargo transportation [9].

In order to develop the Arctic territories of Russia, about 150 different projects are currently being implemented. The amount of funding for these projects is about 5 trillion rubles. More than 50 projects from the general list are aimed at the development of the transport and logistics infrastructure and shipbuilding. The main priorities are the development of the Northern Sea Route and the modernization and development of all levels of the Arctic transport system. Special attention is paid to the creation and development of icebreaking fleets and coastal infrastructure (ports, hydrometeorological equipment, means of communication, etc.).

Today, Russia has 18 Arctic ports. These are the ports of Murmansk, Arkhangelsk, Kandalaksha, Vitino, Onega, Mezen, Varandey, Naryan-Mar, Sabetta, Dudinka, Dickson, Khatanga, Tiksi, Anadyr, Pevek, Provideniya, Egvekinot and Beringovsky [10]. The development strategy of the sea port infrastructure of Russia until 2030 provides for the reconstruction of the ports of Murmansk, Arkhangelsk, Kandalaksha, Vitino and Varandey. In addition to this, it is planned to create new deep-water seaports in Indiga and Harasaveya. It is also planned to build at least three new shipping offshore terminals and four container terminals in various ports of the Northern Sea Route [11]. In conjunction with the creation of new and reconstruction of existing railway, automotive and airport networks, as well as the development of a complex of logistic and "Arctic services" (icebreaking, pilotage, navigation safety, navigation in the difficult conditions of the Arctic, etc.), these measures will allow turning existing ports into large transport and logistics hubs that meet the latest international requirements.

The three most significant projects for the development of Arctic transport logistics in the Arctic zone of the Russian Federation, which we used for analysis in our work are considered in more detail below. These are the projects “Northern latitudinal way”, “Belkomur”, “Integrated development of the Murmansk transport hub”.

The project “Northern latitudinal way” is today the largest project for the development of transport infrastructure in the Arctic zone of the Russian Federation. The project includes the construction of the Obskaya-Salekhard-Nadym-Pangody-Novy Urengoy-Korotchaevo railway [12]. The total length of the railway will be 707 km. The value of the project increases significantly due to the fact that this railway will allow to connect the Northern and Sverdlovsk railways into a single network. Due to this, cargo flows of the Ural region will be optimized. The transport accessibility of vast industrial areas will be improved. Stable access to oil and gas fields will be provided, the development of which without this railway was impractical. In addition, the Northern latitudinal way in the future will provide access of the transport system of Russia to the Northern Sea Route through the Yamal Peninsula.

The project is implemented on the principles of concession and public-private partnership. The amount of financing of the project exceeds 270 billion rubles. In particular, the concession agreement includes the implementation of the following objects: [13]

- the right-bank railway section to the combined bridge across the Ob River from the town of Salekhard;
- the left-bank railway section to the combined bridge across the Ob River from the Obskaya station (not including the Obskaya station);
- combined bridge crossing over the Ob River (railway part);
- bridge over the river Nadym (railway part);
- Salekhard linear section (inclusive) - Nadym (Chorey) (inclusive).

More than 19 explored hydrocarbon deposits are located in the “Northern latitudinal way” zone. The main cargo base of the railway under construction will be represented by hydrocarbon raw materials (gas condensate, a wide fraction of light hydrocarbons, mining by-products, etc.), as well as crude oil and polyethylene produced at the Novy urengoy gas chemical complex. In the east direction, it is assumed the direction of the cargo of development (reinforced concrete products, pipes, materials and equipment, etc.).

It should also be noted that in order to increase the efficiency of the project, it was proposed to build a new Bovanenkovo-Sabetta railway line 170 km long. The Bovanenkovo - Sabetta line will provide access of the Northern Latitudinal Route railway infrastructure to the Northern Sea Route through the port of Sabetta, which significantly increases the investment attractiveness of the project as a whole.

One of the most important elements of the Arctic transport logistics are seaports operating in the Arctic. One of these ports is the port of Murmansk - the largest port in the world, located beyond the northern Arctic Circle. The role of the Murmansk port in the development of the Northern Sea Route and the entire Arctic transport and logistics system is unique. In terms of cargo volumes transported, the port of Murmansk ranks first among all Arctic ports. It accounts for about 70% of cargo from the total cargo turnover of the Northern Sea Route and about 14% of all cargo of the country transported by sea.

The Murmansk port includes three large parts - a sea trading port, a sea fishing port, a sea passenger terminal, and is capable of receiving vessels with a deadweight of up to 300 thousand tons, with a draft of up to 15.5 and a length of more than 265 m. In addition, almost all the infrastructure necessary for receiving large ships, including ship repair, as well as the base of the nuclear icebreaking fleet, is located in the port of Murmansk, which allows icebreaking and ice pilotage of ships throughout the Northern Sea Route [14].

In addition, there are other objective prerequisites for the development of the Murmansk transport hub: free access to the open ocean with relatively low navigation intensity, proximity to the European and American markets, reliable transport links with industrial regions of Russia. Therefore, for the development of the Arctic transport infrastructure and the Northern Sea Route, the project "Integrated Development of the Murmansk Transport Hub" was developed. The goal of the project is to create a world-class modern deepwater sea hub on the basis of the port in Murmansk. This hub will be integrated into international transport corridors and will be able to function all year round. The hub will include centers for the processing of containerized cargo, oil cargo, transshipment of coal and mineral fertilizers. The project is implemented on the basis of a public-private partnership, and the volume of project financing amounts to about 145 billion rubles.

The project includes a set of works for the creation and modernization of the transport and logistics infrastructure on the western and eastern shores of the Kola Bay. The most significant of which are:

- construction and reconstruction of port infrastructure facilities on the coast of the Kola Bay, including the construction of terminals for the transshipment of coal, oil and petroleum products on the west bank;
- construction of the container terminal on the east bank, reconstruction of the coal terminal, construction of a warehouse and distribution zone associated with the container terminal;
- development of railway infrastructure, including the construction of the Vykhodnoy-Lavna railway line (Vykhodnoy station - a bridge over the Tuloma river - Murmashi 2 station - Lavna station);
- construction of 10 railway stations and parks, reconstruction of additional railway lines at four stations, reconstruction of railway lines (from Volkhovstroy station);
- development of road infrastructure, including the development of the road network of the city of Murmansk;
- dredging works in the water area of the Kola Bay;
- creating a logistics center;
- reconstruction of the highway "Kola" [15].

Thus, the modernization of the seaport of Murmansk will make it one of the largest transport and logistics hubs of trunk and international transport, which is especially important for the development of the Northern Sea Route and the development of the shelf of the Arctic seas.

The Belkomur project (acronym for the words Beloye More (White Sea), Komi (Komi Republic), Ural (Ural region)) is included in the Strategy for the Development of Railway Transport in the Russian Federation until 2030 and in the Transport Strategy of the Russian Federation for the period until 2030 [16, 17] and directly affects the interests of the four regions of Russia (Arkhangelsk region, Murmansk region, Perm region and the Komi Republic). The project involves the construction of the missing and reconstruction of the existing railway sections for the launch of a through railway communication along the rectifying route Arkhangelsk - Syktyvkar - Perm (Solikamsk) with a length of 1161 km. More than 700 km of them is the construction of new railways, and about 450 km is the repair, modernization and reconstruction of already existing sections of railways [18]. The new railway will link the industrial areas of Western Siberia with the seaport in Arkhangelsk, as well as shorten the route for goods delivered by rail from China and Kazakhstan to Europe by 800 km. The amount of project financing exceeds 170 billion rubles. The project will be implemented on the terms of a concession and the principles of public-private partnership [19].

The main cargo flows will be formed by such goods as coal, mineral fertilizers, oil and timber cargo, ores, construction materials, containers in the direction of the northern ports of Russia - to the ports of Murmansk, Arkhangelsk, Belomorsk, St. Petersburg. The implementation of the Belkomur project together with such projects as the Northern Latitudinal Route, the Integrated Development of the Murmansk Transport Hub, the development of the Arkhangelsk Sea Port and the Sabetta port on Yamal contributes to the integrated socio-economic development of the Arctic zone of the Russian Federation. The Belkomur project is also crucial for the development of the Northern Sea Route and the protection of Russia's interests in the Arctic. Integrated implementation of these projects will provide a synergistic effect for the development of the Arctic transport and logistics infrastructure of the Arctic [20].

We used various natural language documents for each of the above projects for analysis. A preliminary analysis showed that due to various financial crises, problems of public-private partnerships, withdrawal from projects of some participants and other reasons, the deadlines for the implementation of the main stages of projects often shift, and the financing necessary for their implementation increases. In this regard, the actual task is to conduct operational monitoring and visualization of the current status of the project implementation. Using for this purpose our information technology for extraction and visualization of geo-data from natural language texts for the automatic construction of cartographic interfaces, it is possible to make a more accurate assessment and forecasting of the time and financial costs of project implementation. Information technology is described in more detail in the next section.

3 Methods

The information technology for extraction and visualization of geo-data from natural language texts, which is described in this work, provides automatic construction of cartographic interfaces based on existing geoservices. The cartographic images allow visual analysis of the data presented in natural language documents and promptly make changes to the image by adding new documents to the corpus of documents. The technology includes three main stages, which are divided into smaller steps.

The first stage involves the preparation of a corpus of documents for further analysis. In our work we carry out the formation of a thematic corpus of documents dedicated to the development of the largest transport and logistics projects for the development of the Arctic zone of the Russian Federation (AZRF). The thematic corpus is based on texts containing the three main projects mentions:

1. The "Integrated Development of the Murmansk Transport Hub" project (ID MTH);
2. The "Belkomur" project;
3. The "Northern latitudinal way" project (NLW).

The quality of the results of all subsequent stages of the technology directly depends on how complete and relevant the formed thematic corpus of documents is. At the same time, it should be noted that the presented information technology for extraction and visualization of geo-data from natural language texts can also function in an iterative mode. In other words, if any inconsistencies or the unsatisfactory quality of the map interface associated with the quality of the data are detected, or when new thematic documents appear, it is possible to supplement the corpus with new texts and analyze them.

The second stage involves the analysis of texts and the processing of the results. As mentioned above, various methods are used for analysis. A distinctive feature of this information technology is that the analysis of the thematic corpus of documents is carried out in parallel, both on the basis of syntax and morphological, and on the basis of semantic methods of text analysis. This approach is used to improve the accuracy of identifying toponyms and other objects of the real world that are geographically referenced in natural language texts.

The third stage of the technology is directly related to geocoding and visualization of a plurality of identified objects that are spatially referenced, and the construction of a cartographic interface based on existing geo-information services.

Thus, the input of the information technology for extraction and visualization of geo-data from natural language texts, describing the studied project of the development of Arctic transport logistics, and the output is a geo-image or a digital map of the studied project.

The general scheme of the information technology for extraction and visualization of geo-data from natural language texts is presented in Fig. 1. The figure shows the main stages and steps of the technology, as well as input and output data for each step.

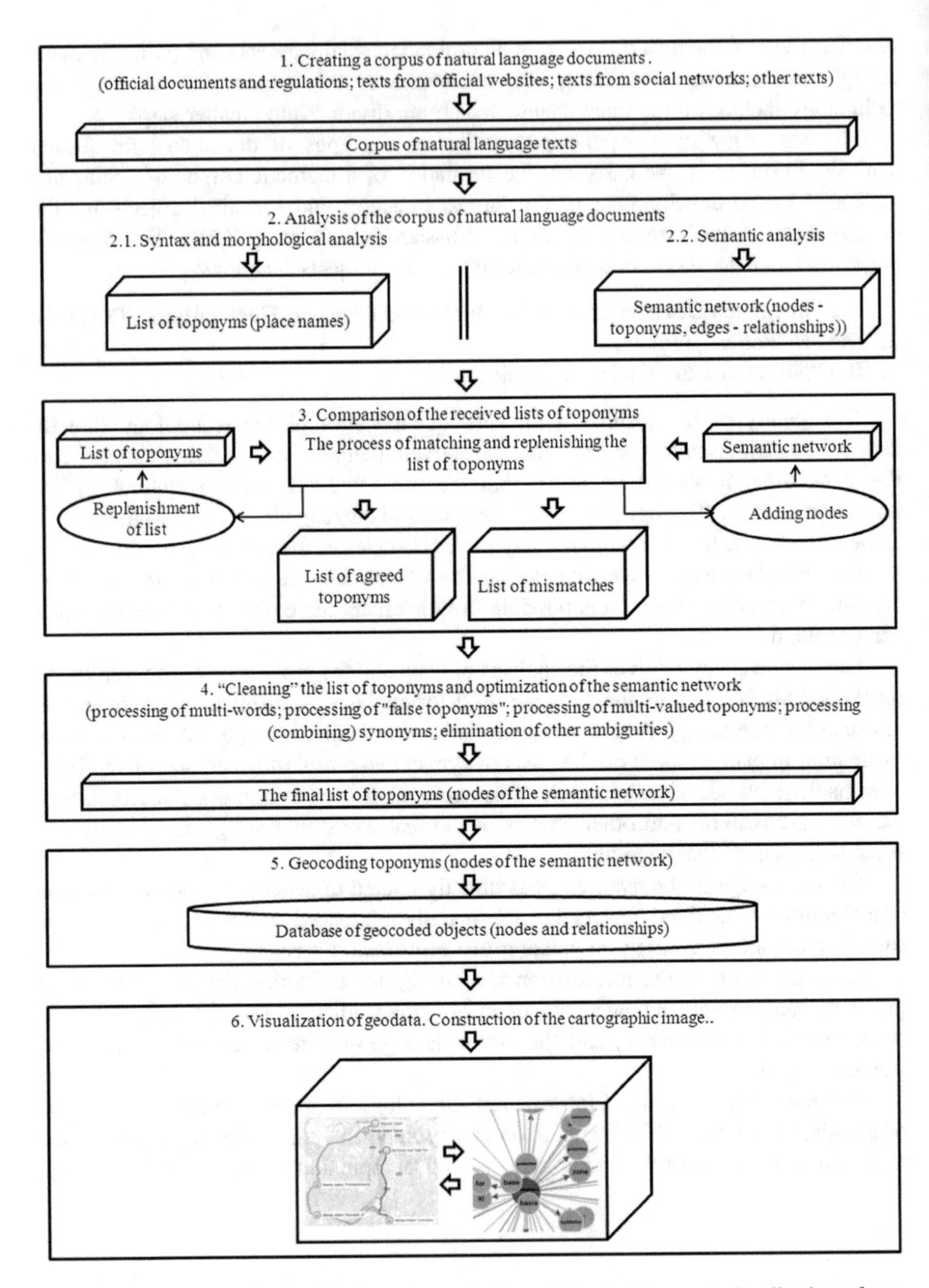

Fig. 1. The general scheme of information technology for extraction and visualization of geodata from natural language texts.

3.1 The Technology of Semantic Text Analysis

In problem of text analyzing are separates subtasks are distinguished: Entity recognition, feature extraction, definition of relationships between objects. Combining the solution of all these tasks allows to extract formalized knowledge.

In this paper, we produce automatic extraction of objects and their properties through multi-level analysis of thematic texts. Particular attention is paid to geographic objects, infrastructure objects and their names (toponyms) that are important for spatial geovisualization in the field of transport logistics. These objects are often expressed by neologisms and compound words, and the problem is the dynamics of the description of infrastructure projects, which limits the use of dictionary-based approaches. The multilevel analysis consists in a combination of statistical and linguistic approaches with the aim of mutually refining the results of the analysis of each stage.

At the first level, we form an associative semantic network based on the methods of distributive semantics, namely, we count the number of shared words in one sentence. We form frequency characteristics to evaluate stable phrases and determine of each word's denotations. To work on small sets of texts, as well as to clarify the result, the resulting pairs of words of each associative relationship are characterized by the distributive semantic proximity index expressed by the cosine proximity between word vectors obtained by the Word2Vec model [21].

At the second level, the formation of syntactic trees of the source texts, and the subsequent integration of the obtained trees into the model of a semantic network with multiple links, is carried out. The syntax analysis of the words in the sentence is done by the Syntaxnet library [22]. Integration of syntactic trees into a network is carried out by calculating the integral assessment of the semantic proximity of concepts based on the similarity of concepts, related concepts, as well as their morphological and syntactic properties. This level of analysis implies more computational costs, therefore it is performed on the same set of texts asynchronously with the first stage (Fig. 2).

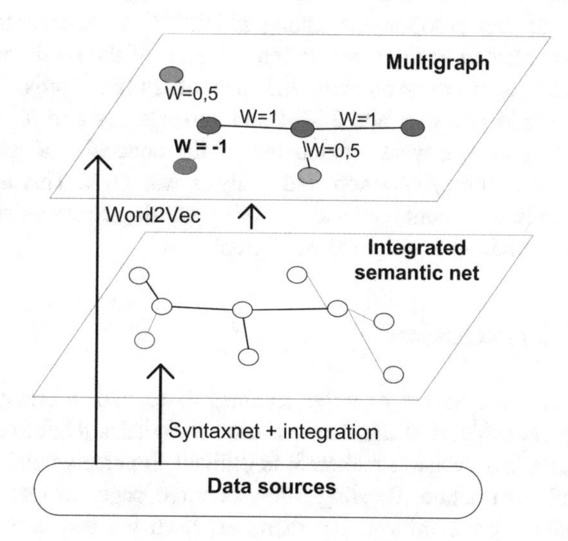

Fig. 2. The combination of statistical and linguistic approaches

The result is an integrated semantic model in the form of a weighted multigraph. At the initial stage, for automated extraction of toponyms, the syntactic, morphological, and statistical characteristics of the concepts of a multigraph containing the circumstances of the place in the analyzed text are expertly determined. The assumption is that concepts that perform one role in the text have similar characteristics in the resulting multigraph. The result of such a filter on the text is a list of concepts that reflect the circumstances of the place. The following is an example that reflects some of the concepts derived from the analysis of Wikipedia articles (Table 1).

Table 1. The example of toponyms extracted from Wikipedia articles

Galaxy	Moscow	Culture	Russia
the city	of Ukraine	the set	Africa
inhabit	Ukraine	Russia	Africa
Europe	traction	linux	temperature
plagiarism	titles	empire	Ottawa
homer	magazines	succession	Ossetia
Thuringia	flow	France	land
federation	north	France	Great Britain
federation	article	Kyrgyzstan	the border
federation	Russia	Kyrgyzstan	the border
USA	the middle	Ethiopia	Kamchatka
Europe	the middle	states	Kamchatka
plagiarism	diseases	states	stamp
Athens	of Ukraine	Europe	Rus

Among the concepts thus obtained, the number of defined circumstances of the place is 73%, the number of toponyms among which is 61%. In order to determine the list of toponyms contained in the development projects of the northern territories, we conducted an analysis using texts with the size of 24322 words. We obtained a multigraph from 15345 concepts and 32424 relationships. As a result of applying the proposed filter, 35 concepts were singled out of the concepts, of which 23 were a toponym. Thus, the accuracy of automated analysis was 66%. This indicator can be increased by taking into account new and clarifying existing characteristics of concepts and relations when filtering a weighted multigraph.

4 Results and Discussion

Currently, there are free access to huge volumes of textual information devoted to various development projects of the Arctic transport logistics. Unfortunately, even for well-trained experts and decision makers it is difficult to perceive and analyze a large amount of textual information. Reading a few hundred pages of text, identifying the meaning and analyzing the information extracted from the text is a complex, time-consuming task that takes a lot of time and effort. Thus, the cognitive load on the person is constantly increasing, which can lead to a distortion of the perception of information and incorrect analysis of the facts.

On the other hand, modern computing systems can handle huge amounts of text documents relatively easily. However, information systems are not always able to correctly identify and take into account the relationship of facts and objects. Therefore, information systems cannot compete in decision making with a person.

The developed information technology for extraction and visualization of geo-data from natural language texts solves the problem of presenting a large amount of information in a form that is most convenient for the end user (decision maker). The result of this information technology is a cartographic interface. This interface includes an interactive geo-image, additional information about geographical objects and visualization management tools.

The result of the work of the information technology for extraction and visualization of geo-data from natural language texts for the automatic construction of cartographic interfaces for the corpus of documents devoted to the "Integrated Development of the Murmansk Transport Hub" project is presented in Fig. 3.

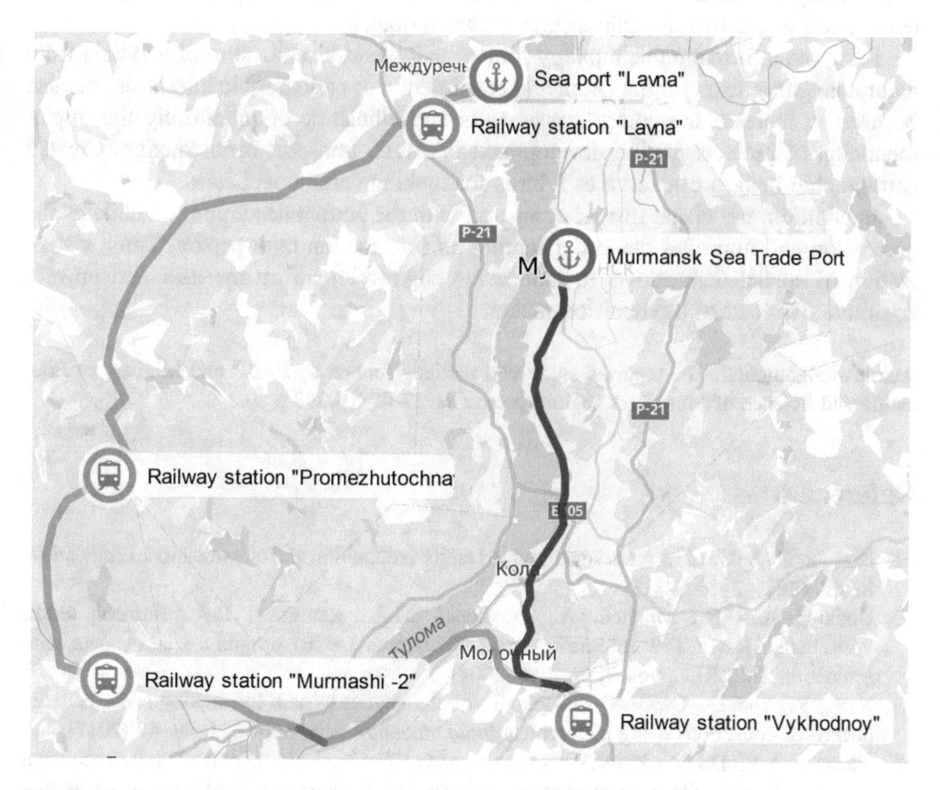

Fig. 3. Cartographic interface for the "Integrated Development of the Murmansk Transport Hub" project.

Using this cartographic interface, a decision maker has the ability to perform interactive visual analysis of the geo-image to identify hidden relationships between geographic objects that are not obvious when analyzing texts, as well as to quickly understand the "geographical meaning" of the objects being studied and the relationships between them.

5 Conclusions

The paper describes an approach to analyzing texts for extracting geodata and creating cartographic images based on this data. This approach allows to significantly simplify the process of extracting, analyzing and visualizing geodata from natural language texts. It is shown that the use of automated text analysis and geocoding provides the ability to create visualization of geodata found in natural language texts.

The proposed information technology allows users to quickly generate interactive map images for interactive visual analysis by experts and decision makers. The analysis of geo-images created on the basis of text analysis provides an opportunity to establish new values of unobvious relationships between objects.

Also, the information technology for extraction and visualization of geo-data from natural language texts for the automatic construction of cartographic interfaces can also be used to increase the comprehensiveness and efficiency of monitoring the implementation of projects for the development of Arctic transport infrastructure. Created cartographic images can serve as a basis for making certain decisions.

In addition, the visual image, compared with the textual description, facilitates the perception and improves the understanding of information by a person. Thus, visualization of spatial data allows to reduce the cognitive load on the user and prevent cognitive overload of the decision maker.

Acknowledgements. The reported study was partially funded by RFBR and Ministry of Education and Science of Murmansk region (project № 17-47-510298 p_a).

References

1. Nadeau, D., Sekine, S.: A survey of named entity recognition and classification. Lingvisticae Invest. **30**(1), 3–26 (2007)
2. Doddington, G.R., Mitchell, A., Przybocki, M.A., Ramshaw, L.A., Strassel, S.M., Weischedel, R.M.: The automatic content extraction (ACE) program - tasks, data, and evaluation. In: LREC, vol. 2, pp. 1–4 (2004)
3. Yang, Y., Yao, Q., Qu, H.: VISTopic: a visual analytics system for making sense of large document collections using hierarchical topic modeling. Vis. Inform. **1**, 40–47 (2017)
4. Vicentiy, A.V., Dikovitsky, V.V., Shishaev, M.G.: The semantic models of arctic zone legal acts visualization for express content analysis. In: Advances in Intelligent Systems and Computing, vol. 763, pp. 216–228 (2018)
5. Li, C., Dong, X., Yuan, X.: Metro-Wordle: an interactive visualization for urban text distributions based on wordle. Vis. Inform. **2**, 50–59 (2018)

6. Hahmann, S., Burghardt, D.: How much information is geospatially referenced? Networks and cognition. Int. J. Geograph. Inf. Sci. **27**, 1171–1189 (2013)
7. Purves, R.S., Clough, P., Jones, C.B., Hall, M.H., Murdock, V.: Geographic information retrieval: progress and challenges in spatial search of text. Found. Trends Inf. Retr. **12**, 164–318 (2018)
8. Rietveld, P.: Climate change adaptation and transport: a review. In: Smart Transport Networks: Market Structure, Sustainability and Decision Making, vol. 1, pp. 29–48 (2013)
9. Shpak, A.V., Serova, V.A., Biev, A.A.: Modern problems of the transport infrastructure of the Russian Arctic regions. North and Market: The Formation of an Economic Order, **6**, 31–36 (2014). (in Russian)
10. Register of Seaports. http://www.morflot.ru/deyatelnost/napravleniya_deyatelnosti/portyi_rf/reestr_mp.html
11. Development Strategy for the Sea Port Infrastructure of Russia until 2030 (Approved by the Maritime Collegium under the Government of the Russian Federation on 28 September 2012). http://www.rosmorport.ru/media/File/seastrategy/strategy_150430.pdf
12. About the project "Northern latitudinal way". http://rareearth.ru/ru/pub/20170127/02936.html
13. Order of the Government of the Russian Federation, No. 1663-p, 8 August 2018. http://static.government.ru/media/files/1XA6VqKeHz83g8ATkA4XZBbiAi9ARtro.pdf
14. Bolshakov, Ya.A., Fridkin, V.N.: Opportunities and prospects of the Murmansk region in the development of transport of the Arctic region. Vestnik MSTU **2**, 363–371 (2016). (in Russian)
15. Description of the project "Integrated development of the Murmansk transport hub". http://ppp-transport.ru/ru/o-retu/proekty-retu/kompleksnoe-razvitie-murmanskogo-transportnogo-uzla/opisanie/
16. The strategy of development of railway transport in the Russian Federation until 2030. https://www.mintrans.ru/documents/2/1010
17. Transport strategy of the Russian Federation for the period up to 2030. https://www.mintrans.ru/documents/2/1009
18. The speech transcript of the head of the Komi Republic. Electronic resource. https://komiinform.ru/news/106894
19. Belkomur project site. http://www.belkomur.com
20. Tarakanov, M.A.: Transport projects in the Arctic: synchronization, complexity. Bull. Kola Sci. Cent. Russ. Acad. Sci. **1**, 80–85 (2014)
21. Word2Vec. https://code.google.com/archive/p/word2vec/
22. Syntaxnet. https://opensource.google.com/projects/syntaxnet

Multi-agent System in Smart Econometric Environment

Jan Tyrychtr[1(✉)], Martin Pelikán[1], Roman Kvasnička[2], Veronika Ander[1], Tomáš Benda[1], and Ivan Vrana[1]

[1] Department of Information Engineering, Faculty of Economics and Management, Czech University of Life Sciences in Prague, Prague, Czech Republic
tyrychtr@pef.czu.cz

[2] Department of Systems Engineering, Faculty of Economics and Management, Czech University of Life Sciences in Prague, Prague, Czech Republic

Abstract. The complexity of current economic phenomena and their analysis requires increasingly complex methods of both statistical and mathematical nature. In this article, we present the idea of using a multi-agent paradigm to process an econometric analysis more efficiently. We present a conceptual model of multi-agent system (MAES) to support econometric tasks. This concept is shown on the procedure of verification of the prognostic properties in a multi-agent system. We consider three agents, which calculate the normalized deviations for each time series. In our case study, the agent sends the entire pivot table to the decision component in the multi-agent system and the result of the prognostic properties verification of its econometric model. Our MAES proposal has four additional components: econometric data component, sensing data component, data processing agent, decision-making component and environment.

Keywords: Multi-agent system · Econometric intelligence system · Ambient intelligence · Smart environment · OLAP · Decision support

1 Introduction

The basic idea behind Smart Environments [4, 5, 7] is in enriching a real environment with technology such as sensors, networks, artificial intelligence, or autonomous agents. Such a system processes real-time data, evaluates it to support users in these smart environments. This concept is sometimes enriched by provision of the ubiquitous computing [9, 11, 12] (unobtrusive presence of computing devices everywhere), ubiquitous communication (access to network and computing devices everywhere) and intelligent user adaptive interfaces (intelligent interfaces between users and computers which adapts to user preferences). In this case, we speak of Ambient Intelligence [1, 3, 6, 10, 18] (sometimes referred to as Smart Intelligence).

The crucial components of these smart environments are multi-agent systems. Intelligent agent studies are roughly divided into three categories [2]: (1) those seeking to design agents similar to human behaviour; (2) those seeking to implement intelligent

R. Silhavy et al. (Eds.): CoMeSySo 2019, AISC 1046, pp. 434–442, 2019.
https://doi.org/10.1007/978-3-030-30329-7_38

autonomous software; and (3) those which focus on coordination or the collaboration of many agents (called multi-agents).

This article focuses on the use of the last category of agent in smart econometric environments. For example, the market price forecast for a product that may be various for different divisions of multi-national corporation (depending on the market situation) may represent a significant computing and workload at the highest level to find a common econometric model. It would be necessary to work with multiple models, different economic data and variables for such forecasts. This can be achieved by the concept of multi-agent systems [14, 15, 17, 19], which represent a group of freely linked autonomous systems (agents) working together to achieve a common goal (analysis, forecasting, calculation of indicators, etc.). Each agent works autonomously and can calculate econometric tasks and it sends results directly to the system decision-making module independent of other agents. Using these multi-agent systems brings similar benefits like a teamwork among people. These advantages are primarily to automate data analysis and reduce computing performance to processing these analyses that are handled by independent agents. For example, stock price developments require processing of big data that change in real time. By distributing tasks to individual agents by econometric dimensions, they do not overload the system.

In this article, we present a conceptual model of a multi-agent system to support econometric tasks. We show the procedure of verification of the prognostic properties in a multi-agent system on a simple example of normalized deviations calculation.

2 Methods

In this article, we propose a new multi-agent system through conceptual models that take into account the basic components of the multi-agent system. As a case study, we consider three agents processing their own econometric models, which are subsequently verified by the agents for their prognostic character. For this purpose, we use the concept presented in our earlier study [16]. First, we show a simulated example of an econometric model evaluation with normalized deviation calculations. We display calculated results in the pivot tables (within OLAP). Last, we design the conceptual model of multi-agent system.

2.1 Normalized Deviations

The normalized deviation is the ratio between the offset of the actual value and its standard deviation [8]:

$$N_{it} = \frac{\widehat{y}_{it} - y_{it}}{\sigma_{y_i}}, where\, i = (1\ldots g)\, and\, t = (1\ldots n) \tag{1}$$

$\widehat{y}_{it}$ is a theoretical value of the *i-th* endogenous variable at time *t*.
y_{it} is the real value of the *i-th* endogenous variable at time *t*.

σ_{y_i} is the standard deviation of the i-th endogenous variable computed as the square root of the total variance.

If $N_{it} = 1$, the same result can be achieved when the theoretical value of $\widehat{y}_{it}$ was replaced by the average value of $\overline{y}_{it}$. If $N_{it} > 1$, the prognosis is worse than if the value has been replaced an average value. If $N_{it} = 0$, it means that the prognosis value agrees with the reality.

The normalized deviation of the i-th endogenous variable of the model is computed according to the formula:

$$N_i = \sqrt{\frac{1}{n}\sum\nolimits_{t=1}^{n} N_{it}^2}, \; where\; i = (1 \ldots g) \tag{2}$$

The normalized deviation for individual years of the time series is calculated according to the formula:

$$N_t = \sqrt{\frac{1}{g}\sum\nolimits_{i=1}^{g} N_{it}^2}, \; where\; t = (1 \ldots n) \tag{3}$$

The normalized deviation for the whole model is in the form:

$$N = \sqrt{\frac{1}{g}\frac{1}{n}\sum\nolimits_{i=1}^{g}\sum\nolimits_{t=1}^{n} N_{it}^2} \tag{4}$$

To calculate the normalized deviations of the model is, according to the formula (1), most appropriate derive the normalized deviation matrix N_{it}. The matrix's size is g*n. The formulas (2) and (3) represent the quadratic mean of the elements per each row and column. We will use this feature for calculation of a data cube in OLAP.

2.2 Decision Support of Analyses

As in our previous article [16], we use a multidimensional view of the normalized deviations to support the verification of prognostic character. A data cube is composed of the set of n dimensions $D = \{d_1, d_2, \ldots, d_n\}$, where each d_n is the name of n-th dimension and the set of k measures $M = \{m_1, m_2, \ldots, m_k\}$, where each m_k is the name of n-th measure.

The set of points of view on normalized deviations (V_{SD}) is defined as the set of matrices in which each matrix represents the values of normalized deviations of specific econometric models. A V_{SD} is defined as the data cube:

$$V_{SD} = \{[nv_1],[nv_2]\ldots[nv_m]\}, \tag{5}$$

where each $[nv]$ is a 2-dimensional matrix, as follows:

$$[nv] = \begin{array}{c} \\ e_1 \\ e_2 \\ \vdots \\ e_n \end{array} \begin{array}{c} \begin{array}{cccc} t_1 & t_2 & \dots & t_m \end{array} \\ \begin{bmatrix} nv_{11} & nv_{12} & \dots & nv_{1m} \\ nv_{21} & nv_{22} & \dots & nv_{2m} \\ \vdots & \vdots & \ddots & \vdots \\ nv_{n1} & nv_{n2} & \dots & nv_{nm} \end{bmatrix} \end{array} \quad (6)$$

Where each e represents an *endogenous variable* from E to dimension d_n from D, each t represents *time* from T to dimension d_t from D and each nv is a particular calculated value of the *normalized deviation* to values v_i from V. Formula (1) represents the measure m_k from M of the data cube.

3 Results

Now we will show a procedure for evaluating the prognosis model of properties in a multi-agent system on a simple example. We assume that the regular derivation of the prognosis from the econometric model was preceded by the verification of the properties of the individual equations, which can be assessed on the basis of: (a) economic interpretability of the calculated parameters, (b) multicollinearity between explanatory variables, (c) tightness and degree of dependence of endogenous and explanatory variables, (d) the statistical significance of parameters, (e) the autocorrelation of resi-dues and (f) the normalized deviations.

Obviously, a complex multi-agent system would have to deal with a variety of tasks. We will focus on the last point, to analyse the normalized deviations of the individual equations.

We assume that each agent has declared model variables and calculated equations. Each agent also knows real and theoretical values $\mathrm{y_n}$ and total variance S_y^2.

The agent's approach is computationally simple and does not differ from commonly used approaches in econometrics. First, the agent calculates the normalized deviations for the *i-th* endogenous variable (for each equation) according to formula (2). The results of the calculation are shown on a simulated example for five endogenous variables from 2011–2018 (see Table 1).

Table 1. Calculation of normalized deviations N_{it}.

$\mathrm{N_{it}}$	2011	2012	2013	2014	2015	2016	2017	2018
$\mathrm{y_1}$	−0.4439	1.6058	0.6641	0.5682	0.6223	0.2767	0.4211	0.0999
$\mathrm{y_2}$	0.9114	0.2916	0.9001	0.8093	−0.5231	0.2772	0.3948	−1.1195
$\mathrm{y_3}$	0.1395	−0.7777	0.2164	−1.2403	0.7943	0.7223	−0.1515	0.7580
$\mathrm{y_4}$	−0.6424	0.7039	1.5159	0.4567	−0.2605	0.5431	0.9350	0.3673
$\mathrm{y_5}$	0.9826	0.1547	0.1563	0.3900	−0.5200	0.6062	0.7122	−0.4389

Subsequently, the agent calculates the normalized deviations for each time series according to formula (3) and the normalized deviation for the whole model according to formula (4).

Calculated values can be sent to the user in the form of a two-dimensional matrix according to (6). The resulting pivot table can be expertly assessed or further processed by the user (expert) in the econometric environment (see Table 2).

Table 2. Calculation of normalized deviations N_{it}^2.

N_{it}^2	2011	2012	2013	2014	2015	2016	2017	2018	$\sum N_{it}^2$	N_i
y_1	0.1970	2,5786	0.4411	0.3228	0.3872	0.0766	0.1773	0.0100	4,1907	0.7238
y_2	0.8307	0.0851	0.8102	0.6549	0.2737	0.0768	0.1559	1,2534	4,1406	0.7194
y_3	0.0195	0.6048	0.0468	1,5383	0.6310	0.5217	0.0230	0.5746	3,9596	0.7035
y_4	0.4126	0.4955	2,2980	0.2085	0.0679	0.2950	0.8743	0.1349	4,7867	0.7735
y_5	0.9656	0.0239	0.0244	0.1521	0.2704	0.3675	0.5073	0.1926	2,5038	0.5594
$\sum N_{it}^2$	2,4254	3,7879	3,6206	2,8767	1,6302	1,3375	1,7377	2,1655		
N_t	0.6965	0.8704	0.8509	0.7585	0.5710	0.5172	0.5895	0.6581		**1,6427**

In our case study, besides the entire pivot table to the decision-making component in the multi-agent system, the agent sends also the result of verifying the prognostic properties of its econometric model. In this case, the calculated normalized deviation for the whole model is crucial for the automatic component; in our example, value is 1.6427.

In this case, the first agent's calculated value is 1.6427. The decision-making component will interpret the result so that the value of the normalized deviation $N > 1$. The result of prognosis is worse than if he had been replaced by average.

Other agents will do the same as the first agent. They will calculate the normalized deviations for the *i-th* endogenous variable, the normalized deviations for the individual time series, and the normalized deviation for the entire model.

For example, the second agent will calculate the normalized deviation for the entire model as shown in Table 3. Conversely, the third agent calculates the normalized deviation for its econometric model with further different result as shown in Table 4.

Table 3. Agent 2 – calculation of normalized deviations N_{it}^2.

N_{it}^2	2011	2012	2013	2014	2015	2016	2017	2018	$\sum N_{it}^2$	N_i
y_1	0.0673	0.0082	0.0356	0.2428	0.1496	0.0582	0.0816	0.0042	0.6473	0.2845
y_2	0.0058	0.0656	0.0042	0.0088	0.0034	0.0584	0.1291	0.0240	0.2994	0.1935
y_3	0.0108	0.0981	0.0327	0.0760	0.1288	0.0823	0.0350	0.1040	0.5677	0.2664
y_4	0.0061	0.0047	0.0786	0.1774	0.0876	0.2378	0.1596	0.1101	0.8619	0.3282
y_5	0.0217	0.0142	0.0146	0.1257	0.0653	0.1374	0.0766	0.1401	0.5956	0.2729
$\sum N_{it}^2$	0.1116	0.1908	0.1658	0.6307	0.4347	0.5741	0.4819	0.3825		
N_t	0.1494	0.1953	0.1821	0.3552	0.2949	0.3388	0.3105	0.2766		**0.3953**

Table 4. Agent 3 – calculation of normalized deviations N_{it}^2.

N_{it}^2	2011	2012	2013	2014	2015	2016	2017	2018	$\sum N_{it}^2$	N_i
y_1	0.1970	0.3670	0.3650	0.2583	0.3627	0.0659	0.1609	0.0064	1.7832	0.4721
y_2	0.0084	0.0738	0.7746	0.0119	0.2950	0.0661	0.1405	0.0195	1.3898	0.4168
y_3	0.0143	0.0886	0.0386	0.0677	0.5996	0.4932	0.0294	0.5447	1.8761	0.4843
y_4	0.0039	0.4677	0.0876	0.1907	0.0787	0.2531	0.1723	0.1206	1.3746	0.4145
y_5	0.0265	0.0182	0.0186	0.1369	0.0576	0.3436	0.4792	0.2105	1.2911	0.4017
$\sum N_{it}^2$	0.2500	1.0153	1.2843	0.6655	1.3936	1.2220	0.9822	0.9017		
N_t	0.2236	0.4506	0.5068	0.3648	0.5279	0.4944	0.4432	0.4247		**0.8141**

Now we design a new conceptual model of multi-agent system to support econometric tasks. Figure 1 represents the structure of our multi-agent econometric system (MAES). The MAES has four additional components:

- **Econometric data component:** the component contains mainly analytical databases, data warehouses and multidimensional data views. The data are mainly of economic nature and processed for econometric models.
- **Sensing data component:** it gains context values - economic and other phenomena important for econometric models. This may include sensing the number and type of customers, payment transactions, temperature, number of errors, yield on land, quantity of product, etc.
- **Data processing agent:** each agent calculates the form of an econometric model. Agents of our system automatically and independently process data and calculate important indicators for economic decisions. Agents send calculated data to decision-making component. Data flow from agents is provided with so-called inhibition mechanism. The mechanisms of inhibition resolve conflicts between agents who have to send data together for assessment according to a decision component plan.

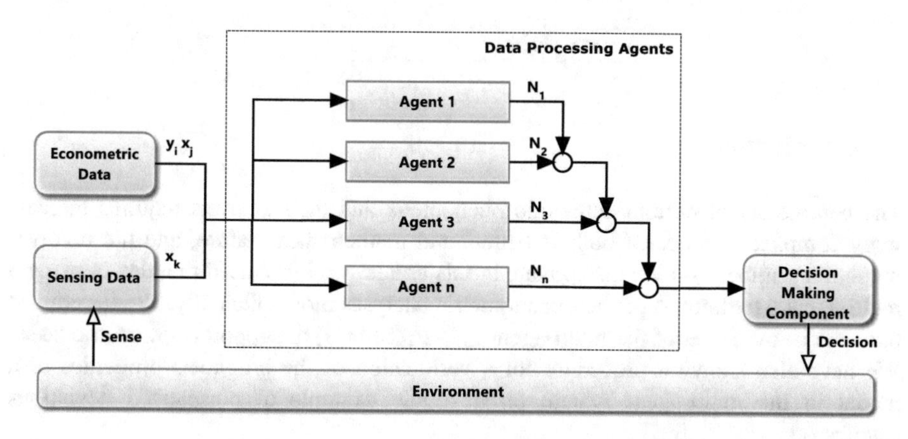

Fig. 1. Conceptual model of multi-agent econometric system

- **Decision-making component:** it chooses the adequate decision based on calculated indicators.
- **Environment:** the agent must take into account the current state of the environment, economic phenomena and user preferences in calculations.

The decision-making component sorts the obtained normalized deviations from the agents by size value and selects the smallest value that characterizes the best prognostic properties of the econometric model of agent. In our case study, the value corresponds to agent n.2 (see Table 5). The sorting function can be very simple, such as Bubble Sort with the complexity $O(n^2)$.

```
1.  function bubbleSort(array a)
2.    for i in 1 -> a.length - 1 do
3.      for j in 1 -> a.length - i - 1 do
4.        if a[j] < a[j+1]
5.          swap(a[j], a[j+1]);
```

[Pseudocode 1: Bubble sort]

Based on this selection of the best value, the system makes a decision. For example, it applies the selected econometric model of the agent to the appropriate economic phenomenon, performs other indicators calculations for selected agents only (econometric models) or based on the results, it attempts to improve parameters of the remaining econometric models.

Table 5. Solutions proposed by three agents.

Agent #	Sorted normalized deviation
Agent 2	$N_2 = 0.3953$
Agent 3	$N_3 = 0.8141$
Agent 1	$N_1 = 1.6427$
…	…
Agent n	N_n

4 Conclusion

The complexity of current economic phenomena and their analysis requires increasingly complex methods of both statistical and mathematical nature, and the involvement of a number of other disciplines. In this article, we introduced the idea of using a multi-agent paradigm to process econometric analyses more efficiently. We introduced the conceptual model of the multi-agent system (MAES) to support econometric tasks. We have also shown a procedure for a verification of the prognosis properties of a model in the multi-agent system on a simple example of normalized deviations calculation.

The results we presented in this article can be further exploited for development of the multi-agent system in smart econometric environments [13]. Such systems can be developed for a number of real-life problems. Econometrics works with a variety of methods that require a human element to decide on a number of attributes of the created models (e.g., correlation, multicollinearity, dummy variable creation, etc.). All these decisions are very difficult to automate. The development of multi-agent systems represents great potential in this area.

Acknowledgements. This work was conducted within the project *Ambient intelligence in decision-making problems in uncertainty conditions* (2019B0008) funded through the IGA foundation of the Faculty of Economics and Management, Czech University of Life Sciences Prague and within the project *Smart Environments - Modelling and Simulation of Complex Intelligent Systems (SEMSCIS)* funded through the Czech Science Foundation, Czech Republic, grant no. 20-12412S.

References

1. Aarts, E., Wichert, R.: Ambient Intelligence, pp. 244–249. Springer, Heidelberg (2009)
2. Augusto, J.C., Nakashima, H., Aghajan, H.: Ambient intelligence and smart environments: a state of the art. In: Handbook of Ambient Intelligence and Smart Environments, pp. 3–31. Springer, Boston (2010)
3. Cook, D.J.: Multi-agent smart environments. J. Ambient Intell. Smart Environ. **1**(1), 51–55 (2009)
4. Cook, D.J., Das, S.K.: Smart Environments: Technology, Protocols and Applications. Wiley, Hoboken (2004)
5. Cook, D.J., Das, S.K.: How smart are our environments? An updated look at the state of the art. Pervasive Mob. Comput. **3**(2), 53–73 (2007)
6. Cook, D.J., Augusto, J.C., Jakkula, V.R.: Ambient intelligence: technologies, applications, and opportunities. Pervasive Mob. Comput. **5**(4), 277–298 (2009)
7. Das, S.K., Cook, D.J.: Designing smart environments: a paradigm based on learning and prediction. In: Pattern Recognition and Machine Intelligence, pp. 80–90. Springer, Heidelberg (2005)
8. Gergelyi, K., Kolek, J., Šujan, I.: Komlexné prognózy v socialistickom hospodárstve: určene prac. federálnych a nár. hosp. orgánov, vedecko-výskum. prac. a štud. vys. škôl ekon., ako aj prísluš. smerov na iných školách. Alfa, t. Svornost' (1973)
9. Lyytinen, K., Yoo, Y.: Ubiquitous computing. Commun. ACM **45**(12), 63–96 (2002)
10. Remagnino, P., Foresti, G.L.: Ambient intelligence: a new multidisciplinary paradigm. Syst. Man Cybern. Part A Syst. Hum. IEEE Transact. **35**(1), 1–6 (2005)
11. Saha, D., Mukherjee, A.: Pervasive computing: a paradigm for the 21st century. Computer **36**(3), 25–31 (2003)
12. Satyanarayanan, M.: Pervasive computing: vision and challenges. Pers. Commun. IEEE **8** (4), 10–17 (2001)
13. Tesfatsion, L.: Agent-based computational economics: modeling economies as complex adaptive systems. Inf. Sci. **149**(4), 262–268 (2003)
14. Tučník, P.: Multicriterial decision making in multiagent systems–limitations and advantages of state representation of behavior. In: Advances in Data Networks, Communications, Computers, pp. 105–110 (2010)

15. Tučník, P., Kožaný, J., Srovnal, V.: Multicriterial decision-making in multiagent systems. In: International Conference on Computational Science, pp. 711–718. Springer, Heidelberg (2006)
16. Tyrychtr, J., Pelikán, M., Štiková, H., Vrana, I.: Multidimensional design of OLAP system for context-aware analysis in the ambient intelligence environment. In: Software Engineering Perspectives and Application in Intelligent Systems, pp. 283–292. Springer, Heidelberg (2016)
17. Van Aart, C.: Organizational Principles for Multi-Agent Architectures. Springer, Heidelberg (2004)
18. Weber, W., Rabaey, J., Aarts, E.H. (eds.): Ambient Intelligence. Springer, Heidelberg (2005)
19. Weyns, D.: Architecture-Based Design of Multi-agent Systems. Springer, Heidelberg (2010)

Organization of Secure Telecommunication Networks in Groups of Mobile Robots Based on Chaos and Blockchain Technologies

Vladimir Shabanov(✉)

Southern Federal University, Rostov-on-Don, Russia
v.b.shabanov@gmail.com

Abstract. Currently, the topic of information exchange in groups of small-sized robots is very popular, the reason for this is the popularization of the use of mobile robots, and therefore the question of the crypto-security of the communication channels of the team becomes. The article discusses the existing approaches to the implementation of information exchange and provides a brief analysis of the features of information exchange, depending on the type of management strategy. The proposed method of organizing the exchange of protected information in a decentralized group of small-sized robots. The proposed method is based on the use of data encryption on the basis of chaotic models for the transmitted information to members of the coalition and the use of the blockchain technology, for entering the obtained information into a distributed database.

Keywords: Group control system · Blockchain · Chaotic algorithms · Information exchange · Limited communications · Dynamic systems · Interaction of robots · Small-sized robots

1 Introduction

The popularity of the use of coalitions of mobile robots can be explained by the achievement of high efficiency groups of robots solving technical challenges [1, 2]. The problem of teamwork becomes urgent. To solve this problem, use is made of the theory of dynamic chaos, which makes it possible to provide an encrypted connection with the team and the blockchain technology, which makes it possible to make a group of decentralized and used to store information about mission objectives and information from sensors.

2 Encoding of Outgoing Data

The foundation of the theory was laid in 1961 by Edward Lorenz in his work – "The Butterfly Effect". Chaotic models are used in many sciences: mathematics, computer science, engineering, finance. Chaotic models were also found in the human brain, where they are used in information processing.

The advantages of using these models include high performance indicators of the model, when solving technically demanding tasks on low-power computing devices.

R. Silhavy et al. (Eds.): CoMeSySo 2019, AISC 1046, pp. 443–448, 2019.
https://doi.org/10.1007/978-3-030-30329-7_39

As applied to the provision of an encrypted communication channel for a coalition of mobile robots, this approach is interesting because of the limited computational resources and small electrical power resources.

The solution to the problem of data transfer between team members and the distributed registry was assigned to dynamic chaotic systems, allowing for secure communication between devices.

The use of chaotic systems to solve the issue of transferring information between devices is not a new topic, there are many options for solving this problem with the ability to communicate both client-client and client-group [3]. The choice of dynamic chaos is due to sensitivity to the initial parameters, the complexity of the structure determined by pseudo-randomness.

Chaotic models have features that are great for organizing information exchange in coalitions of mobile robots:

- Orthogonal, due to the irregularity of chaotic signals, their autocorrelation function has a strong attenuation, which best indicates the use in multi-agent systems, due to the large number of devices on the same frequency
- Crypt resistance, the signals are not regular and have a complex structure, the same generator can produce different signals with minor changes. In view of this, it is not possible to predict signal changes in the long-term and medium-term.
- Broadband, signals are non-periodic and have a wide spectrum.

It is proposed to use a block encryption algorithm, a characteristic feature of which is the use of direct and inverse transforms, with both transformation processes being similar in nature. Encrypted information is divided into blocks of fixed length and their sequential processing takes place.

3 Distributed Registry for Decentralized Data Storage

Blockchain is a decentralized data storage and computing system, originally developed for Bitcoin digital cryptocurrency [4, 5]. In the future, the active introduction of technology into various spheres began: medicine [6], industry [7], IoT [8], robotics [9–11], etc.

The structure of a chain of blocks can be represented as an index of a timestamp, a hash of the previous block and data (see Fig. 1).

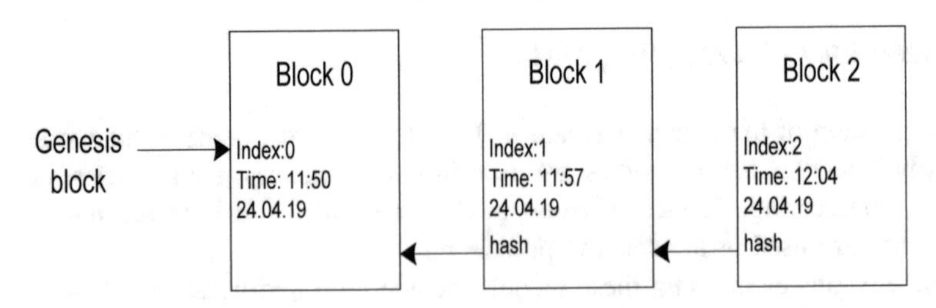

Fig. 1. The architecture of the chain of stored data of the team of robots

Hashing blocks is used to save data, the hash of the previous block used allows you to always know where the information is stored, request the required data at any time. It is also worth noting the significant advantage of hash functions is the ability to check for changes ensuring reliable data storage.

In case of a possible collision of network if devices create at the same time blocks with the same index, the priority of execution will remain behind a chain with a large number of blocks. (see Fig. 2)

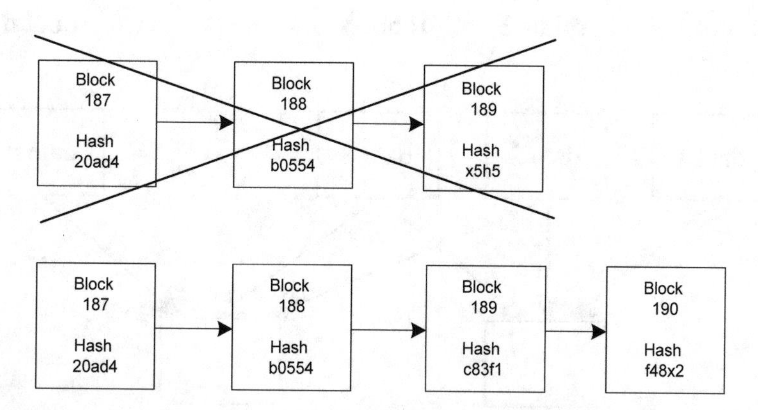

Fig. 2. Result of a collision in a chain of blocks

The exchange of information on the blockchain is an integral part of the system's performance. There are several rules for maintaining network synchronization by nodes.

- A message about the generation of a new unit by the node immediately prior to creation
- When a node is connected to a new peer, information about the last generated block is requested.
- When finding a block node with an index greater than it, either a full block chain request is made, or a block is added to its target.

With regard to the management of groups of small-sized robots, it is advisable to use the 2 synchronization option. It provides a number of advantages related to the scalability of the network and the refinement of the collective tasks of the joining robot.

The hashing function [12] is responsible for protecting the stored information in the blockchain; it prevents the addition of non-valid blocks to the chain. It also regulates the computational complexity of the network. The use of various algorithms allows you to configure the protection of the information block chain in different ways. We can single out several popular algorithms [13] and adapt them for use in groups of small-sized robots, for example:

- cryptonight [14], which uses the ring signatures protocol, allows you to achieve anonymity in a transaction by using a user group, in which one of them signs a transaction from the whole group, and it is not certain which user. With regard to group management, this allows you to create a secure system in which, even when gaining control over a system element, you cannot calculate the source of information in the block, which ultimately protects other team members. (see Fig. 3)

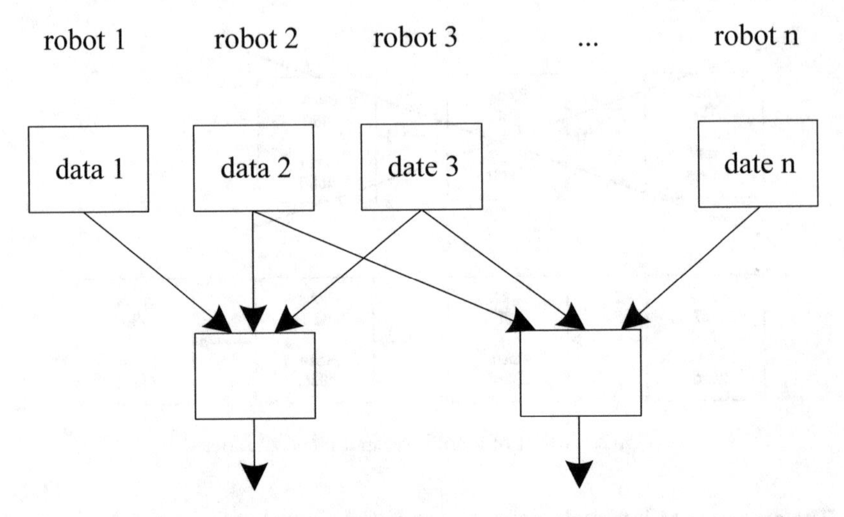

Fig. 3. Circuit protocol of ring signatures.

For the difficulty of analyzing the chain of blocks, except for ring signatures, one-time addresses are used as recipient addresses, indicating the right to dispose of one of the exits, but not allowing the sender to use the output. From the user's point of view, such a transaction can be used as an entry point to any transaction to which it refers. Increasing transaction fees increases anonymity by expanding the mix.

- ethereum [15–17], the algorithm is based on smart contracts. There are 2 types of accounts: external, controlled by private keys and contractual accounts, controlled by the code specified in the terms of the contract.

For external accounts, the ability to send messages to other external and contractual accounts is provided. To do this, create a new transaction using the private key.

Contract accounts cannot independently initiate a transaction, they are used to respond to already created transactions.

All actions in the ethereum blockchain occur only due to transactions initiated by external accounts. (see Fig. 4). Created transactions exist to connect external requests and the internal state of the platform.

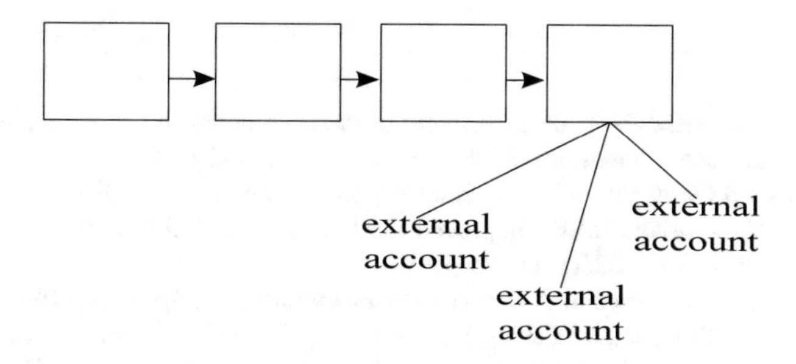

Fig. 4. The ethereum blockchain.

It is also worth noting one of the advantages of the ethereum platform – the "Ghost protocol" [18], which allows to preserve the stability of the system during the periods of collision creation by choosing the chain of blocks with the largest number of calculations.

4 Research Results

Analyzing the existing projects related to the blockchain and robotics, it should be noted that recently the topic is gaining momentum. Most projects are based on the ethereum algorithm [10], which is leading due to the speed of creating blocks of 15 s.

Using the blockchain technology when managing groups of small-sized robots, it is possible to solve the problems of lack of knowledge, even in conditions of fragmentation of the group. Due to protocol consistency and network replication, the blockchain does not create a single point of failure.

The main problems of the technology worth mentioning the possibility of "attacks 51%" [19], as well as the resource requirements to maintain the operation of the chain. At the same time, in view of the use of chaotic algorithms based on a block network for encrypting transmitted messages, the possibility of "attack 51" in the network is negligible. Regarding the issue of resource consumption in POW algorithms, it is necessary to determine in advance which information can be processed locally on the robot, and which information should be transferred to the network.

It is also worth noting a side function of the blockchain – keeping a protected audit log of sensors onboard a small-sized robot. Recorded data can be further analyzed even in case of loss of the analyzed team member.

Acknowledgment. The reported study was funded by RFBR according to the research project №17-29-07054 and №18-58-00051.

References

1. Dorigo, M.: SWARM-BOT: an experiment in swarm robotics. In: Proceedings 2005 IEEE Swarm Intelligence Symposium, 2005. SIS 2005, pp. 192–200 (2005)
2. Intel Lights Up the Night with 500 'Shooting Star' Drones, Intel (2016)
3. Milanovic, V., Zaghloul, M.E.: Improved masking algorithm for chaotic communications systems. Electron. Lett. **32**(1), 11–12 (1996)
4. Nakamoto, S., et al,: Bitcoin: A Peer-to-peer Electronic Cash System (2008)
5. Narayanan, A., Bonneau, J., Felten, E., Miller, A., Goldfeder, S.: Bitcoin and Cryptocurrency Technologies: A Comprehensive Introduction. Princeton University Press, Princeton (2016)
6. Azaria, A., Ekblaw, A., Vieira, T., Lippman, A.: Medrec: using blockchain for medical data access and permission management. In: 2016 2nd International Conference on Open and Big Data (OBD), pp. 25–30 (2016)
7. Sikorski, J.J., Haughton, J., Kraft, M.: Blockchain technology in the chemical industry: Machine-to-machine electricity market. Appl. Energy **195**, 234–246 (2017)
8. Samaniego, M., Deters, R.: Virtual resources & blockchain for configuration management in IoT. J. Ubiquit. Syst. Pervasive Netw. **9**(2), 1–13 (2017)
9. Ferrer, E.C., Hardjono, T., et al.: RoboChain: A secure data-sharing framework for human-robot interaction. arXiv Prepr. arXiv:1802.04480 (2018)
10. Strobel, V., Dorigo, M.: Blockchain technology for robot swarms: a shared knowledge and reputation management system for collective estimation (2018)
11. Ferrer, E.C.: The blockchain: a new framework for robotic swarm systems. In: Proceedings of the Future Technologies Conference, pp. 1037–1058 (2018)
12. Rogaway, P., Shrimpton, T.: Cryptographic hash-function basics: definitions, implications, and separations for preimage resistance, second-preimage resistance, and collision resistance. In: International Workshop on Fast Software Encryption, pp. 371–388 (2004)
13. Luntovskyy, A., Guetter, D.: Cryptographic technology blockchain and its applications. In: The International Conference on Information and Telecommunication Technologies and Radio Electronics, pp. 14–33 (2018)
14. Noether, S.: Ring signature confidential transactions for monero. IACR Cryptol. ePrint Arch. **2015**, 1098 (2015)
15. Dannen, C.: Introducing Ethereum and Solidity. Springer, Heidelberg (2017)
16. Wood, G., et al.: Ethereum: a secure decentralised generalised transaction ledger. Ethereum Proj. Yellow Pap. **151**, 1–32 (2014)
17. Buterin, V., et al.: A next-generation smart contract and decentralized application platform. White Pap. **3**, 37 (2014)
18. Atzei, N., Bartoletti, M., Cimoli, T.: A survey of attacks on Ethereum smart contracts. IACR Cryptol. ePrint Arch. **2016**, 1007 (2016)
19. Bastiaan, M.: Preventing the 51%-attack: a stochastic analysis of two phase proof of work in bitcoin (2015). http://referaat.cs.utwente.nl/conference/22/paper/7473/preventingthe-51-attack-astochastic-analysis-of-two-phase-proof-of-work-in-bitcoin.pdf

Markdown Optimization—A Taxonomy and Realization of the Economic Decision Problem

Felix Weber(✉) and Reinhard Schütte

University of Duisburg-Essen, Universitätsstraße 9, 452239 Essen, Germany
{felix.weber, reinhard.schuette}@icb.uni-due.de

Abstract. The process of taking a markdown on products with a difference in the anticipated sales volume is common practice for a diverse range of businesses. The present decision problems include both the amount and the timing of markdowns. In this article, a taxonomy that structures the markdown optimization problem based on a state-of-the-art literature review is developed as groundwork for further scientific research.

Keywords: Markdown · Optimization · Retail · Taxonomy · Price · Simulation

1 Motivation

The share of markdowns within the overall number of sales is continuously growing. This trend is especially apparent in fashion retailing where the high overall markdown volume even results in positively-changed consumer behavior. An analysis of three years of data from the US found that consumers are only willing to pay 76% of the full price of the examined women's wear categories [1]. This change in consumer behavior can be linked with the growing occurrence of markdowns and shows how badly markdowns are carried through, in general. The same development, a growing share of price reductions, is also present for promotion management. This may have long-term effects on customers, like the updating of reference prices or a decreased willingness to pay for products at the regular price [2]. For a regular customer, the distinction between a markdown and a price promotion is not present and therefore these two aspects interlink with each other. On a company-specific level the economic potential of enhancing the markdown process can be illustrated by the US department store Macy's that is testing the employment of RFID technology to all its stock. By that, Macy's is able to monitor the exact inventory level in real-time. Due to the better data availability in the already rolled out stores, Macy's was able to increase the full-price sales in the women's shoe department by 2.6% [3]. Another example [4] shows how price optimization leads to an improvement of the margin by 0.2 % while reducing the costs for related processes by 25%.

To understand the underlying problem, the markdown process needs to be regarded as a whole. The initial operating activities of deciding on a future assortment and setting the sales price in line with the desired profit margin is based on hypothetical

R. Silhavy et al. (Eds.): CoMeSySo 2019, AISC 1046, pp. 449–460, 2019.
https://doi.org/10.1007/978-3-030-30329-7_40

assumptions of a demand curve, availability, predefined reactions of the competition, and the anticipation of the appeal or attractiveness on the customer side.

This finally results in an order of a sales quantity that is derived from the above-mentioned considerations. The ordered products are then produced, distributed to central warehouses, and finally, delivered to single retail outlets. Starting with this initial sales quantity and value, the sales period begins. At any point of this sales period —but mostly at the end—the difference between the expected sales volume and the effectively realized sales volume at that point of time determines whether a write-off or markdown is needed.

We define the markdown optimization problem as the determination of optimized timing (when?) and reduction amount (how much?) for each relevant retail outlet or warehouse of a company.

In this context, markdown models and considerations about the optimization of markdowns has been a well-researched subject in the past. Most of this work emerged from the domains of marketing, retailing, and business administration, but there have also been contributions from several other research fields like mathematics, data science, operation research, tourism, and production research. This has resulted in a vast body of literature that is spread over a wide variety of domains, publication venues, and starting positions. However, a comprehensive overview of what has been investigated in the past and what are the open issues for the future are is lacking.

In the following text, we develop a classification to describe the phenomena related to this topic. This is intended to accomplish a better understanding of the area and will act as groundwork to build upon in further research. Throughout the development process, we include the relevant aspects derived from an extensive literature review and enhance these findings with our own perspectives by questioning them from logical and practical viewpoints.

2 Markdowns – Problem Description and Taxonomy

As thematic placement of the process of markdown optimization, the marketing domain is appropriate. According to [5], markdown optimization can be reduced to the "price" component of the 4Ps-classification. The markdown of a certain product is a type of price differentiation (also called price discrimination), as the same product is offered for a varying price. With the price change being on a temporal basis and limited to a short period of time—the sales period—it can be further defined as a temporal price differentiation. Temporal price differentiation is characterized by different prices that are dependent on time and purchase [2, 6, 7]. However, this definition matches that of markdowns as well as promotions, which calls for a clear separation of these two kinds of discount. The delimitation of both can be illustrated by the product lifecycle. While a promotion can be applied at any point of a product's lifecycle, it is most commonly done at the introduction and maturity phase, while markdowns are only applied at the end of the sales period. Here, two different price change patterns can be emphasized. Due to the nature of promotions, the price will be reduced (P_r), and then, after the promotion period, the price will go back to the regular price (P_n).

In a markdown scenario, the price is reduced multiple times, but is not raised again. An initial reduction is most likely followed by further one if the stock does not sell out. For products without a well-defined product lifecycle, for example, consumer goods like "Nutella Spread", the need for a markdown can result from an approaching best before date or the decreasing willingness to pay on the customer side. Fast-moving consumer goods (FMCG), consumer packaged goods (CPG) goods, and all other products that have an infinite or at least a not precisely definable life span, will be on sale for more than one period. Markdown can be grouped by two characteristics regarding the individual interval of the markdown (Fig. 1 left side).

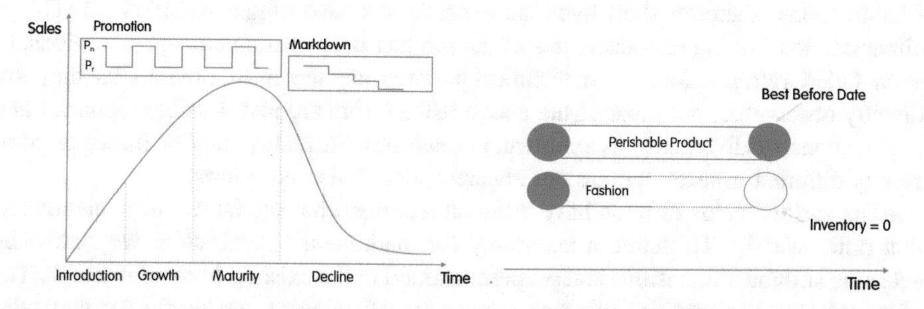

Fig. 1. Delimitation towards a promotion (left) and the two different kinds of markdown (right)

While any perishable goods in grocery retailing will have a fixed markdown time span with a closed interval (the best before date), fashion and most non-food products have an open interval that only ends when the last product is sold.

In general, the need for markdowns is a discrepancy between the expected or forecasted demand and the actual demand over the set sales period. This mismatch can have various causes, with the initial miscalculation of the sales price being the most obvious one.

To avoid a pricing error in advance and to reduce the need for markdowns further on, the willingness to pay can be tested in advance, and then the appropriate (possibly lower) price can be set right from the beginning of the sales period. Particularly when there is a broad range of branches spread over a wide geographical area, local assessment of the expected sales amount with the initial price is needed.

The reasons for markdown can further be divided into miscalculations in price setting and changed conditions during the sales period. In addition to the aforementioned price error, the errors made prior to the start of the selling period include buying mistakes, like overestimating the sales potential and ordering too much stock; misjudgment of the customers' preferences; and a poor timing in the order or the distribution. A common scenario valid for fashion retailing is where most of the products are ordered months in advance and are manufactured overseas. This means that quick adjustments are not possible, especially when the timespan between the order and delivery of a sweater, even for Walmart, takes 6 months [8]. Contractual obligations for guaranteed minimum volumes with the wholesale partners or manufacturers can lead to oversupply. The other potential error in the pre-sales period can be caused by pricing mistakes. Overestimating the customers' willingness to pay, underestimating the pricing of the competition, or general miscalculations should also be mentioned at this point.

During the selling period, external as well as internal factors make a markdown necessary. External conditions can change differently to expected, for example, different seasonality, weather conditions, unexpected (promotional) actions of the competition, discontinued product lines, or successor and replacement products by the same or a different manufacturer. Internal errors also lead to discrepancies. These include (as missed sales opportunities cannot be compensated later on) poor stock-keeping, assortment errors, non-optimal shelf placement, and uninformed sales staff. For perishable goods, a shorter shelf-live than expected can also trigger a markdown. This is either due to differing characteristics of the product itself, wrong storage conditions, or even failed refrigeration. Such technical failures are the most obvious as they are directly observable, but most of the above listed errors are just possible scenarios and explanations of divergent environmental conditions. In reality, any of these, or completely different aspects that are not apparent, could also be causes.

The various fields of trade have different realities, characteristics, rules, and effects that differ widely. To define a taxonomy for markdown optimization, we started by selecting suitable dimensions and properties found in the existing body of research. The aforementioned papers and research articles are all relevant and useful for particular situations and the presented specific circumstances, but none are sufficiently specific and systematic enough to act as an overarching classification of the subject as a whole. Thus, we define a taxonomy to cover all of the existing work in this field (Table 1). This includes several aspects that were not mentioned in the examined literature but seem valid from a practical and logical perspective.

Table 1. Taxonomy of markdown optimization

Characteristics	Values			
Position in Supply Chain	Feedstock supplier	Manufacturers	Wholesales	Retail
Group of merchandises	Fashion	Consumer electronics		Grocery
Nature of sale	Regular	Promotional		Season
Market structure	Monopoly	Oligopoly		Perfect Competition
Possibility of markdown	No possibility		N-possibilities	
Intended effect	Depreciation	Revenue	Storage costs	Image effects
Kind of demand	Deterministic/constant		Uncertain	
Cause for markdown	Unsaleable at end of sales period		Declining willingness to pay	
Objective function	Simple		Complex	

The presented taxonomy consists of nine dimensions that characterize the specific situation: the position of the considered subject within the supply chain, the merchandise group, the sales nature, the overall market structure, the possibility of making a price reduction, the intended effect, the kind of demand, the cause that makes the markdown necessary, and the resulting objective function. Each dimension is presented in detail in the following section.

The first differentiating dimension is the position within the supply chain (Shen et al. 2016). This can either be in the early phase, where the production of goods is the subject matter [9], or in the later phases of trading, including the wholesale and retailing levels. The early stages of the supply chain include the feedstock suppliers of the initial raw material and the precursors for the manufacturers. An important aspect here is the consideration of the effects of any markdown within the supply chain [9].

Any price reduction might or might not be passed onto the other participants. However, the desired effect of increased sales can be realized only when a supplier-introduced price markdown is passed onto the customer; otherwise, the subtrahend will be seeped away as additional profit for the manufacturers or the wholesale companies. Generally, wholesale involves the sale of goods to retailers, in particular, and all non-natural persons, in general. This definition of wholesale also covers multistage trading companies that operate on both levels, like the German EDEKA group, which has a central wholesale and distribution operation as well as local independent retailers. Retailing is the downstream process of selling goods to the end customer for consumption.

The second characteristic dimension is the merchandise group, which mainly differentiates the assortment and range of sold goods. According to the focus in the current literature, we organize this dimension into fashion [10], consumer electronics [11], and grocery [12]. Here, it is important to note that we implicitly use this as a superset of product groups to represent the current research focus and also to simplify the taxonomy. For example, the grocery group would include all conceivable subsets of products: fruits and vegetables, dairy products, meat products, frozen foods, beverages, and dry foods. For grocery retailers, in particular, supermarkets and food stores, the selling period differs vastly between fresh and durable foods. For fresh foods, like fruits, vegetables, meat, and fish, the selling period itself, and particularly the period of markdown, is very short. If the best before date of an apple is around two weeks, the period in which it can be sold before its appearance changes, making the product unsellable, is even shorter. The competitive pressure of maintaining high product availability might even lead to situations where a certain number of leftovers is needed, as non-availability directly influences attitudes towards the retailer in a negative way. In the fashion sector, the main paradigm is the seasonality of products. Each season introduces a new set of fashions (colors and designs), and the products from the last season become unsellable or are removed for the purpose of pressuring the customers to regularly rebuy. For electronics, the selling period can be characterized similarly to the fashion industry with regular introduction of new product successors that replace the old products and make them less desirable. Especially for brown goods, the needed shelf space is much larger with higher storage costs. The trigger for sale discrepancies is technological development, rather than aspects related to appearance.

All groups mentioned above can have a different sales nature. For "regular" products, we assume that all kinds of sales are conducted over a longer period of time of at least one to two years. This group of products is also sold over more than one period and is restocked on a frequent basis. Mainly perishable goods (foods) are considered here, as each individual product (e.g., an individual pack of carrots) has a limited life time, but the product class (here, carrots in general) itself will still be for purchase, even after the individual product's spoilage. The relevance of regular sales is the highest for most businesses, as it represents the main source of revenue. Additionally, even if this aspect is not, or is only vaguely, discussed, all kind of business include such groups of regularly sold products. Even pure fashion retailers include a range of regular sold products, here mostly called fashion "basics", like lingerie. In contrast, seasonal sales are limited to a specific and predefined time range, and although after that period, all products could technically and legally still be sold, the customers' willingness to pay decreases sharply with the start of a new season or the launch of replacement products. This is the case for electronic resellers, as successive products are released in regular and short intervals. For example, personal computers have a sales period that is limited to a maximum of 12 to 16 months, as, in accordance to Moore's law, a new generation of processors is released after this period of time. Considering the number of different components that a personal computer consists of, this period is even shorter and covers only 2 to 4 months [13]. For fashion, a typical season is three to six months. The business goal behind this kind of sales is to create a need for regular repurchases and to constitute a special kind of branding. The period of seasonal sales is mostly limited to a single one, and replenishment is mostly not possible due to limited production or replenishment periods longer than the current sales period. Promotional sales are defined by an even shorter sales period. A well-known example here is the weekly changing advertisements within grocery retailing, like "ALDI Aktuell" from Germany's leading discounter. Here, special products, mostly non-food, are stocked additionally to the normal grocery assortment (regular sales). These products can still be sold after the end of the initially planned sales period (generally one week), but the shelf space is blocked for the successor products of the upcoming promotion. As shelf space is, in general, rare, a tradeoff between a possible loss by the markdown and the requirement of free shelf space for the upcoming week's offers needs to be considered. The business goals of promotional sales are to increase the frequency of customer visits, to reach different customer groups, and to gain additional revenue that otherwise could not have been realized. Here, the sales period is again limited to a single occurrence without the option of replenishment.

The dimensions of the market structure are organized into the different market forms of monopoly [14], oligopoly [15], and perfect competition. According to the structure–conduct–performance paradigm of the industrial organization economics discipline, the market structure affects the behaviors and therefore the performance of the participants. The structure of the market is described as the gap between perfect competition, supply concentration, demand concentration, product differentiation, and market entrance barriers, and is also determined by the nature of the product and the technology available. This directly influences the participants' economic conducts, such as strategic behavior, investments, research or advertising, and pricing approaches. These decisions then have a direct influence on the individual market performances (the efficiency or profitability).

The possibility of taking a markdown reflects the fact that due to regulatory and legal requirements, many products cannot be priced freely. For example, this is the case for pharmaceuticals and books in Germany. This is not mentioned at all in the existing literature, but to provide a holistic review, this practical restriction is an important consideration to be noted and thus added to the taxonomy. Without the legal requirement of price fixing, there is no logical limitation to the number of markdowns that can be made. Many papers have presented an imaginary limitation of only a single markdown, but this is only due to the simplification of the resulting mathematical model and does not correspond to any business restriction in the real world.

The intended effects of a markdown differ: to sell off the product before its depreciation and replacement with a successor product [11], the realization of revenue that otherwise could not fully be realized, and the reduction or avoidance of storage costs [16] to image effects. The depreciation of a product can be caused by spoilage, meaning the product needs to be disposed, or deterioration, where the customers' willingness to pay is decreasing, but the product is still sellable. The replacement by a successor product, in fashion or consumer electronics for example, is caused by the simultaneous decreases in customer demand and willingness to pay. The reduction of storage costs, especially by decreasing instore shelf space, and the concurrent reduction of capital commitment is another common intended effect. The regular refreshment of the assortment is also used to create a positive psychological effect on the customer side.

Demand, as the central mechanism of action for markdown optimization [17], can be grouped into deterministic demand, where the demand function is known or can be derived, and constant demand, which is a conceivable property for convenience goods. In many cases, the demand is uncertain, as either no valid data exists, or the correlations of impact are not known.

The causes for the markdown are either unsaleability at the end of the sales period, the occurrence of a best before date, or the customers' declining willingness to pay [10] over the course of time.

The resulting objective function covers the included effects and objectives. This can either be a simple function that optimizes a single goal or the incorporation of a complex set of goals.

3 Exemplary Markdown Optimization Situations

An important part of the motivation for the discussion of the presented taxonomy is to point out the possible applications in a large class of intended applications, because with the overview resulting from the taxonomy, other groups of applications may be found that are currently completely outside the research focus. In the following text, we present two different markdown optimization situations that illustrate the broad range of characteristics that can be derived from the taxonomy (see Fig. 2). Probably the most evident and discussed application of markdown optimization is present within fashion retailing, which is outlined as the first situation. The other situation presented below refers to a food product within grocery retailing. The aim of presenting these totally different situations is to show the different implications for the resulting decision model.

The elementary business of a pure fashion retailer is the recurring sale of seasonal apparel supplemented by accessories and basic apparel products, like belts or socks. These are different situations and we concentrate on the core aspect, the seasonal products that have a predefined life time that is limited by the season (Fig. 2 left side). The typical division of seasons is analogous to the calendar year: spring, fall, winter, and summer, or combinations of the terms ("spring/summer" and "fall/winter"). The market structure represents an oligopoly, with a few retailers competing for a vast number of consumers. Within the boundaries of legal restrictions (for example, the partial prohibition of sales below cost price), the amounts and number of markdowns can be chosen freely. The intended effects of markdowns range from realizing an enhancement in revenue to maintaining the image effect by continuing to offer the newest trends and designs.

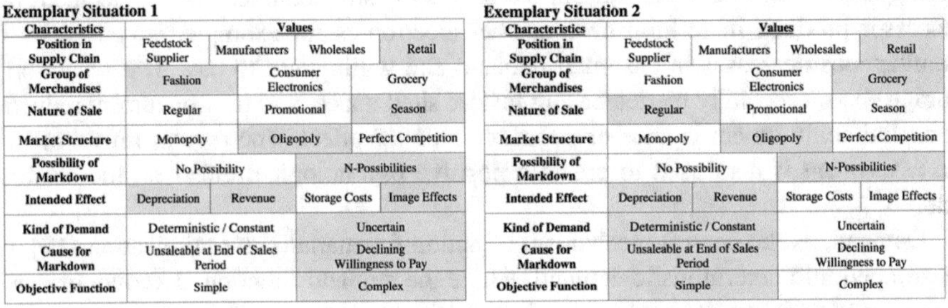

Exemplary Situation 1

Characteristics	Values			
Position in Supply Chain	Feedstock Supplier	Manufacturers	Wholesales	Retail
Group of Merchandises	Fashion	Consumer Electronics	Grocery	
Nature of Sale	Regular	Promotional	Season	
Market Structure	Monopoly	Oligopoly	Perfect Competition	
Possibility of Markdown	No Possibility	N-Possibilities		
Intended Effect	Depreciation	Revenue	Storage Costs	Image Effects
Kind of Demand	Deterministic / Constant	Uncertain		
Cause for Markdown	Unsaleable at End of Sales Period	Declining Willingness to Pay		
Objective Function	Simple	Complex		

Exemplary Situation 2

Characteristics	Values			
Position in Supply Chain	Feedstock Supplier	Manufacturers	Wholesales	Retail
Group of Merchandises	Fashion	Consumer Electronics	Grocery	
Nature of Sale	Regular	Promotional	Season	
Market Structure	Monopoly	Oligopoly	Perfect Competition	
Possibility of Markdown	No Possibility	N-Possibilities		
Intended Effect	Depreciation	Revenue	Storage Costs	Image Effects
Kind of Demand	Deterministic / Constant	Uncertain		
Cause for Markdown	Unsaleable at End of Sales Period	Declining Willingness to Pay		
Objective Function	Simple	Complex		

Fig. 2. Exemplary classification of two different markdown optimization situations

Because each product is tied to the single season in which it is sold in, the demand for the product is uncertain. This effect is intensified with the given nature of the calendrical season, where the length of a season is mostly uncertain. Also, the consumers' preferences cannot be fully predicted. The cause for the markdown is the customers' decreasing willingness to pay for the product over time. The resulting objective function is complex, as several aspects need to be considered. Besides the basic improvement in revenue, the image effects that need to be considered are vast—not only because they can lead to possible changes in consumer behavior, as mentioned above, but also because they effect the overall image of the retail chain and the concerned brand and the maintenance of the brands image (for example, a premium brand might not want to appear as having cheap deals).

The second situation illustrated in the following is a grocery retail company and FMCGs (Fig. 2 right side). The characteristics of its products are that they have a short shelf life, either as a result of high consumer demand or because the product deteriorates rapidly, and the products are sold over a long period of time and can be replenished on a regular basis. Two examples fall under this category: the above already mentioned "Nutella spread" and canned drinks, like orange juice. The significant difference from the first situation is that, here, the markdown is applied on a single product or batch (cans with a best before date of 08.03) and not the product class ("Nutella spread" in general), as the product will be replenished on a regular basis. The market structure for grocery retailing is again characterized as an oligopoly, and there are no mandatory restrictions towards the number of price changes. The intended effect of making a markdown is primarily to cope with the spoilage of the products by selling them off before they need to be disposed. Clearing this stock also implies that at least some kind of positive return is achieved compared to completely writing-off the products. As FMCG are sold over a longer period of time and are frequently subject to promotional price reductions, the demand can either be determined, as enough valid historical data is available, or can be considered to be constant. As presented above, the cause for the markdown is the fact that the product becomes unsaleable at the end of the sales period, as the products cannot legally be sold after the best before date. The objective function here is simple, as the main objective is to reduce any losses accompanying the spoilage.

It is evident that both of the above-described situations are unique and therefore need individual consideration when making markdown optimization decisions.

4 Solving the Decision Problem

4.1 Heuristics for Solving Ill-Structured Decision Problem

With the above outlined taxonomy, it is possible to structure the decision problem and its individual components. Decisions, the selection from a range of different choices, are a central practical task and are therefore a crucial aspect for scientific research within the economic science field. The components of solving a decision problem include the structure, the formulation, and the mathematical resolution to create an explicit model of the underlying problem. A decision model is the combination of the following components [18]: the set of environmental situations, the set of the possible action alternatives, the set of the results of the action alternatives (E), the set of target amounts, and the set of resulting utility values.

Based on this, the decision model is created by linking these components. By matching each action alternative within each environmental situation to the resulting action alternatives, the effect function is created. The evaluation function, as the second linkage within the decision model, is created by assigning each effect function to a target amount. To complete the model, the evaluation function and effect function are connected with the utility values.

Although this approach appears simple in theory, there are problems with its implementation the real world. To constitute such a decision model, the set of all possible environmental situations, action alternatives, results of the action alternatives, target amounts, and utility values need to be known. The needed data is extending the currently available data [19] by far. Furthermore, the dependencies and interdependencies between the functions need to be unequivocally definable. Lastly, an approach to solve the resulting model needs to be available. Rieper [18] defines a well-suited decision problem by the following properties:

– The interdependency between the different alternatives and outcomes are certain and can be defined. The to-be-achieved goals can be listed and sorted accordingly to the resulting utility values
– Within each environmental situation, all possible alternatives can be rated
– A one-dimensional objective function is given, and therefore, the resulting utility values can be calculated
– An efficient solution method exists that can solve the resulting model within an acceptable time span

Hence, any decision problem that lacks any of these four properties is an ill-structured decision problem. By reviewing the above-presented cases of different markdown situations, it becomes clear that both situations lack at least one of the properties and thus represent ill-structured decision problems.

An opportunity to bypass the complexity of creating an adequate decision problem is the deployment of simple rule-based heuristics in a mathematically advanced optimization model. Otherwise complex Machine Learning model can be used, but are currently not in use by many retailers due to, inter alia, technological obstacles [20]. PricewaterhouseCoopers [21], for example, suggests that retailers with more than $500 million in annual returns can benefit most from using markdown optimization software, whereas smaller retailers can achieve good results with heuristic approaches. The use of heuristics to determine the markdown amount and timing is common practice in most retail scenarios. The most important formulas used are alterations of the following [22]: The outdate represents the last day of the sales period, and the "weeks of sales period" (WoSP) represents the number of weeks left from today until the end of the sales period. The units sold in a business week from Monday to Saturday $\left(\sum_{t=1}^{t_n} Units_{sold}\right)$ and the units on the shelf at the beginning of the week $\left(Units_{t-1}\right)$ are the key variables here.

The "weeks of supply" (WoS) represents the number of weeks of inventory left and is calculated on a weekly basis, where

$$\boldsymbol{WoS} = \frac{Units\ in\ stock_{tn+7}}{\sum_{tn}^{t_n+6} Units\ sold} \tag{1}$$

Based on these simple analytics the markdown amount is then determined through fixed guidelines, for example as the following formulas [2–4] illustrate:

$$\text{WoS - WoSP} < 1,\ \text{no markdown is taken} \tag{2}$$

$$\text{WoS - WoSP} > 4,\ \text{results to a } 20\%\ \text{markdown} \tag{3}$$

$$\text{WoS - WoSP} > 8,\ \text{results to a } 50\%\ \text{markdown} \tag{4}$$

The exact markdown percentage is defined by gut feeling of the responsible staff or general guidelines within the retail organization. To measure the effectiveness of the performed markdown two simple analytics can be captured the weeks before and after the markdown: the sell through percentage (ST) represents the pace of the sales:

$$ST = \frac{\sum_{tn}^{t_n+6} Units\,sold}{Units\,in\,stock_{tn+7}} \quad (5)$$

As the WoS and ST have an inverse relationship to each other, together they give a good indication when to take a markdown, the success of a markdown and when the order of the replacement product can be triggered.

5 Conclusion

This article proposes a comprehensive taxonomy to classify and characterize the current state-of-the-art-research efforts in this area with the addition of newly identified aspects. The term "markdown" was delimited to the also common practice of promotional price reductions. Considering the specifics of the time intervals of markdowns (open and closed intervals), two different kinds of markdowns were defined.

The research revealed several patterns, like a strong focus on the fashion domain, and gaps in the existing literature. The key challenge of developing a suitable decision model was underlined as well as the option of bypassing this by applying a heuristic approach. Researchers and practitioners can use the taxonomy to reveal new groups of applications related to markdown optimization that are currently completely outside of the research focus.

We believe that the results of our review will help to advance the needed research within this area and that the taxonomy in particular will be useful for the development and assessment of new research directions.

References

1. Zaczkiewicz, A.: Pricing Report Shows the Negative Impact of Apparel Markdowns (2016). http://wwd.com/business-news/other/retail-apparel-pricing-report-first-insight-li-fung-10690630/. Accessed 14 May 2017
2. Diller, H.: Preispolitik. 3. Auflage (ed.). Kohlhammer, Stuttgard (2007)
3. McKevitt, J.: Macy's RFID effort boosts sales, fulfillment (2017). Accessed 12 May 2017
4. Weber, F., Schütte, R.: Digital technologies for pricing problems-a case study on increasing the level of digitization at a leading German retail company. In: The 10th International Multi-Conference on Complexity, Informatics and Cybernetics 2019, Orlando, Florida, USA: IIIS (2019) https://doi.org/10.13140/rg.2.2.18730.47042
5. McCarthy, E.: Basic Marketing: A Managerial Approach. Irwin, Indiana (1960)
6. Meffert, H., Burmann, C., Kirchgeorg, M.: Marketing: Grundlagen marktorientierter Unternehmensführung Konzepte - Instrumente – Praxisbeispiele, 12 edn. pp. 357–768. Springer Fachmedien Wiesbaden Wiesbaden (2015)
7. Schmalen, H.: Preispolitik. Fischer, Stuttgart (1995)
8. Clark, J.: Fashion Merchandising: Principles and Practice. Palgrave Macmillan, Basingstoke (2014)
9. Chung, W., Talluri, S., Narasimhan, R.: Price markdown scheme in a multi-echelon supply chain in a high-tech industry. Eur. J. Oper. Res. **215**(3), 581–589 (2011)

10. Namin, A., Ratchford, B.T., Soysal, G.P.: An empirical analysis of demand variations and markdown policies for fashion retailers. J. Retail. Consum. Serv. **38**, 126–136 (2017)
11. Chung, W., Talluri, S., Narasimhan, R.: Optimal pricing and inventory strategies with multiple price markdowns over time. Eur. J. Oper. Res. **243**(1), 130–141 (2015)
12. Namin, A.: Do consumer demographics affect dynamic price markdowns of seasonal goods? Arch. Bus. Res. **3**(5), 70–77 (2015)
13. Scheimann, T.: Produktlebenszyklen - Immer schneller neuer (2011). http://www.tagesspiegel.de/wirtschaft/produktlebenszyklen-immer-schneller-neuer/4041756.html. Accessed 11 June 2017
14. Dong, J., Wu, D.: Two-period pricing and quick response with strategic customers. Int. J. Prod. Econ. **215**, 165–173 (2017)
15. Gupta, D., Hill, A.V., Bouzdine-Chameeva, T.: A pricing model for clearing end-of-season retail inventory. Eur. J. Oper. Res. **170**(2), 518–540 (2006)
16. Caro, F., Martínez-de-Albéeniz, V.: The effect of assortment rotation on consumer choice and its impact on competition. In: Tang, C.S., Netessine, S. (eds.) Consumer-Driven Demand and Operations Management Models: A Systematic Study of Information-Technology-Enabled Sales Mechanisms, pp. 63–79. Springer US, Boston, MA (2009)
17. Ni, G., et al.: Optimal online markdown and markup pricing policies with demand uncertainty. Inf. Process. Lett. **115**(11), 804–811 (2015)
18. Rieper, B.: Betriebswirtschaftliche Entscheidungsmodelle : Grundlagen. NWB : Betriebswirtschaft, Herne [u.a.]: Verl. Neue Wirtschafts-Briefe (1992)
19. Kari, M., Weber, F., Schütte, R.: Datengetriebene Entscheidungsfindung aus strategischer und operativer Perspektive im Handel. HMD Praxis der Wirtschaftsinformatik (2019)
20. Weber, F., Schütte, R.: State-of-the-art and adoption of artificial intelligence in retailing. Digit. Pol. Regul. Governance. https://doi.org/10.1108/dprg-09-2018-0050
21. PricewaterhouseCoopers: Pricing it right - In Brief: Markdown Effectiveness. (2010).https://www.pwc.com/ca/en/retail-consumer/publications/markdown-effectiveness-2011-03-24-en.pdf. Accessed 27 Mar 2017
22. Tepper, B.K., Greene, M.: Mathematics for Retail Buying, 8th edn. Bloomsbury, New York (2016)

AAC Intervention on Verbal Language in Children with Autism Spectrum Disorders

Pavel Zlatarov(✉), Galina Ivanova, and Desislava Baeva

University of Ruse, Ruse, Bulgaria
{pzlatarov, dbaeva}@uni-ruse.bg,
givanova@ecs.uni-ruse.bg

Abstract. The paper presents an alternative/augmentative software solution for children with impaired communication skills, focusing on learners with autistic spectrum disorders. Using a series of images and sounds to represent words, the solution aims to help children build sentences in order to express desires and preferences towards objects and actions. The paper explores the idea of using a custom language model based on a subset of the Bulgarian language and defines rules which can be used to automate checking entered sentences for correctness. Technical and methodical aspects of the proposed system are also discussed. The system is designed to store rules in a central database and information about learners' progress in a personalized profile, making progress easily trackable and analyzable.

Keywords: Sentence · Communication · AAC · Language · Speech

1 Introduction

Speech and language disorders are among the most widespread disorders during early childhood years. They correlate with the development of basic mental processes, with communication skills, as well as any subsequent abnormalities in learning and comprehension, and with the specifics of the emotional and behavioural functions. A considerable number of children with special education needs (mainly related to the autistic spectrum or cerebral palsy) have a reduced or absent communication potential. This might be attributed to either psychological factors or motor impairments. Any improvements in the functional communication makes a great impact on the quality of life of individuals with severely impaired communication skills; it's possible to reduce the feeling of frustration, anxiety, depression, and aid them in understanding and interacting with their social surroundings.

With its fast-paced development, information and communication technology is transforming into an ever more powerful resource, helping children, parents, speech therapists and resource teachers.

A specialized software product, which, using a series of images to aid the child in building a sentence, which in turn can be used to express a preference towards an action or object, can provide valuable help in communication with these children. When installed on a mobile device, it can serve as a personal assistant and would aid

R. Silhavy et al. (Eds.): CoMeSySo 2019, AISC 1046, pp. 461–469, 2019.
https://doi.org/10.1007/978-3-030-30329-7_41

teachers and parents in the classroom and everyday life. With its help, more accurate diagnostics of the intellectual capabilities of the child can be a plausibility.

Led by the knowledge that people use different channels to relay information during communication in a social context, a team of specialists from the University of Ruse (Bulgaria) is developing and perfecting a software system, which aims to facilitate the organization of the educational process, as well as help children with impaired language skills communicate.

The fact that speech development is vastly different for each child with language disorders at every stage has been taken into account during the analysis and design of the software system; this imposes the creation of a particularly adaptive and flexible user interface, backed by a database of personally-tailored information for each user; this will ensure that their progress is objectively trackable and analyzeable.

2 Background

A wide array of products that have proven their usefulness and functionality have been developed by various teams worldwide. A particular property of this type of software is that they just store rules and properties for languages that they have been programmed for, which makes them difficult to directly translate into a new language.

The P.E.C.S (Picture Exchange Communication System) is a methodical approach that has been designed to aid children and adults with autistic spectrum disorders to build functioning communication skills. A key component of the PECS system is the emphasis on communication initiation [1], an ability that is often impaired in people with autism [1]. Built on a board (similar to a board game), the system does not allow control of the correctness of the sentences its users can build. The PECS methodology has been the base of the design of a number of software existing on the market.

NikiTalk is an alternative and augmentative communication (AAC) application based on creating phrases using images (with associated text and audio) that are grouped into various categories. The app also includes a keyboard and a drawing pad function to allow its users to create custom, hand-drawn symbols. A web service allows parents to create pages/categories, to upload images, and access to 70 different voices. It has been developed for children, teenagers and adults with impairments, who find explaining what they have done or what they plan to do, difficult; the application helps them explain by using images, text and speech. It has proven especially useful to people with communication difficulties and behavioural issues, helping them increase the level of autonomy and participation [2]. The application only supports a handful of languages out of the box, making it less flexible.

Make Sentences is an interactive tool available on mobile devices and recommended for English speakers for young learners in pre-school or older (5+). The tool can also be useful for learners of English as a foreign language. Their main task is to put scattered words in an order depending on what they hear. Learners can then build simple or more complex phrases and sentences by putting words in the correct order [3].

LinGo Play is an interesting and effective language learning app; in its essence, it is a vocabulary trainer built to help learn words and phrases through flashcards and mini-games. The LinGo language course includes a variety of topics; examples include: Education, Business, People, Home, Nature, Animals, Science, Sports and Tourism, Art, Food, Appliances, Furniture, Beauty and Health, Medicine, and many more. While it includes over 5172 flashcards of 4141 words, 373 phrases, and has over 600 English lessons, it is only available in and applicable for the English language [4].

LOGOPED 2.0 is a web-based application, the main goal of which is to enable virtual consulting and enhance therapy sessions of people with communication disorders. This system stores data about the sessions and experiments conducted with its use [5].

3 Description of the System's Action

3.1 Theoretical Setup

Building the proposed system is based on the theoretical model created by Noam Chomsky [6]. According to him, the ability to construct phrases is inherent (congenital) and corresponds to the existence of a so-called language apparatus. This apparatus interacts with other systems and includes at least two components: a cognitive system for information storage and an executive system that serves to access and process that information. Notably, the executive system is not necessarily tied to a particular language [6].

The cognitive system interacts with two subsystems of the executive system: the articulatory-perceptual (A-P) and conceptual-intentional (C-I). This corresponds to the introduction of two interfaces with two representational modes: a phonetic mode for the A-P interface and a logical mode for the C-I interface.

These structural principles are largely subconscious, as are many other biological and cognitive properties of humans. Chomsky's universal grammar postulates that there exist inherent basic principles of the language apparatus, which are of vital importance for all languages. The grammar constitutes a theory of the initial state of the language apparatus [6].

This approach has been motivated by the exceptional speed at which children are able to learn languages, the similarities in the steps which children around the world take in this process, as well as the fact that children make characteristic mistakes during learning their first language.

In 1957 Noam Chomsky publishes "Syntactic Structures". His idea revolves around using strict mathematical patterns to describe the syntax of the natural language. In his work, Chomsky conceives generative-transformative grammars as the first grammatical model emphasizing on formality and clarity. The transformative rules allow for building and interpreting sentences. With a finite set of similar rules, and a finite set of words, humans are capable of constructing sentences from an infinite set.

In his effort to determine if formal grammars can express key properties of human languages, Chomsky divides them into classes with an increasing expressive power. He then concludes that modeling of different aspects of a given language requires different classes of grammars [7].

A Chomsky grammar (G) has the following form:

$$G = (T, N, \rightarrow, S) \tag{1}$$

where:

- T is a set of *terminal* symbols
- N is a set of *non-terminal* symbols
- $\rightarrow$ is a finite set of replacement *rules*
- S is an element from N called a *start symbol* or axiom.

The hierarchy of grammars determines corresponding types of formal languages for which it is known that $L_3 \subset L_2 \subset L1 \subset L_0$, meaning that all regular languages are contextless, all contextless languages have a context and all context languages are recursively countable, while none of the types are equal to each other.

3.2 Custom Language Model

With any natural language, certain words and phrases are specific or unique to a particular domain and context. If we assume that the vocabulary of children with autistic spectrum disorders is limited and strictly formalized, and the sentences that they create have a maximum length of 5–6 words, it can be assumed that, theoretically, the structure of their speech can be described with the rules of a contextless generative grammar, according to Chomsky's hierarchy.

Applying the generative-transformative method entails that word combinations and phrases are formally represented via their structures, while taking into account the relationships of subordination and composition between their individual constituents. Knowing the dependencies of the structure of a given type of phrase or sentence, a particular word combination or a specific sentence can be generated.

Creating a functioning, high-quality algorithm for sentence construction imposes maintaining a vocabulary of a large enough set of words and their corresponding images. Words are divided into groups according to their purpose.

Within the structural representation of the sentence, the following symbols are introduced: Noun, Adj (adjective), Verb, P (preposition), Adv (adverb), Conj (conjugation), Pron (pronoun), Num (numeric), S (sentence). The projections are represented as a tree, containing nodes representing different syntactical categories.

In the case of correctly provisioned rules for sentence construction, the generated information can be checked by means of automation. As such, each sentence can be described using the following rules (transitions):

- Sentence (S) $\rightarrow$ Noun phrase (NP) and Verb Phrase (VP)
- Noun Phrase (NP) $\rightarrow$ Adjective (Adj) and Noun Phrase (NP)
- Noun Phrase (NP) $\rightarrow$ Noun
- Verb Phrase (VP) $\rightarrow$ Verb or Noun Phrase (NP)

Word groups formed around a central word and its subordinates are called syntactic phrases. The conjugated verb serves to connect separate groups of words to form a sentence. A structure of such a sentence is represented by a syntactic tree, an example of which is shown on Fig. 1.

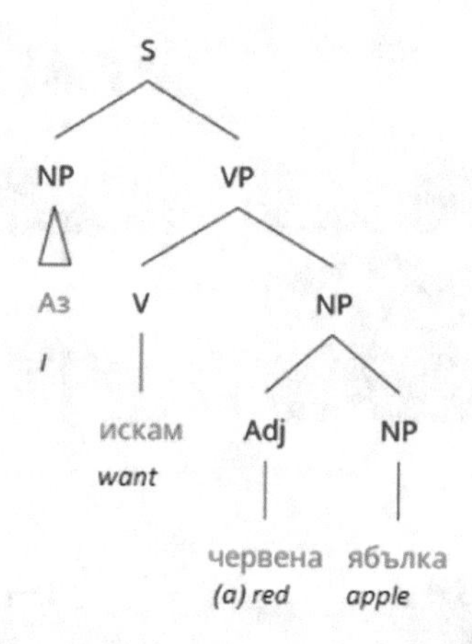

Fig. 1. A sample sentence generated using the language rules.

According to the theory, each sentence needs to have a subject [8, 9]. However, in the Bulgarian language there exist subject-less sentences where the subject position is vacant. This situation is not common in the speech of the target user group; this imposes the mandatory usage of pronouns such as "I", or a more specific addressing as a third person instead of the speaker addressing themselves (e.g. "John wants…").

3.3 Technical Requirements and Description

The software system requires the usage of a database to store the rules to be applied. In the database, each word's entry contains its string representation, a corresponding image (e.g. a JPEG file), a sound file (e.g. a waveform or MP3 file) with its pronunciation, and related metadata such as gender, number, tense, etc. The verbs used are initially planned to only be in the present tense.

During their preparation to work with a given student, speech therapists have the ability to enter new words which should be included in sentences during that particular session. They can also substitute the images and sound files in the included vocabulary with their own. This way, the system can be improved and enriched over time.

An activity diagram of the menu allowing the creation of a phrasal structure or a sentence is presented on Fig. 2. These structures can be entered into the system in advance by an authorized party (e.g. a specialist or administrator). Once they're stored in the database, these structures can be used to check if the word order of sentences entered by users is correct.

Every addition or change to the information needs to be reviewed and approved by the system administrator to ensure the integrity of the database.

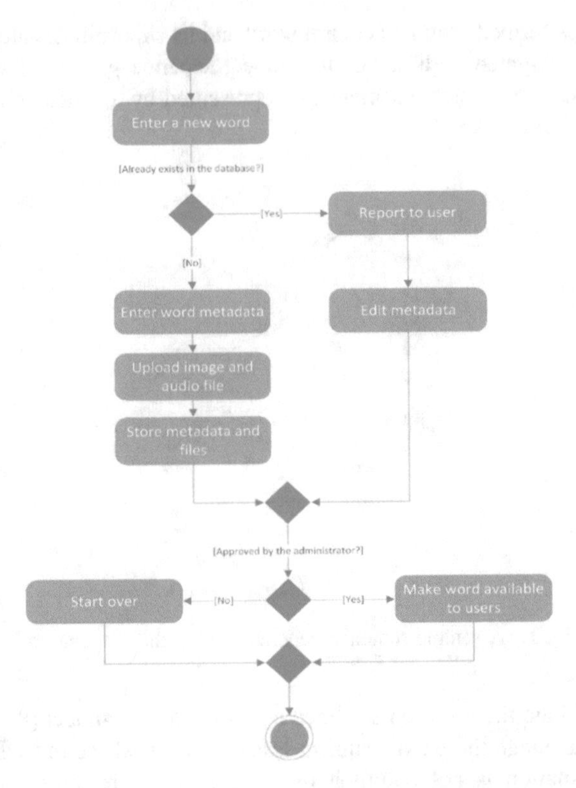

Fig. 2. Activity diagram of the process of adding new words to the database

Fig. 3. The user interface for entering a sentence

A picture of the user interface used to add phrasal structures and sentences to the database is shown on Fig. 4. Since the administrative console of the application is web-based, it can be accessed using any device with access to the Internet. The user interface for learners can also be accessed, mobile apps for the major platforms. Since images are being stored along with each word, the interface for learners is focused around these images. It is shown on Fig. 4.

The information about sentences is presented using a code matrix, where every part of the speech, gender and number have a numeric equivalent; words in the vocabulary are assigned an incrementing number, which also depends on the group they belong to.

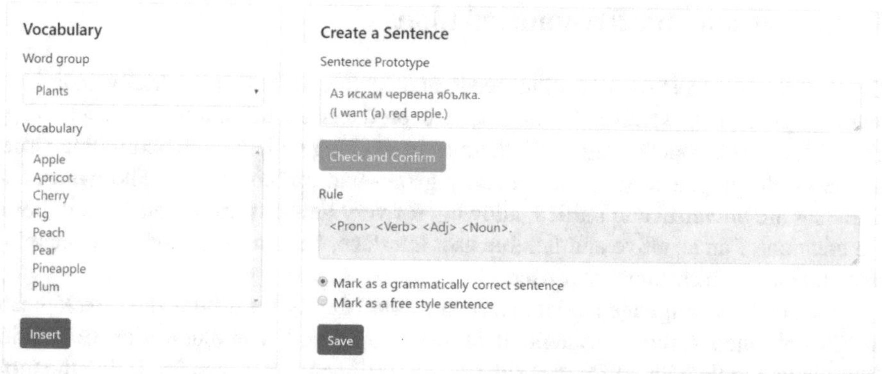

Vocabulary
Word group
Plants
Vocabulary
Apple
Apricot
Cherry
Fig
Peach
Pear
Pineapple
Plum
Insert

Create a Sentence
Sentence Prototype
Аз искам червена ябълка.
(I want (a) red apple.)
Check and Confirm
Rule
<Pron> <Verb> <Adj> <Noun>.
Mark as a grammatically correct sentence
Mark as a free style sentence
Save

Fig. 4. The user interface used by specialists and administrators to create sentences

The presented software system allows users to build specific, contextless languages, which are tailored to the learners' ability to communicate. It is an indispensable resource in the creation of an environment for experimenting in the field of speech development in children with autistic spectrum disorders.

Even though the presented method for building sentences is powerful, it has some limitations; however, some possible extensions have been planned. It is also possible to support several user languages simultaneously.

4 Methodical Aspects

As an initial step, the program is based on simple, imperative communication (e.g. demanding for objects) instead of the idea of expressing or communicating abstract topics. These expressions are done by the child, who presents the picture of the desired object to their teacher or parent. The system then gradually moves on to including symbols and using these symbols in forming functional sentences, where, for example, an action can be conveyed similarly to an object, and emotions can be expressed as well.

The ease of use of touch screen displays enables users who might have motor impairments to more easily make use of the system [10]. A good functional solution is to be able to use the application and all of its content on more than one device (e.g. a laptop, a tablet computer, and a smartphone), while all of the application variants collect data about the learner's achievements and store them in a central knowledge base [11].

A custom vocabulary and statistics about the development are stored and kept for each child within their own personal profile, which allows tracking and analyzing their eventual progress in a correct and objective manner.

5 Conclusion and Recommendations

Every child has their own specific needs and require following a strictly individual learning plan [12], which a lot of software products do not allow for. Technology geared towards compensating for difficulties in working with children exhibiting issues in speech development are predominantly game- and exercise-based. The games and exercises are universal and unified, allowing for very little customization. This imposes the creation of an adaptive and flexible user interface, backed by a database containing information on each individual user.

A specialized language model has been created during the design and development of the presented software system. It allows people with communication difficulties (especially people with ASD) to build grammatically correct sentences. Using the tools provided by the system, the information that users with speech impairments need to convey when communicating with others, can be correctly identified.

This way, every specialist will have the ability to create and modify different modes (depending on progress and or/therapy) that can be used during their work with the learners – e.g. emphasizing on words such as foods, animals, home, etc. Supporting an individual vocabulary leads to a more personal experience; this way, if the child works with different specialists, all the work done with the learner will be accumulated and unified into a common database, with the ability to monitor the overall progress. At home, parents can switch into the free style communication mode, where every word learned so far can be used.

The ability to constantly improve and enrich the word models and language rules would lead to the creation of an extremely powerful, flexible resource that is adaptable to a considerable number of languages. Furthermore, the models and information collected around them can be the base for more powerful algorithms and modules (e.g. artificial intelligence powered systems), which would extend the system's capabilities and, with each iteration, get better and better for its primary user group.

Acknowledgements. The study was supported by contract of University of Ruse "Angel Kanchev", № BG05M2OP001-2.009-0011-C01, "Support for the development of human resources for research and innovation at the University of Ruse 'Angel Kanchev'". The project is funded with support from the Operational Program "Science and Education for Smart Growth 2014 – 2020" financed by the European Social Fund of the European Union.

References

1. Bondy, A., Sulzer-Azaroff, B.: The Pyramid Approach to Education in Autism. Pyramid Educational Products, Newark (2002)
2. NikiTalk Homepage. http://www.nikitalk.com/Default.aspx. Accessed 01 June 2019
3. Make Sentences app (Apple iTunes store). https://itunes.apple.com/us/app/make-sentences/id587265511?mt=8. Accessed 22 Apr 2019
4. LinGo Play app (Apple iTunes store). https://apps.apple.com/us/app/learn-languages-lingo-play/id969976197. Accessed 22 Apr 2019

5. Sivakova, V., Totkov, G., Terzieva, T.: LOGOPED 2.0: software system for e-consulting and therapy of people with communicative disorders. In: Proceedings of the International Conference on Computer Systems and Technologies and Workshop for PhD Students in Computing, p 73. ACM (2009)
6. Chomsky, N.: On certain formal properties of grammars. Inf. Control **1**, 91–112 (1959)
7. Chomsky, N.: Three models for the description of language. IRE Trans. Inf. Theory **2**, 113–124 (1956)
8. Leafgren, J.: A concise Bulgarian grammar. Center for Slavic, Eurasian, and East European Studies, Duke University (2011)
9. Osenova, P., Simov, K.I.: Формална граматика на българския език. (A Formal Grammar of the Bulgarian Language). Bulgarian Academy of Sciences (2007)
10. Chen, K., Savage, A., Chourasia, A., Wiegmann, D., Sesto, M.: Touch screen performance by individuals with and without motor control disabilities. Appl. Ergon. **44**(2), 297–302 (2013)
11. Levin, M.: Designing Multi-device Experiences: An Ecosystem Approach to User Experiences Across Devices. O'Reilly Media Inc., Sebastopol (2014)
12. Iovannone, R., Dunlap, G., Huber, H., Kincaid, D.: Effective educational practices for students with autism spectrum disorders. Focus Autism Other Dev. Disabil. **18**(3), 150–165 (2003)

Using Autoregressive Integrated Moving Average (ARIMA) for Prediction of Time Series Data

Dmitrii Borkin, Martin Nemeth(✉), and Andrea Nemethova

Faculty of Materials Science and Technology in Trnava, Institute of Applied Informatics, Automation and Mechatronics, Slovak University of Technology in Bratislava, Bratislava, Slovakia
{dmitrii.borkin,martin.nemeth, andrea.peterkova}@stuba.sk

Abstract. The data analysis and data mining is broad topic nowadays. One important area and also a step of the data mining process is understanding the data. Proper understanding of the investigated data is not only knowing the structure of the data, but also the behavior and characteristics of the data. Statistical analysis provides multiple methods to analyze and understand the nature of the data. The aim of this paper is to analyze our time series dataset and to assess if the ARIMA model can be successfully used to predict future values of the time series data.

Keywords: ARIMA model · Time series · Energy prediction · Machine learning

1 Introduction

In this paper we aim at the time series data gathered from a thermal plant. These data are in detail described in the article: Comparison of methods for time series data analysis for further use of machine learning algorithms. The main characteristics of these time series data are, that the original dataset consists of two parameters. First, there is the timestamp information and the second parameter is the thermal power output and it is given in megawatt units. The dataset consists of 50000 records throughout 2 years. The thermal plant is a complex process and it is generating huge amount of data with great knowledge potential. The benefit of analyzing the data can be seen in the optimization of the process and in prediction of the future state of the system. The data analysis is also a complex task and first we need to understand the nature and behavior of the data. In this article we present method to test the time series data for the stationary and we are also testing the ARIMA model for prediction the future values of the time series.

R. Silhavy et al. (Eds.): CoMeSySo 2019, AISC 1046, pp. 470–476, 2019.
https://doi.org/10.1007/978-3-030-30329-7_42

2 The Initial Data Analysis

The effectivity is important aspect in the data mining and data analysis process. Our dataset consists of 50000 records. Plotting this amount of data into a histogram would lead to disarranged plot. For better overview of the data we have decided to use the mean values of the thermal power output during each week. When plotting only mean values we are able to see trending and seasoning in the data much clearer. The loss of the data is in this case not significant, and for training the prediction model we were using the whole dataset (Fig. 1).

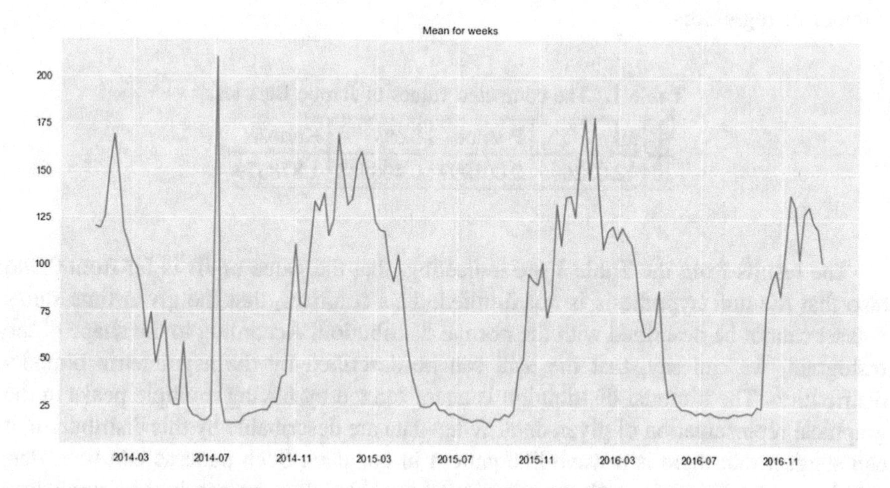

Fig. 1. Simplified overview of the time series data using mean values

2.1 Computation of Coefficient of Variation

After preparing and plotting the data, we have determined the coefficient of variation of our time series data. This coefficient is a measure of relative variability. It can be presented as a ratio of the standard deviation to the average, or mean. The computation of the coefficient of variation is given as follows [4, 5]:

$$Coefficient\ of\ variation = \frac{\sigma}{\mu}, \tag{1}$$

where σ is the standard deviation and μ is the mean.

In our case the coefficient of variation has the value of k_var = 0.70.

2.2 Determining the Probability Distribution of Time Series Data

Next task was to assess the probability distribution of the time series dataset. It is important to know the probability distribution of given data, because it can suggest the character of the investigated process. Some probability distributions are describing failures in the system and others for example the time needed to repair the fault.

For this purpose, we have decided to perform the Jarque-Bera test. This test is used to assess if the skewness and kurtosis of the given data are matching the normal distribution. The result of this test is so called test statistic, referred to as JB. The value of JB is always nonnegative and the further is from 0 the stronger is the signal that the data do not have the normal distribution. The JB value is computed as follows [1–3]

$$JB = \frac{n-k+1}{6}\left(S^2 + \frac{1}{4}(C-3)^2\right) \quad (2)$$

where n is the number of observations, S is the skewness, C is the kurtosis and k is the number of regressors.

Table 1. The computed values of Jarque-Bera test

	JB	P-value	Skew	Kurtosis
0	12.432606	0.001997	0.400988	1.878679

The results from the Table 1 are indicating, that the value of JB is far from 0 and also that the null hypothesis is not fulfilled. This confirms, that the given time series dataset cannot be described with the normal distribution. According to the shape of the histogram, we can say, that the data can be described by the asymmetric bimodal distribution. The bimodal distribution is characterized by having multiple peaks in the graphical representation of given data. When data are describable by this distribution, it can suggest that there is a wave-like pattern in our data. Such patterns can be understood as some kind of repetitive behavior of the data, thus we can build a prediction model to predict future values of the target variable of the investigated process (Fig. 2).

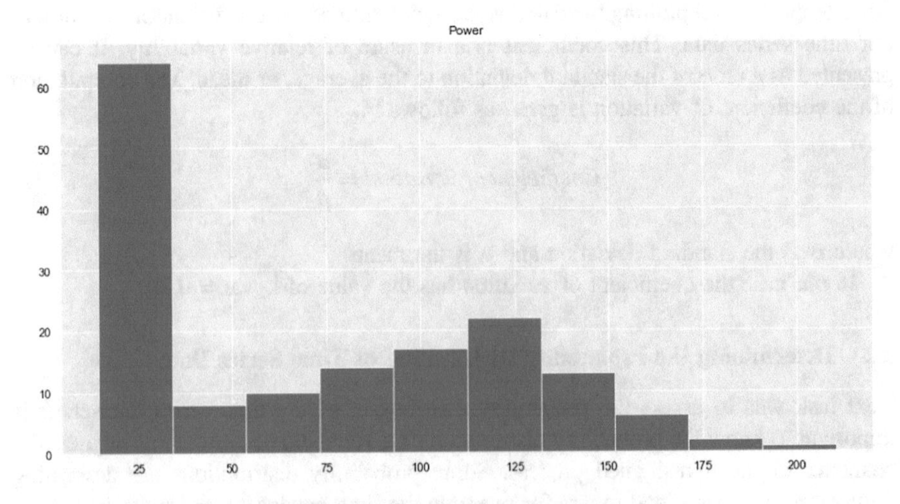

Fig. 2. The distribution of the time series data.

2.3 Assessing the Stationarity of the Time Series Data

We have subsequently decided to test the time series data on stationary. For this analysis, we have performed the Dickey – Fuller test. This test also tests the null hypothesis that a unit rook is present in an autoregressive model. The autoregressive model is a representation of a random process which is time-varying.

In our case the p value is equal to 0,0006. That means that the hypothesis of unstationarity does not apply to the given time series dataset. Thus we can say that the time series data is stationary.

In previous section of this paper, we showed, that our time series data can be described by bimodal distribution. Even though that the P value indicated the stationarity, we assume that building a prediction model using the ARIMA method will be more accurate and effective if we can modify our time series to even higher degree of stationarity.

To do this transformation, we have used the numpy library and the transformation was performed in Python with the diff function from this library. Following figure shows the time series data after the transformation (Fig. 3).

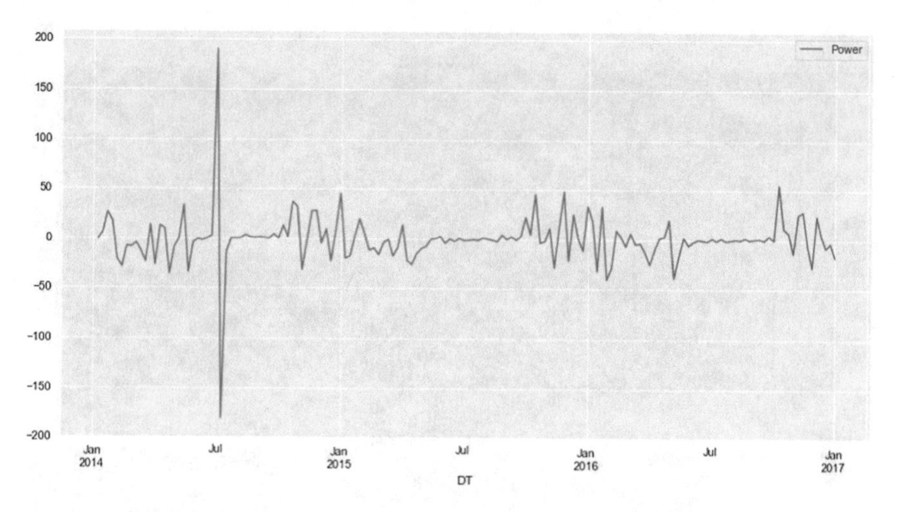

Fig. 3. The data after transformation with the numpy library.

3 ARIMA

The main aim of this paper is to assess whether it is appropriate to use the ARIMA method (Autoregressive integrated moving average) on our time series data. The ARIMA is an autoregressive model mostly used for time series data. The purpose of this model is to better understand the data and can be also used to predict future values in the series. The ARIMA model is generally defined as ARIMA (p, d, q) (P, D, Q)m, where m is the number of periods in every season. The uppercase version of parameters P, D, Q are the autoregressive, differencing and moving average of the seasonal part of given model. The lower case parameters in a form of non-negative integers, where p is the order of the model, d is the degree of differencing and q is the order of the moving average model [6, 7].

It is needed to construct the autocorrelation function and the partial autocorrelation function to be able to determine the parameters mentioned above.

The autocorrelation is telling us how correlated is the time series with the past values of it. The autocorrelation function is the plot of the correlation. The x-axis shows the correlation coefficient and the y axis shows the number of lags. On the other hand, the partial autocorrelation can be understood as a summary of the relationship between observations in a time series with prior time steps observations.

Based on the autocorrelation function and the partial autocorrelation function we can determine the d, p and q parameters. In our case the d parameter is equal to 1. This parameter represents the order of the time series. As it is shown in the figure X we can assume the parameter d to be equal to 1. The p parameter is also equal to 1 based on the partial autocorrelation function plot, where we can see that the first peak is significantly beyond the boundary. The q parameter is according to the autocorrelation function also equal to 1 based on the first peak, which is reaching the value 1 (Fig. 4).

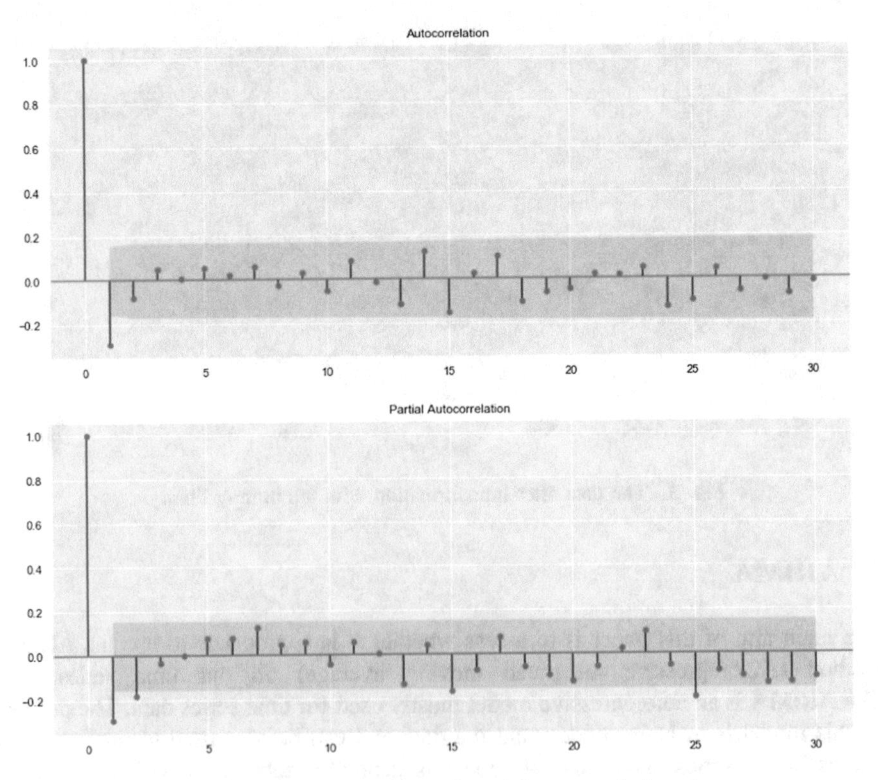

Fig. 4. The autocorrelation and partial autocorrelation plot

3.1 Evaluation of the ARIMA Model

In this section of the paper we would like to assess the fitness of the use for the ARIMA model on our time series data. First of all we have decided to perform the Ljung-Box Q test. The purpose of this test was to assess if the values in our time series dataset are random. The test statistics is defined as follows [8, 9]:

$$Q = n(n+2)\sum_{k=+}^{h}\frac{\rho_k^2}{n-k} \tag{3}$$

Where n is the sample size, ρ_k is the sample autocorrelation at lag k and h is the number of lags. For this test there are two hypotheses the H0 stands for that the data are independently distributed, and the H1 stands for that the data are not independently distributed.

The following table shows the results of the Ljung-Box Q test on our data. These results are confirming that the data values can be understood as a white noise (Table 2).

Table 2. The computed values LJung-Box Q test

Q	P_value
0.000369	0.984670
0.220838	0.895459
0.356030	0.949168
0.676012	0.954254
1.551993	0.906993

To conclude the evaluation, we have computed the coefficient of determination. In this step we have trained the ARIMA model on our dataset, which we have divided into training set and test set. The process of training and computing the coefficient of determination was also performed using Python.

If $\widehat{y}_\imath$ is the predicted value of the i-th sample, and y_i is the corresponding true value for total n samples, the estimated coefficient of determination R^2 is defined as follows [10–12]:

$$R^2(y, \widehat{y}_\imath) = 1 - \frac{\sum_{i=1}^{n}(y_i - \widehat{y}_\imath)^2}{\sum_{i=1}^{n}(y_i - \overline{y})^2} \tag{4}$$

Where $\overline{y} = \frac{1}{n}\sum_{i=1}^{n} y_i$ and $\sum_{i=1}^{n}(y_i - \widehat{y}_\imath)^2 = \sum_{i=1}^{n}\epsilon_i^2$

After computing the coefficient of determination was $R^2 = 0.00896$.

4 Conclusion

In our paper we have presented methods of assessing the stationarity of the time series data and we have also assessed the fitness of use of the ARIMA model on our data set to predict future values of our target variable. In general, the ARIMA model is appropriate for the use of prediction time series data. We have proved that our time series dataset is stationary in first sections of our paper. According to the results of the concluding tests, we can say that the ARIMA model is not appropriate for predicting stationary time series data. This paper is devoted to the stage of choosing the proper method for building a prediction model for our time series data. As it is clear from our result, the importance of understanding the character of the data is crucial in choosing right methods in later stages of datamining.

Acknowledgments. This publication is the result of implementation of the project: "UNIVERSITY SCIENTIFIC PARK: CAMPUS MTF STU - CAMBO" (ITMS: 26220220179) supported by the Research & Development Operational Program funded by the EFRR.

This publication is the result of implementation of the project VEGA 1/0673/15: "Knowledge discovery for hierarchical control of technological and production processes" supported by the VEGA.

References

1. Brillinger, D.R.: Time Series: Data Analysis and Theory. Siam (1981)
2. Granger, C.W.J.: Some properties of time series data and their use in econometric model specification. J. Econom. **16**(1), 121–130 (1981)
3. Hoeffding, W.: A class of statistics with asymptotically normal distribution. In: Breakthroughs in Statistics, pp. 308–334. Springer, New York (1992)
4. Abdi, H.: Coefficient of variation. Encycl. Res. Des. **1**, 169–171 (2010)
5. Searls, D.T.: The utilization of a known coefficient of variation in the estimation procedure. J. Am. Stat. Assoc. **59**(308), 1225–1226 (1964)
6. Contreras, J., et al.: ARIMA models to predict next-day electricity prices. IEEE Trans. Power Syst. **18**(3), 1014–1020 (2003)
7. Zhang, G.P.: Time series forecasting using a hybrid ARIMA and neural network model. Neurocomputing. **50**, 159–175 (2003)
8. Burns, P.: Robustness of the Ljung-Box test and its rank equivalent (2002)
9. Ljung, G.M., Box, G.E.P.: On a measure of lack of fit in time series models. Biometrika **65**(2), 297–303 (1978)
10. Nagelkerke, N.J.D., et al.: A note on a general definition of the coefficient of determination. Biometrika. **78**(3), 691–692 (1991)
11. Ozer, D.J.: Correlation and the coefficient of determination. Psychol. Bull. **97**(2), 307 (1985)
12. Nakagawa, S., Johnson, P.C.D., Schielzeth, H.: The coefficient of determination R^2 and intra-class correlation coefficient from generalized linear mixed-effects models revisited and expanded. J. R. Soc. Interface **14**(134), 20170213 (2017)

An Approach to the Models Translation Intelligent Support for Its Reuse

Maxim Polenov(✉), Artem Kurmaleev, Alexander Gorbunov, and Vladimir Pereverzev

Institute of Computer Technologies and Information Security, Southern Federal University, Taganrog, Russia
{mypolenov, avgorbunov, vapereverzev}@sfedu.ru, art.kurmaleev@gmail.com

Abstract. In this paper, the approach to reuse and translation of models' source code from programming and modeling languages were discussed and reviewed. The current state of research for the reuse and translation of the code was considered and analyzed. Intelligent tools were analyzed in terms of application to the task of models' translation. The expert system was chosen as a tool, and the choice was described and reviewed.

Keywords: Modeling tools · Model reuse · Translation of models · Multitranslator · Distributed Storage of Models · Expert system

1 Introduction

Since the development of the first modeling tool, engineers and researches have created various models using corresponding simulation tools and modeling languages or visual interfaces. Some tools were abandoned due to modern packages releases with new features. That overall turned out in wasting of models for those tools and time costs of creating new ones for the new modeling package each time.

All above lead us to the development of the translation tools that will allows to avoid spending time on recreating our models for new or necessary modeling packages. Due to specifics and requirements it is Multitranslator (MT) [1]. MT can translate the model's code from input to output language in required format. But to get this conversion it is required to create appropriate translation module [1] (from input to output language) written in inner MT languages for grammars and generated code description [2].

And so, it is the standalone model and programming languages translation tool with GUI that require individual computer installation to work and a researcher that understands the specifics of processed languages and MT itself. It was fully capable to translate many of our models that reduce time costs at creation those models.

At first, MT was decided to transfer into client-server model [3] with a different version of GUI focused only on translation of models on a client and to move MT to a server part. It came up to us with a problem of usage and sharing of our models since the architecture was split, and so the models' databases on both sides were added. Also, it shifted the processing to server that reduced time costs again on a big model code

R. Silhavy et al. (Eds.): CoMeSySo 2019, AISC 1046, pp. 477–483, 2019.
https://doi.org/10.1007/978-3-030-30329-7_43

processing due to server to workstation power comparison. And additionally, it reduced time costs for new users for learning MT with its complex GUI. These tools have been developed and called the Distributed Storage of Models [4].

However, there were still some uncertain cases [2] in model's translation procedure that require the model's developer to choose between the possible directions of target code generation. With a variety of translation modules of MT and languages, it can become very unhandy to solve such issues on our own or by finding an expert for each case.

For a while we were looking for the appropriate solution that would both be appropriate and minimal at time costs of development and support for our case. We ended up going through the intelligent application and its tools since an expert would use its own intelligence and knowledge to make a right choice.

With a revision of nowadays tools and after some discussions, we have chosen the expert system and knowledge base as the best solution [5, 6] for this specific task. The development of alpha version of an expert system has proved the decrease of time costs based on the results of the first tests [7].

More of the interest is what we have discussed and analyzed to make a correct choice and what impact it causes on the resulting architecture of our multilanguage translation system.

2 State-of-the-Art of Model's Language Translation Research

Before the talk about the intelligent tools, it is necessary to review current approaches for the model's reuse and translation.

One approach considers models as knowledge artifacts that can serve as a base for a future of science and engineering [8] where our approach is the same. Also, it states for the matter of models reuse automation as we are. Researches divide models reuse based on the representation of the simulation into four areas:

1. Multi-formalism, multi-scale modeling;
2. Reuse across communities of interest and the implementation spectrum;
3. Exploitation of model and simulation web services;
4. Quality-centric approaches to component evaluation.

Where on the first point researchers claim that to implement reuse of this area it is necessary to use new approaches for a different applications and components.

On the second point, it is claimed that the higher level of abstraction is for reuse, the simpler and the better results are achievable [8]. Especially it depends on the application approaches. For example, if we take (as we are trying to achieve in this paper) the lowest level of achievability, the level of programming languages, it also depends on paradigms used in input and output languages. For example, if source code language is made using Object Oriented Approach, and target language does not support such paradigm then the generated code is impossible to be reused in such way. So, it is claimed that complete reuse is only possible by using new methodologies, approaches and techniques as well for this point.

Web services claim that component-based development could solve this problem, but there is another problem of compatibility on both software and hardware levels in coup with an overloaded code and lack of standards with lack of organization and documentation. So, the researchers [8] directly call web services the only solution for reuse so they have to be supported.

And the last point describes that in the current state application can be reused on conditions of being both:

- substantiated to be enough for the intended uses for which it is created;
- match the intended uses of the simulation model where it will be integrated.

The authors also claim that it is critically important to process the reuse with quality-centric approach instead of accuracy-centric approach. In explanation it is told that there are also other important indicators that are not included in accuracy-centric approach, such as adaptability, composability, extensibility, interoperability, maintainability, modifiability, openness, performance, reusability, scalability, and usability [8].

The research ends up using big data analysis and Data Information and Knowledge Management (DIKM) Pyramid with utilizing of knowledge management as a solution of models reuse. Except for deep review of the pyramid itself, the research offers few solutions for the different tasks that can be encountered while solving the problem of reuse:

- Machine learning. Can be used to monitor and check the running simulations, sorting out models and creating model bases with labels. Also, it can check for basic syntactic and semantics of models' code to know if a reused model is written according to the language basics;
- Context Management. Paper states that there might be a list of context attributes in a contextual framework that appears to hold those attributes such as assumptions, constrains, intentions etc. [8]. This should help the researches to make a correct decision for each model on reuse;
- Domain Knowledge Extension for Collaboration and Enhanced Decision Making. It states that validity for the reuse can be viewed as a two-part process, assessment of the results and inferring of validity of the results that can improve the output models code;
- Reuse Through Model Discovery. For this task, Discovery [9] can be achieved only through consistent and relevant metadata that is only possible when properly labeled and marked up.

Another research also applies knowledge management instead of unitizing the knowledge that has to be extracted from the given data [10]. In this approach, physical model's data that are based on Modelica language is represented as RSHP [11] universal knowledge model [10].

It forms necessary models from code for future reuse, what MT does too, but in the internal specific format. It is applied to web applications and programming language reuse with specific output language [10].

This overall gives us a picture of non-existing formal languages translation for models in current researches with mainly knowledge reuse approach based mostly on the knowledge about models themselves and language that was used to code it. In addition, no one is focused on automation for generation of the required models. Even when there is already a vast number of coded models in many languages.

All above lead us into further improvement of our system for the future works and researches.

3 Analysis of Intelligent Tools Application

3.1 Language Analysis

Models and code are always written in a formal language [12] that is constructed with a set of rules with a set of words.

To start with, let us consider the main properties of formal languages:

1. Generated by some formal grammar;
2. Described or matched by a regular expression;
3. Accepted by some automaton [13];
4. Accepted by decision procedures [13].

Last point gives us another tip to the next sections but before that we have to review what we have in current intelligent tools to choose from and what is the best choice and why.

3.2 Intelligent Tools and Analysis

Due to a variety of scientific problems already solved, there is a list of approaches and tools that have been used to solve the problems. They are already applied in practice from video games and art to automatic car driving and diagnosis of different illnesses and diseases as well as robotic systems.

Moreover, the tools are as wide as the list of applications and it has to be considered if it suits for the current task.

Search and Optimization
It consists of mathematical optimization, search algorithms and evolutionary computation.

Optimization part could solve the problem by formalizing everything into mathematical functions that has to find the best value of the according variability to fit the result, but it is definitely very complex to formalize and develop. In addition, it requires doing the process repeatedly each time we change our input or output language, since the function is tied to the exact set of languages to support.

Search algorithms could find the answers for us, but it takes a lot of calculation resources and even specific search algorithms to develop for each specific pair of languages as for the optimization algorithms due to rules change.

Evolutionary algorithms require a lot of computation due to a big variety of languages with a different criterias. Each language on its own has to be processed and if any small change ruins the result as it is for the all of those three approaches.

Neural Networks, Classifiers and Statistical Learning Methods
They behave as a human brain and can only be applied to the tasks that can be formalized as a weight functions and have to learn from lots of examples. As a result, we have to remake networks for each specific pair of languages. And our task is very determined by the alphabet and the rules so there is no need in determining what makes learning process accurate but useless since we already know the rules. Also, it can theoretically lead to random inaccurate results in different cases.

The classifiers could do the job since they are trained in statistical and machine learning approaches on the given data set. Also, the main job of classification and translation of our task is already done by MT. Moreover, machine learning is a type of neural network so all above applies to it as well.

Probabilistic Methods for Uncertain Reasoning
Tools for this group come from probability theory since the main case of usage is uncertainty. Again, we have determined rules of formal languages for input and output what makes probabilistic algorithms a very complex and not compatible solution at the same time.

It is necessary to say that Bayesian networks also perform mostly worse than classifiers for most of the applied data [14, 15] so the choice between these two categories won't ever fall on this one.

Logic Programming and Automated Reasoning
This subject is as wide as the second in this section and includes huge variety of tools. Logic appears to fit the predefined rules of the formal languages since it is not needed to determine the correct data when it is already given.

Also, logic can process the data in automated way if there's a set of rules by the process of decision making, that fits all our criteria at the same time.

3.3 Choice Overhaul

In the chosen subject, there are many different tools to choose from but before we choose, we have to remind ourselves who usually do the job we are willing to pass.

Currently in predetermined tasks like translation, there are always some professionals that know more than other do. What if we could gather all those experts so could we cover all the rules that were ever know? It would be their knowledge combined all together. It is called the knowledge base and a tool will be the expert system that processes all the rules, stored in the knowledge base.

4 Conclusion

The paper appears as the review, the research and the analysis for the current state of models reuse problem. It starts from in-deep review of the correspondent papers including researches and theses. In this review, it becomes clearer that problem is still on the stage of approach development and even old knowledge can be applied to resolve it using special tools.

The analysis of current intelligent problems is revised step-by-step and each tool is analyzed to apply for the task of models' code translation with props and cons in a systematical way. All above led to the choice of the exact tool based on expert system that fit for all of criteria for the solution.

The results of this research make the base for the future development and researches for the application of the intelligent tools for the modeling languages translation.

Acknowledgments. The reported study was funded by Russian Foundation for Basic Research (RFBR) according to the research project No. 19-07-00936.

References

1. Chernukhin, Yu., Guzik, V., Polenov, M.: Multilanguage Translation for Virtual Modeling Environments. Publishing house of Southern Scientific Center of Russian Academy of Sciences, Rostov-on-Don (2009). (in Russian)
2. Chernukhin, Yu., Guzik, V., Polenov, M.: Multilanguage translation usage in toolkit of modeling systems. WIT Trans. Inf. Commun. Technol. **58**, 397–404 (2014)
3. Tanenbaum, E., Van Sten, M.: Distributed Systems: Principles and Paradigms, 2nd edn. Prentice-Hall, Upper Saddle River (2006)
4. Polenov M., Guzik V., Gushanskiy S., Kurmaleev A.: Development of the translation tools for distributed storage of models. In: Proceedings of 9th International Conference on Application of Information and Communication Technologies (AICT 2015), pp. 30–34. IEEE Press (2015)
5. Polenov, M., Guzik, V., Gushansky, S., Kurmaleev, A.: Intellectualization of the models translation tools for distributed storage of models. In: Informatics, Geoinformatics and Remote Sensing (Proceedings of 16-th International Multidisciplinary Scientific Geoconference (SGEM 2016)), vol. 1, pp. 255–262. STEF92 Technology (2016)
6. Polenov, M., Gushanskiy, S., Kurmaleev, A.: Synthesis of expert system for distributed storage of models. In: Software Engineering Perspectives and Application in Intelligent Systems. Advances in Intelligent Systems and Computing, vol. 575, pp. 220–228. Springer (2017)
7. Polenov, M., Kurmaleev, A., Gushanskiy, S.: Synthesis of intellectual tools to support models translation for mobile robotic platforms In: Advances in Intelligent Systems and Computing, vol. 763, pp. 282–291. Springer (2019)
8. Fujimoto, R., Bock, C., Chen, W., Page, E., Panchal, J.H.: Research Challenges in Modeling and Simulation for Engineering Complex Systems. Simulation Foundations, Methods and Applications Series. Springer (2017)
9. Gustavson, P., Daehler-Wilking, R., Blais, K., Rutherford, H.: M&S asset discovery: services, tools and metadata. In: 2011 Interservice/Industry Training, Simulation, and Education Conference (I/ITSEC) (2011)
10. Gallego, E., Álvarez-Rodríguez, J.M., Llorens, J.: Reuse of physical system models by means of semantic knowledge representation: a case study applied to Modelica. In: Proceedings of the 11-th International Modelica Conference, pp. 747–757 (2015)
11. Llorens, J., Morato, J., Genova, G.: RHSP: an information representation model based on relationship. In: Studies in Fuzziness and Soft Computing, vol. 159, pp. 221–253. Springer (2004)

12. Sander M.: Developing Interacting Domain Specific Languages, Universiteit Utrecht (2007)
13. Rozenberg, G.: Handbook of Formal Languages. Springer, Heidelberg (1997)
14. Russell, S.J., Norvig, P.: Artificial Intelligence: A Modern Approach, 3rd edn. Prentice Hall, Upper Saddle River (2009)
15. Van der Walt, C.M., Barnard, E.: Data characteristics that determine classifier performance. SAIEE Afr. Res. J. **98**(3), 87–93 (2006)

A Review of Use Case-Based Development Effort Estimation Methods in the System Development Context

Ho Le Thi Kim Nhung(✉), Huynh Thai Hoc, and Vo Van Hai

Faculty of Applied Informatics, Tomas Bata University in Zlin, Nad Stranemi 4511, 76001 Zlin, Czech Republic
{lho, huynh_thai, huynh_thai}@utb.cz

Abstract. Software Effort Estimation – (further only SEE), is a critical factor in the early phase of the software development life-cycle and hence - the success or failure of a software project depends on the accuracy of the estimated effort. In recent years, Use Cases for Software Effort Estimation has gained wide-ranging popularity. It is suitable for Effort Estimation in the early stages of software development since it helps project managers to bid on projects, and to efficiently allocate resources. It has attracted many researchers' interest in Use Case-based approaches due to the promising results obtained - including their early applicability. In this article, we look into a systematic review of previously published materials in order to summarise various Software Effort Estimation – (further only SEE), models and developments, based on Use Case Point. The study also provides insights into the effects of all factors that contributed to the Use Case size as an estimation for effort. Apart from this, the paper also provides standard criteria to evaluate the models' accuracy and effectiveness.

Keywords: Software effort estimation · Use Case Point Method

1 Introduction

Software Project Development has become a dynamic and competitive industry that requires the ability of high-level human resources. Software products are getting ever more complicated, unpredictable, and challenging to control than ever. Recent decades have witnessed many research projects in the software field designed to move software development processes into regular rules, (that are) more manageable, and predictable. To complete a project done, in-time, and deliver it to the customer as scheduled, project managers must estimate the cost of the software product - as well as the resources, effort and time needed [27]. Due attention is paid to software measurement problems; such as the prediction of project duration - or defect density. These problems show that the Project Management role has become (much) more significant [24].

Effort Estimation – (further only EE), plays a crucial role in the success of the overall solution delivery. Early EE in the initial software development life cycle phase is a critical factor in order to avoid any failures in a project. Our task provides an inside view of software products in support of Budgeting, Scheduling and Planning activities.

R. Silhavy et al. (Eds.): CoMeSySo 2019, AISC 1046, pp. 484–499, 2019.
https://doi.org/10.1007/978-3-030-30329-7_44

As well as Project Bidding, Human Resources Allocation, and Risk Minimisation. EE is essential for many reasons [28]. Firstly, before starting the project - it helps to make informed decisions about how to manage resources. Then, designing the project plan which provides knowledgeable choices about how to control and plan the project and how to deliver the project on time, on schedule, and within budget. When supervising the project's development, it is essential to make sure that the right amount of effort is allocated to various activities. Therefore, this has led many researchers to study software estimation for a more accurate SEE [1, 3, 11].

However, software estimation is an excellent example of a difficult task – based on Requirements Specifications. As regards using inaccurate models, such estimation decisions may be a recipe for disaster. Many software project failures are the most obvious example of problems in managing complex, distributed software systems [9]. The 2018 Standish Group CHAOS report indicated that many software companies still put forward no practical software costs, or work within strict schedules - and finished their projects behind schedule and over-budget - (48%–65%); or failed to complete them at all - (48%–56%) in 2018 [10]. The results pointed out that the actual efforts and schedules for most projects are overruns when compared to the estimations. If the software cost is underestimated, this would bring inefficiencies to the project - and the actual price would surely be superseded. Lastly, even if completed on schedule, these overestimated projects typically expand and spend more resources than planned; while the functionality and quality of these undervalued projects are reduced to be suitable for the plan [11]. These can lead to losing the bid, or wasting time, money, staff and other resources, and conclude in a financial loss; or even, the organisation's bankruptcy. Consequently, Effort Estimation Methods – (further only EEM), represent actual problems that are vital for Software Project Planning, (SPP) [12].

In the requirements stage of the software life-cycle, Use Cases can be useful in measuring estimated SE early on in a software project – prior to acquiring the requisite information [22]. According to Sommerville [24], engineering process requirements principally include four interleaved steps - as shown in Fig. 1, and software estimation tasks can be conducted at any stage within the process.

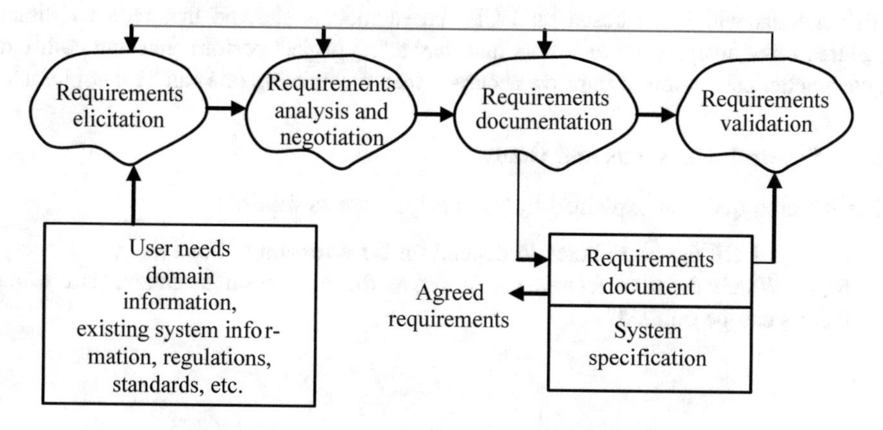

Fig. 1. Engineering process requirements [24].

Use Cases are expected to offer a reliable estimation of the SE of the corresponding future system. A survey – by Neil et al. [7], concentrated on the techniques used for Requirements Elicitation, Description, and Modelling. They indicated that over 50% of these software projects used Use Cases in the early phase. As a result of which, the use of Use Cases for SEE has become widespread.

Researchers have expressed interest in the Use Case-based approaches - along with their initial applicability. Many methods have been proposed in the review. Some methods suggested that, adding more complexity levels for the Use Cases Weight, the Actor Weight -, or both [18, 22, 23] would help. A few methods recommended discretising existing complexity levels into more detailed options [4, 16, 25]. Other methods propose the calibration of complexity weights into different complexity levels [5, 8, 19, 20, 26]. Hence, the Use Case Point - (UCP) method, [2], can be used to make estimations - Khatibi et al. [13], declared that the UCP method could be employed in order to predict Software Development Effort and to demonstrate that the UCP method is good enough. Silhavy et al. [3] showed that the values of the unadjusted UCP weights - which represent many use cases, are a critical element in deciding the Estimated Effort of the software. Apart from this, all other variables (e.g. Unadjusted Actor Weights, Technical Complexity Factors and Environmental Complexity Factors) derived from the UCP method, also influence EE [25].

In this article, we look into a systematic review of Use Case-based EE methods. The paper focuses on analysing the use of complexity levels and weights, and the software development environment factors of each approach for new Use Case driven projects. The structure of the paper is organised as follows. Section 1, describes the introduction. Section 2, defines the research questions and outlines the research objectives. Section 3, describes the previous work completed in modifying the UCP method for better EE. The paper ends with a discussion and future work.

2 Problem Statement

In this research paper, we focus on an overview of various software effort estimation (EE) models which are based on UCP. These models showed that with additional features - like analysis, more factors that may affect project performance and ability to give a better estimation to improve accuracy than the existing one can be incorporated.

2.1 Research Questions and Goals

The research question explained by this analysis are as follows:

RQ1: Is UCP the best choice to depend on for estimating effort?
RQ2: Which decisive factors can improve the estimation accuracy, and which factors can be omitted?

2.2 Evaluation Criteria

In EE, there are different criteria that needed to evaluate the effectiveness of models, like standard evaluation. The EE's accuracy in terms of the Mean of Magnitude of Relative Error (MMRE) and Percentage of Prediction within x% (PRED(x)) [27, 30, 33] are the two most common metrics used in software engineering that have been proposed [32]. Both parameters are based on the quantity called Magnitude of Relative Error (MRE), which is described by Eq. 1, where y_i is the known real value, and $\hat{y}_i$ is the predicted value.

$$MRE_i = \frac{|y_i - \hat{y}_i|}{y_i} \tag{1}$$

The MMRE measures the sample mean of MRE, and PRED(x) the percentage of the estimates that are within x% of the actual efforts (in terms of MRE). MMRE and PRED(x) are shown in Eqs. 2 and 3 - respectively, where n is the number of observations.

$$MMRE = \frac{1}{n}\sum\nolimits_{i=1}^{n} MRE_i \tag{2}$$

$$PRED(x) = \frac{1}{n}\sum\nolimits_{i=1}^{n} \begin{cases} 1, & \text{if } MRE_i \le x \\ 0, & \text{otherwise} \end{cases} \tag{3}$$

Low MMRE values and high PRED(x) values are required. These values help to determine the estimation accuracy and model stability. According to [32, 33, 37], an accurate effort prediction model should have an MMRE ≤ 0.25 (to ensure that the mean estimation error should be less than 25%) and PRED(25) ≥ 0.75 (meaning no less than 75% of the predicted values with MRE are lower than 0.25. Some authors also take into consideration PRED(30). Therefore, a model can be regarded to be accurate when it satisfies any of the following cases: MMRE ≤ 0.25 and PRED(25) ≥ 0.75 or PRED(30) ≥ 0.70 [34].

Another metric that is also used regularly in EE model evaluation [31] is the Mean Magnitude of Relative Error relative to the Estimate (MMRE). And the MER value is comparable to MRE, but the denominator of MER is the predicted effort instead of the actual effort. The formulae for MER and MMER are shown in Eqs. 4 and 5, respectively.

$$MER_i = \frac{|y_i - \hat{y}_i|}{\hat{y}_i} \tag{4}$$

$$MMER = \frac{1}{n}\sum\nolimits_{i=1}^{n} MER_i \tag{5}$$

According to the simulation study of Foss et al. [34], they proved that both metrics - MMRE and MMER might lack accuracy in evaluation in some cases. In the case of the actual effort being small, then MMRE would be high. Or, if the predicted effort is low, then MMER would also be high. Therefore, the authors recommend that MMRE should not be used when comparing EE models and that using the standard deviation would be better. The metric was first proposed by Karl in 1894 [35], as shown in Eq. 7, where $\overline{x}$ is the mean error for each observation i as presented in Eq. 6.

$$\overline{x} = \frac{1}{n}\sum_{i=1}^{n} (y_i - \widehat{y}_i) \tag{6}$$

$$SD = \sqrt{\frac{1}{n-1}\sum_{i=1}^{n} (x_i - \overline{x})^2} \tag{7}$$

The mean error with a standard deviation represented, is shown in Eq. 8.

$$\overline{x} \pm SD \tag{8}$$

Another critical determinant to estimation results is the number of observations that are needed to get meaningful results. When the observations rise, then the standard error (SE) value must be reduced [29]. The SE can be calculated as shown in Eq. 9, where $SD_{(MRE \leq x)}$ means the standard deviation of the indicator values of the MRE values are less than x (i.e., 1 for an MRE less than x, 0 for otherwise) for n observations.

$$SE_{PRED(x)} \approx \frac{SD_{(MRE \leq x)}}{\sqrt{n}} \tag{9}$$

3 Methodologies Used

Based on UCP [2], researchers became interested in the Use Case-based approach along with their early applicability. Many methods have been suggested, like Use Size Points (USP) [5], Fuzzy Use Case Size Points (FUSP) [4], Use Case Reusability (UCR) [8], extended Use Case Points (e-UCP) [16], Revised Use Case Points (Re-UCP) [18], Use Case Points - Activity Based Costing (UCPabc) [19], Adapted UCP [20], UCA [22], Enhanced UCP [23], fuzzy-UCP [25], and SUCP [26]. In common with the benefits of using these methods, various problems and concerns about these approaches have also been put forward. A summary of the EE methods based on Use Case, is shown in Table 6.

The problems primarily focus on four issues. Firstly, complexity in the Use Case models - both explicit and written, and implicitly in the required realisation. Secondly, how to adjust the Technical Complexity Factors (TCFs) and Environmental Complexity Factors (ECFs). For example, some TCFs do not influence the overall project [14]. These then multiply with the weight of the use cases and actors and the impact is such that they do influence things, e.g., technical factor T_1 reflects the requirement to be able to install the system easily. The more massive a system is, the more time-consuming it will be to write its installation procedure. Then, how does one classify the Use Cases. It is complicated to define the metrics, e.g., many variations of use case specification style and formality measure the length of a use case. Diev [21], argued that defining the UCP model's elements - (actors and use cases), and assigning weights to them across and within projects are the main problems to obtaining a reasonably accurate estimate. Lastly, how to make the right choice in a productivity factor whose units are person-hours (ER value). This value is a crucial operator for the estimated effort from UCP size measurements - especially when the historical dataset is not available. For example, if the UUCP is 90, the estimation result then is $90 * 20 = 1800\,\text{hr}$, (by using an ER of 20). Or the outcome will change to 2700 hr, (by using an ER of 30). The ER value significantly influences the final EE.

EE models based on use case can be categorised in many different ways, using many different criteria. In particular, Wen et al. [36], classified estimation models based on machine learning such as, for example, Case-Based Reasoning, Artificial Neural Networks, Decision Trees, Bayesian Networks, Support Vector Regression, Genetic Algorithms, Genetic Programming, and Association Rules. According to Qi et al. [17], EE models are classified into three main groups of approaches that focus on UCP factors. To answer RQ2, we used the categorisation of Qi et al. The models involve adding more complexity levels for the use cases weight, the actor weight - or both of them; discretising existing complexity levels into more detail options, and empirically calibrating complexity weights to the different complexity levels.

3.1 Adding Extra Complexity Levels

The Revised Use Case Point method (Re-UCP) [18], is a modified approach based on the UCP method of EE. Re-UCP was adjusted by adding the actor and use-case weighting - as shown in Tables 4 and 5 and by supplying a new parameter in both Technical (T14) - as in "Scalability", and Environmental Complexity Factors (E9) - "Project Methodology". Critical Actors and Critical Use Cases were added to the actor and use cases types. The principal purpose of the Re-UCP model is to connect the gap between Approximated and Actual software efforts in the rapidly changing software development industry. The performance of Re-UCP has revealed improvements in the estimation of the efforts, with minimum trends in deviation from the practical efforts when compared with UCP on 14 projects.

Minkiewicz, also suggested Use Cases with Actors (UCP Sizing) [22], to add one extra rating level to the Use Case weighting system (Table 5), and to determine the relationship for a UCA with Unadjusted Function Points.

Nassif et al., proposed an Enhanced UCP model [23], also known as Stochastic Gradient Boosting (SGB) to estimate software effort based on three variables, namely UCP, productivity, and complexity. The SGB approach helps the model to improve the estimation accuracy by employing the function in a series, and appending the output of all the individual functions. Besides this, the model extends the use case complexity weights to 20, 25, and 30 points. They also constructed a fuzzy model to predict team productivity - based on the weighted sum of environmental complexity factors, and used MMRE, PRED, MSE, MdMRE for evaluation purposes. However, this model is only suitable for those projects that have about 2475 UCP. In conclusion, the authors showed that the Enhanced UCP model has greater accuracy than both the Regression model and the UCP model.

3.2 Discretising Existing Complexity Levels

The interactive Use Case Points (iUCP) method [6] proposed by Nunes et al., modifies the UCP method to make it suitable for the agile development of interactive software. It differs from UCP in the complexity of the assessment and weighting of UAW. Based on the interaction complexity, iUCP identifies six actor weights, which are assigned to differences in classification - as shown in Table 4. The significant contribution of the method is to help software developers and interaction designers to apply heuristics that are suitable for interactive applications and that work consistently within projects. Apart from having no bias results, iUCP ignored the estimation of technical and environmental factors. These factors are inclined to bias, affected by the experience of the expert [15].

The Extended Use Case Points (EUCP) method [4] was proposed by Wang et al. By integrating Fuzzy Set Theory and Bayesian Belief Networks with the UCP model, the method extends use case complexity levels from three to five classes - as shown in Table 5.

EUCP is useful for a manager to control the risk of overruns since it is easy to capture the difference between use cases and show the complexity of change when updating a use case. The EE of the method did not involve the productivity factor. In the end, the authors compared the UCP and EUCP methods and analysed the proposed effort with two case studies.

The extended Use Case Points (e-UCP) method [16], is based on the UCP and focused on the intimate details of the use cases. Periyasamy et al., proposed using a Use Case Narrative with six parameters, as shown in Table 1. Besides this, the e-UCP changed actor complexity from three to seven levels and classified use case complexity into four levels, as shown in Table 5. The method reuses the TCF, ECF weights, and the productivity factor - as given in the original UCP method [2]. The results of the model were closer to matching the actual time taken in completing projects.

Table 1. Use Case Narrative weight classification [16].

Use Case Narrative parameters	Weight factor
Input parameter	0.1
Output parameter	0.1
A predict in pre-condition	0.1
A predict in post-condition	0.1
An action in a successful scenario	0.2
An exception	0.1

Nassif et al., proposed a model using Fuzzy Logic and Neural Networks - (fuzzy-UCP) [25] to improve the accuracy of the model, and the results grew to 22% over Karner's model. The model discretises the Use Case complexity levels into ten categories - according to the number of transactions in a Use Case, as shown in Table 5. The authors also constructed the productivity factors, based on the weighted sum of TCFs.

3.3 Empirically Calibrating Complexity Weights

Sholiq, proposed the UCPabc model [19], which utilises an activity-based costing technique. The method uses all of the standardised factors - as suggested by the UCP method, except for the fact that the constant of productivity factors is 8.2 person-hours. In the end, Sholiq concluded that the UCPabc model provides satisfactory results, i.e., the Cost Estimation method is close to the real cost.

Another adjustment to the UCP is called Adapted UCP (AUCP) [20]. The method is suitable for incremental development estimations for large-scale projects. AUCP follows the standards of the UCP method, but with significant variations. Initially, all actors are classified as average, and all Use Cases are assumed to be complicated. All the technical and ECFs are replaced by three factors – the Adaptation Adjustment Factor (AAF), the Equivalent Modification Factor (EMF), and the Overhead Factor (OF). The constant of productivity factors is 36 person-hours. In the case where the OF is not used, the use is 72 person-hours. Braz et al., proposed the first, named the Use Case Size Points (USP) [5] as a size metric which calibrates the internal structure of the Use Case to enhance the precision of effort estimation. A detailed Use Case classification includes the Actors classification, pre-condition classification, and post-condition classification, main scenarios, alternate scenarios, exception classification, and the adjustment factor. Additionally, the USP also reused 14 technical complexity factors and five environment factors of the UCP model. A productivity factor of 26 person-hours is used - as per the estimations. The total of all these factors gives the Unadjusted Use Case Size Points (UUSP), which is subsequently multiplied by the difference between the Technical Complexity Factor and the Experience Factor. The results are compared with Function Points and UCP metrics. The second one is called Fuzzy Use Case Size Points (FUSP). The method considers Fuzzy Set Theory concepts to create gradual classifications that deal with uncertainty better. The primary factors discussed in this metric are the same USP. The results are compared with USP, and the model's results are slightly better than the USP.

Qi et al. proposed the Use Case points using Bayesian Analysis (Bayesian UCP) [17]. Bayesian UCP uses the Bayesian approach to modify the Use Case complexity weights of the UCP method to improve effort estimation accuracy. The model involves three steps. Firstly, calculating the a priori means and variances of the Use Case complexity weights from UCP [2], EUCP [4], Re-UCP [18], UCP Sizing [22], and Enhanced UCP [23] as the precursor information. Secondly, calculating Use Case complexity weights and their variances using MLR, based on the observational dataset. Lastly, calculating the Bayesian Weighted Average using the outcomes of the two steps. The performance of the Bayesian approach for estimating Use Case complexity weights is evaluated by comparing it with the "a priori" method (A-Priori), regression method (Regression), UCP method [4], and the Bayesian-based method (Bayesian). In conclusion, the Bayesian UCP evaluated the out-of-sample estimation accuracy by 10-fold cross-validation in terms of MMRE, PRED(.15), PRED(.25), and PRED(.50). The Bayesian-UCP exceeds the other estimators by around 17% in terms of MMRE.

SUCP [26] is proposed based on the UCP model and two sets of adjustment factors relating to the ECFs and TCFs. The method used for realising the goal is the Cross-Validation Process, which analyses various variants of UCP, with and without specific factors. In the end, the method suggested new TCFs and ECFs. The new TCFs are defined as follows: Efficiency (T'1), Operability (T'2), Maintainability (T'3), and Interoperability (T'4). The new ECFs are assigned to follow Team Experience (E'1) and Team Cohesion (E'2).

The Use Case Reusability (UCR) [8] is proposed for the projects that reuse artifacts developed earlier in preceding projects with the same range. The model presents a new classification of Use Case, based on their reusability; TCFs, and ECFs are also redefined by EE experts - as shown in Tables 2 and 3. In conclusion, the UCR method showed improvements of estimated effort results as compared with the UCP model. Namely, that the absolute values of MRE and MMRE for UCR method are lower than the absolute values of MRE and MMRE for the UCP model, and the PRED(20) for UCR is higher than the PRED(20) for the UCP model.

Table 2. UCR – technical complexity factors [8].

UCR – TCFs	Weight factor		
	Low	Medium	High
UCR-T1 software architecture	0.8	1.0	1.2
UCR-T2 portable	0.83	1.0	1.17
UCR-T3 new system integration	0.87	1.0	1.13

Table 3. UCR – environment complexity factors [8].

UCR – ECFs	Weight factor		
	Low	Medium	High
UCR-E1 application experience	1.11	1.0	0.89
UCR-E2 team members experience	1.11	1.0	0.89
UCR-E3 maturity of requirements	1.07	1.0	0.93
UCR-E4 team co-location	1.05	1.0	0.95
UCR-E5 team size	1.05	1.0	0.95
UCR-E6 maintenance of project documentation	1.05	1.0	0.95
UCR-E7 team cohesion	1.02	1.0	0.98

Table 4. Actor types in estimation methods.

Method	Actor type	Weight factor
iUCP [12]	Simple system	1
	Average system	2
	Simple human-being	3
	A complex system	3
	Average human	4
	Complex human	5
e-UCP [16]	Very simple	0.5
	Simple	1
	Less than average	1.5
	Average	2
	Complex	2.5
	Very complex	3
	Most complex	3.5
Re-UCP [18]	Simple	1
	Average	2
	Complex	3
	Critical	4

Table 5. Use case type in estimation methods.

Method	Use case type	Weight factor
EUCP [4]	Very simple	5
	Simple	10
	Average	15
	Complex	20
	Very complex	25
e-UCP [16]	Simple	0.5
	Average	1
	Complex	2
	Critical	3
Re-UCP [18]	Simple	5
	Average	10
	Complex	15
	Critical	20
UCP Sizing [22]	Simple	5
	Average	10
	High	15
	Very high	20

(*continued*)

Table 5. (*continued*)

Method	Use case type	Weight factor
Enhanced UCP [23]	Very low	5
	Low	10
	Normal	15
	High	20
	Very high	25
	Extra high	30
Fuzzy-UCP [25]	1–2 transactions	5
	3 transactions	6.45
	4 transactions	7.5
	5 transactions	8.55
	6 transactions	10
	7 transactions	11.4
	8 transactions	12.5
	9 transactions	13.6
	10 transactions	15
UCR [8]	UC low	5
	UC medium	10
	UC high	15
	UC complex	20
	UC new (reusability)	1.0
	UC similar (reusability)	0.7
	UC identical (reusability)	0.6

Table 6. Summary of the effort estimation methods based on use case with UCP [4].

Method	Actor weights	Use case weights	TCFs	ECFs	ER (Person-hours)
iUCP [6]	Identifying six actor weights with differences in classification	Agree with the standardised use case weights by the UCP method	All TCFs are discarded	All ECFs are discarded	A productivity factor of 20 person-hours was used
EUCP [4]	Agree with the standardised actor weights by the UCP method	Extending use case complexity levels from three to five classes	Including 13 TCFs	Including 8 ECFs	No involvement of the productivity factor

(*continued*)

Table 6. (*continued*)

Method	Actor weights	Use case weights	TCFs	ECFs	ER (Person-hours)
e-UCP [16]	Use case narrative with six parameters and changing actor complexity from three to seven levels	Classifying use case complexity into four levels	Including 13 TCFs	Including 8 ECFs	A productivity factor of 20 person-hours was used
Re-UCP [18]	Adding the actor weighting	Adding the use case weighting	Adding a new T14 – like "Scalability"	Adding a new E9 - like "Project Methodology"	The method and the fixed productivity factor at 20 person-hours per use case point
UCPabc [19]	Agree with the standardised actor weights by the UCP method	Agree with the standardised use case weights by the UCP method	Including 13 TCFs	Including 8 ECFs	The constant of productivity factors is 8.2 person-hours
AUCP [20]	All Actors are classified as average initially	All use cases are assumed to be complicated initially	All TCFs are discarded	All ECFs are discarded	A productivity factor of 36 person-hours per use case is used for AAF, EMF, and OF. The use of 72 person-hours in the case that OF is not used
UCP Sizing [22]	Agree with the standardised actor weights by the UCP method	Adding one rating level to the use case weighting system	Including 13 TCFs	Including 8 ECFs	A productivity factor of 20 person-hours was used
Enhanced UCP [23]	Agree with the standardised actor weights by the UCP method	Extending the use case complexity weights to 20, 25, and 30 points	Including 13 TCFs	Including 8 ECFs	Based on the weighted sum of ECFs

(*continued*)

Table 6. (*continued*)

Method	Actor weights	Use case weights	TCFs	ECFs	ER (Person-hours)
Fuzzy-UCP [25]	Agree with the standardised actor weights by the UCP method	Discretising the use case complexity levels into ten categories - according to the number of transactions	Including 13 TCFs	Including 8 ECFs	Based on the weighted sum of TCFs
USP [5]	Actor weights are included as per the specific use case classification	Use case weights are included as per the specific use case classification	Including 13 TCFs	Including 5 ECFs	The productivity factors constant is 8.2 person-hours
FUSP [5]	Actor weights are included as per the specific use case classification	Use Case weights are included as per the specific use case classification	Including 13 TCFs	Including 5 ECFs	The productivity factors constant is 26 person-hours
SUCP [26]	Discarding actor weights	Agree with the standardised use case weights by the UCP method	Rejecting 9 TCFs	Rejecting 6 ECFs	Based on the counted total of ECFs
UCR [8]	Agree with the standardised actor weights by the UCP method	Adding the classification of use case for subsequent projects based on their reusable components	Redefining TCFs that impact on the calculation of effort for initial and subsequent projects	Adding a new ECF factor	The productivity factors constant for the initial project is 10 person-hours, for any subsequent project, this is 5.5 person-hours
Bayesian UCP [17]	Agree with the standardised actor weights by the UCP method	The weights suggested in the prior information and calibrated by multiple linear regression	Including 13 TCFs	Including 8 ECFs	The productivity factors constant is 20 person-hours

4 Discussion and Conclusion

EE in software development is a complicated and challenging activity. There are no metrics or techniques in all cases. Each technique has its corresponding benefits and drawbacks. Due to its early application lifecycle, when enough information is not available to estimate size using more traditional measures, Use Case-based models have gained broad usage recently, and have been proven to produce promising outcomes.

In RQ1, the answer depends on the proper evaluation and inquiry into these approaches. Many proposed methods have addressed these issues suitably - and several of them have improved many problems as well.

In RQ2, the common factors of UCP are studied in the literature are presented in Table 6. The review provides a brief look at the main factors to be considered in the software effort estimation. Actor weights [2, 4–6, 8, 17–20, 22, 23, 25] that actors are categorized as simple, average or complex based on the level of interaction with the system. Several others discard [26] or modify the weighted actors [16]. Use Case weights [2, 4–6, 18, 19, 26] that Use Cases are categorized into simple, average or complex based on transactions within the Use Case. More recently, many others modify or add more rating level of interaction with the system [8, 16–18, 20, 22, 23, 25]. TCFs [2, 4, 5, 16, 17, 19, 22, 23, 25] that thirteen TCFs are analyzed in the majority of the methods whereas some methods discard few TCFs [6, 20, 26], some methods add more TCF depending on the real project [18], or redefining TCFs that impact the calculation of effort for initial and subsequent projects [8]. ECFs [2, 4, 5, 16, 17, 19, 22, 23, 25] that eight ECFs are considered in the majority of the methods whereas several methods discard few ECFs [6, 20, 26], some others add ECFs depending on the real project [8, 18]. Productivity Factor [6, 16–18, 22, 25, 26] that is a factor whose units are person-hours (ER value). Originally, Karner proposed a 20 person-hours productivity factor per UCP [2]. More lately, various others have productivity factors of 36 person-hours [20], 8.2 person-hours [5, 8, 19], based on the weighted sum of ECFs [23, 25], or do not use this factor [4] depending on the character of the development project. The problem is to find the optimum number of factors that are to be considered while estimating effort accuracy. Many methods agree with the standardized thirteen TCFs and the eight ECFs as proposed by the UCP method. As such, we cannot recommend any method to be the best in terms of this attribute. It depends on the project context. It is crucial for the software project manager to recognize main factors related to estimate the software effort and situations, whereas estimation method will be suitable. No existing method can predict with a high degree of accuracy; hence, the study of software effort estimation is necessary to improve on estimation accuracy.

In future research of this study, we will explore the potential of data mining machine learning, which was applied for estimating software development effort. This could provide the estimation accuracy of the model and decreasing the margin of its prediction error.

Acknowledgment. This work was supported by the Faculty of Applied Informatics, Tomas Bata University in Zlín, under Project RO30196021025 and under Project IGA/CebiaTech/2019/002.

References

1. Azzeh, M., Nassif, A.B.: Project productivity evaluation in early software effort estimation. J. Softw.: Evol. Process (2018)
2. Karner, G.: Resource estimation for objector projects. Object. Syst. (1993)
3. Silhavy, R., Silhavy, P., Prokopova, Z.: Analysis and selection of a regression model for the use case points method using a stepwise approach. J. Syst. Softw. **125**, 1–14 (2017)
4. Wang, F., Yang, X., Zhu, X., Chen, L.: Extended use case points method for software cost estimation. IEEE (2009)
5. Braz, M.R., Vergilio, S.R.: Software effort estimation based on use cases. In: Proceedings of the 30th Annual International Computer Software and Applications Conference (COMPSAC 2006). IEEE (2006)
6. Nunes, N.J., Constantine, L.: iUCP: estimating interactive software project size with enhanced use case points. IEEE Softw. **23**, 64–73 (2011)
7. Neil, C.J., Laplante, P.A.: Requirements engineering: the state of the practice. IEEE Softw. **20**(6), 40–46 (2003)
8. Rak, K., Car, Z., Lovrek, I.: Effort estimation model for software development projects based on use case reuse. J. Softw. Evol. Process (2019)
9. Charette, R.N.: Why software fails. IEEE Spectr. **32**(9), 42–49 (2005)
10. The Standish Group: CHAOS Chronicles. Technical report, The Standish Group International, Inc. (2018)
11. Trendowicz, A., Munch, J., Jeffery, J.: State of the practice in software effort estimation: a survey and literature review. In: Software Engineering Techniques, CEE-SET (2008)
12. Silhavy, R., Silhavy, P., Prokopova, Z.: Evaluating subset selection methods for use case points estimation. Inf. Softw. Technol. **97**, 1–9 (2018)
13. Khatibi, V., Jawawi, D.N.: Software cost estimation methods: a review. Emerging Trends Comput. Inf. Sci. **2**, 21–29 (2010)
14. Cohn, M.: Estimating with use case points. Methods and Tools (2005). https://www.mountaingoatsoftware.com/articles/estimating-with-use-case-points. Accessed 20 June 2019
15. Morgenshtern, O., Raz, T., Dvir, D.: Factors affecting duration and effort estimation errors in software development projects. Inf. Softw. Technol. **49**, 827–837 (2007)
16. Periyasamy, K., Ghode, A.: Cost estimation using extended use case point (e-UCP) model. In: International Conference on Computational Intelligence and Software Engineering (2009)
17. Qi, K., Hira, A.: Calibrating use case points using Bayesian analysis. In: ESEM (2018)
18. Kirmani, M.M., Wahit, A.: Revised use case point (Re-UCP) model for software effort estimation. Int. J. Adv. Comput. Sci. Appl. **6**, 65–71 (2015)
19. Sholiq, Dewi, R.S., Subriadi, A.P.: A comparative study of software development size estimation method: UCPabc vs. function points. In: 4th Information Systems International Conference (ISICO) (2017)
20. Mohagheghi, P., Anda, B., Conradi, R.: Effort estimation of use cases for incremental large-scale software development. ACM (2005)
21. Diev, S.: Use cases modeling and software estimation: applying use case points. ACM Softw. Eng. Notes **31**, 1–4 (2006)
22. Minkiewicz, A.: Use case sizing. PRICE Systems, L.L.C (2015)
23. Nassif, A.B.: Software size and effort estimation from use case diagrams using regression and soft computing models. Western University (2012)
24. Sommerville, I.: Requirements engineering processes (2013)

25. Nassif, A.B., Capretz, L.F., Ho, D.: Enhancing use case points estimation method using soft computing techniques. Global Research in Computer Science (2016)
26. Ochodek, M., Nawrocki, J., Kwarciak, K.: Simplifying effort estimation based on use case points. Inf. Softw. Technol. **53**, 200–213 (2010)
27. Boehm, B.W.: Software Engineering Economics. Prentice-Hall, Englewood Cliffs (1981)
28. Boehm, B.W., Madachy, R., Steece, B.: Software Cost Estimation with COCOMO II. Prentice Hall, Upper Saddle River (2000)
29. Port, D., Korte, M.: Comparative studies of the model evaluation criterions MMRE and PRED in software cost estimation research (2009)
30. Briand, L.C., Emam, K.E., Surmann, D., Wieczorek, I., Maxwell, K.D.: An assessment and comparison of common software cost estimation modeling techniques. In: ICSE 1999, pp. 313–322 (1999)
31. Kitchenham, B.A., Pickard, L.M., MacDonell, S.G., Shepperd, M.J.: What accuracy statistics really measure. IEE Proc. Softw. **148**, 81–85 (2001)
32. Conte, S.D., Dunsmore, H.E., Shen, V.Y.: Software Engineering Metrics and Models. Benjamin-Cummings Publishing, Redwood City (1986)
33. Jørgensen, M.: Experience with the accuracy of software maintenance task effort prediction models. IEEE Trans. Softw. Eng. **21**(8), 674–681 (1995)
34. Foss, T., Stensrud, E., Kitchenham, B., Myrtveit, I.: A simulation study of the model evaluation criterion MMRE. IEEE Trans. Softw. Eng. **29**, 985–995 (2003)
35. Karl, P.: On the dissection of asymmetrical frequency curves. Philos. Trans. R. Soc. **185**, 71 (1894)
36. Wen, J., Li, S., Lin, Z., Hu, Y., Huang, C.: Systematic literature review of machine learning based software development effort estimation models. Inf. Softw. Technol. **54**, 41–59 (2012)
37. Dolado, J.J.: On the problem of the software cost function. Inf. Softw. Technol. **43**, 61–72 (2001)

Author Index

R. Silhavy et al. (Eds.): CoMeSySo 2019, AISC 1046, pp. 501–502, 2019.
https://doi.org/10.1007/978-3-030-30329-7

Printed in the United States
By Bookmasters